Serine Proteases and Their Serpin Inhibitors in the Nervous System

Regulation in Development and in Degenerative and Malignant Disease

NATO ASI Series

Advanced Science Institutes Series

A series presenting the results of activities sponsored by the NATO Science Committee, which aims at the dissemination of advanced scientific and technological knowledge, with a view to strengthening links between scientific communities.

The series is published by an international board of publishers in conjunction with the NATO Scientific Affairs Division

A	Life Sciences	Plenum Publishing Corporation
B	Physics	New York and London
C	Mathematical and Physical Sciences	Kluwer Academic Publishers
		Dordrecht, Boston, and London
D	Behavioral and Social Sciences	
E	Applied Sciences	
F	Computer and Systems Sciences	Springer-Verlag
G	Ecological Sciences	Berlin, Heidelberg, New York, London,
H	Cell Biology	Paris, and Tokyo

Recent Volumes in this Series

Volume 185—Modern Concepts in *Penicillium* and *Aspergillus* Classification
edited by Robert A. Samson and John I. Pitt

Volume 186—Plant Aging: Basic and Applied Approaches
edited by Roberto Rodríguez, R. Sánchez Tamés, and D. J. Durzan

Volume 187—Recent Advances in the Development and Germination of Seeds
edited by Raymond B. Taylorson

Volume 188—Evolution of the First Nervous Systems
edited by Peter A. V. Anderson

Volume 189—Free Radicals, Lipoproteins, and Membrane Lipids
edited by A. Crastes de Paulet, L. Douste-Blazy,
and R. Paoletti

Volume 190—Control of Metabolic Processes
edited by Athel Cornish-Bowden and María Luz Cárdenas

Volume 191—Serine Proteases and Their Serpin Inhibitors
in the Nervous System:
Regulation in Development and in Degenerative and
Malignant Disease
edited by Barry W. Festoff

Series A: Life Sciences

Serine Proteases and Their Serpin Inhibitors in the Nervous System

Regulation in Development and in Degenerative and Malignant Disease

Edited by

Barry W. Festoff

University of Kansas and
Kansas City Veterans Affairs Medical Centers
Kansas City, Missouri

Associate Editor
Daniel Hantaï
INSERM
Paris, France

Plenum Press
New York and London
Published in cooperation with NATO Scientific Affairs Division

Proceedings of a NATO Advanced Research Workshop on
Regulation of Extravascular Fibrinolysis in
Nervous System Development and Disease,
held July 2–8, 1989,
in Maratea, Italy

Library of Congress Cataloging in Publication Data

NATO Advanced Research Workshop on Regulation of Extravascular Fibrinolysis in Nervous System Development and Disease (1989: Maratea, Italy)
    Serine proteases and their serpin inhibitors in the nervous system: regulation
in development and in degenerative and malignant disease / edited by Barry W.
Festoff associate editor, Daniel Hantai.
        p.   cm.—(NATO ASI series. Series A, Life sciences; v. 191)
    "Proceedings of a NATO Advanced Research Workshop on Regulation of Ex-
travascular Fibrinolysis in Nervous System Development and Disease, held July
2–8, 1989, in Maratea, Italy"—T.p. verso.
    "Published in cooperation with NATO Scientific Affairs Division."
    Includes bibliographical references.
    Includes index.

    ISBN 978-1-4684-8359-8   ISBN 978-1-4684-8357-4 (eBook)
    DOI 10.1007/978-1-4684-8357-4

    1. Serine proteinases—Congresses. 2. Serpins—Congresses. 3. Fibrinolysis—
Congresses. 4. Nervous system—Growth—Congresses. 5. Nervous system—
Pathophysiology—Congresses. I. Festoff, Barry W. II. Hantai, Daniel. III. North
Atlantic Treaty Organization. Scientific Affairs Division. IV. Title. V. Series.
    [DNLM: 1. Fibrinolysis—congresses. 2. Nervous System—growth & develop-
ment—congresses. 3. Nervous System Diseases—pathology—congresses. 4.
Plasminogen Activators—congresses. 5. Serine Proteinases—physiology—
congresses. 6. Serpins—physiology—congresses. QU 136 N279s 1989]
QP609.S47N37   1989
616.8'0471—dc20
DNLM/DLC   90-7453
for Library of Congress   CIP

© 1990 Plenum Press, New York
A Division of Plenum Publishing Corporation
233 Spring Street, New York, N.Y. 10013

Softcover reprint of the hardcover 1st edition 1990

To Shea and Mara
—B.W.F.

THE MARATEA CONFERENCE

Organization Committee

Barry W. Festoff, M.D.
Director

University of Kansas Medical Center at the
Veterans Administration Medical Center,
Kansas City, MO 64128 USA

Georgia Barlovatz-Meimon, Ph.D.

Université Paris XII (Val de Marne)
Créteil, FRANCE

Robin Carrell, M.D.

University of Cambridge Clinical School
Cambridge CB2 2QL UNITED KINGDOM

Daniel Hantaï, M.D., Ph.D.

I.N.S.E.R.M. Unité 153,
75005 Paris, FRANCE

Gustave Moonen, M.D., Ph.D.

Université de Liège
Liège, BELGIUM

SPONSORS

NATO DIVISION OF SCIENTIFIC AFFAIRS
ABBOTT LABORATORIES
ALS MND RESEARCH FOUNDATION
AMVESTORS FINANCIAL CORPORATION
ARMY MEDICAL RESEARCH COMMAND
ASSOCIATION FRANÇAISE CONTRE LES MYOPATHIES
ATHENA NEUROSCIENCES
BAYER AG
BOEHRINGER INGELHEIM AG
CIBA-GEIGY AG
CNS RESEARCH, INC.
DELTA BIOTECHNOLOGY (UK)
E.I. DUPONT DE NEMOURS & CO., INC.
LILLY RESEARCH LABORATORIES
FONDAZIONE SIGMA TAU
GENENTECH, INC.
GLAXO RESEARCH LABORATORIES
INVITRON CORPORATION
KABI, FRANCE
MARATEA E BASILICATA AZIENDA TURISMO
MARION LABORATORIES, INC.
NEUROLOGICAL & PSYCHIATRIC RESEARCH & TREATMENT FOUNDATION
PFIZER INC.
SANDOZ, FRANCE
SCHERING AG
SCHERING CORPORATION
SIGMA TAU
SWINGSTERS, INC
THE UPJOHN COMPANY
U.S. DEPARTMENT OF VETERANS AFFAIRS

"Now *here*, you see, it takes all the running *you* can do, to keep in the same place. If you want to get somewhere else, you must run at least twice as fast as that!"

THE RED QUEEN TO ALICE
LEWIS CARROLL
THROUGH THE LOOKING GLASS

SERINE PROTEASES AND THEIR INHIBITORS IN THE NERVOUS SYSTEM

When we decided, in the Fall of 1988, to hold a NATO Advanced Research Workshop on the topic of the possible roles and regulation of serine proteases and their high molecular weight inhibitors, the serpins, in the nervous system we had no idea just how fast this fledgling field was developing. Just six months before, the simultaneous publication of the presence of a Kunitz-type protease inhibitor domain in the β amyloid precursor protein occurred. The isolation and purification of this molecule had only been published in 1984, the cloning of the gene just three years later. It was only in 1983 that the glial-derived neurite promoting factor was shown by Denis MONARD and his colleagues to be a fibrinolytic protease inhibitor and only in 1986 that it was similar to protease nexin (re-named protease nexin I). Protease nexin I itself was first reported by Joffre BAKER and Dennis CUNNINGHAM and their colleagues, to be made by human foreskin fibroblasts in 1980 and by our lab in clonal mouse muscle cells in 1981. It was not until 1987 that the cloning of the cDNA for the glial factor showed that it was identical to protease nexin I except for three amino acids and that protease nexin I was almost as good a promoter of neurite outgrowth as was the glial factor. In 1988 it was confirmed by Michael McGROGAN, Randy SCOTT and BAKER that the two molecules were virtually identical and that two forms of protease nexin I were produced by fibroblasts, the α and β forms, while only the β form was synthesized by astrocytes.

In the meantime, reports of the activity and possible roles of serine proteases in neural development, both in the periphery as well as in the central nervous system, were appearing. The remodelling that occurs in developing and regenerating tissues was not absent from the nervous system and several laboratories, notably those of

Nicholas SEEDS, Nurit KALDERON, Gustave MOONEN, Randall PITTMAN and our own, endeavored to show the effects of such enzymes in these tissue kinetic situations. Urokinase and tissue plasminogen activator levels were estimated and relative specificity for either the PNS, both nerve and neuromuscular, or CNS were found. The possible involvement of these proteases, along with their cell-surface receptors was just being explored in other remodelling and motility conditions and was being sought for in neural remodelling and development.

Fibrinolysis, and its control, had received little attention until recently in the pathogenesis of human primary brain tumors. This area has just begun to be of interest to clinicians, primarily neurosurgeons, and is likely to receive continued attention in the future. Raymond SAWAYA and his associates, and Jasti RAO in our laboratory, have focused on the balance of serine proteases and inhibitors in glioma pathogenesis and possible treatment.

The initial idea for the Workshop had its seed planted on both sides of the Seine during a delightful and productive sabbatical year we spent in Paris. It actually began one cold and wet December night in 1984 at the Brasserie Lipp when, under the influence of our then not-quite-three-year-old daughter, Mara, I met Professor H. Coen HEMKER of the University of Maastricht. He, in turn, introduced me to those delightful, and always helpful, sisters Jeannette and Claudine SORIA. Together with my then-new colleague in Professor Michel FARDEAU'S I.N.S.E.R.M. Unité 153, Daniel HANTAÏ, we extended our studies on the serine protease:serpin balance in the neuromuscular system, which continue on both sides of the Atlantic today.

We sought to organize the program of the meeting to bring together experts in the biochemistry and molecular biology of fibrinolysis and protease inhibition with those in neurobiology who might not have had previous occasions to come together. Some of our critics, perhaps influenced by other claims of fusion then extant in the lay press, said it was unlikely to be accomplished. After the first day, with numerous arrival mishaps, including the perennial strikes on the Italian rail system and others, we were not sure, ourselves, whether it would come off right. However, in another 24 hours, the ambience of the place and the hospitality of the staff of the Hotel Villa del Mare, began to work their magic on the participants so that, ultimately, most came away from THE MARATEA CONFERENCE with a truly remarkable feeling that they had experienced something new, unique and exciting. We are now convinced of this achievement and hope that the final product of this unique research workshop, this book, will convey those sentiments to the reader and, at the same time, stand as a reference book for some time chronicling, in fact, the launching of this field of protease regulation in neurobiological situations.

As editor, I am indebted to the members of the Organizing Committee, especially to Daniel HANTAÏ, who also served as associate editor, and to Robin CARRELL, who saw the "wisdom" of proceeding with such a workshop at this time. We are grateful to all the speakers at the Workshop for their timeliness in providing their manuscripts and to their secretaries and typists for their willingness in sending both hard copies as well as floppy disks *in usable* formats. By so doing we have been able to edit, correct and print this book in the shortest possible time. The credit for this book goes to them, while if any imperfections should be found, we take total responsibility for all. Thanks are due to our word processor assistant, Joyce CAPPS and to our editorial assistants, Julie ALEXANDER and Pamela GILFORD in Kansas City. I'd also like to thank Drs. Jasti RAO and Rajendra REDDY and Riichiro SUZUKI, of the Neurobiology Research Lab, for their untiring support throughout the entire process.

We gratefully acknowledge the financial support of all those organizations listed on the SPONSORS page, without which we could not have accomplished this Workshop. Dr. Craig SINCLAIR, of NATO's ARW Programme, was very supportive, along with Dr. BARTOLOMUCCI of FONDAZIONE SIGMA TAU and Dr. Friederich SCHUMANN of BAYER AG, who helped to defray some of the costs of preparing the final edited manuscript for publication. We wish to thank Melanie

YELITY and Gregory SAFFORD of PLENUM PUBLICATIONS in New York for their advice and assistance. On behalf of all the participants, we finally wish to thank Sr. A. GUZZARDI, his son, Giacomo, and the staff of the Hotel Villa Del Mare along with Madame T. IANNINI and her staff of the AZIENDA AUTOMNA SOGGIORNO E TURISMO DE MARATEA for transmuting a research meeting in their quaint seaside village into a truly memorable experience we will always treasure.

Barry W. Festoff
Kansas City, Fall 1989

CONTENTS

Francesco BLASI, M. Vittoria CUBELLIS, Ann Louring ROLDAN,
Maria Teresa MASUCCI, Niels BEHRENDT, Vincent ELLIS, Ettore APPELLA and
Keld DANØ

SECTION I

Biochemistry and cell biology of serine proteases and serpins

THROMBIN STRUCTURAL REGIONS IN DETERMINING BIOREGULATORY FUNCTIONS

JOHN W. FENTON II

Wadsworth Center for Laboratories & Research, New York State Department of Health, Empire State Plaza, Box 509, Albany, NY 12201, and Department of Physiology, The Albany Medical College of Union University, Albany, NY 12208, USA

INTRODUCTION

Thrombin (EC 3.4.21.5) is the activation product of its blood-circulating or cellular-sequestered zymogen, prothrombin. Unlike the majority of enzymes and activated proteins of the blood-coagulation, fibrinolytic, and complement systems, thrombin has multiple bioregulatory functions, which are central in hemostasis, wound healing, and perhaps developmental, as well as certain disease processes.[1-6]

Hemostasis occurs at three levels, which are: i) plasma proteins (e.g., coagulation factors V, VIII, and IX; proteins C and S, complement components), ii) blood cells (e.g., platelets, monocytes, neutrophils), and iii) blood vessels (e.g., endothelium, smooth muscles). Moreover, thrombin functions at all three levels of hemostasis, while the majority of cell types known are responsive to greater or lesser extents to thrombin. Thrombin furthermore functions both as a proteolytic enzyme with arginine- or lysine-directed specificities or as a nonenzymic activated protein (hormone-like) involving receptor occupancy on certain cells.[1-6]

In contrast to other proenzymes of coagulation and fibrinolytic serine proteinase, prothrombin has an unique gene structure indicating that its gene is more highly evolved than those of related proenzymes.[7] In this regard, the amino-acid sequence of the thrombin B chain (serine proteinase moiety of prothrombin) can be aligned to that of chymotrypsin, where the insertion peptides in the thrombin sequence occur at chymotrypsin gene exon junctions.[3] However, these insertions do not correspond as closely to the exon junction of prothrombin.[8] In further contrast to most closely related serine proteinase proenzymes,[7] prothrombin notably lacks a disulfide linkage connecting the majority of the propiece activation fragment (fragment F 1.2) to the enzyme moiety (the thrombin A and B chains). This has the consequence that thrombin is freed from the propiece upon activation of prothrombin. It also has the consequence that thrombin functions must derive biological specificities entirely from the thrombin moiety of prothrombin, whereas other related proteinases (e.g., plasmin; tissue plasminogen activator) have specificities [e.g., fibrin(ogen) binding] residing in

Serine Proteases and Their Serpin Inhibitors in the Nervous System
Edited by B. W. Festoff
Plenum Press, New York, 1990

their retained activation fragments. Thus, the unique amino-acid sequence of thrombin (e.g., the carbohydrate attachment insertion peptide; the arginine-and lysine-rich sequences; the arginyl-glycyl-aspartyl peptide) are potentially implicated in thrombin functions.[4-6]

The propiece fragment of prothrombin contrastingly has specificities pertaining to thrombin generation, which involve formation of the prothrombin activation complex (activated factors V and X, phospholipids, and Ca^{2+}). Prothrombin in association with this complex is not activated as a true classical enzymic process but rather as a "burst reaction" with slow turnover. This is the result of product inhibition of the prothrombin propiece fragment and the limiting role of activated factor V (factor Va). Thrombin, moreover, activates factor V much more rapidly than factor Xa. This amplification mechanism greatly magnifies the availability of thrombin to carry out other functions and is the critical step in regulating thrombin generation.[9]

Adjacent to the catalytic site in thrombin (the active-site histidine and serine) and corresponding to the residues immediately preceding thrombin-susceptible bonds in peptides or proteins, thrombin has an apolar binding site.[10] This site accounts for thrombin specificity with various peptides[11] and chromogenic tripeptide substrates[12] but does not explain thrombin specificity or interactions with fibrin (ogen).[13,14] Fibrinogen recognition involving the Aα cleavage requires an additional site. This site is destroyed or lost upon conversion of α- to β- and/or γ-thrombin by autoproteolytic or limited tryptic digestion.[15-17] Since these thrombin forms, which lack clotting activities, retain essentially full activities with synthetic substrates, the destroyed site is an exosite independent of the classical active site consisting of the catalytic site and adjacent regions.[12-14] Furthermore, since the exosite is responsible for binding of α-thrombin to negatively charged resin[14,18] and clotting activity is inhibited by various anionic substances [e.g., heparin and derivatives;[18,19] the anionic tail fragments of hirudin],[19-22] it is appropriately an "anion-binding exosite."[14] This exosite is attributed to a cluster of arginines and lysins where the β-thrombin is cleaved. It is also implicated in α-thrombin binding to various cells; this activity is blocked by hirudin, the anticoagulant from the European medicinal leech.[23-25] Hirudin also blocks chemotactic activity with monocytes[26] and neutrophils.[27] Such activity is an example of nonenzymic activity and has been ascribed to the carbohydrate attachment insertion peptide,[28] although hirudin does not affect the accessibility of the carbohydrate side chain.[18] In addition, thrombin has an arginyl-glycyl-aspartyl sequence, which may be involved in certain cellular interactions.[4-6]

During the thrombotic process, α-thrombin, as generated from prothrombin, participates in its amplification, then activates platelets, transforms surface properties of endothelium, and subsequently converts fibrinogen into clottable fibrin. With fibrin deposition and thrombus formation, α-thrombin becomes actively incorporated into the clot along with other plasma proteins (e.g., factor XIII; plasminogen) and various blood cells (e.g., neutrophils). Thus, the thrombus is a storage reservoir for thrombin, which can be subsequently released.[29] Sequestering and activation of neutrophils are the hallmarks of thrombin-induced pulmonary edema and lung injury,[30,31] and presumptively other thrombotic episodes follow similar courses of maturation. With neutrophil activation occurs oxygen radical production,[30] as well as the release of proteinase, which begins thrombus degradation.[32] This releases thrombin from within the thrombus to presumably further stimulate endothelial or other cells to secrete tissue plasminogen activator and other substances.[33] Thus, thrombin not only causes the formation of thrombi but may also be the initiator of fibrinolysis.[34]

Thrombin further induces albumin transport across endothelial cell monolayers;[35] this is presumptively an important process in edema (tissue swelling) and the cotransport of substrates to the subendothelium. Thrombin itself causes smooth muscle contracture,[36] is a potent mitogenic substance for several cell types,[24,25] as well as being an angiogenic substance[5] and inducing bone resorption.[37] Therefore, functions of thrombin do not stop with clotting but appear to continue through the entire healing process.

Because thrombin appears to function largely at the cellular level, it would seem to have its evolutionary origins as a developmental proteinase. The predominance of thrombin activities at the cellular level is illustrated by the ability of γ-thrombin, which lacks clotting activity, to induce pulmonary edema without causing clotting.[34] Cellular responses to thrombin include morphological rounding or shape change, intracellular structural rearrangements, and secretory processes. The majority of these cellular effects require catalytically active forms of thrombin,[4-6] although some cells need both high-affinity binding to the cell (receptor occupancy), as well as catalytically active thrombin.[38] Only a few activities, such as leukocyte chemotaxis, are known to occur with catalytically inactive thrombin forms alone.[26,27]

Although the blood barrier of the central nervous system should exclude prothrombin or thrombin, pore changes, such as those with edema or other injury, could allow passage into nerve tissue. The brain (and presumptively other nervous tissues) has long been recognized as a very potent source of thromboplastin for prothrombin activation.[39] The presence of protease nexin-1 in such tissue suggest it may serve to scavenge thrombin or similar proteinases.[40,41]

Thrombin is known to bind to brain tissue and to stimulate metabolic processes in neuroblastoma cells in tissue culture.[42] More recently, thrombin has also been found to cause neurite retraction,[40,41] a process suggestive of the rounding in shape changes of other cells. Present evidence suggests that prothrombin is not actively synthesized in brain tissue,[43] but whether its cousins or distant relatives are so remains to be determined. Most likely, prothrombin has developed from a proenzyme with developmental functions. This would help explain the diverse and underlying functions of thrombin in hemostasis, wound healing, and certain disease processes, where these functions are largely targeted at the cellular level.

ACKNOWLEDGMENTS

I would like to thank Dr. Barry Festoff for inviting me to participate in this symposium and Christine Bradley for secretarial assistance. Studies on human thrombin have been supported in part by NIH Grant HL13160-18.

REFERENCES

1. J.W. Fenton II, Thrombin specificity, **Ann. N.Y. Acad. Sci.** 370:468-495 (1981).
2. J.W. Fenton II, Thrombin. **Ann. N.Y. Acad. Sci.** 485:5-15 (1986).
3. J.W. Fenton II, and D.H. Bing, Thrombin active-site regions., **Semin. Thromb. Hemostas.** 12:200-208 (1986).
4. J.W. Fenton II, Structural regions and bioregulatory functions of thrombin, in: "Cell Proliferation: Recent Advances," vol. II, A.L. Boynton and H.L.Leffert, eds., pp 133-151, Academic Press, New York, NY (1987).
5. J.W. Fenton II, Regulation of thrombin generation and functions. **Semin. Thromb. Hemostas.** 14:234-240 (1987).
6. J.W. Fenton II, Thrombin bioregulatory functions, **Adv. Clin. Enzymol.** 6:186-193 (1988).
7. D.M. Irwin, K.A. Roberston, and R.T.A. MacGillivray, Structure and evolution of the bovine prothrombin gene, **J. Mol. Biol.** 200:31-45 (1988).

8. D.H. Bing, R.J. Feldmann, and J.W. Fenton II, Structure-function relationships of thrombin based on the computer-generated three-dimensional model of the B chain of bovine thrombin. **Ann. N.Y. Acad. Sci.** 485:104-119 (1986).

9. F.A. Ofosu, J. Hirsh, C.T. Esmon, G.J. Modi, L.M. Smith, N. Anvari, M.R. Buchanan, J.W. Fenton II, and M.A. Blajchman, Unfractionated heparin inhibits thrombin-catalyzed amplification reactions of coagulation more efficiently than those catalyzed by factor Xa, **Biochem. J.** 257:143-150 (1989).

10. S.A. Sonder and J.W. Fenton II, Proflavine binding with in the fibrinopeptide of human α-thrombin, **Biochemistry** 23:1818-1823 (1984).

11. J.-Y. Chang, Thrombin specificity. Requirement of apolar amino acid adjacent to the thrombin cleavage site of polypeptide substrates, **Eur. J. Biochem.** 151:217-224 (1985).

12. S.A. Sonder and J.W. Fenton II, Thrombin specificity with tripeptide substrates comparison of human and bovine thrombins with and without fibrinogen clotting activities, **Clin. Chem.** 32:934-937 (1986).

13. S.D. Lewis, L. Lorand, J.W. Fenton II, and J.A. Shafer, Catalytic competence of human α- and γ-thrombin in the activation of fibrinogen and factor XIII, **Biochemistry** 26:7597-7603 (1987).

14. J.W. Fenton II, T.A. Olson, M.P. Zabinski, and G.D. Wilner, Anion-binding exosite of human α-thrombin and fibrin(ogen) recognition, **Biochemistry** 27:7106-7112 (1988).

15. J.W. Fenton II, M.J. Fasco, A.B. Stackrow, D.L. Aronson, A.M. Young, and J.S. Finlayson, Human thrombins, production, evaluation and properties of α-thrombin **J. Biol. Chem.** 252:3587-3598 (1977).

16. J.W. Fenton II, B.H. Landis, D.A. Walz, and J.S. Finlayson, Human thrombins, in: "**Chemistry and Biology of Thrombin**", R.L. Lundblad, J.W. Fenton II, and K.G. Mann, eds., pp. 43-70, Ann Arbor Science Publishers, Ann Arbor, MI (1977).

17. J.W. Fenton II, B.H. Landis, D.H. Bing, R.D. Feinmann, M.P. Zabinski, S.A. Sonder, L.J. Berliner, and J.S. Finlayson, Human thrombin: preoperative evaluation, structural properties, and enzymic specificity, in: "**The Chemistry and Physiology of the Human Plasma Proteins.**" D.H. Bing, ed., pp. 151-183, Pergamon, New York, NY (1979).

18. T.A. Olson, S.A. Sonder, G.D. Wilner, and J.W. Fenton II, Heparin binding in proximity to the catalytic site of human α-thrombin, **Ann. N.Y. Acad. Sci.** 485:96-103 (1986).

19. J.W. Fenton II, J.I. Witting, C. Pouliott, and J. Fareed, Thrombin anion-binding exosite interactions with heparin and various polyanions, **Ann. N.Y. Acad. Sci.**, 556:158-168 (1989).

20. J.L. Krstenansky and S.J. Mao, Antithrombin properties of C-terminus of hirudin using synthetic unsulfated N^α - acetyl-hirudin 45-65, **FEBS Lett.** 211:10-16 (1987).

21. S.J.T. Mao, M.T. Yates, T.J. Owen, and J.L. Krstenansky, Interaction of hirudin with thrombin: identification of a minimal binding domain of hirudin that inhibits clotting activity, **Biochemistry** 27:8170-8173 (1988).

22. J.M. Maraganore, B. Chao, M.L. Joseph, J. Jablonski, and K.L Ramachandran, Anticoagulant activity of synthetic hirudin peptides, **J. Biol. Chem.** 264:8692-8698 (1989).

23. S.W. Tam, J.W. Fenton II, and T.C. Detwiler, Dissociation of thrombin from platelets by hirudin. Evidence for receptor processing, **J. Biol. Chem.** 254:8723-8725 (1979).

24. K.C. Glenn, D.H. Carney, J.W. Fenton II, and D.D. Cunningham, Thrombin active-site regions required for fibroblast receptor binding and initiation of cell division, **J. Biol. Chem.** 255:6609-6616 (1980).

25. J.F. Perdue, W. Lubenskyi, E. Kivity, S.A. Sonder, and J.W. Fenton II, Protease mitogenic response of chick-embryo fibroblasts and receptor binding/processing of human α-thrombin. **J. Biol. Chem.** 256:2767-2776 (1981).

26. R. Bar-Shavit, A. Khan, G.D. Wilner, and J.W. Fenton II, Monocyte chemotaxis: stimulation by specific exosite region in thrombin, **Science** 220:728-731 (1983).

27. R. Bizios, L. Lai, J.W. Fenton II, and A.B. Malik, Thrombin-induced chemotaxis and aggregation of neutrophils, **J. Cell. Physiol.** 128:485-490 (1986).

28. R. Bar-Shavit, A. Kahn, M.S. Mudd, G.D. Wilner, K.G. Mann, and J.W. Fenton II, Localization of chemotactic domain in human thrombin, **Biochemistry** 23:397-400 (1984).

29. G.D. Wilner, M.P. Danitz, M.S. Mudd, K.-H. Hsieh, and J.W. Fenton II, Selective immobilization of α-thrombin by surface-bound fibrin. **J. Lab. Clin. Med.** 97:403-411 (1981).

30. A.B. Malik, Pulmonary microembolism, **Physiol. Rev.** 63:1114-1207 (1983).

31. J.E. Kaplan and A.B. Malik, Thrombin-induced intravascular coagulation: role in vascular injury, **Semin. Thromb. Hemost.** 13:398-415 (1987).

32. C.W. Francis and V.J. Marder, Degradation of cross-linked fibrin of human leukocyte proteases, **J. Lab. Clin. Med.** 107:342-352 (1986).

6

33. E.G. Levin, D.M. Stern, P.P. Nawroth, R.A. Marlar, D.S. Fair, J.W. Fenton II, and L.A. Harker, Specificity of the thrombin-induced release of tissue plasminogen activator from cultured human endothelial cells. **Thromb. Haemost.** 56:115-119 (1988).

34. R.R. Garcia-Szabo, J.W. Fenton II, S.K. Lo, M. Hussain, and A.B. Malik, Increased transvascular protein transport in lungs induced by nonthrombogenic γ-thrombin, **Am. J. Physiol.** 256:H1690-1696 (1989).

35. J.G.N. Garcia, A. Siflinger-Birnboim, R. Bizios, P.J. Del Vecchio, J.W. Fenton II, and A.B. Malik, Thrombin-induced increases in albumin permeability across the endothelium, **J. Cell. Physiol.** 128:96-104 (1986).

36. D.A. Walz, G.F. Anderson, R.E. Ciaglowski, M. Aiken, and J.W. Fenton II, Thrombin-elicited contractile responses of aortic smooth muscle, **Proc. Soc. Expl. Biol. Med.** 180:518-526 (1985).

37. O. Hoffman, K. Kalushofer, K. Koller, M. Peterlik, T. Mavreas, and P. Stern, Indomethacin inhibits thrombin, but not thyroxin-stimulated resorption of fetal limb bones, **Prostaglandins** 31:601-608 (1986).

38. D.H. Carney, G.J. Herbosa, J. Stiernberg, J.S. Bergmann, E.A. Gordon, D. Scott, and J.W. Fenton II, Double signal hypothesis for thrombin initiation of cell proliferation, **Semin. Thromb. Hemost.** 12:231-240 (1986).

39. R. Biggs, **"Human Blood Coagulation, Haemostasis and Thrombosis,"** 2nd ed. Blackwell Scientific Publications, Oxford (1976).

40. D. Gurwitz and D.D. Cunningham, Thrombin modulates and reverses neuroblastoma neurite outgrowth, **Proc. Natl. Acad. Sci. USA.** 85:3440-3444 (1988).

41. D.D. Cunningham and D. Gurwitz, Proteolytic regulation of neurite outgrowth from neuroblastoma cells by thrombin and protease nexin-1, **J. Cell. Biochem.** 39:55-64 (1989).

42. R.M. Snider, M. McKinney, and E. Ritchelson, Thrombin binding and stimulation of cyclic guanosine monophosphate formation in neuroblastoma cells, **Semin. Thromb. Hemost.** 12:253-262 (1986).

43. D.D. Cunningham and S.L. Wagner, personal communication (1989).

REGULATION AND CONTROL OF THE FIBRINOLYTIC SYSTEM

H. ROGER LIJNEN AND DÉSIRÉ COLLEN

Center for Thrombosis and Vascular Research
Campus Gasthuisberg, O & N
Herestraat 49
B-3000 Leuven, Belgium

INTRODUCTION

Mammalian blood contains an enzymatic system, called the fibrinolytic system, that is capable of dissolving blood clots. This system comprises an inactive proenzyme, plasminogen, which can be converted to the active enzyme plasmin, that will degrade fibrin into soluble fibrin degradation products. Two immunologically distinct types of physiological plasminogen activators have been identified: the tissue-type plasminogen activator (t-PA) and the urokinase-type plasminogen activator (u-PA). Inhibition of the fibrinolytic system may occur either at the level of the plasminogen activators, by plasminogen activator inhibitors (PAI-1 and PAI-2), or at the level of plasmin, mainly by α_2-antiplasmin. Plasminogen activation may also be induced by an "intrinsic" pathway involving several proteins such as Factor XII, high molecular weight kininogen (HMWK) and prekallikrein.

MAIN COMPONENTS OF THE FIBRINOLYTIC SYSTEM

Plasminogen

Human plasminogen is a single-chain glycoprotein with a molecular weight of 92,000, containing 790 amino acids and 5 homologous triple-loop structures or "kringles."[1] The normal plasma concentration is about 1.5 to 2 μM. Native plasminogen has NH_2-terminal glutamic acid ("Glu-plasminogen") but is easily converted by limited plasmic digestion to modified forms with NH_2-terminal lysine, valine, or methionine, commonly designated "Lys-plasminogen." This conversion occurs by hydrolysis of the Arg67-Met68, Lys76-Lys77, or Lys77-Val78 peptide bonds. The hydrodynamic properties of both types have been reviewed elsewhere.[2]

A 2.7 kb insert of a cDNA clone for human plasminogen containing the complete coding region has been sequenced. The amino acid sequence predicted from this cDNA is close to the published protein sequence and differs only in 4 amide assignments (Glx, Asx) and in the presence of an extra isoleucine at position 85, giving

Serine Proteases and Their Serpin Inhibitors in the Nervous System
Edited by B. W. Festoff
Plenum Press, New York, 1990

a total of 791 amino acids for human plasminogen.[3] The plasminogen gene was mapped to the long arm of chromosome 6 at band q26 or q27.[4]

Plasminogen is converted to plasmin by cleavage of a single Arg-Val peptide bond[5] corresponding to the Arg560-Val561 bond based on the 790 amino acid numbering system. The two-chain plasmin molecule is composed of a heavy chain or A-chain, originating from the NH$_2$-terminal part of plasminogen and a light chain or B-chain constituting the COOH-terminal part. The B-chain was found to contain an active site similar to that of trypsin,[1] composed of His602, Asp645, and Ser740. Investigation of the activation pathways of plasminogen with the use of monoclonal antibodies specific for Lys-plasminogen, has revealed that activation of Glu-plasminogen in human plasma occurs by direct cleavage of the Arg560-Val561 peptide bond without generation of Lys-plasminogen intermediates.[6]

The plasminogen molecule contains structures, called lysine binding sites, which are located in the plasmin A-chain. These lysine binding sites mediate its interaction with fibrin and with α_2-antiplasmin. Therefore, it has been suggested that the lysine binding sites play a crucial role in the regulation of fibrinolysis.[7,8]

Physiological Plasminogen Activators

Two types of physiological plasminogen activators have been identified; one related to the activator found in tissues (tissue-type plasminogen activator, t-PA) and the other related to the activator found in urine (urokinase-type plasminogen activator, u-PA). u-PA may be obtained as a two-chain molecule (tcu-PA) or as a single-chain form (scu-PA or pro-urokinase).

Tissue-type Plasminogen Activator (t-PA). The plasminogen activator found in blood represents vascular plasminogen activator which is synthesized and secreted by endothelial cells[9] and is now called "tissue-type plasminogen activator" (t-PA) or alteplase. t-PA antigen levels in normal plasma are about 5 ng/ml.[10]

t-PA has been purified from the culture fluid of a stable human melanoma cell line (Bowes, RPMI-7272) in sufficient amounts to study its biochemical and biological properties.[11] t-PA for clinical use is presently produced by recombinant DNA technology (Activase®, Genentech Inc. or Actilyse®, Boehringer Ingelheim GmbH).

Native t-PA is a serine proteinase with a M_r of about 70,000, composed of a single polypeptide chain of 527 amino acids with Ser as NH$_2$-terminal amino acid.[12] t-PA is converted by plasmin to a two-chain form by hydrolysis of the Arg275-Ile276 peptide bond (numbering based on a total of 527 residues). The NH$_2$-terminal region is composed of several domains with homologies to other proteins: residues 1-43 are homologous to the "finger domains" (F) in fibronectin,[13] residues 44-91 are homologous to human epidermal growth factor (E), and residues 92-173 (K$_1$) and 180-261 (K$_2$) are both homologous to the "kringle" regions of plasminogen.[12] The region comprising residues 276-527 is homologous to that of other serine proteinases and contains the catalytic site, which is composed of His322, Asp371, and Ser478.[12]

t-PA has a specific affinity for fibrin. The structures involved in the fibrin-binding of t-PA are fully comprised within the A-(heavy) chain. Evidence obtained with deletion mutants suggests that binding of t-PA to fibrin is mediated both via the finger domain and via the second kringle region.[14,15] A lysine-binding site (LBS) is involved in the interaction of the kringle-2 domain with fibrin (abolished by the lysine analogue 6-aminohexanoic acid), but not in the interaction of the finger domain with fibrin.[16] The structures required for the enzymatic activity of t-PA are fully comprised within the B-chain.

The activation of plasminogen by t-PA, both in the presence and the absence of fibrin, follows Michaelis-Menten kinetics. Although different kinetic constants were obtained by several investigators,[17-20] there is a consensus that the presence of fibrin enhances the efficiency of plasminogen activation by t-PA with 2 to 3 orders of magnitude. Three groups[14,15,21] have reported that t-PA mutants lacking the finger domain are equally well stimulated by fibrin as intact t-PA, whereas mutants lacking the second kringle structure but containing the finger domain show significantly reduced stimulation, suggesting that interaction of K_2 with fibrin would primarily be responsible for fibrin stimulation.

In contrast, Gething et al.[22] reported that t-PA mutants lacking both kringles are not stimulated by fibrin. However, mutants containing only one kringle (either K_1 or K_2) are indistinguishable from one another and from wild-type t-PA, suggesting that K_1 and K_2 would be equivalent in their ability to mediate stimulation of the catalytic activity of t-PA by fibrin.

The gene coding for human t-PA[23] has been localized to chromosome 8. The t-PA gene has been characterized in detail.[24-26] Thirteen intervening sequences divide the gene into 14 coding regions. The assembly of the t-PA gene is an example of the "exon shuffling" principle. This is suggested by the findings that the different structural domains on the heavy chain (F, E, K_1, K_2) are encoded by a single exon or by two adjacent exons.[27] Because of the striking correlation between the intron-exon distribution of the gene and the putative domain structure of the protein, it was suggested that these domains would be autonomous structural and/or functional entities ("modules").[13,28]

Urokinase-type Plasminogen Activator (u-PA). Two-chain urokinase (tcu-PA), a trypsin-like serine proteinase composed of two polypeptide chains (M_r 20,000 and 34,000), has been isolated from human urine and from cultured human embryonic kidney cells. Over the last years, several groups have isolated a single-chain form of urokinase (scu-PA) from urine, plasma or from conditioned cell culture media. Recently, it was also obtained by recombinant DNA technology and prepared from the translation product in *E. Coli* of an expression plasmid coding for human urokinase.[29]

scu-PA is a single chain glycoprotein with M_r 54,000 containing 411 amino acids.[30] Upon limited hydrolysis by plasmin or kallikrein of the Lys^{158}-Ile^{159} peptide bond, the molecule is converted to a two-chain derivative. The catalytic center is located in the COOH-terminal chain and is composed of Asp255, His204 and Ser356. The NH$_2$-terminal chain contains regions homologous to human epidermal growth factor (residues 5-49) and one region homologous to the plasminogen kringles. A low M_r two-chain urokinase (M_r 33,000) can be generated with plasmin by hydrolysis of the Lys135-Lys136 peptide bond.[31]

The cDNA of u-PA has been isolated and the nucleotide sequence determined.[30,32] The human u-PA gene is located on chromosome 10; it contains 11 exons and the intron-exon organization of the gene closely resembles that of the t-PA gene.

tcu-PA activates plasminogen directly following Michaelis-Menten kinetics. Initially, refolded scu-PA expressed in *E. Coli* was found to be fully active. However, subsequently preparations obtained from natural sources or expressed in mammalian cell systems were found to have approximately 1% of the activity of tcu-PA.[33] A low intrinsic activity of scu-PA was confirmed by others,[34] whereas some authors have claimed that scu-PA has no enzymatic activity and is a genuine proenzyme.[35]

Inhibitors of the Fibrinolytic System

Inhibition of the fibrinolytic system may occur at the level of plasmin, mainly by α_2-antiplasmin, or at the level of plasminogen activators, mainly by plasminogen activator inhibitor-1 (PAI-1).

Alpha$_2$-antiplasmin. Alpha$_2$-antiplasmin is a single-chain glycoprotein with a molecular weight of 70,000 consisting of 452 amino acids. Alpha$_2$-antiplasmin belongs to the serine protease inhibitor protein family (serpins).[36-38] The concentration of the inhibitor in normal plasma is about 7 mg/100 ml (about 1 μM).[39]

α_2-Antiplasmin forms a 1:1 stoichiometric inactive complex with plasmin. This occurs in two successive reactions: a fast reversible second-order reaction followed by a slower irreversible first-order transition. The model can be represented by:

$$P + A \underset{k_{-1}}{\overset{k_1}{\rightleftharpoons}} PA \overset{k_2}{\longrightarrow} PA'$$

in which P is plasmin, A is α_2-antiplasmin, PA a reversible inactive complex and PA' an irreversible inactive complex. The reactive site peptide bond in α_2-antiplasmin cleaved by plasmin[36] consists of Arg364-Met365. A bond is formed between the active site seryl residue in plasmin and a specific arginyl residue in the inhibitor. The second-order rate constant[40] of the reaction k_1 is 2 to 4 x 10^7 M^{-1}s^{-1}. This rate is strongly decreased when the lysine-binding sites and the active site in the plasmin molecule are not available such as when plasmin is digesting fibrin. Thus, the half-life of plasmin molecules on the fibrin surface is estimated to be 2 to 3 orders of magnitude longer than that of free plasmin.[8]

Plasminogen Activator Inhibitor-1 (PAI-1). PAI-1 was first identified in conditioned media of human endothelial cells and rat hepatoma cells. It has subsequently been identified in plasma, platelets, placenta and conditioned media of fibrosarcoma cells and hepatocytes.[41] It is a single chain glycoprotein with M_r about 52,000 consisting of 379 amino acids. The cDNA has been sequenced, revealing that PAI-1 is a member of the serine protease inhibitor (serpin) family.[42,43] Its reactive site consists of Arg346-Met347. The PAI-1 gene is approximately 12.2 kb in length and consists of nine exons and eight introns. It has been mapped[44] to chromosome 7, bands q21.3-q22.

In healthy individuals, highly variable plasma levels of both PAI activity and PAI-1 antigen were observed. PAI activity ranges from 0.5 to 47 U/ml with the majority of values (80%) below 6 U/ml. PAI-1 antigen ranges between 6 ng/ml in 10% of the plasmas to 85 ng/ml (geometric mean: 24 ng/ml).[45] PAI activity is very rapidly cleared from the circulation via the liver; a half-life of 7 minutes in the rabbit has been reported.[46]

PAI-1 is stabilized by binding to a plasminogen activator inhibitor binding protein identified as S-protein or vitronectin.[47] PAI-1 is the primary inhibitor of t-PA and u-PA in human plasma.[48] It reacts with single chain and two chain t-PA and with tcu-PA, but not with scu-PA or with streptokinase. The second order rate constant for the inhibition of single chain t-PA by PAI-1 is in the order of 10^7 M^{-1}s^{-1}, while inhibition of two chain t-PA and tcu-PA is even more rapid. PAI-1 is released during platelet aggregation, resulting in a 10-fold increased level.[49]

PAI-1 is very labile and is found in tissue culture fluid in a "latent" form which can be partly reactivated by treatment with denaturing agents such as urea, guanidine hydrochloride and dodecyl sulfate.[50] Production and/or release of PAI-1 from various cells is regulated by several stimuli, including thrombin, dexamethasone, endotoxin, interleukin-1, transforming growth factor ß and tumor necrosis factor.[41] PAI-1 levels are elevated in several disease states (see below, section on pathophysiology).

REGULATION AND CONTROL OF PHYSIOLOGICAL FIBRINOLYSIS

Molecular Interactions between the Components of the Fibrinolytic System

Plasminogen interacts specifically but weakly with fibrin through structures called the "lysine-binding sites".[51] The presence of t-PA increases the binding of plasminogen to a fibrin clot.[52] t-PA has a specific affinity for fibrin; the dissociation constant of the activator-fibrin complex was estimated to be 0.14 μM.[17] Neither tcu-PA or scu-PA bind specifically to fibrin.[53]

α_2-Antiplasmin interacts weakly with Glu-plasminogen through the lysine-binding site(s) ($K_d = 4$ μM).[54] It reacts very rapidly with plasmin (second order rate constant 2-4 x 10^7 $M^{-1}s^{-1}$); the rate of this inhibition is strongly dependent on the availability of free lysine-binding sites and a free active site in the plasmin molecule.[40] α_2-Antiplasmin is cross-linked to the fibrin α chain when blood is clotted in the presence of calcium ions and activated coagulation factor XIII, thereby inhibiting endogenous fibrinolysis.[55]

Mechanism of Action of t-PA

t-PA is a poor enzyme in the absence of fibrin, but the presence of fibrin strikingly enhances the activation rate of plasminogen.[17] The kinetic data of Hoylaerts et al.[17] support a mechanism in which fibrin provides a surface to which t-PA and plasminogen bind in a sequential and ordered way yielding a cyclic ternary complex. Fibrin essentially increases the local plasminogen concentration by creating an additional interaction between t-PA and its substrate. The high affinity of t-PA for plasminogen in the presence of fibrin thus allows efficient activation on the fibrin clot, while no efficient plasminogen activation by t-PA occurs in plasma. However, others have claimed that fibrin influences both the K_m and k_{cat} of the activation of plasminogen by t-PA.[18]

Plasmin formed on the fibrin surface has both its lysine-binding sites and active site occupied and is thus only slowly inactivated by α_2-antiplasmin (half-life of about 10-100 s); free plasmin, when formed, is rapidly inhibited by α_2-antiplasmin (half-life of about 0.1 s).[40] The fibrinolytic process thus seems to be triggered by and confined to fibrin.

It was proposed[16] that initial binding of t-PA to fibrin would be governed by the finger domain, and that following partial degradation of fibrin, newly exposed carboxyl-terminal lysine residues result in enhanced binding of t-PA via kringle 2. Higgins & Vehar[56] have confirmed that, when fibrin is degraded by plasmin, new t-PA binding sites with markedly lower (2-4 orders of magnitude) dissociation constants are formed. Thus, early fibrin digestion by plasmin could accelerate fibrinolysis by increasing the binding of both t-PA and plasminogen.

Mechanism of Action of u-PA

tcu-PA has no specific affinity for fibrin and activates fibrin-bound and circulating plasminogen relatively indiscriminately. Extensive plasminogen activation and depletion of α_2-antiplasmin may occur following treatment of thromboembolic diseases with tcu-PA, leading to degradation of several plasma proteins including fibrinogen, Factor V and Factor VIII.

scu-PA, in contrast to tcu-PA, has significant fibrin-specificity. A hypothetical mechanism of action has been proposed[53] on the basis of the observation that scu-PA has some intrinsic plasminogen activating potential, that scu-PA does not significantly activate plasminogen in plasma in the absence of fibrin and that fibrin reverses the inhibition exerted by plasma, although not via direct binding of scu-PA to fibrin.

These findings suggest that plasma components interfere with plasminogen activation and that the fibrin-specificity of the activation of plasminogen by scu-PA is due to reversal by fibrin of the inhibited state.

Alternatively, the presence of fibrin may enhance the activation rate of fibrin-bound plasminogen.[57] In clot lysis systems in human plasma *in vitro*, conversion of scu-PA to tcu-PA has been shown to constitute a major positive feed-back mechanism for fibrinolysis.[33] The molecular interactions which regulate the fibrin-specific activation of the fibrinolytic system by scu-PA remain to be further detailed.

PATHOPHYSIOLOGICAL ASPECTS OF FIBRINOLYSIS

Excessive Fibrinolysis

α_2-Antiplasmin Deficiency and Bleeding. Patients with congenital α_2-antiplasmin deficiency may have a clinically severe bleeding tendency.[58-63] This bleeding tendency may be due to premature lysis of hemostatic plugs, because in the absence of α_2-antiplasmin, plasmin molecules generated on the fibrin surface of the hemostatic plug would persist much longer than their normal half-life. In contrast to the first published families with α_2-antiplasmin deficiency, which were discovered through a homozygous propositus with severe bleeding, several cases have now been reported where heterozygotes have no discernible bleeding symptoms.[58,62,63] In contrast, 6 out of 16 heterozygotes in the Dutch families[59] and 2 out of 5 heterozygotes in the American families[64] presented with mild bleeding tendencies. The α_2-antiplasmin levels in all heterozygotes described thus far are very similar, i.e. 40 to 60% of normal.

Decreased levels of α_2-antiplasmin have been observed in liver diseases and appear to be an important factor in the increased fibrinolytic activity. Decreased levels of α_2-antiplasmin have also been reported in patients with disseminated intravascular coagulation and some forms of renal disease.[2] α_2-Antiplasmin levels may be significantly reduced in patients undergoing thrombolytic therapy, as a result of systemic activation of the fibrinolytic system.[65]

An abnormal α_2-antiplasmin that is associated with a serious bleeding tendency has been found in two siblings in a Dutch family, and is referred to as α_2-antiplasmin Enschede. These individuals have 3% of normal functional activity and 100% of normal antigen levels.[66] The ability of the protein to reversibly bind plasmin or plasminogen was not affected, but the abnormal α_2-antiplasmin is converted from an inhibitor of plasmin to a substrate. The molecular defect of α_2-antiplasmin Enschede consists of the insertion of an extra alanine residue (GCG insertion) somewhere between amino acid residues 353 and 357 (4 Ala residues), seven to ten positions on the aminoterminal side of the P_1 residue (Arg364) in the reactive site of α_2-antiplasmin.[67]

Excess t-PA Levels and Bleeding. Excessive fibrinolysis due to increased t-PA levels may cause bleeding complications. A life-long hemorrhagic disorder associated with enhanced fibrinolysis due to increased levels of circulating plasminogen activator has been described. No deficiency of any known inhibitor of fibrinolysis could be detected.[68] A similar case of (inherited) increased fibrinolytic potential due to an excess of t-PA was reported in a Spanish family.[69]

Impairment of Fibrinolysis

PAI Activity and Thrombosis. Increased levels of PAI activity have been reported in several thrombotic disease states. Increased levels of PAI were observed in coronary heart disease[70,71] and acute myocardial infarction.[72,73] In patients who had suffered a myocardial infarction, the depressed fibrinolytic activity was found to be due to an increased PAI activity;[74,75] increased levels of PAI activity in these patients were associated with an increased risk of reinfarction.[76] Plasma PAI activity was found to be increased in patients with a recent episode of thrombosis.[77,78] In prospective studies, however, no difference was found in PAI activity in patients with or without deep vein thrombosis (DVT).[79,80] In patients undergoing hip replacement, preoperative PAI-activity was significantly elevated in those patients who would develop post-operative DVT.[81] Such relationship was however not observed in patients undergoing major abdominal surgery.[79,82] These discrepancies suggest either that increased levels of PAI-activity may constitute a risk factor for the development of DVT or that PAI actually increases as a result of DVT. In patients with recurrent DVT, a defective fibrinolytic response to venous occlusion is often observed, as a result of either a decreased release of t-PA or a high level of PAI.[83-85] Juhan-Vague et al.[84] have investigated 120 patients with spontaneous or recurrent DVT and observed 3 groups based on their response to venous occlusion. A poor fibrinolytic response (less than 2-fold increase) to venous occlusion occurred in 35 percent of these patients, one quarter of them (12 patients) with deficient t-PA release and high risk for recurrent DVT and three quarters (32 patients) with normal t-PA release but increased level of PAI-1. Mellbring et al.[86] reported that patients subjected to major abdominal surgery, who have enhanced fibrinolytic activity preoperatively (as measured from the levels of plasmin-α_2-antiplasmin complex), have less tendency to develop postoperative deep vein thrombosis.

Plasminogen Deficiency and Thrombosis. Congenital plasminogen deficiency as a cause of thrombosis, characterized by a parallel decrease of functional and immunoreactive plasminogen, has been reported in only a few cases.[87-91]

Reduced levels of plasminogen have been observed in several clinical conditions, including liver disease[92] and sepsis (25 to 45% of normal activity).[95] Possible mechanisms for acquired plasminogen deficiency in severe liver disease are depression of the synthesis as well as increased consumption.[94] Degradation of plasminogen into low M_r-plasminogen by leukocyte elastase has been suggested as a possible explanation for the reduction observed in septic patients.[93]

Abnormalities in the plasminogen molecule (dysplasminogenemia) resulting in defective activation to plasmin have been described by several authors. The abnormality may consist of a defect in the active site[95-98] or a defect in the activation mechanism.[99,100] In plasminogen Tochigi I, Tochigi II and Nagoya the molecular defect was identified as a single amino acid substitution (Ala600 to Thr) disturbing the charge relay system of the active center.[96,97] Wohl and coworkers[99,100] have characterized three human plasminogen variants (Chicago I, II and III), identified in young males with a history of recurring deep vein thrombosis. Both homozygote plasminogens Chicago I and II have an activation defect characterized by a higher Michaelis constant and impaired plasminogen activator binding but normal cleavage of the Arg560-Val561 peptide bond.[100] The homozygote plasminogen Chicago III has both an impaired affinity for plasminogen activators as evidenced by an increased Michaelis constant, and an impaired cleavage of the Arg560-Val561 peptide bond.[99]

Characteristic of all these dysplasminogenemias is a lowered ratio of functional to immunoreactive plasminogen.

Plasminogen Activator Deficiency and Thrombosis. Impairment of fibrinolysis due to deficient synthesis and/or release of t-PA from the vessel wall or to increased levels of PAI-1 is associated with a tendency to thrombosis.

Isacson & Nilsson[101] found a defective release of t-PA from the vessel wall during venous occlusion and/or a decreased t-PA content in walls of superficial veins in about 70% of a large series of patients with idiopathic recurrent venous thrombosis. This association of recurrent venous thrombosis with a defect in the fibrinolytic system is more frequent than with any other known disturbance of hemostasis. Korninger et al.[102] followed a group of 121 patients with a history of venous thrombosis and/or pulmonary embolism for 56 ± 19 months and observed a significantly lower recurrence rate in patients with a post-occlusion euglobulin lysis time shorter than 60 min (4.8% per year) as compared to patients with a post-occlusion euglobulin lysis time longer than 60 min (10.3% per year). In these two studies[101,102] it was not evaluated whether the impaired fibrinolytic activity was related to a deficient release of t-PA from the vessel wall or to an increased rate of inactivation or elimination. Subsequent studies[84,103] have shown that both mechanisms may contribute to a deficient fibrinolytic response.

CONCLUSION

Regulation and control of the fibrinolytic system depends on specific molecular interactions between its main components. Disorders of the fibrinolytic system may result either from impairment of fibrinolysis (thrombotic complications) or from excessive activation of the fibrinolytic system (bleeding tendency).

REFERENCES

1. L. Sottrup-Jensen, T.E. Petersen, and S. Magnusson, *in*: "Atlas of protein sequence and structure", vol. 5, Suppl. 3, M.O. Dayhoff, ed. National Biomedical Research Foundation, Washington DC, 91 (1978).
2. D. Collen, and H.R. Lijnen, The fibrinolytic system in man. **CRC Crit. Rev. Hemat. Onc.** 4: 249 (1986).
3. M. Forsgren, B. Raden, M. Israelsson, et al., Molecular cloning and characterization of a full-length cDNA clone for human plasminogen. **FEBS Lett.** 213: 254 (1987).
4. J.C. Murray, K.H. Buetow, M. Donovan, et al., Linkage disequilibrium of plasminogen polymorphisms and assignment of the gene to human chromosome 6q26-6q27. **Am. J. Hum. Genet.** 40: 338 (1987).
5. K.C. Robbins, L. Summaria, B. Hsieh, and R.J. Shah, The peptide chains of human plasmin. Mechanism of activation of human plasminogen to plasmin. **J. Biol. Chem.** 242: 2333 (1967).
6. P. Holvoet, H.R. Lijnen, and D. Collen, A monoclonal antibody specific for Lys-plasminogen. Application to the study of the activation pathways of plasminogen *in vivo*. **J. Biol. Chem.** 260: 12106 (1985).
7. D. Collen, On the regulation and control of fibrinolysis. **Thromb. Haemost.** 43: 77 (1980).
8. B. Wiman, and D. Collen, Molecular mechanism of physiological fibrinolysis. **Nature (London)** 272: 549 (1978).
9. E.G. Levine, and D.J. Loskutoff, Cultured bovine endothelial cells produce both urokinase and tissue-type plasminogen activators. **J. Cell Biol.** 94: 631 (1982).

10. D.C. Rijken, I. Juhan-Vague, F. De Cock, and D. Collen, Measurement of human tissue-type plasminogen activator by a two-site immunoradiometric assay. **J. Lab. Clin. Med.** 101: 274 (1983).

11. D. Collen, D.C. Rijken, J. Van Damme, and A. Billiau, Purification of human tissue-type plasminogen activator in centigram quantities from human melanoma cell culture fluid and its conditioning for use *in vivo*. **Thromb. Haemost.** 48: 294 (1982).

12. D. Pennica, W.E. Holmes, W.J. Kohr, et al., Cloning and expression of human tissue-type plasminogen activator cDNA in *E. coli*. **Nature** 301: 214 (1983).

13. L. Banyai, A. Varadi, and L. Patthy, Common evolutionary origin of the fibrin-binding structures of fibronectin and tissue-type plasminogen activator. **FEBS Lett.** 163: 37 (1983).

14. A.J. van Zonneveld, H. Veerman, and H. Pannekoek, Autonomous functions of structural domains on human tissue-type plasminogen activator. **Proc. Natl. Acad. Sci. USA** 83: 4670 (1986).

15. J.H. Verheijen, M.P.M. Caspers, G.T.G. Chang, et al., Involvement of finger domain and kringle 2 domain of tissue-type plasminogen activator in fibrin binding and stimulation of activity by fibrin. **EMBO J.** 5: 3525 (1986).

16. A.J. van Zonneveld, H. Veerman, and H. Pannekoek, On the interaction of the finger and kringle-2 domain of tissue-type plasminogen activator with fibrin. **J. Biol. Chem.** 261: 14214 (1986).

17. M. Hoylaerts, D.C. Rijken, H.R. Lijnen, and D. Collen, Kinetics of the activation of plasminogen by human tissue plasminogen activator. Role of fibrin. **J. Biol. Chem.** 257: 2912 (1982).

18. W. Nieuwenhuizen, M. Voskuilen, A. Vermond, et al., The influence of fibrin(ogen) fragments on the kinetic parameters of the tissue-type plasminogen-activator-mediated activation of different forms of plasminogen. **Eur. J. Biochem.** 174: 163 (1988).

19. M. Rånby, N. Bergsdorf, B. Norrman, et al., Tissue plasminogen activator kinetics, *in:* "Progress in Fibrinolysis", vol. 6, Davidson, Bachmann, Bouvier and Kruithof, eds., Churchill Livingstone, Edinburgh, 182 (1983).

20. D.C. Rijken, M. Hoylaerts, and D. Collen, Fibrinolytic properties of one-chain and two-chain human extrinsic (tissue-type) plasminogen activator. **J. Biol. Chem.** 257: 2920 (1982).

21. G.R. Larsen, K. Henson, and Y. Blue, Variants of human tissue-type plasminogen activator. Fibrin binding, fibrinolytic, and fibrinogenolytic characterization of genetic variants lacking the fibronectin finger-like and/or the epidermal growth factor domains. **J. Biol. Chem.** 263: 1023 (1988).

22. M.-J. Gething, B. Adler, J.-A. Boose, et al., Variants of human tissue-type plasminogen activator that lack specific structural domains of the heavy chain. **EMBO J.** 7: 2731 (1988).

23. B. Rajput, S.F. Degen, E. Reich, et al., Chromosomal locations of human tissue plasminogen activator and urokinase genes. **Science** 230: 672 (1985).

24. M.J. Browne, A.W.R. Tyrrell, C.G. Chapman, et al., Isolation of a human tissue-type plasminogen activator genomic DNA clone and its expression in mouse L cells. **Gene** 33: 279 (1985).

25. S.J.F. Degen, B. Rajput, and E. Reich, The human tissue plasminogen activator gene. **J. Biol. Chem.** 261: 6972 (1986).

26. T. Ny, F. Elgh, and B. Lund, The structure of the human tissue-type plasminogen activator gene: correlation of intron and exon structures to functional and structural domains. **Proc. Natl. Acad. Sci. USA** 81: 5355 (1984).

27. L. Patthy, Evolution of the proteases of blood coagulation and fibrinolysis by assembly from modules. **Cell** 41: 657 (1985).

28. H. Pannekoek, C. de Vries, and A.J. van Zonneveld, Mutants of human tissue-type plasminogen activator (t-PA): structural aspects and functional properties. **Fibrinolysis** 2: 123 (1988).

29. H.R. Lijnen, D.C. Stump, and D. Collen, Single-chain urokinase-type plasminogen activator: mechanism of action and thrombolytic properties. **Sem. Thromb. Hemost.** 13: 152 (1987).

30. W.E. Holmes, D. Pennica, M. Blaber, et al., Cloning and expression of the gene for pro-urokinase in *Escherichia coli*. **Biotechnology** 3: 923 (1985).

31. G.J. Steffens, W. Günzler, F. Ötting, et al., The complete amino acid sequence of low molecular mass urokinse from human urine. **Hoppe-Seyler'Z. Physiol. Chem.** 363: 1043 (1982).

32. A. Riccio, G. Grimaldi, P. Verde, et al., The human urokinase-plasminogen activator gene and its promoter. **Nucleic Acids Res.** 13: 2759 (1985).

33. H.R. Lijnen, B. Van Hoef, F. De Cock, and D. Collen, On the mechanisms of plasminogen activation and fibrin dissolution by single chain urokinase-type plasminogen activator (scu-PA) in a plasma milieu *in vitro*. **Blood** 73: 1864 (1989).

34. V. Ellis, M.F. Scully, and V.V. Kakkar, Plasminogen activation by single-chain urokinase in functional isolation. A kinetic study. **J. Biol. Chem.** 262: 14998 (1987).

35. L.C. Petersen, L.R. Lund, L.S. Nielsen, et al., One-chain urokinase-type plasminogen activator from human sarcoma cells is a proenzyme with little or no intrinsic activity. **J. Biol. Chem.** 263: 11189 (1988).

36. W.E. Holmes, L. Nelles, H.R. Lijnen, and D. Collen, Primary structure of human α_2-antiplasmin, a serine protease inhibitor (serpin). **J. Biol. Chem.** 262: 1659 (1987).

37. H.R. Lijnen, W.E. Holmes, B. Van Hoef, et al., Amino-acid sequence of human α_2-antiplasmin. **Eur. J. Biochem.** 166: 565 (1987).

38. H.R. Lijnen, B. Wiman, and D. Collen, Partial primary structure of human α_2-antiplasmin. Homology with other plasma protease inhibitors. **Thromb. Haemost.** 48: 311 (1982).

39. B. Wiman, and D. Collen, Purification and characterization of human antiplasmin, the fast-acting plasmin inhibitor in plasma. **Eur. J. Biochem.** 78: 19 (1977).

40. B. Wiman, and D. Collen, On the kinetics of the reaction between human antiplasmin and plasmin. **Eur. J. Biochem.** 84: 573 (1978).

41. E.K.O. Kruithof, Plasminogen activator inhibitors. A review. **Enzyme** 40: 113 (1988).

42. T. Ny, M. Sawdey, D.A. Lawrence, et al., Cloning and sequence of a cDNA coding for the human β-migrating endothelial cell-type plasminogen activator inhibitor. **Proc. Natl. Acad. Sci. USA** 83: 6776 (1986).

43. H. Pannekoek, H. Veerman, H. Lambers, et al., Endothelial plasminogen activator inhibitor (PAI): a new member of the serpin gene family. **EMBO J.** 5: 2539 (1986).

44. K.W. Klinger, R. Winqvist, A. Riccio, et al., Plasminogen activator inhibitor type 1 gene is located at region q21.3-q22 of chromosome 7 and genetically linked with cystic fibrosis. **Proc. Natl. Acad. Sci. USA** 84: 8548 (1987).

45. E.K.O. Kruithof, A. Gudinchet, and F. Bachmann, Plasminogen activator inhibitor 1 and plasminogen activator inhibitor 2 in various disease states. **Thromb. Haemost.** 59: 7 (1988).

46. M. Colucci, J.A. Paramo, and D. Collen, Generation in plasma of a fast-acting inhibitor of plasminogen activator in response to endotoxin stimulation. **J. Clin. Invest.** 75: 818 (1985).

47. P.J. Declerck, M. De Mol, M.C. Alessi, et al., Purification and characterization of a plasminogen activator inhibitor-1 binding protein from human plasma. Identification as a multimeric form of S protein (Vitronectin). **J. Biol. Chem.** 263: 15454 (1988).

48. E.K.O. Kruithof, C. Tran-Thang, A. Ransijn, and F. Bachmann, Demonstration of a fast-acting inhibitor of plasminogen activators in human plasma. **Blood** 64: 907 (1984).

49. E.K.O. Kruithof, G. Nicoloso, and F. Bachmann, Measurement of plasminogen activator inhibitor 1: development of a radioimmunoassay and observations on its plasma concentration during venous occlusion and after platelet aggregation. **Blood** 70: 1645 (1987).

50. C.M. Hekman, and D. Loskutoff, Endothelial cells produce a latent inhibitor of plasminogen activators that can be activated by denaturants. **J. Biol. Chem.** 260: 11581 (1985).

51. B. Wiman, and P. Wallen, The specific interaction between plasminogen and fibrin. A physiological role of the lysine binding site in plasminogen. **Thromb. Res.** 10: 213 (1977).

52. C. Tran-Thang, P. Wyss, E.K.O. Kruithof, et al., Tissue-type plasminogen activator increases the binding of plasminogen to clots. **Haemostasis** 14: 17 (Abstract) (1984).

53. H.R. Lijnen, C. Zamarron, M. Blaber, et al., Activation of plasminogen by pro-urokinase. I. Mechanism. **J. Biol. Chem.** 261: 1253 (1986).

54. B. Wiman, H.R. Lijnen, and D. Collen, On the specific interaction between the lysine-binding sites in plasmin and complementary sites in α_2-antiplasmin and in fibrinogen. **Biochim. Biophys. Acta** 579: 142 (1979).

55. A. Ichinose, T. Tamaki, and N. N. Aoki, Factor XIII-mediated cross-linking of NH_2-terminal peptide of α_2-plasmin inhibitor to fibrin. **FEBS Lett.** 153: 369 (1983).

56. D.L. Higgins, and G.A. Vehar, Interaction of one-chain and two-chain tissue plasminogen activator with intact and plasmin-degraded fibrin. **Biochemistry** 26: 7786 (1987).

57. R. Pannell, J. Black, and V. Gurewich, Complementary modes of action of tissue-type plasminogen activator and pro-urokinase by which their synergistic effect on clot lysis may be explained. **J. Clin. Invest.** 81: 853 (1988).

58. N. Aoki, H. Saito, T. Kamiya, et al., Congenital deficiency of α_2-plasmin inhibitor associated with severe hemorrhagic tendency. **J. Clin. Invest.** 63: 877 (1979).

59. C. Kluft, E. Vellenga, E.J.P. Brommer, and G. Wijngaards, A familial hemorrhagic diathesis in a Dutch family: an inherited deficiency of α_2-antiplasmin. **Blood** 59: 1169 (1982).

60. C. Kluft, E. Vellenga, and E.J.P. Brommer, Homozygous α_2-antiplasmin deficiency. **Lancet** 2: 206 (1979).

61. K. Koie, K. Ogata, T. Kamiya, et al., α_2-plasmin inhibitor deficiency (Miyasato disease). **Lancet** 2: 1334 (1978).

62. H. Stormorken, G.O. Gogstad, and F. Brosstad, Hereditary α_2-antiplasmin deficiency. **Thromb. Res.** 31: 647 (1983).

63. A. Yoshioka, H. Kamitsuji, T. Takase, et al., Congenital deficiency of α_2-plasmin inhibitor in three sisters. **Haemostasis** 11: 176 (1982).

64. L.A. Miles, E.F. Plow, K.J. Donnelly, et al., A bleeding disorder due to deficiency of α_2-antiplasmin. **Blood** 59: 1246 (1982).

65. D. Collen, H. Bounameaux, F. De Cock, et al., Analysis of coagulation and fibrinolysis during intravenous infusion of recombinant human tissue-type plasminogen activator in patients with acute myocardial infarction. **Circulation** 73: 511 (1986).

66. H.K. Nieuwenhuis, C. Kluft, G. Wijngaards, et al., α_2-Antiplasmin Enschede: an autosomal recessive hemorrhagic disorder caused by a dysfunctional α_2-antiplasmin molecule. **Thromb. Haemost.** 50: 170 (abstract 528) (1983).

67. W.E. Holmes, H.R. Lijnen, L. Nelles, et al., α_2-Antiplasmin Enschede: alanine insertion and abolition of plasmin inhibitory activity. **Science** 238: 209 (1987).

68. N.A. Booth, B. Bennett, G. Wijngaards, and J.H.K. Grieve, A new life-long hemorrhagic disorder due to excess plasminogen activator. **Blood** 61: 267 (1983).

69. J. Aznar, A. Estellés, V. Villa, et al., Inherited fibrinolytic disorder due to an enhanced plasminogen activator level. **Thromb. Haemost.** 52: 196 (1984).

70. J. Mehta, P. Mehta, D. Lawson, and T. Saldeen, Plasma tissue plasminogen activator inhibitor levels in coronary artery disease: correlation with age and serum triglyceride concentrations. **J. Am. Coll. Cardiol.** 9: 263 (1987).

71. J.A. Páramo, M. Colucci, and D. Collen, Plasminogen activator inhibitor in the blood of patients with coronary artery disease. **Br. Med. J.** 291: 573 (1985).

72. L.-O. Almér, and H. Öhlin, Elevated levels of the rapid inhibitor of plasminogen activator in acute myocardial infarction. **Thromb. Res.** 47: 335 (1987).

73. F.W.A. Verheugt, J.W. ten Cate, A. Sturk, et al., Tissue plasminogen activator activity and inhibition in acute myocardial infarction and angiographically normal coronary arteries. **Am. J. Cardiol.** 59: 1075 (1987).

74. A. Hamsten, M. Blombäck, B. Wiman, et al., Haemostatic function in myocardial infarction. **Br. Heart J.** 55: 58 (1986).

75. A. Hamsten, B. Wiman, U. De Faire, and M. Blombäck, Increased plasma levels of a rapid inhibitor of tissue plasminogen activator in young survivors of myocardial infarction. **N. Engl. J. Med.** 213: 1557 (1985).

76. A. Hamsten, U. De Faire, G. Walldius, et al., Plasminogen activator inhibitor in plasma; risk factor for recurrent myocardial infarction. **Lancet** ii: 3 (1987).

77. I. Juhan-Vague, B. Moerman, F. De Cock, et al., Plasma levels of a specific inhibitor of tissue-type plasminogen activator (and urokinase) in normal and pathological conditions. **Thromb. Res.** 33: 523 (1984).

78. B. Wiman, and J. Chmielewska, A novel fast inhibitor to tissue plasminogen activator in plasma, which may be of great pathophysiological significance. **Scand. J. Clin. Lab. Invest.** 45 (Suppl 177): 43 (1985).

79. C. Kluft, A.F.H. Jie, G.D.O. Lowe, et al., Association between postoperative hyper-response in t-PA inhibition and deep vein thrombosis. **Thromb. Haemost.** 56: 107 (1986).

80. G. Mellbring, S. Dahlgren, B. Wiman, and O. Sunnegårdh, Relationship between preoperative status of the fibrinolytic system and occurrence of deep vein thrombosis after major abdominal surgery. **Thromb. Res.** 39: 157 (1985).

81. J.A. Páramo, M.J. Alfaro, and E. Rocha, Postoperative changes in the plasmatic levels of tissue-type plasminogen activator and its fast-acting inhibitors. Relationship to deep vein thrombosis and influence of prophylaxis. **Thromb. Haemost.** 54: 713 (1985).

82. G. Mellbring, S. Dahlgren, and B. Wiman, Plasma fibrinolytic activity in patients undergoing major abdominal surgery. **Acta Chir. Scand.** 151: 109 (1985).

83. M. Jorgensen, and V. Bonnevie-Nielsen, Increased concentration of the fast-acting plasminogen activator inhibitor in plasma associated with familial venous thrombosis. **Br. J. Haemat.** 65: 175 (1987).

84. I. Juhan-Vague, J. Valadier, M.C. Alessi, et al., Deficient t-PA release and elevated PA inhibitor levels in patients with spontaneous or recurrent deep vein thrombosis. **Thromb. Haemost.** 57: 67 (1987).

85. M. Stalder, J. Hauert, E.K.O. Kruithof, and F. Bachmann, Release of vascular plasminogen activator (v-PA) after venous stasis: electrophoretic-zymographic analysis of free and complexed v-PA. **Br. J. Haemat.** 61: 169 (1985).

86. G. Mellbring, S. Dahlgren, S. Reiz, and B. Wiman, Fibrinolytic activity in plasma and deep vein thrombosis after major abdominal surgery. **Thromb. Res.** 32: 575 (1983).

87. A. Girolami, F. Marafioti, M. Rubertelli, and M.G. Cappellato, Congenital heterozygous plasminogen deficiency associated with a severe thrombotic tendency. **Acta Haemat.** 75: 54 (1986).

88. D.K. Hasegawa, B.J. Tyler, and J.R. Edson, Thrombotic disease in three families with inherited plasminogen deficiency. **Blood** 60: 213 (1982).

89. R. Lottenberg, F.R. Dolly, and C.S. Kitchens, Recurring thromboembolic disease and pulmonary hypertension associated with severe hypoplasminogenemia. **Am. J. Hematol.** 19: 181 (1985).

90. P.M. Mannucci, C. Kluft, D.W. Traas, et al., Congenital plasminogen deficiency associated with venous thromboembolism: therapeutic trial with stanozolol. **Br. J. Haemat.** 63: 753 (1986).

91. J.W. ten Cate, M. Peters, and H. Buller, Isolated plasminogen deficiency in a patient with recurrent thromboembolic complications. **Thromb. Haemost.** 50: 59 (Abstract) (1983).

92. R.D. Davis, and R.C. Picoff, Low plasminogen levels and liver disease. **Am. J. Clin. Path.** 59: 175 (1969).

93. L.C. Kordich, V.P. Porterie, O. Lago, et al., Mini-plasminogen like molecule in septic patients. **Thromb. Res.** 47: 553 (1987).

94. M. Verstraete, J. Vermylen, and D. Collen, Intravascular coagulation in liver disease. **Ann. Rev. Med.** 25: 447 (1974).

95. M. Kazama, C. Tahara, Z. Suzuki, et al., Abnormal plasminogen, a case of recurrent thrombosis. **Thromb. Res.** 21: 517 (1981).

96. T. Miyata, S. Iwanaga, Y. Sakata, et al., Plasminogens Tochigi II and Nagoya: two additional molecular defects with Ala^{600} → Thr replacement found in plasmin light chain variants. **J. Biochem.** 96: 277 (1984).

97. T. Miyata, S. Iwanaga, Y. Sakata, and N. Aoki, Plasminogen Tochigi: inactive plasmin resulting from replacement of alanine-600 by threonine in the active site. **Proc. Natl. Acad. Sci. USA** 79: 6132 (1982).

98. J. Soria, C. Soria, O. Bertrand, et al., Plasminogen Paris I: congenital abnormal plasminogen and its incidence in thrombosis. **Thromb. Res.** 32: 229 (1983).

99. R.C. Wohl, L. Summaria, J. Chediak, et al., Human plasminogen variant Chicago III. **Thromb. Haemost.** 48: 146 (1982).

100. R.C. Wohl, L. Summaria, and K.C. Robbins, Physiological activation of the human fibrinolytic system. Isolation and characterization of human plasminogen variants, Chicago I and Chicago II. **J. Biol. Chem.** 254: 9063 (1979).

101. S. Isacson, and I.M. Nilsson, Defective fibrinolysis in blood and vein walls in recurrent "idiopathic" venous thrombosis. **Acta Chir. Scand.** 138: 313 (1972).

102. C. Korninger, K. Lechner, H. Niessner, et al., Impaired fibrinolytic capacity predisposes for recurrence of venous thrombosis. **Thromb. Haemost.** 52: 127 (1984).

103. I.M. Nilsson, H. Ljungner, and L. Tengborn, Two different mechanisms in patients with venous thrombosis and defective fibrinolysis: low concentration of plasminogen activator or increased concentration of plasminogen activator inhibitor. **Br. Med. J.** 290: 1453 (1985).

A KEY MOLECULE DICTATING AND REGULATING SURFACE PLASMIN FORMATION : THE RECEPTOR FOR UROKINASE PLASMINOGEN ACTIVATOR

FRANCESCO BLASI, M. VITTORIA CUBELLIS, ANN LOURING ROLDAN, MARIA TERESA MASUCCI, NIELS BEHRENDT[1], VINCENT ELLIS[1], ETTORE APPELLA[2] AND KELD DANØ[1]

Institute of Microbiology
University of Copenhagen
Øster Farimagsgade 2A
1353 Copenhagen K, Denmark;
[1]Finsen Laboratory, Rigshospitalet
Strandboulevarden 49
2100 Copenhagen Ø, Denmark;
and [2]Laboratory of Cell Biology
NCI, National Institutes of Health
Bethesda, MD 20892, USA

INTRODUCTION

Regulation of cell to cell contacts and of cell migration is thought to require the action of protease systems which are needed to sever the ties that link cells to other cells and to the extracellular matrix, and to overcome barriers like the basement membranes. Since cell migration phenomena occur from gametogenesis through embryonic development to adulthood, and since also many pathological disorders are associated with or require tissue destruction and cell migration, the requirement for extracellular proteolytic enzymes is a very widespread phenomenon, which however has not yet been clearly understood on a molecular biological and biochemical basis.[1-3]

PHYSIOLOGICALLY INVASIVE PROCESSES

Examples of physiological, i.e. controlled, invasive processes are numerous. These processes consist of migration of cells through different compartments, and involve a proteolytic destruction and a resynthesis of the basement membrane separating these compartments. The rupture of the ovarian follicle,[4-6] the release of spermatocytes during mammalian spermatogenesis,[7,8] the invasion of the embryonic trophoblast cells in the uterine muscle tissue,[9,10] the migration of granular neurons

Serine Proteases and Their Serpin Inhibitors in the Nervous System
Edited by B. W. Festoff
Plenum Press, New York, 1990

during the development of the nervous system,[11-13] the sprouting of new capillaries during angiogenesis,[14] the outgrowth of keratinocytes during wound healing,[15] the post-lactational involution of the mammary gland[16-18] and muscle involution after denervation,[19] are examples of such systems. The need for proteolytic enzymes in these processes is intuitively apparent, although their role has not yet always been clearly defined or properly demonstrated.

Proteases have also been proposed to be involved in several aspects of the malignant phenotype like neo-angiogenesis, invasion, tissue destruction and metastasis[20,21] (see also ref. 2). These are biochemically similar to the cell migration phenomena mentioned above and to the tissue-destructive processes occurring in other diseases.

The best studied system regulating extracellular proteolysis is that of plasminogen activators. These enzymes transform the inactive zymogen plasminogen into the broad spectrum protease plasmin, which is in turn able to degrade a wide variety of proteins, including those of the extracellular matrix and basement membranes.[1-3] The collagenase system also appears to have an important role in cellular invasion,[22] but will not be discussed in this context. Other protease systems have been proposed to be involved in the same phenomena, but their biochemical and physiological properties are less characterized.

PLASMIN FORMATION

Plasminogen is an inactive and widespread proenzyme present in high concentrations in most body fluids and tissues which are therefore a reservoir of potential proteolytic activity. Plasminogen can be activated by two different serine proteases: urokinase- (uPA) and tissue-type plasminogen activator (tPA).

Plasminogen is a single chain protein which can be cleaved at the Arg 560-Val 561 polypeptide bond by the plasminogen activators. The resulting active serine protease plasmin consists of two polypeptide chains held together by disulfide bonds. The light chain contains the active site that has strong homologies with other serine proteases, like trypsin, chymotrypsin and elastase.

Plasmin is a protease with relatively broad trypsin-like specificity hydrolyzing proteins and peptides at lysyl- and arginyl- bonds. Plasmin is rapidly inhibited by alpha-2 antiplasmin, and at a much slower rate, by alpha-2 macroglobulin, both occurring in plasma. The ability of plasmin to degrade fibrin is of central importance in the dissolution of the fibrin clot.[23] Plasmin, however, is also capable of limited proteolysis of some proteins, like proenzymes, and this may serve the role of regulating their activation. For example, plasmin can activate pro-urokinase plasminogen activator (pro-uPA) generating fully active two-chain enzyme.[24-26] Other proenzymes that can be activated by plasmin include pro-collagenase.[27] Limited proteolysis by plasmin may also have an important function in the post-translational maturation and processing of some growth factors, like TGF-beta (see ref. 28).

Since cells engaged in invasive processes need proteolytic activities to degrade extracellular matrix and basement membrane proteins, it is particularly interesting to note that plasmin substrates include proteins like fibronectin, laminin, proteoglycans and some collagens.[2,28,29]

Thus plasmin has the biochemical properties required for playing a direct role in cellular invasion and tissue degradation. Since plasmin formation appears to be the controlling step of this pathway, the role of plasminogen activators becomes of central importance in regulating extracellular proteolysis and hence all the proteolysis-dependent biological processes.

In this paper we shall confine our discussion to the role and mechanism of urokinase-type plasminogen activator-mediated plasmin formation. uPA has a *Mr* of about 50,000 daltons and occurs in one-chain or two-chain form.[24-26] The gene product is 431 residues long;[30] secreted, single chain pro-uPA is composed of 411 residues. It has little or no plasminogen activating capacity but by limited proteolysis with plasmin it can be converted into its active two-chain counterpart[24,26,31-34] (see ref. 35 for a discussion). In a recent study[33] it was found that a highly purified preparation of pro-uPA (derived from cells that had been maintained under serum free conditions for a prolonged period) had a plasminogen activating capacity that was more than 250-fold lower than that of two-chain uPA. It was not possible to decide whether this low activity was intrinsic or due to contamination.

The proenzyme nature of single chain uPA is also reflected by little or no reactivity with macromolecular[36-38] and synthetic[24-26,39,40] inhibitors and virtually no amidolytic activity with synthetic substrates.[25,31,36] Activation of pro-uPA occurs by cleavage at residue 157, resulting in a disulfide-bonded two chain enzyme.[30,41] The catalytic activity is located in the carboxy-terminal portion (B chain): in fact, Low Molecular Weight (LMW) uPA, a two-chain protein having a short (21 residues) A chain starting at residue 137 and the entire B chain, is fully active.[41]

THE uPA RECEPTOR

First Vassalli et al.[37] and subsequently Stoppelli et al.[42] demonstrated that human monocytes and monocyte-like U937 cells possess specific, high-affinity binding sites for human uPA. Binding is independent of enzymatic activity and in fact both the active site-blocked DFP-uPA and the inactive proenzyme pro-uPA bind to the same sites with nearly equal affinity. Specificity is absolute and no other homologous molecule is able to compete for binding. For instance no competition was detected with tPA, various growth factors, plasminogen or coagulation factors. Although highly homologous to human uPA, mouse uPA does not compete for binding to human receptors, and conversely human uPA does not bind to mouse receptors.[43-44]

Kinetic analysis of the interaction between uPA and its receptor showed that the number of sites per cells varies from a few thousands in some normal blood cells to several hundred thousands in fibroblasts and in some tumor cells. The affinity of the interaction also varies somewhat, between 0.1 and 2 nM in different cells and in different laboratories, although it has not been rigorously excluded that these differences represent the use of different reagents or other possible experimental variables.[45]

Almost all cell lines so far examined possess the uPA receptor; whether this is a faithful representation of its occurrence in the intact organism is not yet known.

Receptor-Binding Region in uPA

As it appears that the receptor is essential for regulating surface proteolysis, a detailed knowledge of the structural interaction between the receptor and its ligand becomes extremely important when attempting to interfere with this interaction. As stated above, binding to the receptor does not require the enzymatic activity of uPA.

Actually it does not require the catalytic moiety of uPA; the enzymatically active low *Mr* uPA (the carboxy terminal region) has no binding capacity, while the amino terminal fragment (ATF), which is completely devoid of plasminogen activating activity, has all of the binding activity.[42] Binding activity has been further localized at the amino acid residues 10 to 45 of uPA, using synthetic peptides and genetically deleted uPA molecules.[44,46] The receptor-binding region is also known as the growth factor domain for its homology to the epidermal growth factor, a sequence, present in many different proteins in which it may have quite different functions, and which has been proposed to be involved in protein-protein interaction.[47] In fact, the residues of uPA that dictate binding specificity are those between amino acid 20 and 32; they have no homology at all with epidermal growth factor nor to any other known protein, but are flanked by two highly conserved sequences.[44,47] This explains the high specificity of the interaction in the presence of high sequence homology: molecules like epidermal growth factor and tPA do not bind the uPA receptor even at a 100 μM concentration despite their overall sequence homology in the binding domain. We view the growth factor domain of uPA as composed of three superimposed loops: the middle one contributing to binding specificity and the other two to structural stability.

Structure of the uPA receptor

The high affinity of the interaction and the possibility to cross-link uPA or its derivatives to the receptor have provided workable methods for identifying and purifying the uPA receptor.[48] This in turn has given partial amino acid sequence information[49] which has been used to isolate a complementary DNA copy of the uPA receptor mRNA which has been sequenced completely and has been therefore decoded into the amino acid sequence of the uPA receptor protein.[50] The mature protein is composed of 313 amino acid residues, contains a potential trans-membrane domain close to the carboxyl terminus and a very short putative intracytoplasmic domain. From these cDNA data it has been proposed that the uPA receptor is a mostly extracellular molecule; at the amino terminus of the coding region there is a typical signal peptide which is not found in the mature molecule and thus is probably removed upon secretion. The receptor is very rich in cysteine and is highly glycosylated, carbohydrates contributing to its apparent *Mr* by about 40%.

UPA RECEPTORS AND THE FOCAL ADHESION SITES

In cultured cells, as human fibroblasts and HT1080 sarcoma cells, uPA is present at the focal adhesion sites, i.e. the closest contacts between the cells and the substratum.[51-53] Such sites provide a trans-membrane connection between cytoskeleton and extracellular matrix with intracellular vinculin and talin connecting the actin cables to the integrins of the plasma membrane. In addition, uPA is also present at discrete sites of cell to cell contacts. At these sites, uPA is bound to its specific receptor.[53,54] It is not known whether this focalization of the receptor also occurs in the absence of uPA.

Glass et al.[55] recently reported that uPA is present in adhesion plaques of cultured rat mesangial cells and that antibodies to uPA totally inhibited a change in shape and loss of adhesiveness in response to induction of c-AMP elevation.

Plasminogen and plasmin bind to the surface of many cell types.[56] In most cases part of this binding is saturable and usually there is a competition between the two molecules. The affinities for binding varies strongly between cell types and laboratories, with K_D values in the range of 1 nM to 5 μM. Also the number of binding sites varies, often being very high with 10^6-10^7 sites per cell. Little is known of the structural identity of the binding sites, candidates being fibrin, thrombospondin and the membrane glycoprotein GPIIb/IIIa (see ref. 57 for a discussion).

U937 cells bind at their surface both pro-uPA to the uPA receptor and plasminogen to as yet unidentified binding sites.[58,59] Concomitant binding of the two proenzymes leads to strong acceleration (up to 16 fold) of plasminogen activation in comparison to that observed with both reactants in solution.[60] The acceleration is abolished by the addition of either the amino-terminal fragment of uPA or 6-aminohexanoic acid, thus demonstrating the requirement for binding of both pro-uPA and plasminogen to the cells. A kinetic analysis of the various reactions suggests that the increased rate of plasmin formation is primarily due to a strong increase in the rate of plasmin catalyzed conversion of pro-uPA to uPA, while the rate of two-chain uPA catalyzed activation of plasminogen is not increased. It is not known whether an initial activity of the pro-uPA preparation observed in this study was due to contamination or to intrinsic activity of pro-uPA.

The human fibrosarcoma cell line HT1080 produces both pro-uPA and the uPA receptor.[26,48] It also binds added plasminogen at its surface.[54,61] Endogenous pro-uPA binds to the receptor and in the absence of added plasminogen stays in the proenzyme form. Upon addition of plasminogen, cell bound plasmin as well as cell bound two-chain uPA is formed.[54] The plasminogen activation occurs at the cell surface and is decreased by preincubation with diisopropyl fluorophosphate-inhibited uPA (DFP-uPA), indicating that a large part, if not all, of the cell surface plasminogen activation is catalyzed by receptor-bound uPA. Likewise the formation of two-chain uPA is inhibited by low concentrations of tranexamic acid, indicating that it is catalyzed by surface bound plasmin. These experiments were performed in the presence of serum, but the cell bound plasmin was inaccessible to serum inhibitors. In contrast, cell bound two-chain uPA was accessible to inhibition by endogenous and added plasminogen activator inhibitors PAI-1 and PAI-2,[54] in analogy with what was observed with U937 cells.[62]

REGULATION OF THE uPA-uPA RECEPTOR INTERACTION

This can be accomplished at several levels, modulating the rate of synthesis of both ligand and receptor, the rate of activation of pro-uPA precursor, the affinity of the receptor for its ligand, and the synthesis and location of plasminogen activator inhibitors.

1. Synthesis of uPA is strictly regulated by many hormones, by cell growth, differentiation and neoplastic transformation, etc., but the pattern of regulation varies strongly between cell types (see ref. 63). The molecular basis of the regulation is only now starting to be elucidated.[64] Very little is known, however, about the regulation of receptor synthesis. The few available data suggest a general pattern of regulation similar to that of uPA. PMA or EGF increase the synthesis of uPA receptors in at

least four different cell lines[43,65,66] (Stoppelli & DeCesare, unpublished). In the case of monocyte-like U937 cells and PMA the effect is time- and concentration-dependent, parallels cell differentiation into macrophages and is blocked by cycloheximide.[65] An early increase of uPA receptor mRNA has been observed upon PMA treatment of U937 cells (L.R. Lund, personal communication).

2. A second level of regulation has been observed: in U937 and HeLa cells, in addition to the increase in receptor number, a 10 to 50 fold decrease in receptor affinity upon PMA stimulation has been reported.[43,65] This is due to a modification of the receptor itself, but the biochemical basis has not yet been clarified. The decrease in receptor affinity is accompanied by slight physical chemical changes: the receptor migrates more slowly in SDS-polyacrylamide gel electrophoresis and separates into three spots in bidimensional electrophoresis.[65] Since the receptor is heavily glycosylated, and since glycosylation increases upon PMA treatment, this structural change appears to be due to glycosylation.[49] Whether glycosylation and functional changes of the receptor are related, remains to be proven.

3. Since the product of the uPA gene is the inactive zymogen pro-uPA, and since pro-uPA is able to bind the uPA receptor with high affinity,[42,58] conversion of receptor-bound pro-uPA to active two chain uPA is another step of the pathway that is amenable to regulation. No information is available yet as to the physiological pro-uPA activating enzyme. However, plasmin is certainly able to cleave receptor-bound pro-uPA into an active PAI-1 binding (see below) two chain uPA.[54,58,62] Therefore the regulation of proenzyme conversion is another fundamental step in the control of cell-surface proteolytic activity and its understanding may prove invaluable in attempts to manipulate extracellular proteolysis and hence cell migration and all biological phenomena that are based on it. The fundamental role of the receptor in uPA activation is indicated by the recent experiment in which it was shown that receptor-binding increased the rate of plasmin generation initiated by pro-uPA; this is apparently due to an approximately 50 fold increase in the rate of pro-uPA activation by the plasmin generated at the cell surface.[60]

4. A fourth level of regulation can be exerted by the plasminogen activator inhibitors (PAI-1 and PAI-2) (see ref. 67 for a review). Since surface-bound plasmin is protected from its inhibitor alpha-2 antiplasmin,[59] one might have expected also receptor-bound uPA to be resistant to the PAIs. However, the structural data showing the independence of the catalytic from the receptor-binding moiety[42] argue against such possibility. And in fact PAI-1 is able to bind and inhibit receptor-bound uPA, and the uPA-PAI-1 complex is also able to bind specifically to the uPA receptor. However, the affinity of the uPA-PAI-1 complex is somewhat lower and its dissociation rate faster than that of uPA alone[62] (Cubellis & Blasi, unpublished data). Thus PAI-1 binding to uPA does not only regulate uPA activity but also the affinity of uPA for the receptor. In HT1080 and in human fibroblasts uPA is found on the cell surface at sites of contact with other cells or with the substratum (focal contacts), while PAI-1 is mostly present in the extracellular matrix.[51-53] It appears therefore that receptors localize uPA at sites where it is needed for cell migration, and that any movement of the cell would dislocate receptor-bound uPA into an area which contains PAI-1; a cell movement would thus be followed by inhibition of uPA activity by the PAI-1 present in the extracellular matrix. Possibly, a redistribution of free uPA receptors at new sites may occur, and these will eventually be filled by new pro-uPA molecules to be activated.

CONCLUSIONS

The data available so far have allowed us to outline a novel functional cycle of activation and inactivation of the proteolytic pathway that takes place at the surface of the cells, since the main components, uPA and plasminogen, are bound to the plasma membrane, the former to a specific receptor. This allows the cells to concentrate and regulate the proteolytic activity at the sites where it is required, i.e. at the moving edges of the cells. The following cycle can be therefore envisaged at this point: pro-uPA is secreted from cells and bound to the receptor[68] and very efficiently activated to uPA.[54,58,60] Surface uPA, located at focal and cell to cell contacts[51] catalyses surface plasmin formation[54] which results in destruction of cell to cell or cell to substratum contacts. The resulting movement of the cell relocates the uPA activity into a new area in which it is then inhibited by PAI-1[62] or other inhibitors. This allows reformation of new cell to cell or cell to matrix interactions.

REFERENCES

1. E. Reich, Activation of plasminogen: a general mechanism for producing localized extracellular proteolysis, in: "Molecular Basis of biological degradative processes". R.D. Berlin et al., Eds. New York: Academic Press, p 155 (1978).
2. K. Danø, P.A. Andreasen, J. Grøndahl-Hansen, P. Kristensen, L.S. Nielsen, and L. Skriver, Plasminogen activators, tissue degradation and cancer. **Adv. in Cancer Res.** 44: 139 (1985).
3. F. Blasi, J.D. Vassalli, and K. Danø, Urokinase type plasminogen activator: proenzyme, receptor and inhibitors. **J. Cell Biol.** 104: 801 (1987).
4. W.H. Beers, S. Strickland, and E. Reich, Ovarian plasminogen activator: relationship to ovulation and hormonal regulation. **Cell** 6: 387 (1975).
5. R. Canipari, and S. Strickland, Plasminogen activator in the rat ovary. Production and gonadotropin regulation of the enzyme in granulosa and thecal cells. **J. Biol. Chem.** 260: 5121 (1985).
6. A.J.W. Hsueh, Y.-X. Liu, S. Cajander, X.-R. Peng, K. Dahl, P. Kristensen, and T. Ny, Gonadotropin-releasing hormone induces ovulation in hypophysectomized rats: studies on ovarian tissue-type plasminogen activator activity, messenger ribonucleic acid content, and cellular localization. **Endocrinology** 122: 1486 (1988).
7. J.A. Hettle, E.K. Waller, and I.B. Fritz, Hormonal stimulation alters the type of plasminogen activator produced by Sertoli cells. **Biol. Reprod.** 34: 895 (1986).
8. K.K. Vihko, P. Kristensen, K. Danø, and M. Parvinen, Immunohistochemical localization of urokinase-type plasminogen activator in Sertoli cells and tissue-type plasminogen activator in spermatogenic cells in the rat seminiferous epithelium. **Dev. Biol.** 125: 150 (1988).
9. S. Strickland, E. Reich, and M.I. Sherman, Plasminogen activator in early embryogenesis: enzyme production by trophoblast and parietal endoderm. **Cell** 9: 231 (1976).
10. S. Strickland, and V. Mahdavi, The induction of differentiation in teratocarcinoma stem cells by retinoic acid. **Cell** 15: 393 (1978).
11. A. Krystosek, and N.W. Seeds, Plasminogen activator release at the neuronal growth cone. **Science** 213: 1532 (1981).
12. A. Krystosek, and N.W. Seeds, Plasminogen activator secretion by granule neurons in cultures of developing cerebellum. **Proc. Natl. Acad. Sci. USA** 78: 7810 (1981).
13. A. Krystosek, and N.W. Seeds, Peripheral neurons and Schwann cells secrete plasminogen activator. **J. Cell Biol.** 98: 773 (1984).
14. J.L. Gross, D. Moscatelli, and D.B. Rifkin, Increased capillary endothelial cell protease activity in response to angiogenic stimuli in vitro. **Proc. Natl. Acad. Sci. USA** 80: 2623 (1983).
15. J. Grøndahl-Hansen, L.R. Lund, E. Ralfkiær, E. Ottevanger, and K. Danø, Urokinase- and tissue-type plasminogen activator in keratinocytes during wound reepithelialization in vivo. **J. Invest. Dermatol.** 90: 790 (1988).
16. L. Ossowski, D. Biegel, and E. Reich, Mammary plasminogen activator: correlation with involution, hormonal modulation and comparison between normal and neoplastic tissue. **Cell** 16: 929 (1979).

17. L.-I. Larsson, L. Skriver, L.S. Nielsen, J. Grøndahl-Hansen, P. Kristensen, and K. Danø, Distribution of urokinase-type plasminogen activator immunoreactivity in the mouse. **J. Cell Biol.** 98: 894 (1984).

18. N. Busso, J. Huarte, J.-D. Vassalli, A.-P. Sappino, and D. Belin, Plasminogen activators in the mouse mammary gland. Decreased expression during lactation. **J. Biol. Chem.** 264: 7455 (1989).

19. B.W. Festoff, D. Hantaï, J. Soria, A. Thomaïdis, and C. Soria, Plasminogen activator in mammalian skeletal muscle: characteristics of the effect of denervation on urokinase-like and tissue activator. **J. Cell Biol.** 103: 1415 (1987).

20. L. Ossowski, and E. Reich, Antibodies to plasminogen activator inhibit human tumor metastasis. **Cell** 35: 611 (1983).

21. V.J. Hearing, L.W. Law, A. Corti, E. Appella, and F. Blasi, Modulation of metastatic potential by cell surface urokinase of murine cells. **Cancer Res.** 48: 1270 (1988).

22. K. Tryggvason, M. Hoyhtya, and T. Salo, Proteolytic degradation of extracellular matrix in tumor invasion. **Biochim. Biophys. Acta** 907: 191 (1987).

23. D. Collen, On the regulation and control of fibrinolysis. **Thromb. Haemost.** 43: 77 (1980).

24. L. Skriver, L.S. Nielsen, R. Stephens, and K. Danø, Plasminogen activator released as inactive proenzyme from murine cells transformed by sarcoma virus. **Eur. J. Biochem.** 124: 409 (1982).

25. T.-C. Wun, L. Ossowski, and E. Reich, A proenzyme form of human urokinase. **J. Biol. Chem.** 257: 7262 (1982).

26. L.S. Nielsen, J.G. Hansen, L. Skriver, E.L. Wilson, K. Kaltoft, J. Zeuthen, and K. Danø, Purification of zymogen to plasminogen activator from human glioblastoma cells by affinity chromatography with monoclonal antibody. **Biochemistry** 21: 6410 (1982).

27. Y. Eeeckhout, and G. Vaes, Further studies on the activation of procollagenase, the latent precursor of bone collagenase. **Biochem. J.** 166: 21 (1977).

28. O. Saksela, and D.B. Rifkin, Cell-associated plasminogen activation: Regulation and physiological functions. **Ann. Rev. Cell Biol.** 4: 93 (1988).

29. F. Blasi, and M.P. Stoppeli, Molecular basis for plasminogen activation, surface proteolysis and their relation to cancer, in: "Growth Regulation and Carcinogenesis" W.R. Paukowits, Ed., CRC Uniscience (1989).

30. P. Verde, M.P. Stoppelli, P. Galeffi, P.P. DiNocera and F. Blasi, Identification and primary sequence of an unspliced human urokinase poly(A)+ RNA. **Proc. Natl. Acad. Sci. USA** 81: 4727 (1984).

31. R. Pannell, and V. Gurewich, Activation of plasminogen by single-chain urokinase or by two-chain urokinase- A demonstration that single-chain urokinase has a low catalytic activity (pro-urokinase). **Blood** 69: 22 (1987).

32. V. Ellis, M.F. Scully, and V.V. Kakkar, Plasminogen activation by single chain urokinase in functional isolation. **J. Biol. Chem.** 262: 14998 (1987).

33. L.C. Petersen, L.R. Lund, L.S. Nielsen, K. Danø, and L.S. Skriver, One-chain urokinase-type plasminogen activator is a proenzyme with little or no intrinsic activity. **J. Biol. Chem.** 263: 11189 (1988).

34. T. Urano, V.S. de Serrano, P. Gaffney, and F.J. Castellino, The activation of human (Glu^1) plasminogen by human single-chain urokinase. **Arch. Biochem. Biophys.** 264: 222 (1988).

35. K. Danø, N. Behrendt, L.R. Lund, E. Rønne, J. Pöllänen, E.-M. Salonen, R.W. Stephens, H. Tapiovaara, and A. Vaheri, Cell-surface plasminogen activation, in: "Cancer Metastasis" V. Schirrmacher and R. Schwartz-Albiez, eds. Springer-Verlag p 98 (1989).

36. D.L. Eaton, R.W. Scott, and J.B. Baker, Purification of human fibroblast urokinase proenzyme and analysis of its regulation by proteases and protease nexin. **J. Biol. Chem.** 259: 6241 (1984).

37. J.D. Vassalli, D. Baccino, and D. Belin, A cellular binding site for the 55,000 *Mr* form of the human plasminogen activator, urokinase. **J. Cell Biol.** 100: 86 (1985).

38. P.A. Andreasen, L.S. Nielsen, P. Kristensen, J. Grøndahl-Hansen, L. Skriver, and K. Danø, Plasminogen activator inhibitor from human fibrosarcoma cells binds urokinase-type plasminogen activator, but not its proenzyme. **J. Biol. Chem.** 261: 7644 (1986).

39. V. Gurewich, R. Pannell, S. Louie, P. Kelley, R.L. Suddith, and R. Greenlee, Effective and fibrin-specific clot lysis by a zymogen precursor form of urokinase (pro-urokinase). A study *in vitro* and in two animal species. **J. Clin. Invest.** 73: 1731 (1984).

40. S. Kasai, H. Arimura, M. Nishida, and T. Suyama, Proteolytic cleavage of single-chain pro-urokinase induces conformational change which follows activation of the zymogen and reduction of its high affinity for fibrin. **J. Biol. Chem.** 260: 12377 (1985).

41. W.A. Guenzler, G.J. Steffens, F. Otting, G. Buse, and L. Flohé, Structural relationship between human high and low molecular mass urokinase. **Hoppe-Seyler's Z. Physiol. Chem.** 363: 133 (1982).

42. M.P. Stoppelli, A. Corti, A. Soffientini, G. Cassani, F. Blasi, and R.K. Assoian, Differentiation enhanced binding of the aminoterminal fragment of human urokinase plasminogen activator to a specific receptor on U937 monocytes. **Proc. Natl. Acad. Sci.** 82: 4939 (1985).

43. A. Estreicher, A. Wohlwend, D. Belin, and J.D. Vassalli, Characterization of the cellular binding site for the urokinase-type plasminogen activator. **J. Biol. Chem.** 264: 1180 (1989).

44. E. Appella, E.A. Robinson, S.J. Ulrich, M.P. Stoppelli, A. Corti, G. Cassani, and F. Blasi, The receptor binding sequence of urokinase. **J. Biol. Chem.** 262: 4437 (1987).

45. F. Blasi, Surface receptors for urokinase plasminogen activator. **Fibrinolysis** 2: 73 (1988).

46. F. Robbiati, M.L. Nolli, E. Sarubbi, A. Soffientini, M.P. Stoppelli, F. Parenti, G. Cassani, and F. Blasi, A recombinant prourokinase mutant missing the growth factor like domain does not bind to the urokinase receptor. Submitted for publication.

47. E. Appella, I.T. Weber, and F. Blasi, Structure and function of the growth factor-like regions in proteins. **FEBS Lett.** 231: 1 (1988).

48. L.S. Nielsen, G.M. Kellerman, N. Behrendt, R. Picone, K. Danø, and F. Blasi, A 55,000-60,000 Mr receptor protein for urokinase. **J. Biol. Chem.** 263: 2358 (1988).

49. N. Behrendt, E. Rønne, M. Ploug, T. Petri, D. Løber, L.S. Nielsen, W.-D. Schleuning, F. Blasi, E. Appella, and K. Danø, The human receptor for urokinase plasminogen activator. N-terminal amino acid sequence and glycosylation variants. Submitted for publication.

50. A.L. Roldan, M.V. Cubellis, M.T. Masucci, N. Behrendt, L.R. Lund, K. Danø, E. Appella, and F. Blasi, Cloning and expression of the receptor for human urokinase plasminogen activator, a central molecule in cell-surface plasmin-dependent proteolysis. Submitted for publication.

51. J. Pöllänen, O. Saksela, E.M. Salonen, P. Andreasen, L.A. Nielsen, K. Danø, and A. Vaheri, Distinct localizations of urokinase-type plasminogen activator and its type 1 inhibitor under cultured human fibroblasts and sarcoma cells. **J. Cell Biol.** 104: 1085 (1987).

52. J. Pöllänen, K. Hedman, L.S. Nielsen, K. Danø, and A. Vaheri, Ultrastructural localization of plasma membrane-associated urokinase-type plasminogen activator. **J. Cell Biol.** 106: 87 (1988).

53. C. Hébert, and J.B. Baker, Linkage of extracellular plasminogen activator to the fibroblast cytoskeleton: colocalization of cell surface urokinase with vinculin. **J. Cell Biol.** 106: 1241 (1988).

54. R.W. Stephens, J. Pöllänen, H. Tapiovaara, K.-C. Leung, P.-S. Sim, E.M. Salonen, E. Rønne, N. Behrendt, K. Danø, and A. Vaheri, Activation of pro-urokinase and plasminogen on human sarcoma cells: a proteolytic cascade with surface-bound reactants. **J. Cell Biol.** 108: 1987 (1989).

55. W.F. Glass, R.A. Radnik, J.A. Garoni, and J.I. Kreisberg, Urokinase-dependent adhesion loss and shape change after cyclic adenosine monophosphate elevation in cultured rat mesangial cells. **J. Clin. Invest.** 82: 1992 (1988).

56. L.A. Miles, and E.F. Plow, Plasminogen receptors: ubiquitous sites for cellular regulation fibrinolysis. **Fibrinolysis** 2: 61 (1988).

57. K. Danø, N. Behrendt, E. Rønne, V. Ellis, and F. Blasi, The urokinase receptor and regulation of pericellular plasminogen activation, in: "Molecular Biology of the Cardiovascular System" UCLA Symposium on Molecular and Cellular Biology, New Series, vol. 132, R. Roberts and J. Sambrook, Eds., Alan R. Liss Inc., New York, (1989) in press.

58. M.V. Cubellis, M.L. Nolli, G. Cassani, and F. Blasi, Binding of single chain prourokinase to the urokinase receptor of human U937 cells. **J. Biol. Chem.** 261: 15819 (1986).

59. E.F. Plow, D.E. Freaney, J. Plescia, and L.A. Miles, The plasminogen system and cell surfaces: evidence for plasminogen and urokinase receptors on the same cell type. **J. Cell Biol.** 103: 2411 (1986).

60. V. Ellis, M.F. Scully, and V.V. Kakkar, Plasminogen activation initiated by single chain urokinase-type plasminogen activator. Potentiation by U937 monocytes. **J. Biol. Chem.** 264: 2185 (1989).

61. J. Pöllänen, Down-regulation of plasmin receptors on human sarcoma cells by glucocorticoids. **J. Biol. Chem.** 264: 5632 (1989).

62. M.V. Cubellis, P.A. Andreasen, P. Ragno, M. Mayer, K. Danø, and F. Blasi, Accessibility of receptor-bound urokinase to type-1 plasminogen activator inhibitor. **Proc. Natl. Acad. Sci. USA** 86: 4828 (1989).

63. K. Danø, P.A. Andreasen, N. Behrendt, J. Grøndahl-Hansen, P. Kristensen, and L.R. Lund, Regulation of the urokinase pathway of plasminogen activation, in: "Development and function of the Reproductive Organs" vol. II M. Parvinen, I. Huhtaniemi and L.J. Pelliniemi, eds. Ares-Serono Symposia, Rome p 259 (1988).

64. P. Verde, S. Boast, A. Franzé, F. Robbiati, and F. Blasi, An upstream enhancer and a negative element in the 5' flanking region of the human urokinase plasminogen activator gene. **Nucl. Acid Res.** 16: 10699 (1988).

65. R. Picone, E. Kajtaniak, L.S. Nielsen, N. Behrendt, M.R. Mastronicola, M.V. Cubellis, M.P. Stoppelli, K. Danø, and F. Blasi, Regulation of urokinase receptor in monocyte-like U937 by phorbol ester phorbol myristate acetate. **J. Cell Biol.** 108: 693 (1989).

66. D. Boyd, G. Florent, P. Kim, and M. Brattain, Determination of the level of urokinase and its receptor in human colon carcinoma cell lines. **Cancer Res.** 48: 3112 (1988).

67. P.A. Andreasen, B. Georg, L.R. Lund, A. Riccio, and S.N. Stacey, Plasminogen activator inhibitors: hormonally regulated serpins. Molec. Cell. **Endocrinol.** in press.

68. M.P. Stoppelli, C. Tacchetti, M.V. Cubellis, A. Corti, V.J. Hearing, G. Cassani, E. Appella, and F. Blasi, Autocrine saturation of the urokinase receptors. **Cell** 45: 675 (1986).

REGULATION OF TISSUE PLASMINOGEN ACTIVATOR SECRETION FROM HUMAN ENDOTHELIAL CELLS

EUGENE G. LEVIN, KEITH R. MAROTTI,[*]
AND LYDIA SANTELL

Department of Basic and Clinical Research
Scripps Clinic and Research Foundation
10666 North Torrey Pines Road
La Jolla, CA 92037

INTRODUCTION

Tissue plasminogen activator is produced by cultured cells from a variety of sources, and its production can be regulated by hormones,[1-3] tumor promoters,[3-5] serine proteases,[6] and mediators of inflammation.[7] Among the cells producing t-PA is the endothelial cell. One of the proposed functions of the vascular endothelium is to maintain or reestablish a homeostatic environment through a variety of pathways, including the production and release of t-PA. *In vivo*, t-PA levels are elevated following exercise, venous occlusion, or infusion of DDAVP.[8,9] The appearance of t-PA is rapid and transient with the initial increase occurring at or before 30 min and returning to baseline by 5-6 hrs. This pattern of t-PA increase and decrease in blood suggests that it results from a release of stored t-PA. In contrast, stimulation of t-PA release from cultured human endothelial cells by thrombin or histamine is much slower, requiring 4-6 hrs to begin, and is also prolonged, continuing for 12-16 hr and only then declining to baseline levels.[6,10] This sequence of events has also been reported in the stimulation of t-PA antigen secretion from other cells such as rat granulosa cells following administration of leutinizing hormone.[2,3] Thus, this type of response may be characteristic of hormone-induced release of t-PA antigen and may be important in various physiologic and pathologic events.

We have been examining the mechanisms by which tissue plasminogen activator production and secretion are controlled within the cell, i.e., those pathways that are involved with initiating the response following hormone receptor occupancy. To perform these experiments we have employed specific agonists of well-defined intracellular signalling pathways to determine whether the activation of those pathways

*Department of Molecular Biology
The UPJOHN Company
Kalamazoo, MI 49001

Serine Proteases and Their Serpin Inhibitors in the Nervous System
Edited by B. W. Festoff
Plenum Press, New York, 1990

31

results in the induction of t-PA antigen secretion and whether the same pathways mediate the stimulation of t-PA secretion by physiologic agonists.

RESULTS

The addition of increasing concentrations of thrombin to confluent cultures of human endothelial cells for 16 hrs results in a dose-dependent increase in the level of t-PA antigen in the conditioned medium.[6] This increase is first detectable with 0.1 U/ml thrombin and maximum levels reach with 1 U/ml thrombin (12 ng/ml, a 6-fold increase). During the first 6 hrs after thrombin addition, secretion of t-PA does not change (Figure 1). However, between 6 and 16 hrs the rate of t-PA antigen secretion increases 6 to 7-fold and then declines toward baseline levels after 16 hrs. Treatment with a second dose of 1 U/ml thrombin at 24 hrs does not stimulate t-PA secretion any further, suggesting that a period of down-regulation or desensitization occurs with thrombin treatment.

The 6 hr delay that occurs between thrombin addition and enhanced t-PA release suggests that the increase depends upon other metabolic events. To determine whether protein synthesis or RNA synthesis is necessary for the stimulation of t-PA release, cultures were treated with either cycloheximide or actinomycin D at various times after thrombin addition and the amount of t-PA present in the conditioned medium determined. Both inhibitors suppressed thrombin-induced increases in t-PA secretion. The effect of actinomycin D was time-dependent. If this inhibitor was added 3 hrs after thrombin only 20% of the t-PA normally secreted in response to thrombin was released. However if actinomycin D was added 6 hrs after thrombin little inhibitory effect was observed when the t-PA was measured in the culture medium after 24 hrs. Thus, it appears as if RNA synthesis is necessary for increased release of t-PA but that it is only required during the first 6-8 hrs after thrombin addition.

Treatment of endothelial cells with histamine gives identical results as found with thrombin.[10] Histamine increases t-PA secretion in a dose- and time-dependent manner. The time course of release is identical to that of thrombin, including the 4 hr lag, rapid secretion between 4 and 16 hrs, and a decline back to baseline rates after 16 hrs. Secondary treatment of histamine has no stimulatory effect upon t- PA

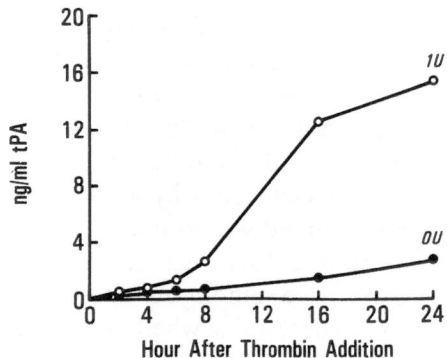

Figure 1. *Time course of thrombin-induced tPA release.* Confluent cultures of endothelial cells were incubated with or without 1 U/ml thrombin for the indicated period. The culture medium was removed, centrifuged, and analyzed for tPA by radioimmunometric assay. Each value represents the average of three experiments involving three separate batches of cells. o 1 U/ml thrombin; •, untreated cells. Reproduced from the Journal of Clinical Investigation 1984, 74, 1988-1995 by copyright permission of The American Society for Clinical Investigation.

secretion after 16 hrs, once again suggesting the onset of desensitized state. Both thrombin and histamine activate the phosphoinositide pathway, leading to the mobilization of calcium and the generation of diacylglycerol, the cellular activator of protein kinase C.[11,12] Thus, it is possible that the mechanism by which both of these agonists induce elevated levels of t-PA secretion involves one or both of these intracellular signalling pathways. Evidence suggesting that histamine and thrombin induce t-PA secretion through similar or identical pathways includes: 1) the lack of a full or even partial additive effect when both are added to endothelial cell cultures simultaneously,[10] and 2) the inability of each to stimulate t-PA secretion once a desensitized state has been established by the other. The generation of diacylglycerol and the onset of desensitized states by both of these agonists suggested that protein kinase C may be involved in the induction of t-PA secretion. Experiments were performed to determine whether the activation of protein kinase C by phorbol esters would lead to stimulation of t-PA secretion followed by desensitization.

To compare the effect of various tumor promoters on t-PA release, dose titration experiments were performed with increasing concentrations of phorbol myristate acetate (PMA), phorbol dibutyrate (PDB), phorbol didecanoate (PDD), and 4-α-phorbol didecanoate (4-α-PDD). PMA and PDB stimulated the largest increases in t-PA concentration after 24 hrs (8-12-fold; $EC_{50} = 13$ and 55 nM, respectively), with PDD stimulating t-PA secretion to about 60% of PMA-treated cultures. The non-tumor promoting 4-α-PDD had no effect at all. Time course studies of t-PA release showed a similar time course to that observed with thrombin and histamine. The accumulation of t-PA began several hours after agonist addition, rose rapidly, and returned to baseline levels. The transient nature of the increase in t-PA release was similar to drug-induced desensitization previously reported to occur in numerous metabolic pathways following treatment with tumor promoting agents. To determine whether these cells became desensitized to phorbol esters, cultures were exposed to PDB for 20 hrs and then treated a second time for 20 hrs. No secondary release of t-PA was observed. Desensitization of phorbol ester-induced t-PA occurs in a dose and time-dependent manner.[10]

If histamine and thrombin were inducing t-PA release via a protein kinase C dependent pathway, then the desensitized state induced by phorbol esters (which results from a loss of protein kinase C activity and immunoprecipitable material) should also result in the loss of responsiveness to histamine and thrombin. Treatment of either thrombin or histamine following a 20 hr incubation with 100 nM PMA (a concentration inducing a fully desensitized state) induced t-PA secretion to only 25% of that occurring when the cells were treated with either thrombin or histamine alone.

Thus, the down-regulation of protein kinase C reduced the ability of either histamine or thrombin to stimulate enhanced t-PA secretion, suggesting a role for protein kinase C in the response to the two physiologic agonists.

It has become apparent that stimulatory effects of secondary messenger pathways can be influenced by simultaneous activation of other messenger pathways.[13] To determine whether activation of adenylate cyclase (increasing the cAMP level) would alter the phorbol ester-induced release of t-PA secretion, cultures were treated with PMA and various compounds which elevate intracellular cyclic AMP levels. In each case, elevation of cAMP alone had no stimulatory effect upon t-PA secretion. However, if cAMP levels were increased simultaneously with PMA treatment, phorbol ester-stimulated release was potentiated (Table 1). In the case of forskolin (an activator of adenylate cyclase) t-PA concentrations increased from approximately 22 ng/ml in 24 hrs (PMA alone) to 122 ng/ml in the same time period (PMA and forskolin). This potentiation was also observed with other tumor promoters (PDB and

Table 1. Potentiation of Phorbol Ester-Induced tPA Release by cAMP. Cultures of endothelial cells were treated with the indicated concentrations of cAMP elevating compounds and/or phorbol esters for 20 hr and tPA concentration determined by radioimmunometric assay. Values are the mean ± S.D. of at least six determinations from two separate experiments performed in triplicate.

Treatment	tPA ng/ml			Fold Increase
None	3.4	±	0.2	
PMA (100 nM)	22.2	±	3.9	6.5
forskolin (100 μM)	3.9	±	0.6	1.2
PMA + forskolin	122.1	±	22.9	35.8
IBMX (100 μM)	1.9	±	0.2	0.6
PMA + IBMX	63.1	±	8.4	18.6
Bt_2cAMP (10 mM)	6.3	±	0.4	1.9
PMA + Bt_2cAMP	94.7	±	8.7	27.8
br-cAMP (10 mM)	2.5	±	0.2	0.7
PMA + br-cAMP	57.0	±	5.0	16.7
PDBu (100 nM)	12.3	±	1.7	3.6
PDBu + forskolin	92.2	±	8.0	27.1
PDD (100 nM)	19.9	±	1.2	5.8
PDD + forskolin	91.9	±	7.5	27.0
4-α-PDD	2.9	±	0.2	0.9
4-α-PDD + forskolin	3.1	±	0.2	0.9

PDD) but not the non-tumor promoter 4-α-PDD. While the rate of secretion of t-PA rose dramatically, no effect on the characteristics of the PMA-induced release occurred. For example, while forskolin caused an upward shift in the amount of t-PA released throughout the entire effective dose range, the concentration of PMA causing half- maximal and maximal release remained the same in the presence or absence of forskolin, and substimulatory concentrations of PMA were still without effect. In addition to the dose-titration curve, cyclic AMP had no effect on the kinetics of the PMA-induced release (Figure 2). As has been seen before with histamine, thrombin, and PMA alone, a 4 hr lag period preceded the onset of enhanced t-PA release (4-16 hrs) and ended with a decline in the rate of release back to baseline levels after 16 hrs. The decline back to baseline levels was followed by a period of desensitization. Maximum potentiation by forskolin was only observed when this compound was added simultaneously with PMA. If addition of forskolin was delayed, the degree of potentiation decreased and was no longer seen if forskolin was added 6 hrs after PMA (Figure 3).

To determine whether the potentiation of this PMA-induced release was a general cellular phenomenon or possibly specific for t-PA, the same experiments were repeated and the effect on plasminogen activator type I (PAI-1) secretion was determined (Table 2). Treatment of cells with phorbol ester stimulated the release of PAI-1 approximately 2- fold while cyclic AMP elevating compounds actually reduced the secretion rate by about 30%. When PMA and forskolin were added simultaneously, the elevation of cyclic AMP attenuated the PMA-induced release of PAI-1 so that instead of a stimulatory response, PAI-1 release rates were actually depressed below control levels. Therefore, in contrast to its potentiative effect on PMA-induced release of t-PA, the elevation of cyclic AMP reverses the stimulatory effect of PMA on PAI-1.

To further study the mechanism by which PMA stimulates t-PA release and PMA potentiates this effect, messenger RNA levels were measured under the various con-

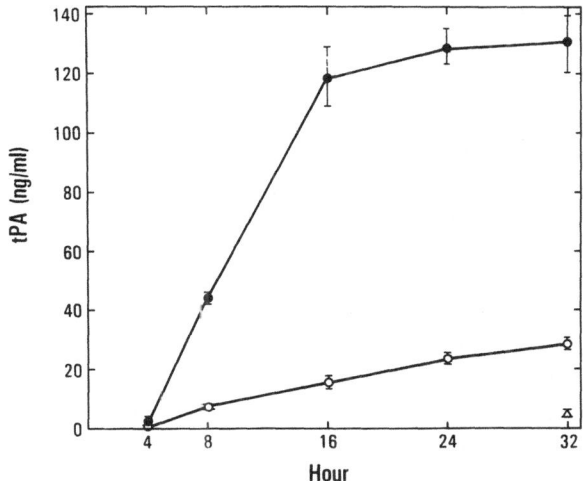

Figure 2. *Time course of PMA and PMA-forskolin-induced release of tPA from endothelial cells.* Cultures were treated with 100 nM PMA in the presence or absence of 100 µM forskolin, and the culture medium removed at the indicated times and assayed for tPA. Data points are the mean and S.D. of six determinations from two separate experiments performed in triplicate. o, PMA alone; •, PMA plus forskolin; Δ, tPA levels in untreated cultures at 32 h. Reproduced from the Journal of Biological Chemistry 1988, 263, 16802-16808 by copyright permission of The American Society for Molecular Biologists and Biochemists.

ditions used previously. In PMA-only treated cultures, t-PA mRNA did not change detectably over the first 4 hrs, increased 3.5-fold by 8 hrs and returned to basal levels by 16 hrs. In the presence of forskolin, little change occurred in 2 hrs, but by 4 hrs the mRNA had increased 12-fold, and by 8 hrs was 25 times that of controls. Once again, levels had returned to control levels by 16 hrs. A 50% decline in the level of t-PA mRNA had already occurred by 12 hrs, indicating that the level did not remain high for a prolonged period of time. Actin transcripts remained stable during this entire time period. Addition of cycloheximide with PMA plus forskolin did not inhibit the induction of t-PA mRNA levels and, in fact, promoted the increase to higher levels (approximately 1.5 times) than in the absence of cycloheximide. The decline in mRNA levels between 8 and 16 hrs also occurred in the presence of cycloheximide.

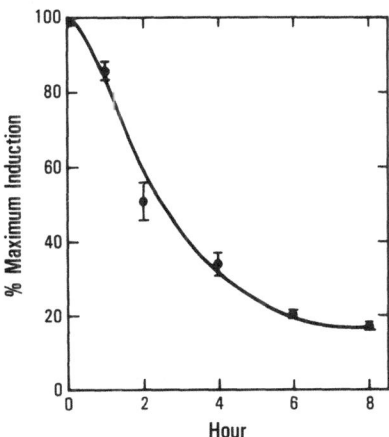

Figure 3. *Effect of delayed addition of forskolin to PMA-treated cultures on the potentiation of tPA release.* Cultures were treated with 100 nM PMA, and 100 µM forskolin was added simultaneously or at various times afterwards. The cells were incubated to 20 h after PMA addition and the tPA concentration determined. Values are the mean and S.D. of six determinations from two experiments performed in triplicate. The tPA concentration in control cultures to which only PMA was added was equal to the 8-h value. Reproduced from the Journal of Biological Chemistry 1988, 263, 16802-16808 by copyright permission of The American Society for Molecular Biologists and Biochemists.

Thus, in contrast to HeLa cells[5] and HT1080 fibrosarcoma cells[14] protein synthesis is not needed for induction or decline in t-PA mRNA levels.

As had been found with t-PA antigen secretion rates, delayed addition of forskolin also limited the degree to which t-PA mRNA increased with PMA treatment. Again, the later the addition the less the potentiation effect by forskolin, and by 6 hrs after PMA no potentiation was observed.

In an attempt to provide further evidence for the association between t-PA mRNA levels and t-PA antigen secretion, the protein kinase inhibitor H-7 was employed.[15] This is an isoquinolinesulfonamide derivative which inhibits several protein kinases in an *in vitro* assay system[15] and has been previously shown to depress phorbol ester-induced responses in a variety of other cell lines.[16,17] Increasing concentrations of H-7 inhibited PMA-induced t-PA antigen secretion in a dose-dependent manner, with 25 μM completely inhibiting secretion. The ID_{50} equalled between 7 and 7.5 μM in both PMA and PMA plus forskolin treated cultures. Neither cell viability or the cell metabolic state was affected by the highest concentration of H-7 employed in these studies. H-7 also inhibited the phorbol ester dependent increase in PAI-1 secretion but had no effect upon the basal rate of secretion, indicating that the inhibitor did not suppress general metabolic activity.

Table 2. Effect of PMA and cAMP on PAI-1 Levels in Conditioned Medium. Cultures were treated for 20 hr and the PAI-1 concentration determined by immunoassay. Values shown are mean ± S.D. of triplicate determinations and is representative of additional experiments.

Treatment	PAI-1 ng/ml	Fold Increase
None	316 ± 6.6	
PMA (100 nM)	536 ± 4.8	1.70
forskolin (100μM)	226 ± 18.0	0.71
forskolin + PMA	198 ± 2.4	0.63
IBMX (250 μM)	216 ± 8.0	0.68
IBMX + PMA	252 ± 4.0	0.79
Bt$_2$cAMP (10 mM)	200 ± 12.0	0.63
Bt$_2$cAMP + PMA	210 ± 2.1	0.66
4-α-PDD (1 μM)	284 ± 11.2	0.89
4-α-PDD + forskolin	206 ± 13.1	0.65

To determine if H-7 inhibition was time-dependent, H-7 was added to cultures at 1, 2, 4, 6, or 8 hrs after PMA or PMA plus forskolin, and the culture medium collected at 24 hrs and analyzed for t-PA antigen. With either PMA or PMA plus forskolin, stimulation was completely attenuated when the inhibitor was added at zero time. However, if H-7 addition was delayed, inhibition of enhanced t-PA secretion diminished, and at 6 hrs after PMA or PMA plus forskolin H-7 had no effect upon the final concentration of t-PA antigen at 24 hrs. The same time course was seen when the effect of H-7 on the accumulation of t-PA mRNA in response to PMA plus forskolin was examined. H-7 inhibited the response to PMA plus forskolin when added with the agonists. When H-7 was added 4 hrs after the agonists message levels were found to be 50% of maximum levels at 8 hrs. H-7 had no effect if added 6 hrs after PMA plus forskolin.

DISCUSSION

We have demonstrated that human endothelial cells in culture produce and release tissue plasminogen activator and that the production of this protein may be regulated by specific intracellular signalling pathways. The activation of protein kinase C appears to be the primary signal which leads to elevation of t-PA mRNA and the subsequent increase in secretion rates. This is suggested by the effect of various phorbol esters on t-PA release and the occurrence of a refractory period following treatment. Protein kinase C has been identified as the major cellular receptor for phorbol esters and the protein through which these compounds exert their effects. [18-21] Also, much evidence exists to associate desensitization with the loss of protein kinase C activity and the level of immunoprecipitable material. Even cAMP dependent potentiation of t-PA secretion appears to be entirely controlled by the events associated with the activation of protein kinase C. The presence of elevated levels of cyclic AMP has no effect on the concentration range of PMA stimulation, substimulatory PMA concentrations remain ineffective, and the presence of forskolin does not eliminate or shorten the delay observed before onset of enhanced t-PA release, the decline in the maximum release rate, or the onset of a refractory period after 16 hrs. Therefore, we have concluded that the signals mediating PMA stimulation continue to define the characteristics of t-PA induction while a cyclic AMP dependent event acts to enhance the magnitude of that response.

The role for t-PA mRNA in regulating t-PA secretion is a common event and is found in HeLa cells,[5] rat granulosa cells,[3,22] and HT1080 fibrosarcoma cells.[14] Hormonal or tumor promoter stimulation of t-PA secretion was associated with increases in the level of t-PA mRNA as we have observed here. However, various differences in the response of t-PA mRNA in these different systems was observed, with the time course and longevity of the increase and the dependence on protein synthesis differing in each system. Thus, we believe that human endothelial cells are distinct from other studied cultured cells with respect to the stimulation of t-PA mRNA levels and antigen secretion.

Underscoring this difference is the effect of protein synthesis inhibition by cycloheximide. In HeLa cells and HT1080 fibrosarcoma cells the inhibition of protein synthesis eliminates the increase in t-PA mRNA levels in response to PMA and dexamethasone, respectively.[5,14] In HeLa cells protein synthesis is not only required for the initiation of the inductive effect but also is necessary for prolongation of the effect since addition at any time after PMA treatment attenuates further increase in mRNA levels. In endothelial cells however we have shown that cycloheximide has no affect on either the increase or the decline in mRNA levels and therefore the regulation of t-PA production in human endothelial cells is independent of protein synthesis.

The result of the H-7 experiments suggests that the response of t-PA mRNA and t-PA antigen secretion to phorbol esters is dependent upon the activation of a protein kinase. Because of the broad inhibitory action of H-7 and the absence of information on how H-7 works in intact cells the conclusion that the protein kinase involved is protein kinase C cannot be established. However, the fact that this protein kinase is the target of phorbol ester action and the positive affect of dioctanoylglycerol, a more specific activator of protein kinase C, do support a role for protein kinase C in the stimulation of t-PA production under the conditions used. The H-7 data also indicate that maximum stimulation of t-PA mRNA is dependent upon the prolonged activation of the event or events that are affected by this inhibitor. This is indicated by the fact that the increased t-PA mRNA levels in response to PMA plus forskolin can be stopped at any time between 0 and 6 hrs after PMA addition and that the affect by cyclic AMP can still be observed anytime during the increase of t-PA mRNA levels. These events could be attributed to continued activation of a protein kinase or the accumulation of an intermediate that is necessary for the increase in t-PA mRNA levels. The necessity of prolonged activation of an event to obtain maximum levels of t-PA antigen secretion and mRNA levels may explain why the response to dioctanoylglycerol, which is unstable and activates protein kinase C only transiently, never attains the level achieved with phorbol esters which maintain activation of protein kinase C for longer periods of time.

While this work has focused on specific intracellular signaling pathways that are activated by non-physiologic agonists, the similarity between the characteristics of these results and those observed with thrombin and histamine suggest that similar if not identical mechanisms occur when cells are treated with these two agonists. Thus, we believe that activation of protein kinase C and possibly the elevation of AMP mediate tPA regulation in endothelial cells.

REFERENCES

1. E. H. Allen, J. A. Hamilton, R. L. Medcalf, M. Kubota, and T. J. Martin, Cyclic AMP-dependent and independent effects on tissue-type plasminogen activator in osteogenic sarcoma cells; evidence from phosphodiesterase inhibition and parathyroid hormone antagonists, **Biochim. Biophys. Acta** 888:199 (1986).
2. M. L. O'Connell, R. Canipari, and S. Strickland, Hormonal regulation of tissue plasminogen activator secretion and mRNA levels in rat granulosa cells, J. Biol. Chem. 262:2339 (1987).
3. T. Ny, Y-X. Liu, M. Ohlsson, P. B. C. Jones, and A. J. W. Hsueh, Regulation of tissue-type plasminogen activator and messenger RNA levels by gonadotropin-releasing hormone in cultured rat granulosa cells and cumulus-oocyte complexes, J Biol Chem 262:11790 (1987).
4. H. Ashino-Fuse, G. Opdenakker, A. Fuse, and A. Billiau, Mechanisms of the stimulatory effect of phorbol 12-myristate 13-acetate on cellular production of plasminogen activator, **Proc Soc Exp Biol Med** 176:109 (1984).
5. E. K. Waller, and W. -D. Schleuning, Induction of fibrinolytic activity in HeLa cells by phorbol myristate acetate: Tissue-type plasminogen activator antigen and mRNA augmentation require intermediate protein biosynthesis, **J. Biol. Chem.** 260:6354 (1985).
6. E. G. Levin, U. Marzec, J. Anderson, and L. A. Harker, Thrombin stimulates tissue plasminogen activator release from cultured human endothelial cells, **J. Clin. Invest.** 74:1988 (1984).
7. M. P. Bevilacqua, R. R. Schleef, M. A. Gimbrone, and D. J. Loskutoff, Regulation of the fibrinolytic system of cultured human vascular endothelium by interleukin 1, J Clin Invest 78:587 (1986).
8. S. Sherry, R. I. Lindemeyer, A. P. Fletcher, and N. J. Alkjaersig, , **J. Clin. Invest.** 38:810 (1959).
9. P. M. Mannucci, M. Abery, I. M. Nilsson, and B. Robertson, , **British J. Haemotol.** 30:81 (1975).
10. E. G. Levin, and L. Santell, Stimulation and desensitization of tissue plasminogen activator release from human endothelial cells, J Biol Chem 263:9360 (1988).
11. T. J. Resink, G. Y. Grigorian, A. K. Moldabaeva, S. M. Danilov, and F. R. Buhler, Histamine-induced phosphoinositide metabolism in cultured human umbilical vein endothelial cells. Association with thromboxane and prostacyclin release, **Biochem. Biophys. Res. Commun.** 144:438 (1987).
12. S. L. Hong, N. J. McLaughlin, C. -Y. Tzeng, and G. Patton, Prostacyclin synthesis and deacylation of phospholipids in human endothelial cells: Comparison of thrombin, histamine and ionophore A23187, **Thromb. Res.** 38:1 (1985).
13. Y. Nishizuka, Studies and perspectives of protein kinase C, **Science** 233:305 (1986).
14. M. L. Medcalf, E. Van den Berg, and W. -D. Schleuning, Glucocorticoid- modulated gene expression of tissue- and urinary-type plasminogen activator and plasminogen activator inhibitor 1 and 2, J Cell Biol 106:971 (1988).
15. T. Matsui, Y. Nakao, T. Koizumi, Y. Katakami, and T. Fujita, Inhibition of phorbol ester-induced phenotypic differentiation of HL-60 cells by 1-(5-isoquinolinylsulfonyl)-2-methylpiperazine, a protein kinase inhibitor, **Cancer Res.** 46:583 (1986).
16. R. M. Burch, A. Lin Ma, and J. Axelrod, Phorbol esters and diacylglycerols amplify bradykinin-stimulated prostaglandin synthesis in swiss 3T3 fibroblasts, J Biol Chem 263:4764 (1988).

17. A. K. Ho, C. L. Chik, and D. C. Klein, Effects of protein kinase inhibitor (1-(5- isoquinolinesul-fonyl)-2-methylpiperazine (H7) on protein kinase C activity and adrenergic stimulation of cAMP and cGMP in rat pinealocytes, Biochem Pharmacol 37:1015 (1988).

18. M. Castagna, Y. Takai, K. Kaibuchi, K. Sano, U. Kikkawa, and Y. Nishizuka, Direct activation of calcium-activated,phospholipid-dependent protein kinase by tumor-promoting phorbol esters, J Biol Chem 257:7847 (1982).

19. J. E. Niedel, L. J. Kuhn, and G. R. Vandenbark, Phorbol diester receptor copurifies with protein kinase C, Proc Natl Acad Sci USA 80:36 (1983).

20. U. Kikkawa, Y. Takai, Y. Tanaka, R. Miyake, and Y. Nishizuka, Protein kinase C as a possible receptor protein of tumor-promoting phorbol esters, J Biol Chem 258:11442 (1983).

21. K. L. Leach, M. L. James, and P. M. Blumberg, Characterization of a specific phorbol ester aporeceptor in mouse brain cytosol, Proc Natl Acad Sci USA 80:4208 (1983).

22. J. A. Ribes, C. A. Francis, and D. D. Wagner, Fibrin induces release of von Willebrand factor from endothelial cells, J. Clin. Invest. 79:117 (1987).

THROMBIN DISINTEGRATES CELL SURFACE UROKINASE FOCAL ADHESION PLAQUES AND DECREASES CELL EXTENSION: IMPLICATIONS FOR AXONAL GROWTH

CAROLINE A. HÉBERT AND JOFFRE B. BAKER

Department of Biochemistry
University of Kansas
Lawrence, Kansas 66045

INTRODUCTION

In addition to its function as the ultimate protease in the coagulation cascade thrombin has several hormone-like effects on cells *in vitro*. This serine protease is a potent mitogen for fibroblastic cells from several species.[1,2] In addition, like other growth factors,[3] thrombin can modulate cellular differentiation. Thrombin or a closely related protease present on neuroblastoma cells antagonizes formation of neurite extensions.[4] Furthermore thrombin has recently been identified as the primary serum inhibitor of neurite extension in neuroblastoma cell cultures.[5] In the present paper we show that thrombin, but not other growth factors tested, causes the disappearance on human fibroblasts of linear strands of cell surface urokinase-type plasminogen activator, strands which have recently been shown to colocalize with cell-to-substratum focal adhesion plaques (FAPs).[6,7] Disappearance or gross alteration of FAPs in fact accompanies the thrombin-mediated dissolution of the cell surfaced urokinase strands.

Cell surface urokinase promotes cell invasion.[8] The presence of FAPs correlates with cell protrusive activity.[9] The observations that thrombin disintegrates FAPs and associated urokinase strands therefore indicates that thrombin is a potential modulator of cell movement and morphology. We show that thrombin causes a 25% decrease in the average length of fibroblasts on coverslips. It is suggested that thrombin may block the differentiation of neuronal cells by destabilizing local cell-to-substratum adhesive contacts and/or by removing cell surface urokinase from sites of contact with the substratum.

MATERIALS AND METHODS

Materials

Human α-thrombin, originally purified in the laboratory of Dr. John Fenton II (New York State Department of Health, Albany, NY), was a gift from Dr. Darrell H. Carney (University of Texas Medical Branch, Galveston, TX). Some of the thrombin used was obtained from American Diagnostica, Inc., Greenwich, CT. Thrombin whose catalytic serine residue was derivatized with diisopropylphosphofluoridate was prepared as described.[10] Transforming growth factor-β (TGFβ) was a gift from Dr. Dan L.

Serine Proteases and Their Serpin Inhibitors in the Nervous System
Edited by B. W. Festoff
Plenum Press, New York, 1990

41

Eaton (Genentech Inc., South San Francisco, CA). Platelet-derived growth factor (PDGF) was purchased from PDGF Inc., Boston, MA. Epidermal growth factor (EGF) was purchased from Collaborative Research Inc., Lexington, MA. Affinity purified antiurokinase antibody was obtained as previously described.[11] Rabbit antivinculin serum was a gift from Dr. Gary Rosenfeld (University of Texas Medical Center at Houston). Rabbit antitalin was a gift from Dr. Keith W.T. Burridge (University of North Carolina, Chapel Hill, NC). Fluorescein-conjugated goat anti-rabbit IgG, yeast tRNA, RIA-grade bovine serum albumin (BSA), sodium azide, saponin, phorbol myristate acetate (PMA), and diisopropylphosphofluoridate were from Sigma Chemical Co., St. Louis, MO. We used 30 mm cell culture dishes (Falcon 3001; Becton Dickinson and Co., Mountain View, CA). Dulbecco's Modified Eagle's Medium (DMEM) was purchased from Hazleton Systems, Inc., Aberdeen, MD. Penicillin-streptomycin solution and fetal bovine serum (FBS) were obtained from Irvine Scientific, Santa Ana, CA. Trypsin and L-glutamine were purchased from KC Biologicals Inc., Lenexa, KS. Paraformaldehyde and trichloroacetic acid were from Fisher Scientific Co., Pittsburgh, PA. Methyle [^{3}H] Thymidine was purchased from Research Products International Inc., Mount Prospect, IL.

Methods

Cell culture. Human foreskin fibroblasts (HF cells) were grown in DMEM supplemented with 7% FBS and L-glutamine as previously described.[12] Culture media contained 200 U/ml penicillin G and 200 mg/ml streptomycin sulfate (pen/strep). Cultures were maintained in a 5% CO_2 atmosphere at 37°C. Unless indicated otherwise, cells were seeded at 2.5 x 10^4 cells/dish, incubated for 10 hr in serum-containing medium then for 2 hr in serum-free medium before addition of PMA, thrombin or other growth factors mentioned. Except where indicated otherwise, the cells were incubated for 12 hr with the various compounds studied diluted in DMEM + 0.1% BSA.

Interference reflection microscopy. Cell-to-substratum contacts in living fibroblasts were observed by interference reflection microscopy as described by C.S. Izzard and L.R. Lochner.[13] At the end of the incubation period, the cell-containing coverslips were placed in an aluminum filming chamber filled with serum-free medium. Focal contacts were observed with a Zeiss IM35 inverted microscope equipped for interference reflection microscopy with a 100X Antiflex objective. Photographs were taken after processing of the image with a DAGE-MTI video camera/Nuvicon 64 tube enhancement system.

Immunofluorescence labeling of urokinase and vinculin. Immunofluorescence labeling was done according to the procedure of M.C. Willingham and I. Pastan.[14] After extensive wash with PBS, the cells were fixed with 3.7% paraformaldehyde in phosphate buffered saline (PBS). Nonspecific antibody binding was blocked by incubating the cells at room temperature with 1% RIA-BSA in PBS. Fixed cells were incubated for 45 min at room temperature with immune or nonimmune IgG (30 mg/ml) and for 15 min at room temperature with fluorochrome antibody conjugate (50 mg/ml). In the case of vinculin and talin localization, 0.1% saponin was added to all the reagents to permeabilize the cells. All coverslips were mounted in 0.05 M Tris-Cl, 50% glycerol, 0.02% sodium azide (pH 8.1) and observed with a microscope (E. Leitz Inc., Rockleigh, NJ) equipped with epifluorescence and a Fluoreszenz 40:1.3 oil immersion objective.

Assay of the growth factors. The mitogenic activity of the growth factors was checked with a [^{3}H]thymidine incorporation assay. HF cells were seeded at 2 x 10^4 cells/cm^2 and incubated for 24 hr in DMEM supplemented with pen/strep and 5% FBS. The cultures were then rinsed twice with serum-free DMEM and incubated for

48 hr in DMEM supplemented with pen/strep, albumin, transferrin and selenium (ATS-DMEM).[6] After rinsing the cells twice with serum-free DMEM, the various growth factors were added to the cultures in ATS-DMEM. Tritiated thymidine was added to the cultures at a concentration of 5 mCi/ml twelve hours after growth factor addition. After a final incubation period of 12 hr, the cells were rinsed 5x with PBS and lysed in 0.5 N NaOH. The pH was readjusted to 7 with 1 N HCl, 1 mg/ml of yeast tRNA (Sigma) was added to the cell lysate, and the nucleic acids were precipitated with 20% trichloroacetic acid (TCA). The precipitates were filtered onto glass fiber filters (Whatman, GF/B), washed extensively with 5% TCA, and redissolved in 0.5 NaOH. The pH of the samples was readjusted with 1 N HCl, 5 ml of scintillation cocktail were added to each sample, and the radioactivity of the samples was measured in a Beckman β-scintillation counter (LS1801). The average radioactivity of triplicate samples was as follows: negative control (no growth factors added): 47,234 cpm; positive control (10% FBS added): 249,446 cpm; 2.5 mg/ml α-thrombin: 145,580 cpm; 10 ng/ml EGF: 122,010 cpm; 10 ng/ml TGFβ: 48,296 cpm; 10 ng/ml EGF + 10 ng/ml TGFβ: 75,673 cpm; 12.5 ng/ml PDGF: 207,933 cpm.

RESULTS

Many cells have surface receptors for urokinase which may function primarily to focus urokinase activity to the pericellular space.[15] The urokinase present on human fibroblasts is distributed in strands that colocalize with FAPs.[6,7] This prompted us to investigate, by immunofluorescence, the distribution of cell surface urokinase on fibroblasts following pretreatment of the cells with agents known to modulate either urokinase production or FAPs. The urokinase strands present on untreated fibroblasts are shown in Figure 1A. Treatment of the cells with the tumor promoter phorbol myristate acetate (PMA) markedly increased the fluorescence brilliance of the urokinase strands (Figure 1B). This change could reflect the well-known property of PMA to vastly stimulate urokinase secretion[16] which would result in greater occupancy of urokinase receptors by urokinase. We have found that the brilliance of the urokinase strands is enchanced by incubating cells in the presence of exogenous urokinase (data not shown). Figure 1B also shows that a fraction of the urokinase strands on PMA-treated cells had a "rosette" configuration, reminiscent of rosette adhesion plaques present in virally transformed cells.[17]

12 to 24 hr incubation of the fibroblasts with epidermal growth factor (EGF), transforming growth factor-β (TGFβ), or platelet derived growth factor (PDGF) at doses that are optimal for mitogenic stimulation, did not notably alter either the intensity or shape of the urokinase strands (Figure 1C-E). Although EGF stimulates human fibroblast secretion of plasminogen activator, it is far less effective in this regard than is PMA.[18] Both TGFβ and PDGF perturb the FAPs of mouse 3T3 cells,[19,20] an action which might be expected to cause the dissolution of cell surface urokinase strands. However, these growth factors do not release the talin component of focal adhesion plaques, and only release the vinculin component transiently (~ 3 min).[19,20] In the present study we looked for longer lasting effects. The photographs shown in Figure 1 were taken 12 hr after growth factor additions.

In contrast to the above factors, thrombin caused the complete disappearance of the fibroblast urokinase strands (Figure 1F). Loss of the urokinase strands at 12 hr. was complete when thrombin was at 71 nM and was detectable when it was at 9 nM (Figure 2). Thrombin is mitogenic in this dose range. Doses of thrombin below ~ 5 nM are not mitogenic because secreted protease nexin I inactivates it.[21] Figure 2 shows that thrombin (71 nM) that had been inactivated by preincubation with diisopropylphosphofluoridate failed to perturb cell surface urokinase strands (Figure 2), demonstrating that this action of thrombin, like its mitogenic action,[22] is dependent

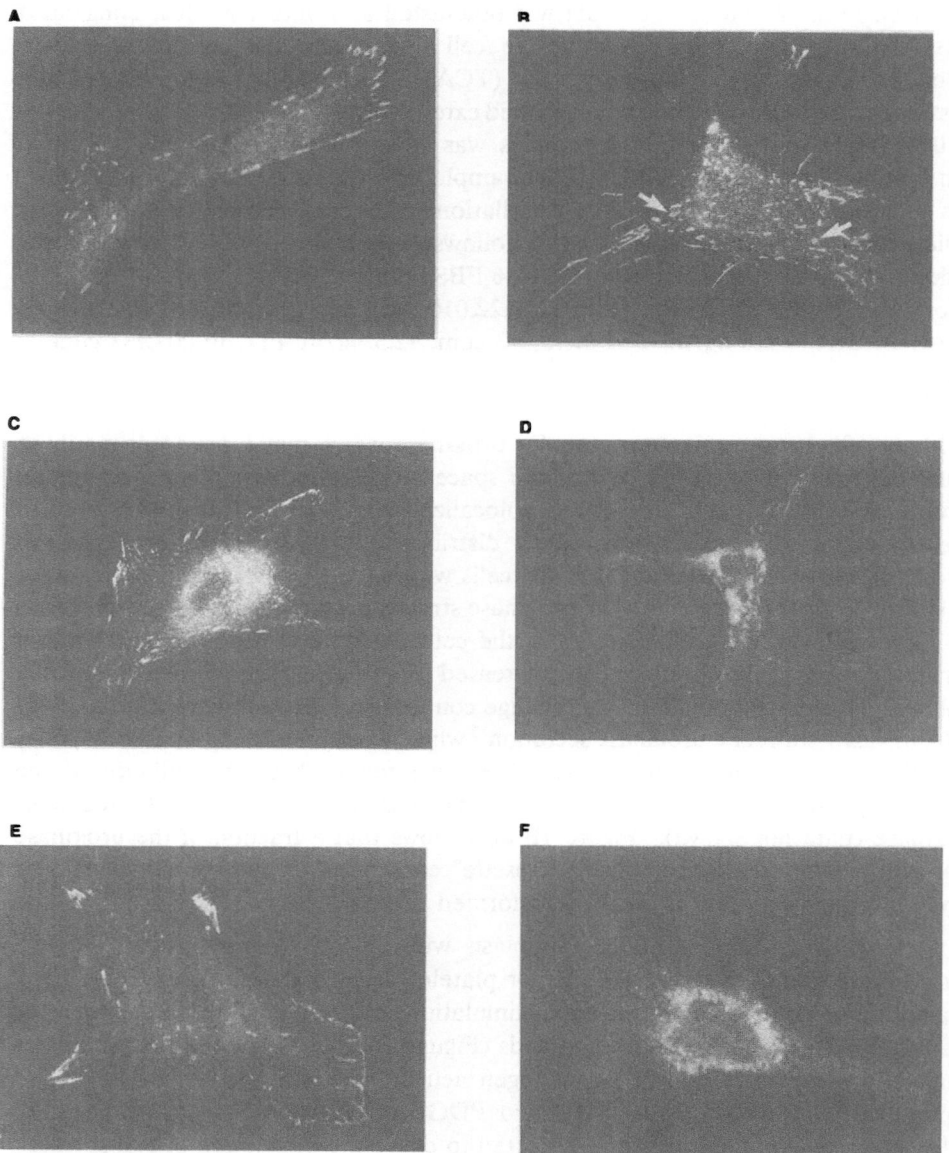

Figure 1. *Effect of various growth promoting substances on the shape and distribution of fibroblast urokinase strands.* Human foreskin fibroblasts were incubated for 12 hr in DMEM + 0.1% BSA + (A) nothing else; (B) 10 nM PMA (arrows point to the "rosettes" of urokinase); (C) 20 ng/ml EGF; (D) 10 ng/ml TGFβ; (E) 25 ng/ml PDGF; (F) 2.5 mg/ml thrombin. The coverslips containing fixed fibroblasts were incubated with affinity-purified rabbit antiurokinase antibody followed by fluorescein-conjugated goat anti-rabbit IgG. Bars 20 μm.

on proteolytic activity. Loss of urokinase strands occurred over the course of several hours of thrombin exposure (Figure 3). Little or no change was observable after 4 hours of treatment. By 8 hours urokinase strands were present but were thinner and elongated compared to those present on untreated cells. The strands became progressively thinner and eventually disappeared by 12 hours of treatment.

Thrombin does not inhibit, but rather modestly stimulates, fibroblast secretion of urokinase.[18] It might cause the disappearance of cell surface urokinase strands by cleaving urokinase.[23] However, fibroblasts that had been incubated for 12 hours with thrombin (71 nM) failed to display urokinase strands even when incubated with exogenous 54 kDa urokinase that had not been exposed to thrombin (data not shown). Thrombin could also eliminate the urokinase strands by redistributing the urokinase receptors. This latter effect would be predicted if thrombin dissolved FAPs, as these structures colocalize with and thus may recruit the urokinase receptors.[6]

FAPs were first examined using immunofluorescence to detect vinculin, an adhesion plaque marker protein. As shown in Figure 4A, control cells contained vinculin in twig structures (characteristic of FAPs)[24] as well as diffuse cytoplasmic vinculin. (Previous studies have shown that the antivinculin antibody used does not stain the cells unless they are permeabilized).[6] Strikingly, the thrombin-treated cells lacked the typical vinculin twig-structures (Figure 4B). In certain cases, vinculin antigen can disappear from FAP structures while talin, another FAP protein, remains in the characteristic FAP pattern.[19] Thrombin-treated cells either contained no discrete concentrations of talin or strands of this protein that were thinner and more elongated than in non thrombin-treated cells (Figure 4C and D). Twelve hour incubation with PDGF or TGFβ (which, as shown above, did not affect urokinase strands) did not change the shapes of FAPs, as indicated by vinculin or talin immunofluroescence (data not shown).

Focal contacts can also be detected by interference reflection microscopy, appearing as dark, tear-shaped structures at the substratum focal plane.[13] Figure 4E shows that, in untreated fibroblasts, these structures were prominent, especially at the cell edges. In thrombin-treated cultures, however, the ventral surfaces of the cells appeared uniformly grey and devoid of focal contacts (Figure 4F). In thrombin-treated cultures, focal adhesion plaques, observed either inculin staining or interference reflection optics disappeared over the same 12 hr time course as did the urokinase strands (see above). While these experiments suggest that thrombin causes redistribution of the urokinase receptors, they do not address the possibility that thrombin might also alter the urokinase binding affinity of the receptors. Relevant binding experiments utilizing [^{125}I]urokinase were prohibited because the thrombin-mediated disintegration of urokinase strands occurred at very low cell densities (vide infra) at which most of the ligand binding is nonspecific. However, it is noteworthy that loss of fluorescent urokinase strands in thrombin-treated cells was accompanied by the appearance of numerous dots of urokinase fluorescence (Figure 1F), suggesting that thrombin caused the redistribution of the receptors without destroying their affinity for urokinase.

During these experiments we noticed that thrombin seemed to decrease the length of the fibroblasts. We therefore measured the cells along their longest axes. Figure 5 shows that, while the cell lengths were distributed over a wide range within both treated and untreated populations, the thrombin-treated cells were, on the average, considerably shorter than the untreated cells (mean of cell lengths; thrombin-treated, 105 mm; control 127 mm). Application of the Kolmogorov-Smirnov test to this data demonstrates that these differences are statistically significant ($P < 0.01$).

After observing thrombin-mediated dissolution of FAPs and colocalized urokinase in approximately 20 successive experiments, we carried out an experiment in which thrombin did not affect these structures. It was noted that, in this case, the cells were present at a slightly higher density than in the previous experiments. We observed

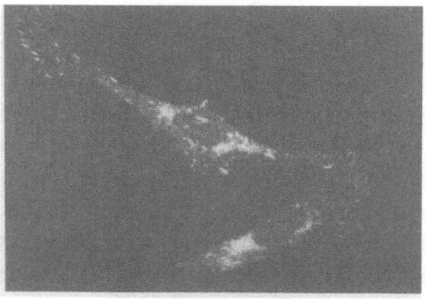

Figure 2. DIP-inactivated thrombin has no effect on the shape and distribution of urokinase strands. Human foreskin fibrolasts were incubated for 12 hr in DMEM + 0.1% BSA + 2.5 mg/ml DIP-thrombin. Bars, 20 mm.

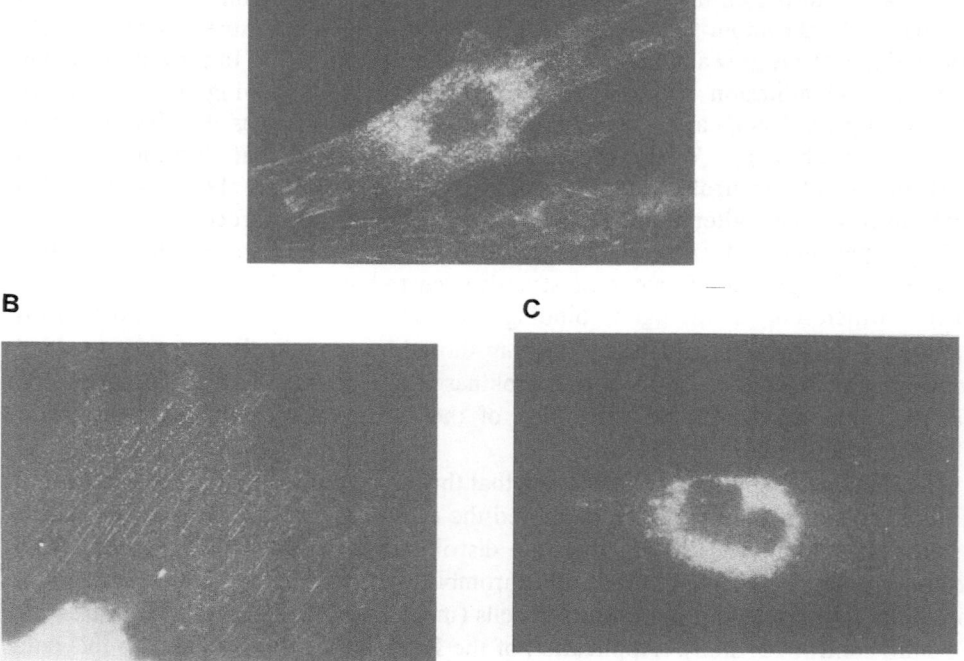

Figure 3. Progressive disintegration of urokinase strands on human foreskin fibroblasts incubated in DMEM + 0.1% BSA + 2.5 mg/ml thrombin for (A) 4 hours; (B) 8 hours; (C) 12 hours. Bars, 20 mm.

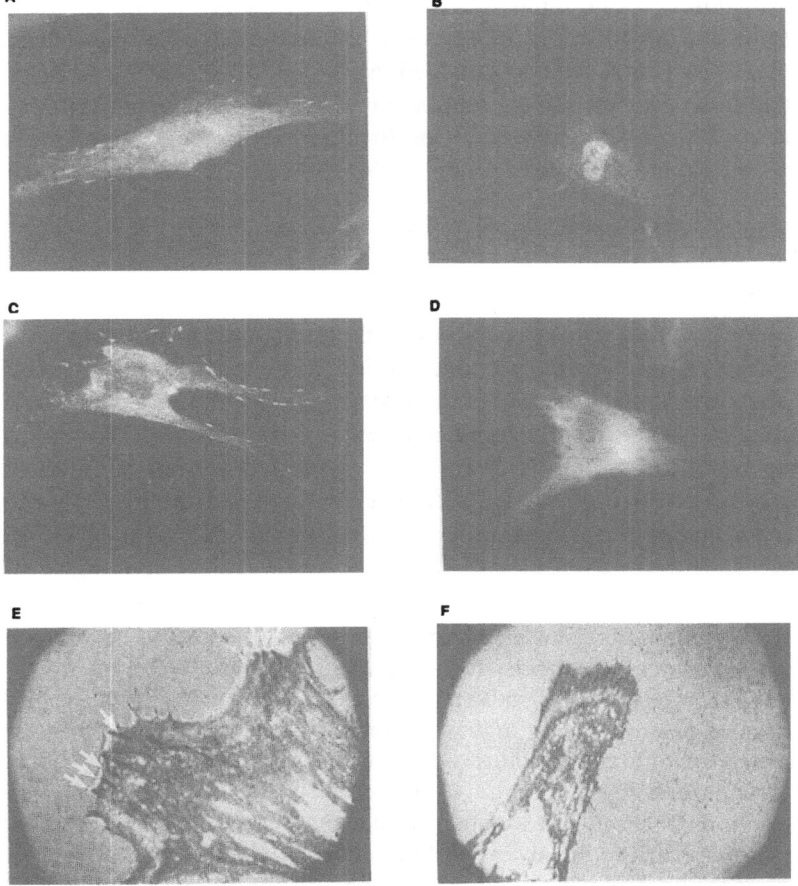

Figure 4. *Effect of thrombin of fibroblast focal adhesion plaques.* Human foreskin fibroblasts were incubated for 12 hr in DMEM + 0.1% BSA + (A,C,E) nothing else; (B,D,F) 2.5 mg/ml thrombin. Coverslips containing fixed fibroblasts were incubated with rabbit antivinculin serum (A and B) or rabbit antitalin IgG (C and D) followed by fluorescein-labeled goat anti-rabbit IgG. (E and F) Focal adhesion plaques (arrows) as observed by interference reflection microscopy. Bars, 20 mm.

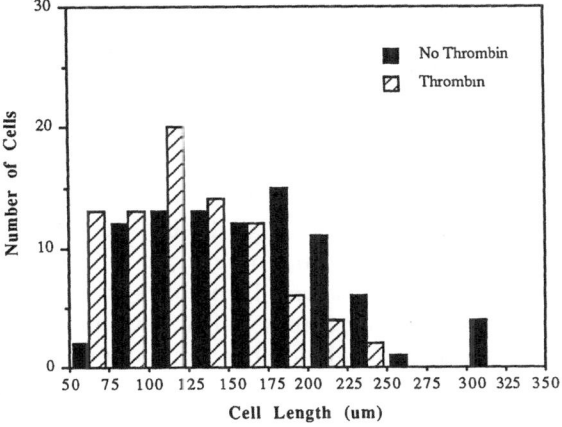

Figure 5. Distribution of cell lengths in thrombin-treated fibroblasts and non-thrombin-treated controls. The cells were treated as described in Figure 1A and 1F and fixed before measurement with an optical micrometer.

that following exposure to thrombin, urokinase strands were present on less than 20% of the cells seeded at 2×10^3 cells/cm^2 but were present on >95% of the cells seeded at 5×10^3 cells/cm^2. This result was obtained in all (3) replicate experiments carried out. This inhibition occurred at a subconfluent cell density. Human fibroblasts reach confluence at 3×10^4 cells/cm^2. High nonspecific fluorescence background prevented the clear determination of the distribution of FAP proteins in confluent cells.

DISCUSSION

The present immunofluorescence study demonstrated that thrombin added to human fibroblasts caused the disappearance of FAP-associated cell-surface urokinase strands and loss or extreme distortion of the FAPs themselves. These events occurred at thrombin doses (>10 nM) which are mitogenic for fibroblasts. However, other modulators of cell growth (EGF, TGFβ, PDGF and PMA) did not cause loss of the urokinase strands, and one of them, PMA, actually increased the intensity of these strands. Our results are not incompatible with findings that PDGF and TGFβ cause disappearance of vinculin from mouse 3T3 cell FAPs.[19] In the latter case talin is unaffected and the loss of vinculin fluorescence is transient, returning to the FAPs thrombin-treated human fibroblasts lasts for many hours. That thrombin alone among the factors tested had these effects may derive from its property to both transmit intracellular second messenger signals (e.g., inositol triphosphate)[25] and to hydrolyze certain cell-surface or extracellular matrix proteins.[26] Clearly the FAPs and associated urokinase are transmembrane structures: the vinculin and talin being internal and the associated urokinase being external. It seems likely that FAPs are stabilized by protein interactions on both sides of the plasma membrane. The stability of these structures evidently varies with even relatively modest changes in the physiological state of the cells, because, as described above, thrombin could not cause their disappearance when cell density increased beyond 5×10^3 cells/cm^2. This cell density is subconfluent.

In view of the wealth of evidence linking urokinase to cell invasion and tissue matrix destruction, it seems likely that cell surface urokinase cleaves tissue matrix proteins with which the cell makes direct contact.[16] The subset of cell surface urokinase that is concentrated at FAPs might function to release cell-to-substratum focal contacts during or preceding cell locomotion or formation of cell extensions. The immunofluorescence intensity of these "urokinase strands" suggests the concentration of urokinase in them is very high. As a consequence, this focused urokinase could have catalytic properties that are not attributed to the soluble form of the enzyme, which has often been considered to exist at low (<nM) concentrations. Interestingly, Gold et al have shown that soluble urokinase at >mM levels cleaves fibronectin, releasing it from extracellular matrix.[27] Available data suggest that most of the urokinase that the cells secrete and therefore load onto their surface receptors is the single chain proenzyme form, although the urokinase receptors bind the active two chain form as avidly as they bind the proenzyme.[28] The nature of the urokinase that is concentrated at FAPs is unknown. Thrombin's ability to eliminate cell-surface foci of urokinase suggests that this protease might have anti-invasive activity. On the other hand there is a negative correlation between numbers of FAPs and the rate of cell locomotion,[29] suggesting that thrombin might accelerate the rate of cell migration.

FAP-like structures are scarce *in vivo*, being present only in certain cells that experience particularly great mechanical stress.[30] Thus, their presence in cultured fibroblasts is likely to be an artifact of attachment to very adhesive plastic substrata. However, it seems probable that the FAP structure has related counterparts that are not identifiable in the light microscope. Further speculation, namely that these sites might also be sensitive to thrombin, is prompted by observations made on neuronal cells, which lack FAPs. Monard et al have shown that a glia-derived thrombin inhibitor and other natural or synthetic thrombin inhibitors stimulate neurite outgrowth in culture.[4] These results suggest that thrombin emanating from either the cells or from a trace of the original serum-containing culture medium suppresses formation of neurite extensions. Several investigators have found that removing serum from neuroblastoma cultures promotes neurite formation.[31,32] Very recently, Gurwitz and Cunningham have provided compelling evidence that the key neurite-suppressing factor in serum is thrombin.[5] The present results suggest that thrombin may act on neuronal cells by modulating cell-to-substratum interactions. In fibroblastic cells there is a positive correlation between FAP number and formation of ruffled membranes and cell extensions (C.A.H.unpublished data).[33,34] Moreover, here we show that loss of FAPs in thrombin-treated fibroblasts is accompanied by decreased extension of the cells on their substratum. Although FAPs are not seen in neuronal cells (C.A.H. unpublished data),[35] FAP-like structures are present in PC12 cells following polyethylene glycol-mediated fusion.[35] Intriguingly, when nerve growth factor is added to these cells to promote the formation of neurites, the pattern of focal adhesion sites changes, with FAPs becoming confined to nascent protrusions and later to the tips of the forming growth cone extensions.

ACKNOWLEDGMENTS

This work was supported by National Institutes of Health grants CA-00886 and CA-29307.

We thank Dr. Carol Erickson and Dr. Roger Leslie for their assistance in obtaining the interference reflection microscopy data, Dr. Gunther Schlager for his advice on the statistical analysis of our data, Philip Heying and Hollis Officer for their help with the photography.

REFERENCES

1. D.H. Carney, K.C. Glenn, and D.D. Cunningham, Conditions which affect initiation of animal cell division by trypsin and thrombin. J. Cell. Physiol., 95:13-22 (1978).
2. L.B. Chen, and J.M. Buchanan, Mitogenic activity of blood components. I. Thrombin and prothrombin. Proc. Natl. Acad. Sci. USA 72:131-135 (1975).
3. M.P. Sporn, and A.B. Roberts, Peptide growth factors are multifunctional. Nature 332:217-219 (1988).
4. D. Monard, E. Niday, A. Limat, and F. Solomon, Inhibition of protease activity can lead to neurite extension in neuroblastoma cells. Prog. Brain Res. 58:359-364 (1983).
5. D. Gurwitz, and D.D. Cunningham, Thrombin modulates and reverses neuroblastoma neurite outgrowth. Proc. Natl. Acad. Sci. USA 85:3440-3444 (1988).
6. C.A. Hébert, and J.B. Baker, Linkage of extracellular plasminogen activator to the fibroblast cytoskeleton: colocalization of cell surface urokinase with vinculin. J. Cell Biol. 106:1241-1247 (1988).
7. J. Pöllänen, K. Hedman, L.S. Nielsen, K. Danø, and A. Vaheri, Ultrastructural localization of plasma membrane-associated urokinase-type plasminogen activator at focal contacts. J. Cell Biol. 106:87-95 (1988).
8. V.J. Hearing, L.W. Law, A. Corti, E. Appella, and F. Blasi, Modulation of metastatic potential by cell surface urokinase of murine melanoma cells. Cancer Res. 48:1270-1278 (1988).
9. J.M. Vasiliev, Spreading of non-transformed and transformed cells. Biochim. Biophys. Acta 780:21-65 (1985).
10. D.A. Low, and D.D. Cunningham, A novel method for measuring cell surface-bound thrombin. J. Biol. Chem. 257:850-858 (1982).
11. D.L. Eaton, R.W. Scott, and J.B. Baker, Purification of human fibroblast urokinase proenzyme and analysis of its regulation by proteases and protease nexin. J. Biol. Chem. 259:6241-6247 (1984).

12. R.W. Scott, D.L. Eaton, N. Duran, and J.B. Baker, Regulation of extracellular plasminogen activator in human fibroblast cultures. The role of protease nexin. J. Biol. Chem. 258:4397-4403 (1983).

13. C.S. Izzard, and L.R. Lochner, Cell-to-substrate contacts in living fibroblasts: an interference reflection study with an evaluation of the technique. J. Cell Sci. 21:129-159 (1976).

14. M.C. Willingham, and I. Pastan, An atlas of immunofluorescence in cultured cells. Academic Press, Inc., Orlando (1985).

15. F. Blasi, J.-D. Vassali, and K. Danø, Urokinase-type plasminogen activator: proenzyme, receptor, and inhibitors. J. Cell Biol. 104:801-804 (1987).

16. K. Danø, P.A. Andreasen, J. Grøndahl-Hansen, B. Kristensen, L.S. Nielsen, and L. Skriver, Plasminogen activators, tissue degradation and cancer. Adv. Cancer Res. 32:146-239 (1985).

17. T. David-Pfeuty, and S.J. Singer, Altered distributions of the cytoskeletal protein vinculin and α-actinin in cultured fibroblasts transformed by Rous sarcoma virus. Proc. Natl. Acad. Sci. USA 77:6687-6691 (1980).

18. D.L. Eaton, and J.B. Baker, Phorbol ester and mitogens stimulate human fibroblast secretions of plasmin-activatable plasminogen activator and protease nexin, an antiactivator/antiplasmin. J. Cell Biol.7:323-328 (1983).

19. B. Herman, and W.J. Pledger, Platelet-derived growth factor-induced alterations in vinculin and actin distribution in BALB/c-3T3 cells. J. Cell Biol. 100: 1031-1040 (1985).

20. B. Herman, M.A. Harrington, N.E. Olashaw, and W.J. Pledger, Identification of the cellular mechanisms responsible for platelet-derived growth factor-induced alterations in cytoplasmic vinculin distribution. J. Cell. Physiol. 126:115-125 (1986).

21. D.A. Low, R.W. Scott, J.B. Baker, and D.D. Cunningham, Cells regulate their mitogenic response to thrombin through secretion of protease nexin. Nature 298:476-478 (1982).

22. K.C. Glenn, D.H. Carney, J.W. Fenton, and D.D. Cunningham, Thrombin active site regions required for fibroblast receptor binding and initiation of cell division. J. Biol. Chem. 225:6609-6616 (1980).

23. A. Ichinose, K. Fujikawa, and T. Suyama, The activation of pro-urokinase by plasma kallikrein and its activation by thrombin. J. Biol. Chem. 261:3486-3489 (1986).

24. B.A. Geiger, A 130K protein from chicken gizzard: its localization at the termini of microfilament bundles in cultured chicken cells. Cell 18:193-205 (1979).

25. D.M. Raben, K. Yasuda, and D.D. Cunningham, Thrombin stimulated metabolism of inositol-containing phospholipids. J. Cell. Biochem. Suppl 0 (10 part A) p. 252.

26. N.H. Teng, and L.B. Chen, Thrombin-sensitive surface protein of cultured chick embryo cells. Nature 259:578-580 (1976).

27. L.I. Gold, R. Schwimmer, and J.P. Quigley, Human plasma fibronectin is a substrate for the plasminogen activator, human urokinase. J. Cell Biol. 105 p.216a (#1221) (1987).

28. M.-P. Stoppelli, C. Tacchetti, M.V. Cubellis, A. Corti, V.J. Hearing, G. Cassani, E. Appella, and F. Blasi, Autocrine saturation of prourokinase receptors on human A431 cells. Cell 45:675-684 (1986).

29. J. Kolega, M.S. Shure, W.-T. Chen, and N.D. Young, Rapid cellular translocation is related to close contacts formed between various cultured cells and their substrata. J. Cell Sci. 54:23-34 (1982).

30. K. Burridge, Substrate adhesions in normal and transformed fibroblasts: organization and regulation of cytoskeletal, membrane and extracellular matrix components at focal contacts. Cancer Rev. 4:18-78 (1986).

31. D. Schubert, S. Humphreys, C. Baroni, and M. Cohn, In vitro differentiation of a mouse neuroblastoma. Proc. Natl. Acad. Sci. USA 64:316-323 (1969).

32. N.W. Seeds, A.G. Gilman, T. Amano, and M.W. Niremberg, Regulation of axon formation by clonal lines of a neuronal tumor. Proc. Natl. Acad. Sci USA 66:160-167 (1970).

33. B. Geiger, Z. Avnur, G. Rinnerthaler, H. Hinssen, and J.V. Small, Microfilament-organizing centers in areas of cell contact: cytoskeletal interactions during cell attachment and locomotion. J. Cell Biol. 99:83-91 (1984).

34. J.V. Small, and G. Rinnerthaler, Cytostructural dynamics of contact formation during fibroblast locomotion in vitro. Exp. Bio. Med. 10:54-68 (1985).

35. S. Halegoua, Changes in phosphorylation and distribution of vinculin during nerve growth factor-induced neurite outgrowth. Dev. Biol. 121:97-104 (1987).

36. J.-D. Vassalli, D. Baccino, and D. Belin, A cellular binding site for the Mr. 55,000 form of the human plasminogen activator, urokinase. J. Cell Biol. 100:86-92 (1985).

STRUCTURE AND FUNCTION OF TISSUE-TYPE PLASMINOGEN ACTIVATOR

ANTON JAN VAN ZONNEVELD, CARLIE DE VRIES
AND HANS PANNEKOEK

Department of Molecular Biology
Central Laboratory of the Netherlands
Red Cross Blood Transfusion Service
1006 AK Amsterdam, the Netherlands

INTRODUCTION

Tissue-type plasminogen activator (t-PA) has a crucial function in fibrinolysis. Its importance as a thrombolytic agent has led to an enormous interest in the structure and function relationships of this molecule. Many studies have been performed both to understand the fundamental molecular mechanisms of plasminogen activation by t-PA and to develop derivatives of the molecule with improved *in vivo* thrombolytic properties. It was found that t-PA can be considered as a mosaic protein composed of autonomous structural and functional domains.

This concept has been substantiated by many studies in which the biological activity of t-PA-deletion mutant proteins or hybrid molecules that are combinations of t-PA domains with other fibrinolytic proteins, was investigated. These analyses led to the identification of the parts of the molecule that interact with the substrate plasminogen and the physiological inhibitor plasminogen activator inhibitor type I (PAI-1). Sites were also located that interact with specific receptors in the liver that are involved in the rapid clearance of t-PA from the bloodstream.

Special interest has been given to the identification of the parts of the molecule that mediate the enhancement of t-PA activity by fibrin. This property is believed to underlie the clot specificity of plasminogen activation by t-PA. The importance of a co-factor like fibrin for t-PA activity is illustrated by the observation that *in vitro* soluble fibrin fragments can accelerate plasminogen activation by t-PA two to three orders of magnitude.

In this review we will try to summarize the recent progress that has been made in the understanding of t-PAs structure and function relationships. We will separately discuss the interactions of the molecule with its substrate plasminogen, the inhibitor PAI-1, the interaction with fibrin and the sites of the molecule which are responsible for the rapid clearance of t-PA by receptors in the liver.

Serine Proteases and Their Serpin Inhibitors in the Nervous System
Edited by B. W. Festoff
Plenum Press, New York, 1990

T-PA is also thought to be important for the activation of plasminogen in extravascular processes, where fibrin is not present. In that setting, a cofactor molecule other than fibrin is likely to be necessary. Thus far, heparin, fibronectin and thrombospondin have been shown to display properties that indicate a potential role of these molecules as a cofactor for t-PA. In the final part of this chapter, the interaction of t-PA with these molecules will be discussed.

BIOLOGY OF t-PA

Biological processes that involve tissue degradation are essential both in differentiation as for the maintenance of the adult organism. The common theme in most of these phenomena is the localized generation of proteolytic activity that subsequently degrades the structural proteins of the particular tissues. Rather then having a specific protease for each and every different structural protein, the organism efficiently employs specific task forces, or proteolytic systems, that lead to the local activation of one or a limited number of "broad specificity" proteases. Plasmin is a typical example of such a protease. It displays a relatively broad trypsin-like specificity, hydrolyzing proteins and peptides at lysyl and arginyl bonds.[1] It is the enzyme that is responsible for the degradation of fibrin and, as such, the main enzyme of the fibrinolytic system (Figure 1).

Due to its broad specificity, plasmin activity should be strictly regulated to prevent aspecific protein degradation. The key feature of this regulation is that plasmin is generated from its inactive zymogen-form plasminogen by specific plasminogen activators (PAs). Since plasminogen is present at a relatively high concentration in the majority of the body fluids,[3] the plasminogen activators are believed to control and to direct the action of plasmin. The two physiological plasminogen activators are urokinase-type plasminogen activator (u-PA) and tissue-type plasminogen activator (t-PA). These molecules are similar in that they are both serine proteases that can convert plasminogen into the active protease plasmin by hydrolyzing a single peptide bond in the plasminogen molecule (arg 560 - val 561). U-PA and t-PA can both be inhibited by plasminogen activator inhibitor type-1 (PAI-1),[2] an inhibitor that is present in platelets and is also deposited in the extracellular matrix of a large number of animal cell cultures.[4] In contrast, the way

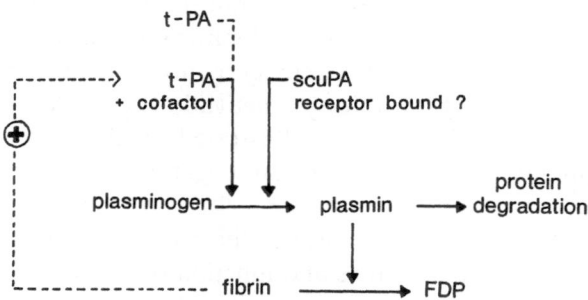

Figure 1   *Schematic representation of plasminogen activation.* Plasminogen activators t-PA or u-PA activate plasminogen by the hydrolysis of the arg 560-val 561 bond. This results in the active protease plasmin that can degrade fibrin or other proteins (reviewed by Bachmann).[2]

their activity is modulated differs. U-PA is synthesized as a single chain pro-enzyme (scu-PA). This pro-enzyme form can be activated by plasmin to u-PA. The localization and the conversion of scu-PA into u-PA (tcu-PA) may well be regulated by the interaction of the molecule with a specific receptor.[5,6] As a consequence, it is thought that u-PA catalyzed plasminogen activation is thought to take place at cell-surfaces. U-PA is involved in a variety of proteolytic processes such as tissue remodelling, tumor invasion, embryogenesis, fertilization and inflammatory processes.[3] T-PA is believed to play its major role in fibrinolysis which leads to the degradation of blood clots. Whereas u-PA is synthesized by a variety of tissues, t-PA is mainly synthesized by the endothelium and secreted as a one-chain molecule (sct-PA).[2,3] The t-PA molecule exhibits a high affinity for fibrin. *In vitro* studies revealed that in the absence of fibrin t-PA is a very poor plasminogen activator. However, in the presence of physiological concentrations of fibrin the activity is greatly enhanced due to a decrease of the Km to plasminogen of about 500-fold.[7] This cofactor role of fibrin can be explained by the observation that both t-PA and plasminogen can bind to fibrin, thereby bringing the enzyme and the substrate together in a ternary complex that results in an enhanced activation rate of plasminogen.

In recent years evidence has accumulated for a number of extravascular roles of t-PA. These processes include spermatogenesis,[8] ovulation,[9] neuron migration,[10,11] development and maintenance of the neuro-muscular junction[12] and wound healing.[13] It is unlikely that fibrin is present in the micro-environment where each of these processes take place. If plasminogen activation by t-PA is important for these remodelling events, some molecule(s) other than fibrin will be necessary to fulfil the cofactor requirement of t-PA activity.

t-PA IS A MOSAIC PROTEIN

The enormous progress in the understanding of the t-PA molecule that will be summarized in this chapter, largely results from the notion that t-PAs fibrin-specific action could be a property that would make the molecule extremely suited for thrombolytic therapy.[14] To facilitate the production of recombinant t-PA and as a tool for further study, t-PA cDNA was isolated and the complete primary structure of the molecule was revealed.[15,16] This structure has been confirmed by the determination of the amino-acid sequence of t-PA isolated from the conditioned medium of Bowes melanoma cells employing the Edman degradation technique.[17] The striking feature of this structure was that the amino-terminal part of the molecule appeared to contain a number of domains, homologous to parts of known plasma proteins. Based on these homologies a model for the secondary structure has been proposed[15,18] (Figure 2). This model and the structure of t-PA has been discussed in detail elsewhere.[20] Briefly, t-PA is synthesized as a pre-pro-precursor protein, including a signal peptide (residues -35 till -13) and a pro-peptide (residues -14 till -1). Next, from the amino-terminus towards the carboxyl terminus, there is a "finger" (F) domain, homologous to the type I fingers of fibronectin, an "epidermal growth factor like" (E) domain and two so-called "kringle domains" (K1, K2), similar to the triple disulfide-bonded structures found in many plasma proteins, including prothrombin and plasminogen. Together these amino-terminal domains constitute the "heavy" (H) chain of the molecule (38 kd). The carboxyl-terminal remainder of the molecule, that carries the serine protease moiety, is called the "light (L) chain" (32 kd). The fully active two chain form of t-PA is derived from the single-chain molecule by cleavage of the arg-275/ile-276 peptide bond by plasmin or trypsin.

When the distribution of the introns and exons of the t-PA gene was resolved,[19] the hypothesis was forwarded that the t-PA gene would have been evolved by a mechanism referred to as "exon shuffling".[21] It was found that the domains of the H chain were coded for by separate exons (signal peptide, pro peptide, F and E domains), or a set of two adjacent exons (the K1 and K2 domains). Some primordial serine protease gene with PA activity might have been provided, by intron recombination, with additional exons coding for structural domains. This way specific functions, such as the affinity for fibrin, could have been added to the plasminogen activator and thus, the mosaic protein t-PA would have been created. If this hypothesis was to hold truth, the domains of the H chain could function as autonomous structural and/or functional 'modules'. By expressing and studying the properties of mutant t-PA proteins that lack one or more domains (based on the exact positions of the introns), we and others have provided evidence that indeed this seems to be the case. For example, two domains, the finger and the K2 domain, were shown to be able to bind to fibrin, both independent of any other part of the molecule, and each in a distinct fashion.[22,23] The "domain theory" has initiated a number of studies with similar t-PA deletion mutant proteins all aimed at assigning specific functions to the different domains. It is possible that these studies will not only contribute to our understanding of the structure and function of the t-PA molecule but ultimately may also lead to "second generation" t-PA derivatives with improved thrombolytic properties. An example of such attempts is the construction of 'chimeric plasminogen activators'.[24-32] In order to combine the fibrin-binding properties of t-PA with the

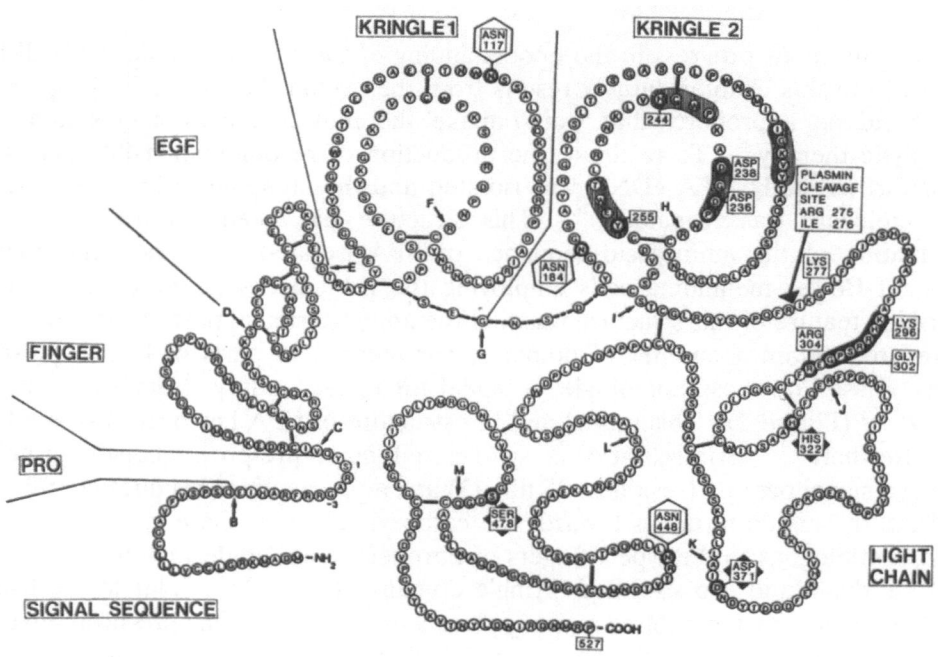

Figure 2 *Model of the secondary structure of the t-PA molecule (adapted after ref. 19).* Small arrows indicate the relative positions of the introns in the t-PA gene. Numbers indicate amino acid positions according to Pennica et al.[15]

54

catalytic domain of u-PA, hybrid u-PA molecules were created carrying the t-PA H chain.[26-32] This *"in vitro* exon shuffling" has supported the domain theory in that the hybrid molecules indeed have fibrin-binding properties. However, the observation that the binding of the hybrid molecules to fibrin is less efficient then that of the native t-PA molecule indicates that the mere presence of binding domains does not suffice to fully express their property.[26,29-32] Other characteristics, such as positioning, will obviously be of additional importance. We will now summarize the information that has resulted from these and other studies on the sites of interaction of t-PA with its substrate plasminogen, the inhibitor PAI-1, the cofactor fibrin and with receptors in the liver, involved in the rapid clearance of the molecule from the bloodstream.

INTERACTION WITH PLASMINOGEN

The property of t-PA to activate plasminogen and the specificity for this substrate is entirely contained within the L chain of the molecule. This was demonstrated using either a L chain preparation isolated from two chain melanoma t-PA[33,34] or with a recombinant L chain t-PA, expressed with a cDNA construct from which the DNA coding for the H chain was deleted.[35] In the absence of fibrin the t-PA L chain is equally efficient in plasminogen activation as the full-length molecule. When fibrin is added, the wild-type molecule is highly stimulated, whereas the activity of the L chain molecule is not altered. Thus, the L chain contains the serine protease moiety and the H chain mediates the fibrin stimulation and, as will be discussed later, the fibrin binding properties.

The L chain of the t-PA molecule represents a chymotrypsin-like serine protease. Such serine proteases have an active site that is composed of three crucial residues.[36] Analogous to chymotrypsin, in t-PA these residues are histidine 322 (his-322), aspartic acid 371 (asp-371), and serine 478 (ser-478). Together they form the so-called 'charge relay network', that functions to transfer a proton from the hydroxyl-group of ser-478, via the imidazole group of his-322, to the carboxylate group of asp-371. This results in the generation of a highly reactive oxygen group on ser-478 that will combine with the substrate to form a tetrahedral intermediate and subsequently an acyl-enzyme intermediate. In chymotrypsinogen, the inactive precursor form, the residue equivalent to asp-371 is not accessible. When trypsin activates chymotrypsinogen by cleavage of its arg-16/ile-17 bond, the newly formed α-amino group of ile-17 forms a salt bridge with an asp residue in the proximity of the ser-residue that is part of the active site triad.[37-39] This event makes the asp-371 equivalent accessible and the 'charge relay network' functional.

In t-PA a similar arg-275/ile-276 peptide bond is split by plasmin to generate two-chain t-PA. Also, an asp-477 residue is in the equivalent position next to the essential serine residue. Surprisingly, in the presence of fibrin single-chain t-PA exhibits plasminogen activator activity similar to that of the two-chain molecule. Furthermore it was shown that in the absence of fibrin the single-chain molecule exhibits a significant plasminogen activator activity and does not behave like a true zymogen. These observations have long been a matter of debate because of the intrinsic complication that the product of the reaction, plasmin, can convert single-chain to two-chain t-PA. The catalytic activity of plasmin for this reaction is very high relative to the other reactions occurring and, even in the presence of the plasmin inhibitor trasylol, single- to two-chain t-PA conversion was observed.[40,41] To circumvent this problem t-PA mutants were expressed that have the arg-275 substituted for, either by glutamic acid (glu-275) or by a 64 glycine residue (gly-275) in order to obtain an uncleavable t-PA variant.[42-44] These studies confirmed that, in the presence of fibrin, indeed, the activity of the single- and the two-chain molecule are nearly identical. Without fibrin the activity of the single-chain molecule is only 20- to 50-fold lower then the two chain molecule. One explanation for the activity of the single-chain molecule is that an ϵ-aminogroup of a properly placed lysine residue within the L chain could take over the function of the amino-terminus (ile-276) that is liberated when t-PA is converted to the two-chain form. Such an ϵ-amino group of a lys-residue could form a salt bridge with asp-477 and thereby induce the active conformation of the t-PA molecule. When fibrin is present this active conformation of the single chain molecule could be stabilized resulting in a fully active t-PA molecule. Peterson et al[45] have tested this hypothesis by mutagenizing two lys-residues that could potentially interact with asp-477 (lys-277,[46,47] lys-416[48]).

The amidolytic activity of one chain t-PA was significantly quenced when lys-416 was mutated whereas mutation of lys-277 had no effect. This observation strongly suggests a role for lys-416 in the activity of sct-PA.

INTERACTION WITH PAI-1

Although t-PA can be inhibited by a number of serine protease inhibitors ("serpins"), both from kinetical considerations, as for reasons of physical localization, PAI-1 may well be the physiological inhibitor of t-PA in the vascular system.[4] T-PA and PAI-1 can form one to one, SDS-stable complexes and the second-order rate constant for this association is extremely high (10^7 $M^{-1}s^{-1}$). Since the mechanism of action of the interaction of serpins with serine proteases is that the inhibitor behaves like an "pseudo substrate",[49] the interaction of t-PA with PAI-1 can be expected to be closely related to the interaction of t-PA with the substrate plasminogen. As expected, the interaction is dependent on the active site[50] and the L chain alone is fully capable of stable complex formation.[51] Recently, some information on the role of the active serine residue at position 478 in the interaction with PAI-1 was obtained.[52] Substitution of the active site residue ser-478 by glycine resulted in a t-PA molecule incapable of activating plasminogen and forming stable complexes with PAI-1. Another substitution mutant with the ser-478 replaced by threonine displayed markedly reduced plasminogen activator activity but the activity was not totally abolished. The residual activity may be due to the fact that the threonine substitution preserves a hydroxyl group at the active site. The capability of forming stable complexes with PAI-1 however, is totally abolished in the thr-478 mutant. It was concluded that the presence of an active site serine residue rather then enzymatic activity per se, is necessary for formation of stable complexes of t-PA with PAI-1. Kinetic studies[53,54] demonstrated that in addition to the interaction with the active site, a second site of interaction appears to be involved in the high rate of complex formation. Di-isopropyl phosphofluoridate-inactivated t-PA inhibited the reaction between PAI-1 and t-PA. A similar inhibitory effect was found on the interaction of PAI-1 with low molecular weight u-PA. Since the latter activator is devoid of the amino-terminal EGF- and kringle domain, the second interaction site is believed to be located on the serine protease domains of the PAs. In addition, Madison et al. demonstrated that it is possible to dissociate substrate and inhibitor binding by making a mutant t-PA molecule that retains 95% of its plasminogen activator activity but is virtually resistant to PAI-1 inhibition[55] By modelling the t-PA - PAI-1 complex analogous to the known three-dimensional structure of trypsin and bovine pancreatic trypsin inhibitor, these investigators predicted the formation of a salt bridge between arg-304 in t-PA and glu-350 of PAI-1. They also noted that compared to the other members of the trypsin-like serine proteases, t-PA contains a seven amino-acid insert (lys(296)-his-arg-arg-ser-pro-gly(302)) adjacent to the predicted PAI-1 interaction site that could be involved in t-PA specificity for PAI-1. Indeed, deletion of residues 296-302 of t-PA or substitution of ser or glu for arg-304 had little effect on the catalytic efficiency, but a significant effect on complex formation with PAI-1. Thus, it was concluded that the area of contact between t-PA and PAI-1 is more extensive than the interaction between the enzyme and its substrate plasminogen.

INTERACTION WITH FIBRIN

As mentioned before, the mechanism underlying the cofactor activity of fibrin is that both t-PA and plasminogen can interact with fibrin.[7,56] When the domain structure of the H chain of t-PA was elucidated, it was speculated that the kringle domains and the finger domain were potential sites for the interaction with fibrin, since in two other fibrin binding plasma proteins, plasminogen (kringles)[57] and fibronectin (fingers),[58,59] similar structures were shown to be responsible for binding to fibrin. In accord, it was found that neither recombinant L chain,[35] nor L chain preparations isolated from two-chain bowes melanoma t-PA[33] had affinity towards fibrin and that the L chain activity could not be stimulated by fibrin. This suggested that the H chain is responsible for fibrin binding. To identify the domains involved in fibrin binding we have employed t-PA deletion mutant proteins that lacked specific domains, with the borders of the domains based on the intron- and exon organization of the t-PA gene.[22] Indeed, we found that the finger domain (F) and the second kringle domain (K2) mediate fibrin affinity and stimulation. This observation raised the question why two domains would be required for binding and stimulation by fibrin. Do they function in a cooperative fashion or do they bind independently of each other, e.g. at different stages in the degradation of fibrin. Similar to the fibrin-binding kringles of plasminogen, the K2 domain of t-PA was found to contain a LBS.[23,60,61] These binding sites are capable of binding carboxyl-terminal lysine and arginine residues, like those created upon degradation of fibrin by plasmin.[62] It has been claimed that both kringles in t-PA contain a LSB, suggesting that K1 might also be involved in fibrin binding.[63] However, most of the data presented in the literature agree with fibrin binding being unique to K2 domain of t-PA.[22,60,61,64-69] Tulinsky et al.[70] have modelled the kringles of plasminogen and t-PA according to the known structure of the prothrombin kringle and proposed the structural requirements that constitute a LSB. The hatched areas in the K2 domain of t-PA (Figure 2) indicate the regions that would contribute to the lysine-binding site. A LBS is characterized by a dipolar surface with the polar parts separated by conserved aromatic residues. In case of K2 of t-PA, the positive charge of ligands such as lysine or e-aminocaproic acid would interact with asp-236 and asp-238. These asp residues are conserved in all LBS-containing kringles. In plasminogen kringle 1 and 4, the negatively-charged group of the lysine-like ligands form ionic interactions with arginine residues at positions equivalent to the thr-252 and val-213 residue of K2 of t-PA. In K2 of t-PA the function of these arg residues might be taken over by his-244 through a positively charged imidazolium ion,[70] or by thr-252 through a hydrogen bond.[66] The aromatic residues that provide the hydrophobic platform include trp-253. This residue is not present in kringle 1 of t-PA and the lack of this residue might be related to the apparent absence of a LBS in the first K1 domain.

To explain the presence of two distinct fibrin-binding sites, one of them being a LBS we proposed a model for the t-PA/fibrin interaction during the progress of fibrinolysis.[23] In the initial phase of fibrinolysis, fibrin is mostly intact and no carboxyl-terminal lysine or arginine residues are present; t-PA then binds specifically to fibrin by its finger domain. Glu-plasminogen is bound to fibrin with a weak affinity, possibly to internal lysine residues by its so-called amino-hexyl (AH) binding site.[71] Plasmin is formed at a low rate which then starts to degrade fibrin and create carboxyl-terminal lys and arg residues. This leads to an increased, LBS mediated, binding of both glu-plasminogen[72] and t-PA and, subsequently, to a higher rate of plasminogen activation. This hypothesis was also based on work of Norrman et al.,[73] who described a kinetic transition in t-PA induced plasminogen activation. A possible physiological role for this transition-mechanism is that it ascertains that no premature lysis of newly formed fibrin takes place. To test this model, we have studied the binding of t-PA and of t-PA mutants to intact fibrin and to fibrin that was partially degraded with plasmin.[68] It was shown that the binding of t-PA increased when fibrin was limitedly digested with plasmin and that part of this augmented binding was dependent on the generation of carboxyl-terminal lysine and arginine residues. In addition, it was shown that a further increase in fibrin binding was likely to be mediated by the finger domain, provided it was correctly exposed in the t-PA mutant proteins used for these studies. Surprisingly, the kringle 2 domain displayed a substantial affinity for intact fibrin suggesting that the LBS could also bind to "internal" lysines. This agrees with a previous report showing the affinity of kringle 2 for amino-hexyl Sepharose,[69] and a study showing the specificity of K2 for different lysine derivatives.[66] These observations indicate that K2 of t-PA harbours a binding site that resembles the 'AH site' of plasminogen. The absence of arginine residues in t-PAs LBS at positions where they could bind the negative charge of lysine could explain the apparently equal affinity for internal and carboxyl-terminal lysine residues.

In contrast to these studies, Higgins and Vehar studied the enhanced t-PA binding to fibrin and concluded that the lysine-binding site was only involved in the binding to intact fibrin.[74] However, in those studies fibrinogen was predigested with plasmin before clotting as opposed to plasmin digestion of preformed fibrin. This difference illustrates the importance of the nature of the 'fibrin' that is used during these type of studies.

INTERACTION WITH RECEPTORS IN THE LIVER

When t-PA is administered to man or animals it is rapidly cleared from the bloodstream by the liver where it is degraded in the lysosomes.[75] The half-life in plasma in different species is very similar: 1-4 minutes in mice, rats, rabbits and dogs, whereas in man a half-life of about 3-6 minutes is observed. The plasma disappearance curves of the I^{125}-labelled t-PA (either one- or two-chain t-PA) appears to be bi-phasic with a fast α-phase ($t\frac{1}{2} \approx 5$ min), followed by a slower ß-phase ($t\frac{1}{2} \approx 10$ min). In some studies a third phase is observed with a half-life of over two hours. To develop a t-PA derivative with a prolonged half-life and possibly with clinical properties superior over that of the wild-type molecule, many studies have been performed to identify the structures on the t-PA molecule that mediate the clearance by the liver. Rijken et al.[76] showed that, in rats, the H chain is cleared from the bloodstream almost 6 fold faster than the L chain, indicating that t-PA is recognized by the liver primarily through the H chain. To identify the cell-types in the liver that are responsible for this specific recognition, Kuiper et al.[77] intravenously injected ^{125}I-t-PA in rats, and subfractionated the liver shortly thereafter. It was found that 80% of the injected t-PA was associated with the liver. Of that fraction 54.5% interacts with the parenchymal cells, 39.5% binds to endothelial liver cells and 6% binds to Kupffer cells. Interestingly, from competition studies, it was concluded that the different cell-types use different recognition systems for the uptake of t-PA. The endothelial cells of the liver take up t-PA mainly by the mannose receptor and thus recognize carbohydrate structures on the molecule. The parenchymal cells or hepatocytes utilize a novel high-affinity system specific for t-PA. The latter finding agrees with the 'in vitro' observation by Bakhit et al.[78] who demonstrated the presence of a carbohydrate-independent uptake system on rat hepatocytes of high affinity (half maximal uptake at 10 nM).

Thus, both carbohydrates- and protein structures are sites of interaction of t-PA with the liver.

The t-PA molecule contains four potential sites for asp-linked glycosylation, but only three of these are used.[17] A high mannose oligosaccharides is present at position 117 and complex oligosaccharide at position 184 and 448. Two natural variants occur, called type I t-PA and type II t-PA, differing in that in type II t-PA the asn 184 is not glycosylated.[17] Several studies have shown that alteration of the carbohydrate structures can decrease the 'in vivo' clearance of the t-PA molecule. Hotchkiss et al.[79] have removed the high-mannose structure at position 117 either enzymatically or by site-directed mutagenesis and subsequently observed a 2.7 fold lower clearance of the altered t-PA molecule as compared to wild-type t-PA. Conversely, when t-PA was expressed in a cell-type that can produce only high-mannose oligosaccharide structures on glycoproteins the clearance was increased by 1.8 fold. These results and a study by Tanswell et al.[80] suggest a role for the carbohydrate at position 117 in the rapid clearance of t-PA. Similar studies have provided evidence that the other carbohydrate structures at position 184[81] and 448[82] might also be involved in recognition by the liver. However, even without carbohydrate structures the t-PA molecule is still rapidly cleared and thus the recognition of the protein backbone by liver receptors is likely to be the major factor for t-PAs rapid clearance.[83]

In search for the protein determinant on t-PA for liver receptors, a number of groups have investigated the clearance of t-PA deletion mutant proteins in animal model systems.[83-89] The consensus of these studies is that the binding to the liver receptor resides in the amino-terminus of the molecule, being either the EGF-domain,[84,85] the finger domain,[83] or both.[86-88] The uncertainty of these experiments is however, that it is apparently difficult to distinguish whether a specific deletion indeed removed the binding site or that the deletion disrupts another site on the molecule essential for receptor binding. In a recent study, Ahern et al.[89] have provided evidence that in fact the region connecting the finger and the EGF-domain between residues 44 and 50 of the finger might harbour the binding site. Since this region containing the putative liver receptor binding site, or the accessibility of this region could be affected both when the EGF-domain is deleted or when the finger domain is deleted, an explanation for the effect of both deletions on the clearance of t-PA could have been provided by this study.

Another approach has been taken by Gething et al.[90] to prolong the 'in vivo' half-life of t-PA. By a mutation located on a possibly exposed "turn" of the EGF-domain (tyr67→asn), a new consensus sequence for N-linked glycosylation was introduced. This variant t-PA was successfully expressed and secreted with an extra carbohydrate group attached to the EGF-domain. Using a cell-binding assay it was found that, in contrast to wild-type t-PA, the mutant t-PA was unable to bind to rat hepatoma cells, presumably because the additional oligosaccharide shields the receptor binding-site on the molecule.

Finally, Morton et al.[91] have suggested that the catabolism of t-PA by the human hepatocyte can be modulated by PAI-1. These investigators employed human Hep-G2 hepatoma cells as an *'in vitro'* model system for t-PA catabolism by the hepatocyte.[92] They have shown that when ^{125}I-t-PA is added to these cells, the t-PA binds to a specific receptor on the surface of these cells and is subsequently internalized and degraded. In a recent study they demonstrated that binding of t-PA to the cells correlates with an increase in the apparent molecular mass of t-PA due to the formation of an SDS-stable complex with PAI-1.[91] Thus, complex formation with PAI-1 may play a role in the uptake of t-PA by the Hep G2-receptor system.

EXTRAVASCULAR COFACTORS

As mentioned in the introductory part of this chapter the role of t-PA in the organism is not restricted to vascular fibrinolysis. We will now summarize the information that has been reported on molecules that potentially could be considered as extravascular cofactors of t-PA that come into play when fibrin is absent. It is important to note that it is not difficult to enhance *in vitro* plasminogen activation by t-PA. Radcliffe and Heinze[93] showed that a variety of denatured proteins are capable of enhancing t-PA induced plasminogen activation in solution. Also, it was reported[94] that if plasminogen is immobilized on plastic surfaces it acquires affinity for soluble t-PA that promotes efficient generation of plasmin, thus, suggesting that immobilization per se is sufficient for enhancement of activation. Even though one can only speculate on the mechanism of these phenomena, they have in common that the lysine-binding sites of plasminogen play an important role. It has been shown[95] that the interaction of ligands with the lysine- binding site of native glu-plasminogen results in conformational changes which lead to an increased rate of plasminogen activation by t-PA or u-PA in the absence of fibrin. Hence, when traces of plasmin are formed, denatured proteins, susceptible to plasmic degradation, could be cleaved and carboxyl-terminal arg- or lys residues could be created. These termini could subsequently bind to the lysine-binding sites of plasminogen and induce the easily activitable "lys-plasminogen-like" conformation, and thus enhance plasmin formation. For example, it has recently been reported that human immunoglobulin G (IgG)[96] binds glu-plasminogen and that glu-plasminogen bound to the IgG molecule can be activated to plasmin. The binding of plasminogen to IgG was lysine-binding site dependent as it was inhibited by ϵ-amino caproic acid. As with fibrin , when the IgG fraction was treated with plasmin the binding of plasminogen to IgG was markedly enhanced. In general, the creation of carboxyl-terminal arg- or lys residues by plasmin in IgG, fibrin or any *in vivo* analogue of fibrin clots, such as denatured or aged proteins, scar tissue or tumors, may lead to a focal immobilization of plasminogen and an enhanced plasminogen activation.[93] Similarly, the lysine binding site of t-PA could also interact with the newly created carboxyl-terminal residues, that could be of additional importance for the enhancement of plasminogen activation. Although the mechanism of a lysine- binding site interactions should not be oversimplified in that they non-specifically bind to any lysine residue,[97] these mechanisms may underlie some of the targeting of the action of plasmin in the organism. The remainder of this chapter will be restricted to some putative extravascular cofactors molecules that can specifically interact with t-PA and stimulate its activity towards plasminogen.

HEPARIN

Heparin is a complex sulfated glycosaminoglycan. Several groups have reported that heparin can bind to t-PA and glu-plasminogen and that heparin can significantly increase the activation of plasminogen by t-PA.[98-102] This activity of the heparin preparation could be abolished by heparinase but not by chondroitinase ABC or AC.[98] Low molecular-weight fragments of heparin where also found to potentiate t-PA-induced thrombolysis in a rabbit jugular vein thrombosis-model.[102] The stimulation of the t-PA activity with heparin was in most cases shown to interfere with fibrin stimulation suggesting that the fibrin- binding sites are identical to the heparin-binding site(s) or, alternatively that they are in close proximity to the heparin binding site(s).[98,99,101] Consequently it was likely that the interaction of t-PA and heparin would be mediated by the H chain of the t-PA molecule. Even though in one report the heparin-binding site of porcine t-PA was mapped to the L chain,[100] Fears[101] showed that the activity of isolated t-PA L chain was not affected by heparin or pentosan polysulphate, and thus suggested that the heparin binding site was indeed located on the H chain of the molecule. Furthermore, an antibody against the K2-domain of t-PA and 6-aminohexanoic acid, both agents that are known to interfere with the binding of fibrin to the second kringle of t-PA, did not affect the stimulation of t-PA by heparin. Thus, the heparin-binding domain was found to be distinct from the fibrin-binding site on K2. Direct evidence that domains of the H chain mediate heparin binding and stimulation of t-PA was provided by Stein et al.[103] From binding studies employing t-PA deletion mutant proteins and heparin agarose it was concluded that the finger domain and to a lesser extend the K2 domain contained affinity for heparin. In addition, it was shown that both domains but particularly the finger-domain, mediate the stimulation of t-PA by heparin. The involvement of the finger domain was not unexpected since the homologous finger from fibronectin also binds to heparin.[59] Using monoclonal antibodies is was demonstrated that, even though fibrin- and heparin-binding sites are present on the same domains in the t-PA molecule, they are likely to be physically distinct structures. Since sulfated polysaccharides are constituents of most cell surfaces, including endothelial cells, heparin could potentially play a role in any tissue-remodelling event that involves cell surfaces. One process where heparin-like structures might play a role as cofactor for t-PA activity is the process of ovulation. This hypothesis was forwarded by Andrade-Gordon et al.[98] Before ovulation, the ovarian follicle fluid is highly viscous due to an abundance of high-molecular weight proteoglycans. These proteoglycans are highly sensitive to plasmin in contrast to e.g. cartilage proteoglycans. Prior to ovulation the ovarian follicle fluid liquifies, allowing the egg to move freely after rupture of the follicle wall. Since both t-PA, produced by granulosa cells in the follicle, and plasminogen are present within the follicle, t-PA might play a role in the degradation of follicular fluid proteoglycans. In the absence of fibrin in this system the heparin oligosaccharides units present on the proteoglycans could function as a cofactor for t-PA activity.

FIBRONECTIN

Immobilized fibronectin, in contrast to soluble fibronectin, is also capable of binding both t-PA and plasminogen.[104] This binding was mapped to the carboxyl-terminal Mr 120.000-140.000 fragment of fibronectin. Furthermore, it was found that the binding of one-chain t-PA was considerable better then the two-chain form of the molecule. T-PA bound to fibronectin retained its ability to activate plasminogen. It was speculated[104] that in the tissues, in the absence of fibrin, the observed interactions between immobilized fibronectin, one-chain t-PA and plasminogen may serve to localize pericellular proteolysis. Barlati et al.[105] have demonstrated the formation of active, SDS-stable, high-molecular weight forms of t-PA that are complexes of t-PA and the 110 kDa central region of the fibronectin molecule. Similar complexes where detected in plasma of patients treated with intravenous infusions of t-PA, following angina crisis or progressive and peripheral ischemia, indicating that the occurrence of these t-PA/fibronectin fragment complexes is not an *in vitro* artefact but may play a role in the processes of site-directed proteolysis.

THROMBOSPONDIN AND HISTIDINE-RICH GLYCOPROTEIN

Thrombospondin (TSP) is a multifunctional 450 KD glycoprotein that is found in the α-granules of platelets and in the extra cellular matrix of a variety of other tissues.[106] Histidine-rich glycoprotein (HGRP) is an α2-glycoprotein of plasma that is also present in a granules of platelets.[107] Silverstein et al.[94] have shown ternary complex formation of t-PA and plasminogen with both TSP and HRGP. The formation of these complexes was specific and dependent on a lysine-binding site and leads to a 40-fold increase in catalytic efficiency of plasminogen activation, due to a decrease of the Km of t-PA for plasminogen. Interestingly, as with fibrin, plasmin associated to TSP or HRGP is protected from inhibition by α2-antiplasmin. The analogy with fibrin was even further extended in the case of TSP-mediated plasminogen activation. In a following paper by the same investigators,[108] it was shown that if t-PA is added to TSP-plasminogen complexes, the binding of plasminogen to TSP is markedly enhanced. This effect appears to be due to the selective proteolysis of TSP by plasmin which generates additional binding sites. A similar increase in plasminogen binding by TSP was found when TSP was pre-incubated with plasmin. When TSP was incubated with elastase or thrombin, no enhanced binding could be demonstrated showing that the effect is plasmin specific. Lys-plasminogen, a plasmin-modified form of native glu-plasminogen (kinetically a better substrate for t-PA than glu-plasminogen), displayed a higher affinity for TSP than glu-plasminogen. The initial production of small amounts of plasmin, from plasminogen immobilized on TSP, could lead to an increase in TSP-plasminogen complex formation as well as to the generation of lys-plasminogen. Together these events could lead to a localized and increased production of plasmin in a fibrin-free micro-environment.

In an immunocytochemical localization study of TSP in the developing mouse embryo O'Shea and Dixit[109] found TSP to be associated with morphogenetic processes of proliferation, migration and intercellular adhesion during differentiation. One striking observation was the close association of TSP with outgrowing nerve processes in the peripheral nervous system. Since t-PA has been found to be similarly present in processes like granule cell migration in the cerebellum,[110] neurite outgrowth,[111] and in the development of the peripheral nervous system,[112] it was hypothesized that TSP could form a nidus for protease generation during these and other types of tissue-remodelling events in early differentiation.

MISCELLANEOUS

Rat oocytes synthesize t-PA when they are stimulated by hormones to initiate meiotic maturation. As mentioned before, it has been suggested that t-PA is involved in the ovulation process, transport of the oocyte through the oviduct, or in implantation of the early embryo.[9] In a recent paper, Bicsak and Hsueh[113] specifically examined the possible existence of a non-fibrin cofactor in the rat oocyte. By studying the activity of t-PA from oocyte preparations in comparison to purified rat t-PA, they observed that the oocyte t-PA activity could not or only slightly be stimulated by poly-lysine or fibrin while purified rat t-PA was normally stimulated by these cofactors. After SDS-PAGE fractionation the oocyte derived t-PA activity could be equally stimulated as the purified rat t-PA, indicating dissociation of t-PA and some molecule with cofactor like activity towards t-PA. It was proposed that the activity of the oocyte t-PA is stimulated by an endogenous factor that could be the physiological cofactor for t-PA activity during ovulation and ovulation related processes. The putative endogenous cofactor was also found to be present in t-PA preparations from the conditioned medium of a rat insulinoma cell line. Machovich and Owen[114] extracted and partially purified a factor from porcine- coronary endothelium that was found to have a capacity to enhance the rate of plasminogen activation by t-PA. Unlike fibrin, whose activity is accounted for, mostly as a lowering of the K_m for plasminogen, this endothelial cofactor increases the Vmax for plasminogen activation without significant effect on K_m. Whether the cofactor is physiologically significant awaits further characterization.

Finally, the activation of plasminogen by u-PA and t-PA can also be enhanced on cell-surfaces. There is a rapidly growing interest in these type of processes that are likely to involve both plasminogen receptors[115] and receptors for u-PA[5] and possibly for t-PA.[116-119] The involvement of receptors in these processes would present a novel dimension to the regulation of plasminogen activation in the mature organism as well as in differentiation.

REFERENCES

1. K.C. Robbins, L. Summaria, and R.C. Wohl, Human plasmin, in: "Methods in Enzymology" (L.Lorand ed.), Academic Press, New York 80: 379 (1981).
2. F. Bachmann, Fibrinolysis, in: "Thrombosis and Haemostasis" (Verstraete, M., Vermylen, J., Lijnen, H.R. and Arnout, J. eds.) International Society on Thrombosis and Haemostasis and Leuven University Press, Leuven p 227 (1987).
3. K. Danø, P.A. Andreasen, J. Grøndahl-Hansen, P. Kristensen, L.S. Nielsen, and L. Sriver, Plasminogen activators, tissue degradation, and cancer. **Adv. Cancer Res.** 44: 139 (1985).
4. D.J. Loskutoff, M. Sawdy, and J. Mimuro, Type I plasminogen activator inhibitor, in: "Progress in Thrombosis and Heamostasis" (Coller, B., ed.) W.B. Saunders, Philadelphia 9: 87 (1989).
5. F. Blasi, Surface receptors for urokinase plasminogen activator. **Fibrinolysis** 2: 73 (1988).
6. W.R. Stephens, J. Pöllänen, H. Tapiovaara, K.C. Leung, P.S. Sim, E.M. Salonen, E. Rønne, N. Behrendt, K. Danø, and A. Vaheri, Activation of pro-urokinase and plasminogen on Human Sarcoma Cells: A proteolytic system with surface-bound reactants. **J. Cell Biol.** 108: 1987 (1989).
7. M. Hoylaerts, D.C. Rijken, H.R. Lijnen, and D. Collen, Kinetics of the activation of plasminogen by human tissue-type plaminogen activator. **J. Biol. Chem.** 257: 2912 (1982).

8. R.J. Coombs, J. Ellison, A. Woods, and N. Jenkins, Only tissue-type plasminogen activator is secreted by immature bovine Sertoli cell-enriched cultures. **J. Endocr.** 117: 63 (1987).

9. A. Tsafriri, T.A. Bicsak, S.B. Cajander, T. Ny, and A. Hsueh, Suppression of ovulation rate by antibodies to tissue-type plasminogen activator and α_2 anti-plasmin. **Endocrinology** 124: 415 (1989).

10. G. Moonen, M. Grau-Wagemans, and I. Selah, Plasminogen activator-plasmin system and neuronal migration. **Nature** 298: 753 (1982).

11. S. Verrall, and N.W. Seeds, Characterization of ^{125}I-tissue plasminogen activator binding to cerebral granule neurons. **J. Cell Biol.** 109: 265 (1989).

12. D. Hantaï, J.S. Rao, and B.W. Festoff, Serine protease and serpins: their possible roles in the motor system. **Rev. Neurol. (Paris)** 144: 680 (1988).

13. J. Grøndahl-Hansen, L.R. Lund, E. Ralfkiaer, V. Ottevanger, and K. Danø, Urokinase- and tissue-type plasminogen activators in keratinocytes during wound reepithelialization *in vivo*. **J. Invest. Derm.** 90: 790 (1988).

14. D. Collen, On the regulation and control of fibrinolysis. **Thromb. Haemost.** 43: 77 (1980).

15. D. Pennica, W.E. Holmes, W.J. Kohr, R.N. Harkins, G.A. Vehar, C.A. Ward, W.F. Bennett, E. Yelverton, P.N. Seeburg, H.L. Heyneker, I.D. Goeddel, and D. Collen, Cloning and expression of human tissue-type plasminogen activator in *E.Coli*. **Nature** 301: 214 (1983).

16. T. Edlund, T. Ny, M. Rånby, L.O. Heden, G. Palm, E. Holmgren, and S. Josefsson, Isolation of DNA sequences coding for a part of human tissue-type plasminogen activator. **Proc. Natl. Acad. Sci. USA** 80: 349 (1983).

17. G. Pohl, M. Kallstrom, N. Bergsdorf, P. Wallén, and H. Jornvall, Tissue plasminogen activator: peptide analysis confirm an indirectly derived amino acid sequence, identify the active site serine residue, establish glycosylation sites, and localize variant differences. **Biochemistry** 23: 3701 (1984).

18. L. Banyai, A. Varadi, and L. Patthy, Common evolutionary origin of the fibrin-binding structures of fibronectin and tissue-type plasminogen activator. **FEBS Lett.** 163: 37 (1983).

19. T. Ny, F. Elgh, and B. Lund, The structure of the human tissue-type plasminogen activator gene: correlation of intron and exon structures to functional and structural domains. **Proc. Natl. Acad. Sci. USA** 81: 5355 (1984).

20. H. Pannekoek, C. de Vries, and A.J. van Zonneveld, Mutants of human tissue-type plasminogen activator (t-PA): structural aspects of functional properties. **Fibrinolysis** 2: 123 (1988).

21. W. Gilbert, Genes in pieces revisited. **Science** 228: 823 (1985).

22. A.J. van Zonneveld, H. Veerman, and H. Pannekoek, Autonomous functions of structural domains on human tissue-type plasminogen activator. **Proc. Natl. Acad. Sci. USA** 83: 4670 (1986).

23. A.J. van Zonneveld, H. Veerman, and H. Pannekoek, On the interaction of the finger and the kringle-2 domain of tissue-type plasminogen activator with fibrin: inhibition of kringle-2 binding to fibrin by ϵ-amino caproic acid. **J. Biol. Chem.** 261: 14214 (1986).

24. K.C. Robbins, I.G. Boreisha, A covalent molecular weight 92000 hybrid plasminogen activator derived from human plasmin fibrin-binding and tissue plasminogen activator catalytic domains. **Biochemistry** 26: 4661 (1987).

25. K.C. Robbins, and Y. Tanaka, Covalent molecular $\approx$92000 hybrid plasminogen activator derived from human plasmin amino terminal and urokinase carboxyl-terminal domains. **Biochemistry** 25: 3603 (1987).

26. L. Nelles, H.R. Lijnen, D. Collen, W.E. Holmes, Characterization of a fusion protein consisting of amino acids 1 to 263 of tissue-type plasminogen activator and amino acid 144 to 411 of urokinase type plasminogen activator. **J. Biol. Chem.** 262: 10855 (1987).

27. J. Pierard, P. Jacobs, D. Gheysen, M. Hoylaerts, B. Andre, L. Topisirovic, A. Cravador, F. de Foresta, A. Herzog, D. Collen, M. De Wilde, and A. Bollen, Mutant and chimeric recombinant plasminogen activators: production in eukaryotic cells and preliminary characterization. **J. Biol. Chem.** 262: 11771 (1987).

28. D. Gheysen, H.R. Lijnen, L. Pierard, F. de Foresta, E. Demarsin, P. Jacobs, M. De Wilde, A. Bollen, and D. Collen, Characterization of a recombinant protein with the finger domain of tissue-type plasminogen activator with a truncated single chain urokinase-type plasminogen activator. **J. Biol. Chem.** 262: 11779 (1987).

29. C. De Vries, H. Veerman, F. Blasi, and H. Pannekoek, Artificial exon shuffling between tissue-type plasminogen activator (t-PA) and urokinase (u-PA): a comparative study on the fibrinolytic properties of t-PA/u-PA hybrid proteins. **Biochemistry** 27: 2565 (1988).

30. S.G. Lee, N. Kalyan, J. Wilhelm, W.-T. Hum, R. Rappaport, S.-M. Cheng, S. Dheer, C. Urbano, R.W. Harztell, M. Ronchetti-Blume, M. Levner, and P.P. Hung, Constructions and expression of hybrid plasminogen activators prepared from tissue-type plasminogen activator and urokinase-type plasminogen activator genes. **J. Biol. Chem.** 263: 2917 (1988).

31. R.H. Lijnen, L. Nelles, B. Van Hoef, E. Demarsin, and D. Collen, Characterization of a chimeric plasminogen activator consisting of aminoacids 1 to 274 of tissue-type plasminogen activator and aminoacids 138-411 of single chain urokinase type plasminogen activator. **J. Biol. Chem.** 203: 19083 (1988).

32. R.H. Lijnen, L. Pierard, M.E. Reff, and D. Gheysen, Characterization of a chimaeric plasminogen activator obtained by insertion of the second kringle structure of tissue-type plasminogen activator (aminoacids 173 through 262) between residues ASP^{130} and SER^{139} of urokinase-type plasminogen activator. **Thromb. Res.** 52: 431 (1988).

33. D.C. Rijken, and E. Groeneveld, Isolation and functional characterization of the heavy and light chain of human tissue-type plasminogen activator. **J. Biol. Chem.** 261: 3098 (1986).

34. I. Dodd, R. Fears, and J.H. Robinson, Isolation and preliminary characterisation of active B-chain of recombinant tissue-type plasminogen activator. **Thromb. Haemost.** 55: 94 (1986).

35. M.E. MacDonald, A.J. Van Zonneveld, and H. Pannekoek, Functional analysis of the human tissue-type plasminogen activator: the light chain. **Gene** 42: 59 (1986).

36. J.J. Kraut, Serine proteases: structure and mechanism of catalysis. **Ann. Rev. Biochem.** 46: 331 (1977).

37. P.B. Sigler, B.A. Jeffery, B.W. Matthews, and D.M. Blow, Structure of crystalline α-chymotrypsin.II. A preliminary report including a hypothesis for the activation mechanism. **J. Mol. Biol.** 35: 143 (1966).

38. B.W. Matthews, P.B. Sigler, R. Henderson, and D.M. Blow, Three-dimensional structure of tosyl-α-chymotrypsin. **Nature (London)** 214: 652 (1967).

39. H.L. Oppenheimer, B. Labouesse, and G.P. Hess, Implication an ionizing group in the conformation and activity of chymotrypsin. **J. Biol. Chem.** 241: 2720 (1966).

40. P. Wallén, M. Rånby, N. Bergsdorf, and P. Kok, Purification and characterization of tissue plasminogen activator: the occurrence of two different forms and their enzymatic properties, in: "Progress in Fibrinolysis" (Davidson, J.F., Nilsson, I.M., Astedt, B. Eds.) Volume 5 Chuchill Livingstone, Edinburgh, p 16 (1981).

41. D.C. Rijken, M. Hoylaerts, and D. Collen, Fibrinolytic properties of one-chain human extrinsic (tissue-type) plasminogen activator. **J. Biol. Chem.** 257: 2920 (1982).

42. K.M. Tate, D.L. Higgins, W.E. Holmes, M.E. Winkler, H.L. Heyneker, and G.A. Vehar, Functional role of proteolytic cleavage at arginine-275 of human tissue-type plasminogen activator as assessed by site-directed mutagenesis. **Biochemistry** 26: 338 (1987).

43. L.C. Petersen, M. Johannessen, D. Foster, A. Kumar, and E. Mulvihill, The effect of polymerised fibrin on the catalytic activities of one-chain tissue-type plasminogen activator as revealed by an analogue resistant to plasmin cleavage. **Biochim. Biophys. Acta** 952: 245 (1988).

44. J.A. Boose, E. Kuismanen, R. Gerard, J. Sambrook, and M.J. Gething, The single chain form of tissue-type plasminogen activator has catalytic activity: studies with a mutant enzyme that lacks the cleavage site. **Biochemistry** 28: 635 (1989).

45. L.C. Petersen, E. Boel, M. Johannessen, and D. Foster, Possible involvement of a lysine residue in establishing the charge relay system responsible for one chain t-PA. **Thromb. Haemost.** 62: 322 (abs.) (1989).

46. P. Wallén, G. Pohl, N. Bergsdorf, M. Rånby, T. Ny, and H. Jornvall, Purification and characterization of a melanoma cell plasminogen activator. **Eur. J. Biochem.** 132: 681 (1983).

47. N. Haigwood, E.P. Paques, G. Mullenbach, G. Moore, L. Desjardins, and A. Tabrizi, Improvement of t-PA properties by means of site directed mutagenesis. **Thromb. Haemost.** 58: 286 (Abstract 1042) (1987).

48. A. Heckel, and K.M. Hasselbach, Prediction of the 3-Dimensional structure of the enzymatic domain of t-PA. **J. Computer-Aided. Mol. Des.** 2: 7 (1988).

49. R. Carrell, and D.R. Boswell, Serpins: The superfamily of plasma serine proteinase inhibitors, in: "Proteinase inhibitors" Barrell, A, and Salvesen, G. (eds.) Elsevier Publishing Co., Amsterdam p 403 (1986).

50. K.O. Kruithof, G. Tran-Tang, A. Ransijn, and F. Bachmann, Demonstration of a fast-acting inhibitor of plasminogen activators in human plasma. **Blood** 64: 907 (1984).

51. A.J. Van Zonneveld, H. Veerman, M.E. MacDonald, J.A. Van Mourik, and H. Pannekoek, On the structure and function of human tissue-type plasminogen activator. **J. Cell. Biochem.** 32: 169 (1987).

52. J.C. Monge, C.L. Lucore, E.T.A. Fry, B.E. Sobel, and J.J. Biladello, Characterization of interaction of active-site serine mutants of tissue-type plasminogen activator with plasminogen activator inhibitor-1. **J. Biol. Chem.** 264: 10922 (1989).

53. J. Chmielewska, M. Rånby, and B. Wiman, Kinetics of the inhibitor of plasminogen activators by the plasminogen-activator inhibitor. **Biochem. J.** 251: 327 (1988).

54. S. Thorsen, M. Philips, J. Selmer, I. Lecander, and B. Asted, Kinetics of inhibitor of tissue-type and urokinase-type plasminogen activator by plasminogen-activator inhibitor type 1 and type 2. **Eur. J. Biochem.** 175: 33 (1988).

55. E.L. Madison, E.J. Goldsmith, R.D. Gerard, M.J.H. Gething, and J.F. Sambrook, Serpin resistant mutants of human tissue-type plasminogen activator. **Nature** 339: 721 (1989).

56. M. Rånby, Studies on the kinetics of plasminogen activator by tissue plasminogen activator. **Biochim. Biophys. Acta** 704: 461 (1982).

57. L. Sottrup-Jensen, H. Klaeys, M. Zajdel, T.E. Petersen, and S. Magnusson, The primary structure of human plasminogen: isolation of two lysine-binding fragments and one "mini"-plasminogen (M.W.38000) by elastase-catalysed-specific limited proteolysis, *in*: "Progress in Chemical Fibrinolysis and Thrombolysis". Davidson, J.F., Rowan, R.M., Smama, M.M. and Desnoyers, P.C. (Raven, New York) 3: 191 (1978).

58. I. Sekiguchi, M. Fukada, and S.I. Hakamori, Domain structure of hamster plasma fibronectin. **J. Biol. Chem.** 256: 6452 (1981).

59. T.E. Petersen, H.C. Thögersen, K. Skorstengaard, K. Vibepedersen, R. Sahl, L. Sottrup-Jensen, and S. Magnussen, Partial primary structure of bovine plasma fibronectin: three types of internal homologies. **Proc. Natl. Acad. Sci. USA** 80: 137 (1983).

60. A. Ichinose, K. Takoi, and K. Fujikawa, Localization of the binding site of tissue-type plasminogen to fibrin. **J. Clin. Invest.** 78: 163 (1986).

61. J.M. Verheijen, M.P.M. Caspers, G.T.G. Chang, G.A.W. de Munk, P.H. Pouwels, B.E. Enger-Valk, Sites in tissue-type plasminogen activator involved in the interaction with fibrin, plasminogen and low molecular weight ligands. **EMBO J.** 5: 3525 (1986).

62. R.F. Doolittle, Fibrinogen and fibrin. **Sci. Amer.** 245(6): 92 (1981).

63. M.J. Gething, B. Adler, J.A. Boose, R.D. Gerard, E.L. Madison, D. McGookey, S. Meldell, L.M. Roman, and J. Sambrook, Variants of human tissue-type plasminogen activator that lack specific structual domains of the heavy chain. **EMBO J.** 7: 2731 (1988).

64. L. Erickson, P.W. Bergum, E.W. Hubert, N.Y. Theriault, E.F. Rehberg, D.P. Palermo, G.A.W. De Munk, J.H. Verheijen, and K.R. Marotti, Enhancements and inhibition of the activity of recombinant analogs of tissue plasminogen activator. **Thromb. Haemost.** 58: 287 (Abstract 1045) (1987).

65. G.R. Larsen, K. Henson, and Y. Blue, Variants of human tissue-type plasminogen activator: fibrin binding, fibrinolytic and fibrinogenolytic characterization of genetic variants lacking the fibronectin finger-like and/or the epidermal growth factor domains. **J. Biol. Chem.** 263: 1023 (1988).

66. S. Cleary, M.G. Mulkerrin, and R.F. Kelley, Purification and characterization of tissue-plaminogen activator kringle-2 domain expressed in *Escherichia Coli*. **Biochemistry** 28: 1884 (1988).

67. S. Urano, A.R. Metzger, and F.J. Castellino, Plasmin-mediated fibrinolysis by variant recombinant tissue plasminogen activators. **Proc. Natl. Acad. Sci. USA** 86: 2568 (1989).

68. C. De Vries, H. Veerman, and H. Pannekoek, Identification of the domains of tissue-type plasminogen activator involved in the augmented binding to fibrin after limited digestion with plasmin. **J. Biol. Chem.** 264: 12604 (1989).

69. G.A.W. De Munk, M.P.M. Caspers, G.T.G. Chang, P.H. Pouwels, B.E. Enger-Valk, and J.H. Verheijen, Binding of tissue-type plasminogen activator to lysine, lysine analogues and fibrin fragments. **Biochemistry** 28: 7318 (1989).

70. A. Tulinsky, C.H. Park, B. Mao, and M. Llinás, Lysine/fibrine binding sites of kringles modeled after the structure of kringle I of prothrombin. **Proteins** 3: 85 (1988).

71. U. Christensen, The AH-site of plasminogen and two C-terminal fragments: a weak lysine-binding site preferring ligands not carrying a free carboxylate function. **Biochem. J.** 223: 413 (1984).

72. E. Suenson, O. Lutzen, and S. Thorsen, Initial plasmin-degradation of fibrin as the basis of a positive feed back mechanism in fibrinolysis. **Eur. J. Biochem.** 140: 513 (1984).

73. B. Norrman, P. Wallén, and M. Rånby, Fibrinolysis mediated by tissue plasminogen activator: disclosure of a kinetic transition. **Eur. J. Biochem.** 149: 193 (1985).

74. D.L. Higgins, and G.A. Vehar, Interaction of one-chain and two chain tissue plasminogen activator with intact and plasmin-degrated fibrin. **Biochemistry** 26: 7786 (1987).

75. J. Krause, Catabolisme of t-PA, its variants, mutants and hybrids. **Fibrinolysis** 2: 133 (1988).

76. D.C. Rijken, and J.J. Emeis, Clearance of the heavy and light polypeptide chains of human tissue-type plasminogen activator in rats. **Biochem. J.** 238: 643 (1986).

77. J. Kuiper, M. Otter, D.C. Rijken, and T.J.C. Van Berkel, Characterization of the interaction *in vivo* of tissue-type plasminogen activator with liver cells. **J. Biol. Chem.** 263: 18220 (1988).

78. C. Bakhit, D. Lewis, R. Billings, and B. Malfroy, Cellular catabolism of recombinant tissue-type plasminogen activator. **J. Biol. Chem.** 262: 8716 (1987).

79. A. Hotchkiss, C.J. Refino, C.K. Leonard, J.V. O'Connor, C. Crowley, J. McCabe, K. Tate, G. Nakamura, D. Powers, A. Levinson, M. Mohler, and M.W. Spellman, The influence of carbohydrate structure on the clearance of recombinant tissue-type plasminogen activator. **Thromb. Haemostas.** 60: 255 (1988).

80. P. Tanswell, M. Schlüter, and J. Krause, Pharmacokinetics and Isolated liver perfusion of carbohydrate modified recombinant tissue-type plasminogen activator. **Fibrinolysis** 3: 79 (1989).

81. D.P. Beebe, and D.L. Aronson, Turnover of t-PA in rabbits influence of carbohydrate moieties. **Thromb. Res.** 51: 11 (1988).

82. D. Lau, G. Kuzma, C.M. Wei, D.J. Livingston, and N. Hsiung, A modified human tissue plasminogen activator with extended half-life *in vivo*. **Biotechnology** 5: 953 (1987).

83. G.R. Larsen, M. Metzger, K. Henson, Y. Blue, and P. Horgan, Pharmaco kinetic and distribution analysis of variants forms of tissue-type plasminogen activator with prolonged clearance in rat. **Blood** 73: 1842 (1989).

84. M.J. Browne, J.E. Carey, C.G. Chapman, A.W.R. Tyrrell, C. Entwisle, M.P. Lawrence, B. Reavy, I. Dodd, A. Esmail, and J.H. Robinson, A tissue-type plasminogen activator mutant with prolonged clearance *in vivo*. **J. Biol. Chem.** 263: 1599 (1987).

85. I. Dodd, B. Nunn, and J.H. Robinson, Isolation, identification and pharmacokinetic properties of human tissue-type plasminogen activator species: possible localisation of a clearance recognition site. **Thromb. Haemostas.** 59: 523 (1988).

86. N.K. Kalyan, S.G. Lee, J. Wilhelm, K.P. Fu, W.T. Hum, R. Rappaport, R.W. Hartzell, C Urbano, and P.P. Hung, Structure-function analysis with tissue-type plasminogen activator (t-PA): Effect of deletion of NH2-terminal domains on its biochemical and biological properties. **J. Biol. Chem.** 263: 3971 (1988).

87. D. Collen, J.M. Stassen, and G. Larsen, Pharmacokinetics and thrombolytic properties of deletion mutants of human tissue-type plasminogen activator in rabbit. **Blood** 71: 216 (1988).

88. K.P. Fu, S. Lee, W.T. Hum, N. Kalyan, R. Rappaport, N. Hetzel, and P.P. Hung, Disposition of a novel recombinant tissue plasminogen activator, 2 - 89 t-PA, in mice. **Thromb. Res.** 50: 33 (1988).

89. T.D. Ahern, G.E. Morris, K.M. Barone, P.G. Horgan, L.B. Angus, K.S. Henson, P.R. Langer-Safer, and G.R. Larsen, Distinguishing the sites in the amino terminal region of tissue-type plasminogen activator (t-PA) required for efficient fibrinolytic activity and rapid clearance from the circulation. **Fibrinolysis** 3 Suppl.1: 5 (1989).

90. M.J. Gething, J. Sambrook, and D. McGookey, Addition of an oligosaccharide side-chain at an ectopic site on the EGF-like domain of t-PA prevents binding to specific receptors on hepatic cells. **Fibrinolysis** 3 Suppl.1: 19 (1989).

91. P.A. Morton, D.A. Owensby, B.E. Sobel, and A.L. Schwartz, Catabolism of tissue-type plasminogen activator by the human hepatoma cell line hep G2. **J. Biol. Chem.** 264: 7228 (1989).

92. D.A. Owensby, B.E. Sobel, and A.L. Schwartz, Receptor-mediated endocytosis of tissue-type plasminogen activator by the human hepatoma cell line hep G2. **J. Biol. Chem.** 263: 10587 (1988).

93. R. Radcliffe, and T. Heinze, Stimulation of tissue-type plasminogen activator by denatured proteins and fibrin clots. A possible additional role for plasminogen activator? **Arch. Biochem. Biophys. Acta** 211: 750 (1981).

94. R.L. Silverstein, R.L. Nachman, L.L.K. Leung, and P.C. Harpel, Activation of immobilized plasminogen by tissue activator. **J. Biol. Chem.** 260: 10346 (1985).

95. L. Banyai, and L. Patthy, Importance of intramolecular interactions in the control of the fibrin affinity and activation. **J. Biol. Chem.** 259: 6466 (1984).

96. P.C. Harpel, R. Sullivan, and T.S. Chang, Binding and activation of plasminogen on immobilized Immunoglobulin G. **J. Biol. Chem.** 264: 616 (1988).

97. P. Bosma, D.C. Rijken, and W. Nieuwenhuizen, Binding of tissue-type plasminogen activator to fibrinogen fragments. **Eur. J. Biochem.** 172: 399 (1988).

98. P. Andrade-Gordon, and S. Strickland, Interaction of heparin with plasminogen activators and plasminogen: effects on the activation of plasminogen. **Biochemistry** 25: 4033 (1986).

99. E.P. Paques, H.A. Stöhr, and N. Heimburger, Study on the mechanism of action of heparin and related substances on the fibrinolytic system: relationship between plasminogen activator and heparin. **Thromb. Res.** 42: 797 (1986).

100. S. Soeda, M. Kakiki, H. Shimeno, and A. Nagamatsu, Localization of the binding sites of pocine tissue-type plasminogen activator and plasminogen to heparin. **Biochem. Biophys. Acta** 916: 279 (1987).

101. R. Fears, Kinetic studies on the effect of heparin and fibrin of plasminogen activators. **Biochem. J.** 249: 7781 (1988).

102. J.M. Stassen, I. Juhan-Vague, M.C. Alessi, F. De Cock, and D. Collen, Potentiation by heparin fragments of thrombolysis induced with human tissue-type plasminogen activator on human single-chain urokinase-type plasminogen activators. **Thromb. Haemostas.** 58: 947 (1987).

103. P.L. Stein, A.J. Van Zonneveld, H. Pannekoek, and S. Strickland, Structural domains of human tissue-type plasminogen activator that confer stimulation by heparin. **J. Biol. Chem.** 264: in press (1989).

104. E.M. Salonen, O. Saksela, T. Vartio, A. Vaheri, L.S. Nielsen, and J. Zeuthen, Plasminogen and tissue-type plasminogen activator bind to immobilized fibronectin. **J. Biol. Chem.** 260: 12302 (1985).

105. S. Barlati, E. Marchina, D. Bellotti, and G. De Petro, Interaction of plasminogen activation with fibronectin fragments *in vitro* and *in vivo*. **Fibrinolysis** 3 Suppl.1: 5 (1989).

106. J.W. Lawler, H.S. Slayter, and J.E. Coligan, Isolation and characterization of a high molecular weight glycoprotein from human blood platelets. **J. Biol. Chem.** 283: 8609 (1978).

107. H. Haupt, and N. Heimburger, Humanserumproteine mit hoher affinitat zu carboxymethylcellulose I. **Hoppe-Seyler's Z. Physiol. Chem.** 353: 1125 (1972).

108. R.L. Silverstein, P.C. Harpel, and R.L. Nachman, Tissue plasminogen activator and urokinase enhance the binding of plasminogen to thrombospondin. **J. Biol. Chem.** 261: 9959 (1986).

109. K.S. O'Shea, and V.M. Dixit, Unique distribution of the extra cellular matrix component thrombospondin in the developing mouse embryo. **J. Cell. Biol.** 107: 2737 (1988).

110. A. Krystosek, and N.W. Seeds, Plasminogen activator secretion by granule neurons in cultures of developing cerebellum. **Proc. Natl. Acad. Sci. USA** 78: 7810 (1981).

111. P.H. Patterson, On the role of protease, their inhibitors and the extracellular matrix in promoting neurite outgrouth. **J. Physiol.** 80: 207 (1985).

112. A. Baron-Van Evercooren, P. Leprince, B. Rogister, P.P. Lefebvre, P. Delrée, I. Selak, and G. Moonen, Plasminogen activators in developing peripheral nervous system, cellular origin and mitogenic effect. **Dev. Brain Res.** 36: 101 (1987).

113. T.A. Bicsak, and A.J.W. Hsueh, Rat oocyte tissue-plasminogen activator is a catalytically efficient enzyme in the absence of fibrin. **J. Biol. Chem.** 264: 630 (1989).

114. P. Machovich, and W.G. Owen, A factor from endothelium facilities activation of plasminogen by tissue plasminogen activator. **Enzyme** 40: 109 (1988).

115. L.A. Miles, and E.F. Plow, Plasminogen receptors: ubiquitous sites for cellular regulation of fibrinolysis. **Fibrinolysis** 2: 61 (1988).

116. D.P. Beebe, Binding of tissues plasminogen activator to human umbilical vein endothelial cells. **Thromb. Res.** 47: 241 (1987).

117. K.A. Hajjar, N.M. Hamel, P.C. Harpel, and R.L. Nachman, Binding of tissue plasminogen activator to cultured human endothelial cells. **J. Clin. Invest.** 80: 1712 (1987).

118. E.S. Barnathan, A. Kuo, H. Van der Keyl, K.R. McCrae, G.R. Larsen, and D.B. Cines, Tissue-type plasminogen activator binding to human endothelial cells. **J. Biol. Chem.** 263: 7792 (1988).

119. K. Deguchi, and S. Shirakawa, Plasminogen activation by tissue-type plasminogen activator in the presence of platelets. **Thromb. Res.** Suppl. VII: 65 (1988).

THE HEPARIN BINDING SITE AND ACTIVATION
OF PROTEASE NEXIN I

DYFED LL. EVANS, PETER B. CHRISTEY AND ROBIN W. CARRELL

Department of Haematology
University of Cambridge
MRC Center, Hill's Road
Cambridge CB2 2QH, United Kingdom

INTRODUCTION

Protease Nexin I (PNI) is a heparin-activatable cell-associated member of the serpin superfamily. It is a 44 kDa thrombin (Th) and urokinase inhibitor which is secreted by cultured human foreskin fibroblasts and some other non-vascular cells.[1,2] In the presence of optimal concentrations of the glycosaminoglycan - heparin, the association constant for the formation of the PNI-thrombin complex increases by a factor of 500-fold;[3] giving a rate constant for the reaction of $1.1x10^9M^{-1}s^{-}$, which is near to the theoretical diffusion-controlled limit.

Two other members of the family also demonstrate heparin enhancement of their interaction with thrombin.

Antithrombin III (ATIII) is the major inhibitor of thrombin in plasma, but also inhibits other proteases of the intrinsic coagulation system.[4] This 58kDa glycoprotein is unique in that it is activated by heparin molecules containing a specific pentasaccharide unit of defined sequence.[5] The maximal rate enhancement for the formation of the ATIII-thrombin complex, is 50000-fold, with the rate constant increasing from $7x10^3M^{-1}s^{-1}$ in the absence of heparin, to a value of $3x10^8M^{-1}s^{-1}$ in the presence of heparin.[6]

Heparin Cofactor II (HCII), a 66kDa serpin which is an inhibitor of thrombin, leukocyte cathepsin G and chymotrypsin.[7,8] The association constant for this reaction is raised maximally in the presence of dermatan sulfate (a factor of 1300 times), whilst in the presence of heparin, there is a 900-fold enhancement.[9]

On the basis of homologous sequence alignments of ATIII and HCII we have recently provisionally identified the structure of the heparin binding site common to both molecules.

Confirmation of this site is presented here, based on sequence alignments and a computer-projected three dimensional model of PNI, based on the known α_1-antitrypsin co-ordinates.[10] Further evidence for the identity of the heparin binding site is also provided by recent mutant studies. Our findings lead to the conclusion that

Serine Proteases and Their Serpin Inhibitors in the Nervous System
Edited by B. W. Festoff
Plenum Press, New York, 1990

the heparin binding site is composed of a zone of positive charge centred on the D-helix and the base of the A-helix.

METHODS

Blocked alignments of PNI and rat glia-derived nexin with other serpin sequences were prepared using the Needleman-Wunsch algorithm with modification to penalise for introducing gaps into regions of defined secondary structure.[11] This demonstrated a conservation of secondary structure, with a tertiary structure superimposable onto that of α_1-antitrypsin.

The work was carried out on a VAX 8600 and an Evans & Sutherland PS300 graphics system. The FRODO molecular modelling system of programs were used for tertiary structure analysis.

In an attempt to deduce the location of the heparin binding site in all three serpins, the following assumptions were made:

a) Heparin is polyanionic and on the basis that the action of heparin can be mimicked by other polysulfated polymers it could be deduced that binding was likely to occur by electrostatic interactions. Thus binding should occur to arginine and lysine residues on the protein surface. (The ionic nature of the initial interaction of heparin and ATIII has been demonstrated by Olson et al.[12]).

b) The nature of the heparin-inhibitor interaction was essentially the same for all three serpins.

c) Certain key residues important for heparin binding were conserved in all three serpins.

RESULTS AND DISCUSSION

Heparin Binding Serpins: Structure and Reactive Centre

The skeletal molecular structure of these three serpins can be confidently projected onto that of the archetypal serpin, α_1-antitrypsin. This is a highly ordered structure, globular in shape with a flattened upper pole containing the reactive centre. The reactive site acts as a pseudosubstrate for target proteases, with the reactive centre of the inhibitor forming a tight fit into the active site of the protease. The complex thus formed, of inhibitor and protease, is then rapidly removed and catabolised.

Our homology alignments indicate that the reactive sites of these serpins are formed thus: Gly-Arg~Ser (392-394) in the case of ATIII, which agrees with cleavage experiments,[13] and previous homology alignments.[14] In HCII, the active site was identified as Pro-Leu~Ser, which also agrees with cleavage experiments.[15] Alignment of PNI with the other serpins indicated that its active site was Ala-Arg~Ser (344-346). This agrees with previous alignments,[16] and our cleavage studies (data not shown). This data therefore lends experimental support for our alignment and modelling approaches.

Protease Nexin I: Identification of the Heparin Binding Site

We present here new evidence that provides confidence in the localisation of the heparin binding site in these three inhibitors to the D-helix and demonstrates the extension of that site to the region of the active centre.

Figure 1 — Alignment of serpin sequences (landscape table).

```
Sequence Number      190             200                210              220             230              240              250               260

SecStruc            !~~~~hC~~~~!                     !~~~~~~hD~~~~~!                  !^^^^sA2^^^^!                       !~~~~hE~~~~!
Hepcof 2            THEQVHSILHFKDVFNA-----SSKYEIT    TIHNLFRKLTHRLFRNF   G--YTLRSVNDLYIQKQFPILLDFKTKVREYYFAE-
AT3                 TLQQLMEVFKFDTISEK--------TSD     QIHFFFAKLNCRLYRKAN  K-SSKLVSANRLFGDKSLTFNETYQDISELVYGAK-
Protease Nexin      TKKQLAMVMRY----------------GVN   GVGKILKKINKAIVSKKN  K--DIVTVANAVFVKNASEIEVPFVTRNKDVFQCE-
Rat GDN             TKKQLSTVMRY--------------NVN     GVGKVLKKINKAIVSKKN  K--DIVTVANAVFVRNGFKVEVPFAARNKEVFQCE-
A1AT Human          THDEILEGLNFN-LTEI---------PEA    QIHEGFQELLRTLNQPDS  Q--LQITGNGLFLSEGLKLVDKFLEDVKKLYHSE-
A1AT Baboon         THSEILEGLNFN-LTEI---------PEA    QVHEGFQELLRTLNKPDS  Q--LQITGNGLFLNKSLKLVDKFLEDVKNLYHSE-
C1INH               LTQLLGAGQNT-KTNL---------ESIL    SYPKEFTCVHQALKGFTT  E--LQLSMGNAMFVKEQLSLLDRFTEDAKRLYGSE-
A1ACT Human         TLTEILKASSSP-HGDL---------LRQ    KFTQSFQHLRAPSISSSD  ----AFRLAARMYLQKGFPIKEDFLEQSEQLFGAK-
A2AP                TLQRLQQVLH---------------AGSGP   CLPHLLSRLCQDLG-PG-  K--DEISTTDAIFVQRDLKLVQGFMPHFFRLFRST-
PAI                 TQQQIQAAMGFK-----------IDDK      GMAPALRHLYKELMGPWN  --VYSFSLASKLYAEEKYPLLPEYLQCVKELYRGG-
OVCHK               TRTQINKVVRFDKLPGFGDSIEAQCGTSV    NVHSSLRDILNQITKPND  E--LELQIGNALFIGKHLKPLAKFLNDVKTLYETE-
TBG                 TQTEIVGTLGFN-LTDT--------PMV     EIQHGFQHLICSLNFPKK  S--LEMTMGNALFLDGSLELLESFSADIKHYYESE-
CBG                 TRAQLLQGLGFN-LTER------SET       EIHQGFQHHQLFAKSDT   QTPLLQSTVVGLFTAPGLRLKQPFVESLGPFTPAIF
Angio Rat           TASQLQVLLGVPVKEGDCTSRLDGH-KVL    TALQAVQGLLVTQGGSSS  QAQLLLSTVVGVFTAPGLHLKQPFVQGLALYTPVVL
Angio Human         TADRLQAILGVPWKDKNCTSRLDAH-KVL    SALQAVQGLLVAQGRADS  --TYSLEIADKLYVDKTFSVLPEYLSCARKFYTGG-
Gene Y Protein      TESQMKKVLHFDSITGAGSTTDSQCGSSE    YVHNLFKELLSEITRPNA  ---VEKEADNDDISFSMNVYGRYSAVFKDSFLRKIG
Cowpox Virus Prot   TAEQLSY-------
```

Key: **R/K** – Positive residues   *D/E* – Negative residues

Figure 1. Alignments of the serpins in the region of the D-helix (boxed), shows clearly the conservation and relative positioning of homologous positively charged residues in the tertiary structure of the four heparin binding serpins; heparin cofactor II (hepcof 2), antithrombin III (AT3), protease nexin I and rat glia-derived nexin (Rat GDN). The other serpins show much more variability in this region, with the whole of the helix being absent in cowpox virus protein. In general they have an equal distribution of positively and negatively charged groups in their D-helices. Thus the conservation of positive charge in this region appears to be an unique feature of the heparin binding members of the family. Abbreviations: A1AT-α_1-antitrypsin; C1INH-C1 inhibitor; A2AP-α_2-antiplasmin; PAI-plasminogen activator inhibitor I; OVCHK-chicken ovalbumin; TBG-thyroxin binding globulin; CBG-cortisol binding globulin; ANGIO-angiotensinogen. (Arbitrary units used for sequence numbering.)

Table 1 : Residues identified as potentially involved in the heparin binding reaction. Those identified in the first analysis were found to be uniquely conserved in the heparin activatable serpins and were termed the prime binding site. A further analysis of each individual serpin was then conducted, to identify extensions of the prime binding site to the N- and C-termini. These extensions from the prime site proved to be far more variable. The largest secondary sites were identified in PNI, whilst the smallest were found in HCII. The existence of secondary sites may explain the different heparin affinities recorded for these molecules. The results obtained are tabulated here.

SERPIN	Residues Identified in First Analysis[†]	Residues Identified in Second Analysis[†]
Antithrombin III	Lys 125 Lys 129 Arg 132 Lys 133 Lys 228	Arg 46 Arg 47 Lys 136 Arg 235 Lys 236 Lys 275
Heparin Cofactor II	Lys 125 Lys 129 Lys 132 Arg 133 Lys 228	Arg 47 Arg 50 Arg 124 Arg 416
Protease Nexin I Rat GDN	Lys 125 Lys 128 Lys 133 Lys 134 Arg 228	Arg 97 Lys 99 Lys 100 Lys 121 Lys 124 Lys 136 Lys 235 Lys 236 Lys 237

[†] Numbering based on Antithrombin III sequence

Analysis of aligned sequences was conducted in two stages. Initially positive residues were sought which were present in essentially homologous positions in the heparin binding serpins, but which were not noticeably conserved throughout other members of the superfamily. As shown in Figure 1, five alignment positions fulfilled these criteria. Four of these residues lie close together on the D-helix. The fifth residue lies 100 residues C-terminal to the D-helix, but on the three dimensional model, appears adjacent to it (Table 1).

The second stage was to analyse each of the three serpins in isolation. Positive residues found not to be conserved between the proteins, but which were contiguous with the prime site, were considered significant. These could conceivably form extended binding sites for molecules of heparin longer than the 22Å primary binding sequence.[17]

When plotted onto a schematic representation of the D-helix, the residues identified in the analysis proved to be on the solvent face. This meant that they would be accessible for heparin binding by charge-charge interactions, as shown in Figure 2.

These basic residues were displayed on the molecular model of PNI. The obvious feature was the presence of a complete positively charged face on the external underside of the D-helix. This feature is uniquely shared with the two other heparin binding members of the family, ATIII and HCII.

The residues involved in this site in PNI; Lys 71, 74, 75, 78, 83 and 84 are in the same, or overlapping positions with homologous basic residues in ATIII and HCII, with one notable exception. Residue Arg 47 which is present in antithrombin and HCII is absent from PNI, which has an asparagine at position 2 in the A-helix. However, PNI has an additional lysine in the D-helix, at position 71 (position 121 in ATIII). On the three dimensional model the Lys 71 is seen to occupy a position nearly superimposeable on that of Arg 47 in antithrombin, and therefore probably serves an analogous function.

This homologous matching on the three dimensional structure, as well as identifying the heparin binding site in PNI, also provides a clear confirmation of the

Internal face

External face

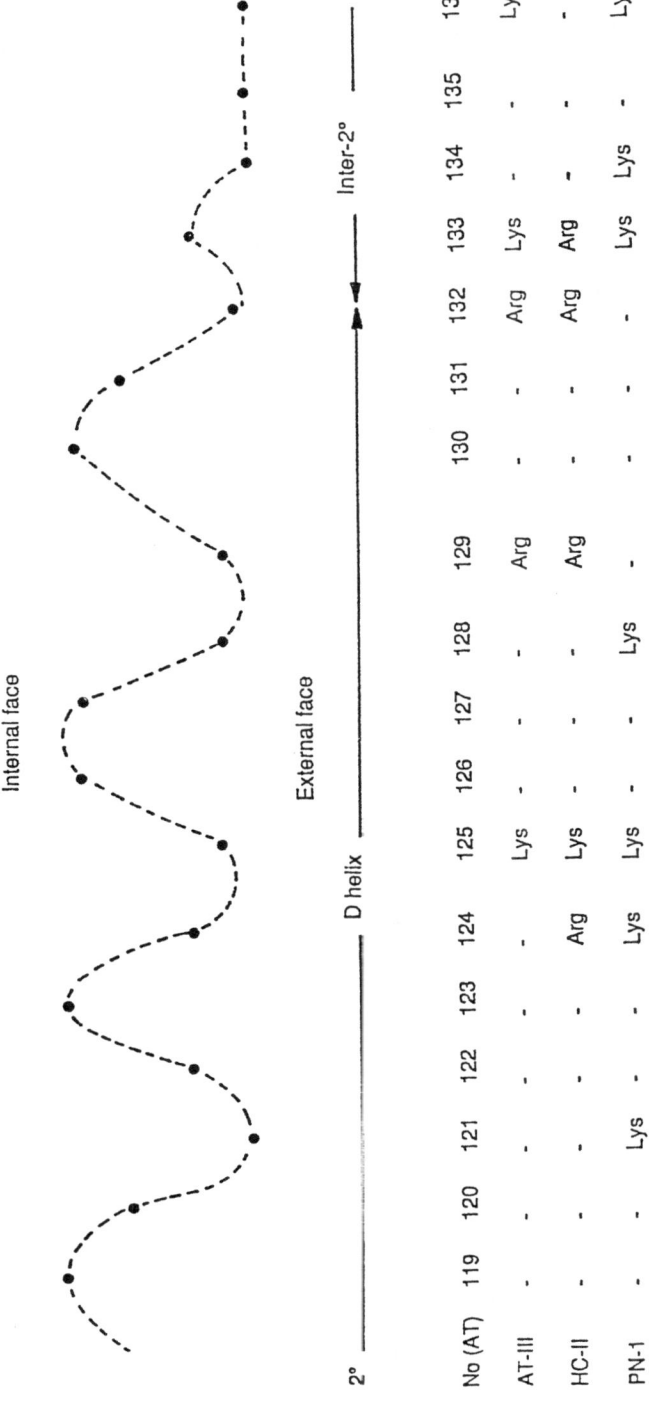

2°						D helix								Inter-2°			
No (AT)	119	120	121	122	123	124	125	126	127	128	129	130	131	132	133	134	135	136
AT-III	-	-	-	-	-	-	Lys	-	-	-	Arg	-	-	Arg	Lys	-	-	Lys
HC-II	-	-	-	-	-	Arg	Lys	-	-	-	Arg	-	-	Arg	Arg	-	-	-
PN-1	-	-	Lys	-	-	Lys	Lys	-	-	Lys	-	-	-	-	Lys	Lys	-	Lys

Figure 2 : Homology alignments of serpins show a feature unique to the heparin binding members, i.e. the presence of a positively charged external face on the D-helix and the adjacent interhelical structure, as shown. (Numbering as in antithrombin.)

73

previously proposed sites in the other two serpins. Thus there is strong evidence from homologies that the common prime binding site is formed by a positive face of the D-helix, extending to the base of the A-helix. (Residue 47 in the case of ATIII and HCII). Furthermore, the comparison of the three structures identifies an additional feature present in each member, though less rigidly conserved. This is an extension of the positively charged binding zone, from beyond the D-helix, around the end of the molecule to above the base of the reactive centre loop. This involves the following residues in PNI (ATIII) : Lys 86 (136), Lys 175 (226), Lys 177 (228) and Arg-Lys 184-185 (235-236).

On the three dimensional model of PNI another extended region of positive charge is also seen to be located near the N-terminus of the molecule. This is composed of the residues Arg 53, Lys 55 and Lys 56. Thus there are three distinct regions of positive charge on the molecule, the first near the N-terminus, the second, (prime site) centered on the D-helix, and the third near the reactive centre.

Evidence from Antithrombin Mutants

Strong supporting evidence as to the exact nature, and extent of the primary heparin binding site is provided by numerous studies of naturally ocurring low heparin affinity mutants and chemically modified forms of antithrombin.

Residues found to be important for, or affected by, heparin binding are shown in Figure 3 on a schematic representation of ATIII projected onto the structure of cleaved α_1-antitrypsin.

Three mutants having decreased affinity for heparin were seen to involve replacements of Arg 47, to Cys in ATIII Toyama,[18] to His in Rouen-1[19] and to Ser in Rouen-II.[20] These mutants clearly demonstrate the key rôle of Arg 47 in heparin binding.

The need for the correct orientation of this arginine also provides the likely explanation for the loss of heparin binding associated with the mutation Pro 41 to Leu in ATIII Basel,[21] and the chemical modification of Trp 49.[22] Both these residues lie close to Arg 47. Any changes occurring in this region would cause displacement of Arg 47 from its native spatial orientation, thus making it inacessible to heparin.

Use of chemical techniques to label lysine residues both in the presence and absence of heparin,[23] identified Lys 107, Lys 125 and Lys 136 as being protected from modification by heparin binding It is likely therefore that these residues are involved in the heparin binding interaction. This study also showed that on heparin binding, Lys 236 becomes more exposed, lending support to the theory that the binding of heparin induces conformational changes in these inhibitors.

In the naturally ocurring high affinity, or ß, form of antithrombin, the loss of a carbohydrate moiety at Asn 135 produces increased heparin affinity.[24] From the three dimensional model, it is clear that the removal of this carbohydrate gives greater access to the heparin binding site. Antithrombin Rouen III is a variant, in which Ile 7 is replaced by asparagine. This introduces a novel glycosylation sequence Asn-Cys-Thr 7-9, with consequent glycosylation of the asparagine, with the position of the peptide being fixed by the S-S bridge of Cys 8 to Cys 128. Thus the new oligosaccharide is placed near the centre of the D-helix and in the model can be seen in close proximity to, and just below, Arg 47. Hence the extent of the site is delineated by two abnormalities of glycosylation. The loss of an oligosaccharide at the upper end of the site at Asn 135 in ß ATIII gives increased access to the site and therefore increased heparin affinity; the gain of an oligosaccharide in Rouen III at a position just below Arg 47 will obscure the site and therefore gives decreased affinity.

It has also been shown,[25] that the isolated ATIII peptide composing residues 114-156 has high affinity for heparin. This is the region of ATIII containing the D-helix, and therefore the major binding site.

The recent characterisation of heparin cofactor II Oslo,[26] which has the mutation Arg 189 (residue 192 in ATIII numbering) to His, leading to reduced affinity for dermatan sulfate also implicated this residue as being in the prime binding site.

Modelling of the Site

With a computer-generated three-dimensional model of PNI, it was possible to carry out matching of the binding site with various heparin configurations. Our projected structure described here does not allow precise fitting, but it does indicate the dimensions involved. The prime binding site, Lys 71 to Lys 83, (≈ 20Å in length) fits within the dimensions of the heparin pentasaccharide (22Å) which is the minimal requirement for activation in antithrombin.

In the case of HCII, maximal activation occurs in the presence of dermatan sulfate, with the minimal binding unit being a high affinity octasaccharide,[27] which fits within the bounds of the prime binding site in HCII.

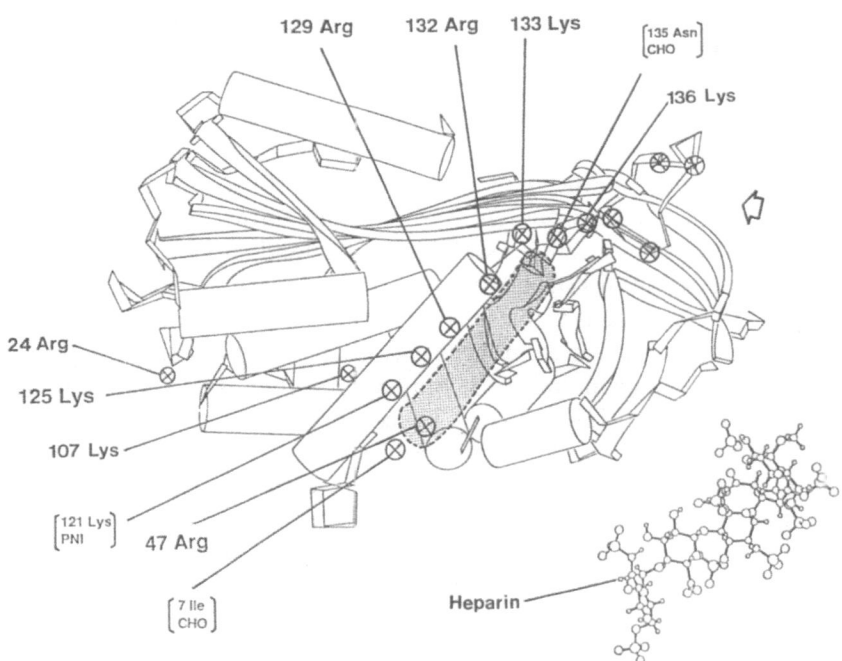

Figure 3 : Schematic view of antithrombin projected onto the structure of cleaved α_1-antitrypsin,[10] showing the prime (pentasaccharide) binding site (shaded) formed by the sidechains of lysines and arginines 47, 125, 129, 132 and 133. The site extends around the molecule to the reactive centre with 136 Lys, 228 Lys, 235 Arg and 236 Lys (shown but not labelled). The estimated region of the reactive centre is arrowed. The prime site is flanked by oligosaccharide attachment points 135 and 7 (Asn in Rouen III). In protease nexin I, the equivalent of 47 Arg is absent, but at position 121 there is a lysine, whose sidechain overlaps that of Arg 47. As shown, the size of the heparin pentasaccharide is slightly larger than the shaded area.

This data would suggest that there is a minimal binding unit in the heparin molecule, which is specific for PNI also. In the case of ATIII and HCII, this minimal binding unit is located in the glycosaminoglycan producing the greatest rate-enhancement effect. For PNI (as in the case of ATIII) the maximal effect is produced by heparin. Comparing the sequence alignments (Figure 1) and the molecular models of PNI and ATIII, it is clear that there are a number of minor differences in their prime binding sites. This would suggest that the minimal binding unit for PNI would not be the heparin pentasaccharide, but it could well be located in another region of the heparin molecule.

CONCLUSION

Sequence comparisons of PNI with ATIII and HCII identify a region of conserved positive charge centred on the D-helix, which is unique to these three proteins and is not conserved in other members of the family (Figure 1). The positive residues in this region seem to be in essentially homologous positions. The characterised ATIII mutants also all lie in regions homologous to those identified as being important for heparin binding in PN1, lending credence to our model.

When projected onto the three-dimensional models, the spatial distribution and orientation of the residues we identified are similar. There are small differences between all three, which almost certainly accounts for their different target glycosaminoglycan specificities. To progress further with the modelling of this region in PNI, it would be useful to see whether a minimal binding unit could be identified and characterised. This could then be modelled onto our proposed structure of the prime binding site in PNI.

The PNI model studies are also of note, as they revealed two other regions of positive charge which were subsequently identified in ATIII and HCII. The region extending from the D-helix to the reactive centre in all three molecules could well be a face of charge which needs to be neutralised before thrombin (which also posesses a charged surface) can gain easy access. This would lend supporting evidence to the ternary complex model of ATIII-Th interaction proposed by Olson & Shore and Danielsson et al.[28,29] This model proposes that a weak ionic complex is initially formed between ATIII and heparin, which induces a conformational change in the inhibitor, producing a 400-fold enhancement in the binding affinity.[30] This increases the rate of formation of the ternary ATIII-Heparin-Th complex. A covalent linkage between protease and inhibitor is then formed, which triggers the loss of the heparin-induced conformational change. This complex binds heparin with only weak affinity, resulting in the loss of the heparin molecule from the complex.

A striking feature of PNI is the presence of an extension of the region of positive charge to the N-terminus of the molecule, away from the active centre. This may well account for PNI's high affinity for heparin. If the homology with ATIII is retained in this region, then this finding provides a basis for constructing a model for the, as yet, undefined N-terminal region of antithrombin.

Antithrombin Rouen IV (J.Y. Borg, personal communication), a recently characterised mutant, implicates the N-terminus of the molecule as being important for heparin binding. In this variant, the substitution 24 Arg to Cys has occured, thus reducing heparin affinity. On the three dimensional model, it is envisaged that this residue could lie in essentially the same spatial position as Arg 53 in PNI.

The model we have is therefore of three heparin binding regions on the PNI molecule. The first lies near the N-terminus and could be the initial attachment point. We then have the prime binding site, which seems in all cases to be the main attachment point. There is then an extended region of charge, leading from the prime site towards the active site. In the case of interaction with thrombin-like molecules, which possess a charged surface near their catalytic sites, attachment of heparin to the extended charge site on the inhibitor is envisaged as being necessary, to prevent charge-charge repulsion between protease and inhibitor, and to allow easier access to the reactive centre. In PNI the production of a net negative charge as a result of heparin binding to the active centre region may explain why the association constant for the formation of the PNI-Th complex reaches the diffusion-controlled limit in the presence of heparin. Charge attraction between heparin and the positive face of thrombin could well bring PNI and Th into more rapid contact, thus giving rise to the observed increase in the association constant.

This hypothesis is supported by evidence from ATIII and HCII. In the case of ATIII, even though the minimal binding unit is a pentasaccharide of heparin, a 14-mer or longer is needed to provide significant acceleration of the ATIII-Th interaction. Wallace et al.,[3] have shown that the interaction of glia-derived nexin and Th follows the same mechanism as that of ATIII and Th. (It has been shown by McGrogan et al.,[16] that glia-derived nexin and PNI primary amino acid sequences are identical). Thus it would be expected that PNI should be activated by a similar mechanism.

For HCII, although the minimal binding unit is an octamer of dermatan sulfate, an 18-mer is needed to give an increase in the rate of inhibition of thrombin. From this it would follow that the dermatan sulfate binds first to the prime site, with the charged face near the active site then being covered, to allow easier access for the protease.

Our results therefore indicate the nature, and extent of the heparin binding site in PNI and also lead to proposals as to the mechanism of heparin binding and activation.

Acknowledgements: A much fuller description of this work is contained in the doctoral thesis of P.B. Christey (University of Cambridge). We gratefully acknowledge the contribution of Dr. J-Y. Borg (Rouen), Dr S.O. Brennan and M.C. Owen (New Zealand) and the support of the MRC of Great Britain and the Wellcome Trust.

REFERENCES

1. J.B. Baker, D.A. Low, R.L. Simmer, and D. Cunningham, Protease-nexin a cellular component that links thrombin and plasminogen activator and mediates their binding to cells. Cell 21: 37 (1980).
2. D.L. Eaton, and J.B. Baker, Evidence that a variety of cultured cells secrete protease nexin and produce a distinct cytoplasmic serine protease binding factor. J. Cell. Physiol. 117: 175 (1983).
3. A. Wallace, G. Rovelli, J. Hofsteenge, and S.R. Stone, Effect of heparin on the glia-derived-nexin-thrombin interaction. Biochem J. 257: 191 (1989).
4. R.D. Rosenberg, Chemistry of the hemostatic mechanism and its relationship to the activation of heparin. Fed. Proc., Fed. Am. Soc. Exp. Biol. 36: 10 (1977).
5. J. Choay, J-C. Lormeau, M. Petiou, P. Sinay, and J. Fareed, Structural studies on a biologically active hexasaccharide obtained from heparin. Ann. N.Y. Acad. Sci. 370: 644 (1981).
6. M. Hoylaerts, W.G. Owed, and D. Collen, Involvement of heparin chain length in the heparin-catalyzed inhibition of thrombin by antithrombin III. J. Biol. Chem. 259: 5670 (1984).
7. K.A. Parker, and D.M. Tollefsen, The protease specificity of heparin cofactor II. Inhibition of thrombin generated during coagulation. J. Biol. Chem. 260: 3501 (1985).

8. F.C. Church, C.M. Noyes, and M.J. Griffith, Inhibition of chymotrypsin by heparin cofactor II. **Proc. Natl. Acad. Sci. USA** 82: 6431 (1985).

9. D.M. Tollefsen, C.A. Pestka, and W.J. Monafo, Activation of heparin cofactor II by dermatan sulfate. **J. Biol. Chem.** 258: 6713 (1983).

10. H. Lobermann, R. Tokuka, J. Deisenhofer, and R. Huber, Human α proteinase inhibitor: crystal structure analysis of two crystal modifications, molecular model and preliminary analysis of the implication for function. **J. Mol. Biol.** 177: 531 (1984).

11. A.M. Lesk, M. Levitt, and C. Chothia, Alignment of the amino acid sequences of distantly related proteins using variable gap penalties. **Protein Eng.** 1: 11 (1984).

12. S.T. Olson, K.R. Srinivasan, I. Björk, and D. Shore, Binding of high affinity heparin to antithrombin III. Stopped flow kinetic studies of the binding interaction. **J. Biol. Chem.** 256: 11073 (1981).

13. H. Jornvall, W.W. Fish, and I. Björk, The thrombin cleavage site in bovine antithrombin. **FEBS Lett.** 106: 358 (1979).

14. R.W. Carrell, D.R. Boswell, S.O. Brennan, and M.C. Owen, Active site of α_1-antitrypsin: homologous site in antithrombin. **Biochem. Biophys. Res. Comm.** 93: 3994 (1980).

15. M.J. Griffith, C.M. Noyes, and F.C. Church, Reactive site peptide structural similarity between heparin cofactor II and antithrombin III. **J. Biol. Chem.** 260: 2218 (1985).

16. M. McGrogan, J. Kennedy, M.P. Li, C. Hsu, R.W. Scott, C.S. Simonssen, and J.B. Baker, Molecular cloning and expression of two forms of human protease nexin I. **Biotechnology** 6: 172 (1988).

17. M. Petitou, P. Duchaussoy, I. Lederman, J. Choay, C.J. Jaquinet, and C.P. Sinay, Synthesis of heparin fragments: a methyl alpha-pentoside with high affinity for heparin. **Carbohydrate Res.** 176: 67 (1987).

18. T. Koide, S. Odani, K. Takahashi, T. Ono, and N. Sakuragawa, Antithrombin III Toyama: replacement of arginine-47 by cysteine in a hereditary abnormal antithrombin III that lacks heparin-binding ability. **Proc. Natl. Acad. Sci. USA** 81: 289 (1984).

19. M.C. Owen, J.Y. Borg, C. Soria, J. Soria, J. Caen, and R.W. Carrell, Heparin binding defect in a new antithrombin III variant: Rouen, 47 Arg to His. **Blood** 69: 1275 (1988).

20. J.Y. Borg, M.C. Owen, C. Soria, J. Soria, J. Caen, and R.W. Carrell, Proposed heparin binding site in antithrombin based on arginine 47. A new variant Rouen II 47 Arg to Ser. **J. Clin. Invest.** 81: 1292 (1988).

21. J-Y. Chang, and T.H. Tran, Antithrombin III Basel. Identification of a Pro-Leu substitution in a hereditary abnormal antithrombin with impaired heparin cofactor activity. **J. Biol. Chem.** 261: 1174 (1986).

22. M.N. Blackburn, R.L. Smith, J. Carson, and C.C. Sibley, The heparin binding site in antithrombin III. Identification of a critical tryptophan in the amino acid sequence. **J. Biol. Chem.** 259: 939 (1984).

23. J-Y. Chang, Binding of heparin to human antithrombin III activates selective chemical modification of Lysine 236. Lys 107, Lys 125 and Lys 136 are situated within the heparin-binding site of antithrombin III. **J. Biol. Chem.** 264: 3111 (1989).

24. S.O. Brenan, P.M. George, and R.E. Jordan, Physiological variant of antithrombin III lacks carbohydrate sidechain at Asn 135. **FEBS Lett.** 219: 431 (1987).

25. J.W. Smith, and D.J. Knauer, A heparin binding site in antithrombin III. Identification, purification and amino acid sequence. **J. Biol. Chem.** 262: 11986 (1987).

26. M.A. Blinder, T.R. Andersson, U. Abildgaard, and D.M. Tollefsen, Heparin cofactor II Oslo: mutation of Arg 189 to His decreases the affinity for dermatan sulfate. **J. Biol. Chem.** 264: 5182 (1989).

27. D.M. Tollefsen, M.E. Peacock, and W.J. Monafo, Molecular size of dermatan sulfate oligosaccharides required to bind and activate heparin cofactor II. **J. Biol. Chem.** 261: 8854 (1986).

28. Å. Danielsson, E. Raub, U. Lindahl, and I. Björk, Role of ternary complexes, in which heparin binds both antithrombin and proteinase in the acceleration of the reaction between antithrombin and thrombin or factor Xa. **J. Biol. Chem.** 33: 15467 (1986).

29. S.T. Olson, and J.D. Shore, Demonstration of a two-step reaction mechanism for inhibition of α-thrombin by antithrombin III and identification of the step affected by heparin. **J. Biol. Chem.** 257: 14891 (1982).

30. S.T. Olson, and J.D. Shore, Transient kinetics of heparin-catalysed protease inactivation by antithrombin III. The reaction step limiting heparin turnover in thrombin neutralization. **J. Biol. Chem.** 261: 13151 (1986).

POLYPEPTIDE CHAIN STRUCTURE OF INTER-α-TRYPSIN INHIBITOR AND PRE-α-TRYPSIN INHIBITOR: EVIDENCE FOR CHAIN ASSEMBLY BY GLYCAN AND COMPARISON WITH OTHER "KUNIN"-CONTAINING PROTEINS

JAN J. ENGHILD, IDA B. THØRGERSEN,
SALVATORE V. PIZZO AND GUY SALVESEN[*]

*From the Department of Pathology, Duke University Medical Center,
Durham, North Carolina 27710*

INTRODUCTION

Proteins structurally related to the proteinase inhibitor aprotinin, systematically known as pancreatic trypsin inhibitor (Kunitz), occur in animals from a variety of orders including mammals, moluscs and coelenterates.[1] The wide distribution of these proteins suggests that the ancestral gene is very old, at least as old as the radiation of multicellular animals.[2]

As is the case in other proteinase inhibitor families (but not the serpins), gene duplications have led to multi-headed proteins, indeed, humans possess single-headed, double-headed and triple-headed aprotinin homologs. For example, the protein comprising the inhibitory domains of human inter-α-trypsin inhibitor (IαI)[**] contains tandem aprotinin-like domains and is, therefore, double-headed.[2] A recent report has suggested the name "bikunin" for this protein, a contraction of its two-domain structure and one of its pseudonyms.[3] We concur with this proposal and we have extended it to encompass the "monokunin" A_4 amyloid peptide precursor-751 (App_{751})[4] and the "trikunin" lipoprotein-associated coagulation inhibitor (LACI).[5]

The bikunin domain of IαI, in common with most of the other kunins, is synthesized as a larger precursor that undergoes proteolytic processing before storage or secretion.[6] However, the bikunin precursor cannot account for the large size of IαI (180-250 kDa). On the contrary, IαI is now thought to be composed of at least three chains that originate from distinct mRNA precursors.[3,6-9] Only one of these mRNAs encodes the inhibitory bikunin domain.

---

[*]To Whom Correspondence Should Be Addressed

[**]Abbreviations used: IαI, inter-α-trypsin inhibitor; PαI, pre-α-trypsin inhibitor; OPA, o-phthalaldehyde; TFMSA, trifluoromethane sulfonic acid; TFA, trifluoroacetic acid; PVDF, polyvinylidene difluoride; PTH, phenylthiohydantoin; SDS, sodium dodecyl sulfate; PAGE, polyacrylamide gel electrophoresis; DCI, 3,4-dichloroisocoumarin; E-64, N-[N-(L-3-trans-carboxyoxiran-2-carbonyl)-L-leucyl]-4-aminobutylguanidine; LACI, lipoprotein associated coagulation inhibitor; APP_{751}, amyloid A4 peptide precursor-751; PN-2, protease nexin2.

Serine Proteases and Their Serpin Inhibitors in the Nervous System
Edited by B. W. Festoff
Plenum Press, New York, 1990

The links that stabilize the complex of the bikunin and the heavier non-inhibitory chains resist dissociation in sodium dodecyl sulfate (SDS) in the presence of reagents that cleave disulfide bonds, leading earlier investigators to conclude that IαI is a single chain glycoprotein.[10,11] Little is known of the nature of these links, although a brief report indicates that bikunin is released from the parent molecule upon treatment with hyaluronidase or chondroitinase ABC.[12] As a prerequisite to investigating the function of IαI we have employed chemical and enzymatic methods to dissociate the molecule for analysis by inhibitory assays and protein sequencing, so that we may define the chain composition of this protein, and others in human blood that contain bikunin.

EXPERIMENTAL PROCEDURES

Materials

1,10 phenanthroline, tetrazotized o-dianisidine (fast blue B salt), N-acetyl-DL-phenylalanine beta-naphthyl ester, N,N-dimethylformamide, alkaline phosphatase conjugated goat anti-rabbit IgG, 5-bromo-4-chloro-3-indolyl phosphate, nitro blue tetrazolium, polyethylene glycol 8000 and Blue Agarose were from Sigma. Bovine trypsin, Staphylococcus aureus V8 proteinase, ovine testicular hyaluronidase, the general serine proteinase inhibitor 3,4-dichloroisocoumarin (DCI), and the general cysteine proteinase inhibitor N-[N-(L-3-trans-carboxyoxiran-2-carbonyl)-L-leucyl]-4-amino-αbutylguanidine (E-64) were from Boehringer Mannheim. O-phthalaldehyde (OPA) was from Pierce. Trifluoroacetic acid (TFA) and trifluoro-methanesulfonic acid (TFMSA) were from Applied Biosystems. Sephacryl S300 HR, DEAE-Sephacel, Superose 6, Superose 12, and MONO Q⁻ columns were from Pharmacia. Antiserum raised against human IαI was from Dakopatts. Fresh frozen human plasma was obtained from the Duke University Medical Center Blood Bank.

Polyacrylamide Gel Electrophoresis (PAGE)

Sodium dodecyl sulfate (SDS)-PAGE was performed on samples treated for 5 min at 100°C in 1% SDS on 5-15 % linear gradient gels (10 x 10 x 0.1 cm) using the glycine/2-amino-2-methyl-1,3-propanediol/HCl system described by Bury.[13] Some gels were stained for trypsin inhibitor activity (see below) and others were stained with Coomassie brilliant blue R-250. Numbers on the side of each gel represent molecular weight of standard proteins in kDa.

Trypsin Inhibitor Counterstained Gels (TIC gels)

The assay was performed essentially as described by Uriel and Berges.[14] Briefly, after electrophoresis, SDS gels were equilibrated in 50 ml 0.1 M sodium phosphate buffer, pH 7.8 for 15 min at 37°C. The buffer was decanted and 50 ml fresh buffer containing 0.04 mg trypsin/mL was added and the gel was incubated at 37°C. After 15 min the gel was rinsed briefly in water and allowed to react with 10 mL of 2.5 mg/mL N-Ac-DL-Phe-beta-napthyl ester in N,N-dimethylformamide and 50 mL 1 mg/mL tetrazotized o-dianisidine in 50 mM NaH_2PO_4, pH 7.8 for 10-60 min at 25°C. Samples from each step of purification were analyzed by this method and "active fractions" are defined as those showing significant inhibition by the TIC procedure.

Protein Purification

All purification steps were done at 4°C. Human plasma (2L) was made 5% in polyethylene glycol 8000, incubated for 1h and the precipitate collected by centrifugation at 10,000 x g for 15 min. The supernatant was made 16% in polyethylene glycol 8000 for 1 h, the pellet was collected by centrifugation and dissolved in 50 mM Tris-HCl, 50 mM NaCl pH 7.4 (column equilibration buffer). The dissolved precipitate was applied to a 5 cm x 30 cm Blue Agarose column with a flow rate of 70 ml/h. When the absorbance at 280 nm was less than 0.02 the column was eluted with equilibration buffer containing 1M NaCl at a flow rate of 40 mL/h. Active fractions were pooled and dialyzed overnight against column equilibration buffer.

The fall through and pooled fractions were applied separately to a 2.5 cm x 30 cm DEAE-Sephacel column in column equilibration buffer. A 2 x 500 mL gradient from 50 to 500 mM NaCl was used to develop the column. Active fractions were pooled and run on a 2.5cm x 150 cm Sephacryl S-300 HR in 50 mM Tris-HCl, 150 mM NaCl, pH 7.4, at 20 mL/h. As a final purification step the active fractions were pooled and run on a MONO Q column using a Pharmacia FPLC system employing a linear gradient from 200 mM to 500 mM NaCl in 50 mM Tris-Cl, pH 7.4 (1 mM/mL) to develop the column.

Protein Fragmentation

Reduced, alkylated PαI (250 μg) was digested with 5 μg of $S.$ $auereus$ V8 proteinase for 24 h in 50 mM NH_4HCO_3 at 23°C. 40 mg of IαI was digested with 1 mg of pepsin for 48 h in 20% formic acid at 37°C. Peptides were separated on Aquapore RP-300 (Brownlee) using a linear gradient (1%/min) from 0.1% TFA to 90% acetonitrile, 0.08% TFA, and a C-18 column (Vydac) using a linear gradient from 0.1% TFA to 90% propan-2-ol. Peptides were detected at 206 or 220 nm and applied directly to polybrene-treated glass fiber filters for Edman degradation.

Amino Acid Sequence Analysis

Automated Edman degradation was carried out in an Applied Biosystems 477A sequencer, using BEGIN-1 and NORMAL-1 cycles, with on line PTH-analysis using an Applied Biosystems 120A HPLC. The instruments were operated as recommended in the user bulletins and manuals distributed by the manufacturer. The different fragments obtained by chemical and enzymatic deglycosylation, as described below, were separated by SDS 5-15% polyacrylamide gel electrophoresis (PAGE). The gels were electroblotted to polyvinylidene difluoride membranes (PVDF-Millipore Immobilon Transfer Membranes) according to Matsudaira.[15] Protein bands were detected by Coomassie Blue staining, excised and placed on polybrene-treated precycled glass fiber filters and sequenced as described above. To block non-proline amino terminals O-phthalaldehyde was employed[16] under conditions specified by Applied Biosystems using the OPA-1 and PRO-1 cycles. Amino acid yields were calculated by subtracting the amount of PTH amino acid in each cycle from the background in the previous cycle.

Chemical Deglycosylation

Chemical deglycosylation using TFMSA was performed on purified protein samples as described by Sojar and Bahl.[17] Lyophilized, salt-free protein was incubated for 30 min at 0°C after which the reaction was stopped by the addition of 50% pyridine in diethyl ether at -40°C, followed by extensive dialysis against 0.1M NH_4HCO_3. TFA was also employed using the following conditions. Samples were dialyzed extensively against water at 4°C, brought to dryness in a Speed Vac Concentrator (Savant) and lyophilized over a pellet of P_2O_5. The samples were placed in a 50°C heating block and flushed with nitrogen. 100 μl of TFA was added and the tubes were sealed. After 30 min the reaction was terminated by freezing the samples in dry ice/ethanol followed by evaporation of the TFA in a Speed Vac. The dry sample was dissolved in 50 μL of ice cold 100 mM NH_4HCO_3 and dried again for 60 min in the Speed Vac. This was repeated twice, to remove all traces of TFA, before the sample was further analyzed.

Enzymatic Deglycosylation

Hyaluronidase was allowed to react with purified proteins in 50 mM Tris-Cl, 100 mM NaCl, 1 mM EDTA pH 7.4 at 37°C. The reaction mixture was made 50 μM in DCI, 10 μM in E-64 and 1 mM in 1,10 phenanthroline in order to hinder proteolytic degradation of the proteins.[18] The concentration of hyaluronidase and the reactions time are indicated in the figure legends.

Zone Electrophoresis of IαI and PαI kDa

Zone electrophoresis was carried out by Duke University Medical Center Hematology/protein laboratory. The electrophoresis was carried out in 2% agarose gels using a Beckman Paragon 6558 and the stained gel was scanned at 600 nm with a Beckman Appraise spectrophotometric densitometer.

RESULTS

Purification of SDS-Stable Trypsin Inhibitors from Human Plasma

Plasma samples from blood drawn from human volunteers 15 min before SDS-PAGE contained SDS-stable trypsin inhibitors of 125,000 Da and 225,000 Da (Figure 1). The trypsin inhibitor counterstain (TIC) assay was used to monitor fractions for SDS-stable trypsin-inhibitory activity during the purification of the human 225,000 Da and 125,000 Da proteins. The purification procedure was optimized to enable simultaneous recovery of both inhibitors from the same batch of plasma, taking advantage of a large differential in affinity of each protein for the blue Sepharose column. 100 mg of pure 225,000 Da protein and 8 mg of pure 125,000 Da protein were obtained from 2 L of plasma. In our hands, use of the proteinase inhibitors E-64, DCI, 1,10-phenanthroline and EDTA, during the purification did not increase the yield or stability of the proteins.

Identity of the Inhibitors

Purified samples of each protein were electrophoresed in agarose gels as described in **Experimental Procedures** (Figure 2). Comparison with standard human plasma samples indicates that the 225,000 Da protein migrates between the α_1 and α_2 zone and is therefore the previously characterized inter-α-trypsin inhibitor (IαI). The 125,000 Da protein migrates before the α_1 zone and may be identical to the prealbumin-like acid-stable trypsin inhibitor reported by Ødum and Ingwersen[19] and the 125,000 Da IαI-like protein detected by Salier et al,[20] although neither of these were characterized. In light of its migration in agarose gels and the relationship to IαI we name the protein pre-α-trypsin inhibitor (PαI).

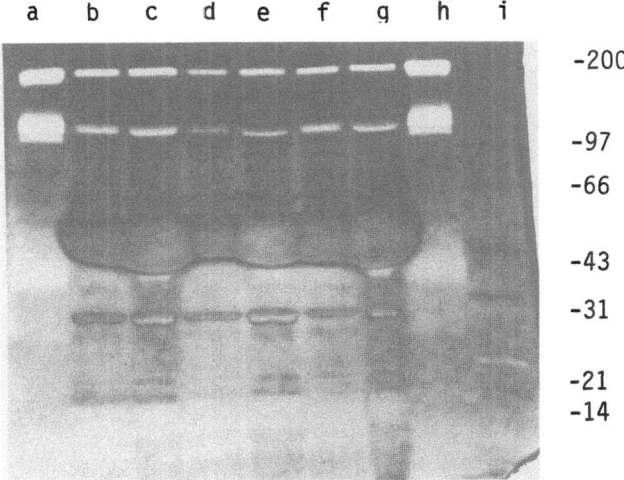

Figure 1. *Occurrence of SDS-stable trypsin inhibitors in human plasma.* Blood was collected from three volunteers from our laboratory. The cells were removed and plasma, (2.5 µl (lanes b,d and f) and 5 µl (lanes c,e and g)) was loaded on SDS-PAGE gels within 15 min of collection. Lanes a and h are purified inhibitors and lane i contains the molecular weight standards. The 21,000 Da inhibitory band in lane i is soybean trypsin inhibitor.

Chain Composition

Fresh plasma samples contain both inhibitors (Figure 1), indicating that PαI is not derived from IαI during plasma storage. Amino terminal protein sequence analysis of the purified proteins revealed the presence of three chains comprising IαI and two chains comprising PαI.[***] Sequence analysis of samples electrophoresed in SDS-

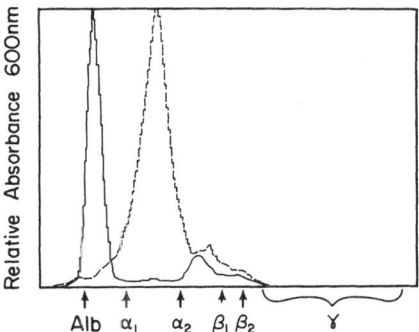

Figure 2. *Zonal agarose gel electrophoresis of the purified proteins.* Zone electrophoresis was performed on samples of purified inhibitors. The gel was stained with coomassie blue and scanned at 600 nm. The broken line is the 225,000 Da protein and the unbroken line is the 125,000 Da protein. The positions of the classic serum fractions are shown for comparison. Alb, albumin.

[***]A more detailed description of results on which a model of the chain structure of IαI and PαI is based may be found in Enghild et al.[21]

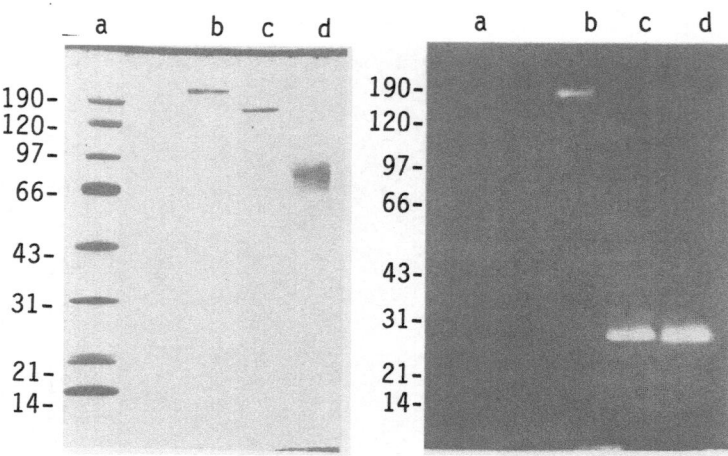

Figure 3. *Dissociation of the 225,000 Da inhibitor protein chains.* SDS-PAGE of standard proteins (lane a), purified inhibitor (lane b), inhibitor treated with 1 μg hyaluronidase per 25 μg inhibitor for 3 h (lane c) and inhibitor treated with TFMSA (lane d). The Left gel was stained with Coomassie blue and the Right gel was TIC stained.

PAGE, after electroblotting to PVDF-membranes, did not show any change in the amino-terminal sequences, indicating that the multiple amino-terminals were not due to loose association of peptides or proteins. Treatment of the proteins with TFMSA as described in **Experimental Procedures**, employing a protocol that is used to deglycosylate proteins, resulted in the production of 30,000 Da and 65-70,000 Da derivatives of IαI (Figure 3, Lane d), and 30,000 and 90,000 Da derivatives of PαI (Figure 4, lane d). Treatment of the proteins with hyaluronidase also resulted in the release of the 30,000 Da derivatives, yet under the conditions used in Figure 3, lane c, the 65-70,000 Da components of IαI remained associated in SDS-PAGE to give a

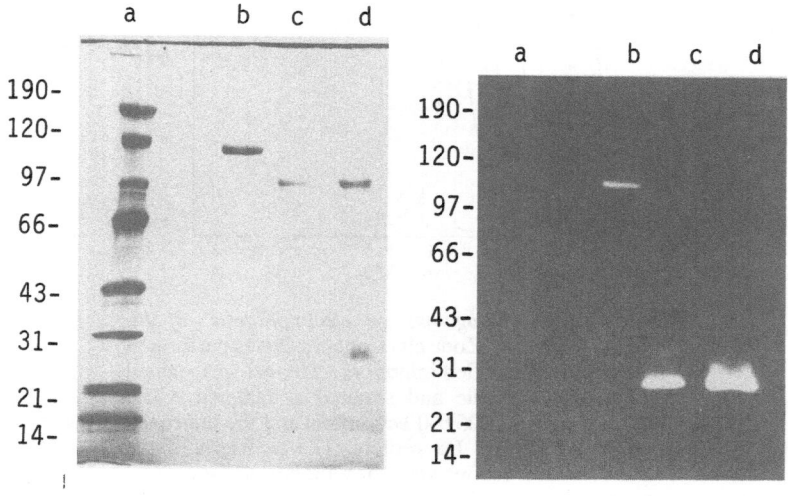

Figure 4. *Dissociation of the 125,000 Da inhibitor.* SDS PAGE of standard proteins (lane a), purified inhibitor (lane b), inhibitor after treatment with 1 μg hyaluronidase per 25 μg inhibitor (lane c) and inhibitor treated with TFMSA (lane d). The Left gel was Coomassie blue stained and the Right gel TIC stained.

derivative of 130,000 Da. The amino-terminal sequences of polypeptides released from IαI or PαI by TFMSA or hyaluronidase treatment were determined after electrophoresis of the products in SDS-PAGE, as described in **Experimental Procedures**. The 30,000 Da polypeptide from IαI was found to be identical to that of PαI and corresponds to the proteinase-inhibitory bikunin domain.[1] The 130,000 Da derivative of IαI produced by hyaluronidase treatment, and the 65-70,000 derivatives produced by TFMSA treatment, contained two amino-terminal sequences in approximately equal yields. One sequence was identical to residues Ser_1 - Asp_{14} of the IαI clone reported by Gebhard et al.[22] The second sequence, inferred by subtraction from the first, does not correspond to sequences in release 20 of the National Biomedical Research Foundation (NBRF) Protein Identification Resource. The 90,000 Da derivative of PαI contained a single sequence that is also not found in this release of the NBRF resource. Therefore, we conclude that IαI possess two separate polypeptides designated heavy chains 1 and 2 (HC1 and HC2), in addition to bikunin, whereas PαI is composed of bikunin and a 90,000 Da polypeptide designated heavy chain 3 (HC3).

The 65,000 Da and 70,000 Da derivatives of IαI were not well separated on SDS-PAGE, making it difficult to distinguish between them by sequence analysis following transfer to PVDF membranes (see **Experimental Procedures**). Nevertheless, we were able to designate the 65,000 Da derivative as heavy chain 1, and the 70,000 Da derivative as heavy chain 2.

Extended amino-terminal sequencing of heavy chains from IαI and PαI was accomplished by treatment of samples with o-phthalaldehyde (OPA) following the second Edman degradation cycle (see **Experimental Procedures**). This procedure prevents reaction of phenylisothiocyanate with amino-terminal residues other than proline and allowed unambiguous identification of extended sequences of the PαI heavy chain (HC3) and the heavy chain of IαI that contains proline in position three (HC2).

Sequence analysis of proteolytic peptides derived from IαI indicates that the heavy chains correspond to cDNA clones isolated by Schreitmüller et al,[7] Salier et al,[8] and Gebhard et al.[3] Sequence analysis of peptides derived from proteolysis of PαI indicates that the heavy chain of this protein corresponds to the cDNA clone recently reported by Diarra-Mehrpour et al.[9] The cDNAs encoding HCl and HC3 are not full length, which explains our inability to match the amino-terminal sequences with the cDNAs. Table I summarizes the amino-terminal sequences of the chains comprising IαI and PαI and relates them to the various published cDNA clones.

Table 1. Amino terminal sequences of the components of IαI and PαI. HI-30, also known as bikunin,[1] contains the trypsin-inhibitory activity. HC2 is identical to the sequence of the cDNA reported by Gebhard et al.[21] The amino-terminal sequence of HC1 has not previously been published; this chain corresponds to the partial cDNAs of Schreitmüller et al[7] and λ Hu HITI-9 of Salier et al.[8] The amino-terminal sequence of HC3 has not previously been published; this chain corresponds to the partial cDNA λ Hu HITI-13 of Diarra-Mehrpour et al.[9]

			Residue Number		
Chain	1	5	10	15	20
HI-30	A V L P Q E E G S G G G Q L V T E V T
HC1	S K S S E K R Q A V D T A V D G T F I A
HC2	S L P G E S E E M M E V D Q V T L Y S
HC3	S L P E G V A N G I E V Y S T K I N S K

Chain Stoichiometry

Since the relation between molecular weight and migration of cross-linked polypeptides in SDS-PAGE is tenuous[23] we were unable to infer the number of individual chains comprising PαI and IαI. However, limited hyaluronidase treatment of IαI resulted in a single molecular weight shift to liberate bikunin, the remaining 130,000 Da derivative contained no inhibitory activity. More extensive digestion with hyaluronidase resulted in conversion of the 130,000 Da derivative into 65,000 Da and 70,000 Da derivatives, similar to TFMSA treatment, with no intermediates. We conclude, therefore, that IαI contains single copies of two distinct 65-70,000 Da heavy chains (HC1 and HC2) and a single bikunin chain. By similar criteria of limited deglycosylation, we conclude that PαI contains a single heavy chain (HC3) linked to a single bikunin.

Composition of the Crosslink(s)

Despite earlier claims that IαI is a single chain protein[10,11] presumably due to the extraordinary stability of inter-chain crosslinks, we present evidence here that IαI is composed of three chains and PαI of two. The first evidence for the composition of the crosslink was presented by Jessen et al[12] who showed that the bikunin could be separated from IαI by treatment of the protein with hyaluronidase and chondroitinase ABC, enzymes that share the ability to degrade chondroitin-sulfate-like glycosaminoglycans (GAGs). However, Jessen et al[12] did not observe dissociation of the 130,000 Da derivative of IαI leading them to conclude that the molecule contains a single heavy chain. We show here that treatment of IαI and PαI with hyaluronidase, or the deglycosylating agents TFMSA and TFA, leads to the liberation of all of the omponent chains of the proteins without detectable proteolysis, confirming the non-protein nature of the crosslink and the chain stoicheometry shown in Figure 6.

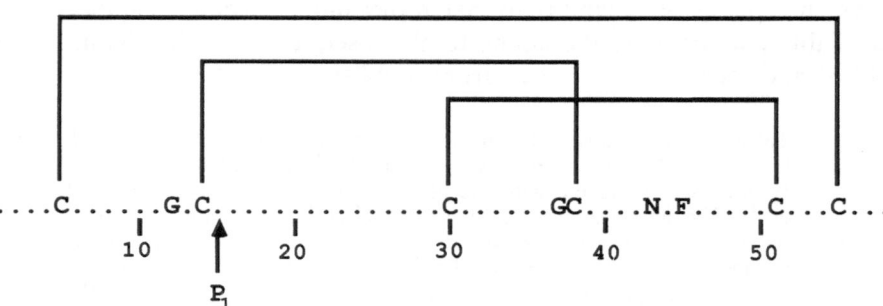

Figure 5: *Consensus sequence of a kunin domain.* The consensus was established following screening of the NBRF protein sequence database, version 20, using bovine aprotinin as the test sequence. A total of 31 kunin domains were identified, to which may be added the three kunins of LACI,[5] the single kunin domains of APP$_{751}$[4] and trypstatin.[24] A kunin domain consists of about 58 amino acids with a high degree of conservation of the residues noted above (standard single letter amino acid code). The pairing of cysteines to form disulfides in aprotinin is likely to be conserved between all kunins.[1] The residue labelled P$_1$ corresponds to the primary specificity site within the reactive site loop of inhibitory kunins.[2] The following proteins show deviations from the consensus. The P$_1$ residue has been deleted, Asn$_{43}$ substituted by Gly, and the β_2-B chain contains Lys in place of Gly$_{12}$ in the β-bungarotoxin -B chain kunins;[25] these kunins are not proteinase inhibitors. Residue Phe$_{45}$ is substituted by Tyr in the horseshoe crab kunin;[26] inhibitory activity is maintained. Bombyx mori kunin contains a single residue insertion somewhere between Cys$_5$ and Gly$_{12}$;[27] inhibitory activity is maintained.

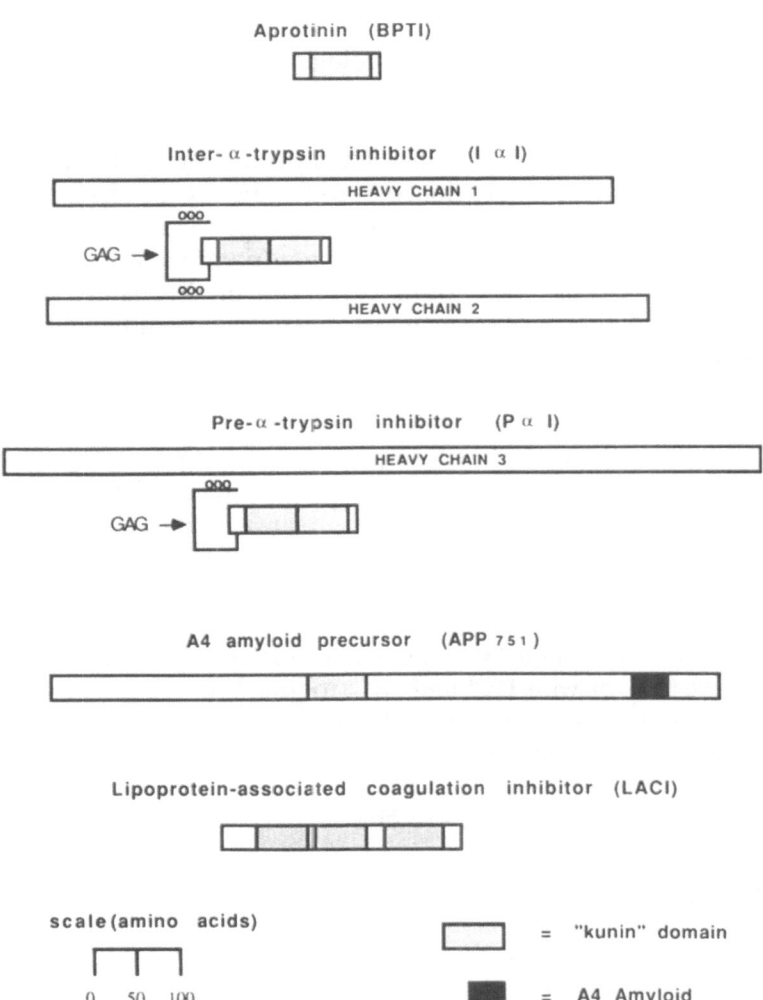

Figure 6. *Proteins containing kunin domains.* The archetypal kunin, bovine aprotinin, contains a single domain (hatched box) with short peptide extensions (open boxes) at the amino- and carboxyl-terminals.[28] IαI is composed of two distinct heavy chains attached to the bikunin domains, probably via a GAG link (thin line). PαI is composed of one heavy chain, different from the two IαI heavy chains, attached to the bikunin via an equivalent GAG-like link. The dots between the GAG chain and the protein chain represent the strong forces that help assemble the complexes. APP$_{751}$, recently shown to be identical to protease nexin 2 (PN-2),[29] contains a single kunin domain and a membrane-spanning domain close to the A4 amyloid segment. APP$_{751}$ may be found in cell-associated and secreted forms.[30] LACI contains three consecutive kunin domains and is therefore a "trikunin". The domains are separated by short connecting peptides.

Location of the Crosslink

Results of limited proteolysis of IαI and PαI followed by protein sequencing indicate that the crosslink originates from Ser$_{10}$ of bikunin.[21] Furthermore, we note that the residues surrounding Ser$_{10}$ conform to a partial concensus (DEXSG) for GAG addition to acceptor protein chains.[31] Ser$_{10}$ of bikunin has been reported to carry a long glycan chain,[32] although these authors did not address the possibility of a GAG-like structure for the chain. Balduyck et al,[33] on the other hand, identified

GAG-like glycan attached to human urinary trypsin inhibitor, a protein thought to be synonymous with bikunin.[1]

Preliminary data based on biosynthetic radiolabeling of IαI in Hep G2 cells in the presence of [^{35}S] sulfate indicate that HI-30, not HC1 or 2, contains sulfated GAG[34] so we reason that the GAG that assambles IαI, and by inference PαI, originates from Ser_{10} of bikunin. Since the IαI heavy chains are separated by extensive hyaluronidase treatment, we speculate that they are linked to each other by the GAG that originates from bikunin. Their greater resistance to hyaluronidase treatment suggests that the inter-heavy chain GAG is less available than the portion of the GAG chain that links bikunin to the heavy chains. We know very little of the nature of this strong interaction between the bikunin GAG and the heavy chains, although Jessen et al[12] suggested that bikunin is linked covalently to heavy chains of IαI by chondroitin sulfate. However, in the absence of any direct evidence for a covalent crosslink we feel that a strong, non-covalent, association of bikunin GAG with regions on HC1, 2 and 3 should also be considered. With respect to this last possibility we note the report of Frenette et al[35] documenting a strong (stable to SDS-PAGE) but non-covalent link between a heparin sulfate proteoglycan and laminin.

We believe our observations clarify the confusion that has surrounded attempts to understand the structure of protein material previously known as "inter-α-trypsin inhibitor." We have shown that the material consists of two distinct complexes, each containing a single trypsin-inhibitory chain that assembles, via a GAG-like glycan, with either HC3 to give PαI, or with HC1 and HC2 to give IαI. An understanding of the structure of the inhibitors should enable the design of experiments to examine the function of the inhibitors and the reasons for their unusual glycan-mediated assembly.

Multiple Forms of Kunin-Containing Proteins

Figure 5 depicts the consensus sequence of a minimal kunin - a highly conserved structure of about 60 residues. The key to the kunin structure is the conservation of cysteine residues with the spacing of C(8) C(15) C(7) C(12) C(3) C where the values in parentheses are the number of residues between each cysteine. Interestingly, residues 186-243 of the human epidermal growth factor receptor fit this cysteine spacing,[36] but this protein contains 4 additional cysteine residues within the matched region. Consequently, it may have a different disulfide arrangement, and it is not known to inhibit proteinases. In the case of aprotinin, disulfides formed between these cysteines play critical roles in the correct folding of nascent protein and the maintenance of inhibitory capacity. By analogy we expect the same to hold for other kunins, yet we note the association of "extra" peptide material with the minimal inhibitory domain of many kunins (Figure 6). It is clear that the members of the kunin superfamily have evolved from a common ancestor, but during the course of this evolution their genes have incorporated coding sequences and the kunin domains have become embedded in larger precursors. For example, aprotinin contains short sequences at the amino- and carboxy- terminal ends of the precursor that are not present in the protein isolated from bovine organs.[28] Similarly, the IαI and PαI bikunin precursor contains α_1-microglobulin at its amino-terminal end. Presumably this is removed before assembly of the complexes. The amino- and carboxy-terminal extensions of the LACI trikunin may not be removed,[5] and we await data on the processing of APP_{751}.[29]

The function of the peptide extensions is an open question. The heavy chains of IαI do not appear to modulate inhibitory activity of the bikunin[37,38] and the APP_{751}

kunin appears to be a good trypsin inhibitor while embedded in the remainder of the precursor.[29]

With the exception of LACI which may inhibit coagulation factor Xa[5] the biological roles of the kunins shown in Figure 6 are unknown. They appear not to be physiological regulators of the well known pancreatic or coagulation serine proteinases[1] and a function in controlling the plasminogen activators and growth factor binding proteins should be considered. Finally, the heavy chains of IαI and PαI, and the extensions of APP_{751}, may target the kunins to certain areas in the body, allowing localization of the inhibitors to certain cells or tissues. An investigation of possible targeting information contained within the heavy chains of IαI and PαI, and the extensions of the APP_{751} precursor, seems to be a pertinent avenue of research to determine the biological roles of those kunin-containing proteins.

SUMMARY

The polypeptide chain composition of protein species referred to in the literature as "inter-α-trypsin inhibitor" was investigated. The material was found to consist of distinct proteins of 125,000 Da and 225,000 Da, each of which contained more than one polypeptide chain. The links that assemble each protein were found to be stable to various strong denaturants, but susceptible to treatment with trifluoromethane sulfonic acid or hyaluronidase, indicating a glycan nature. The 225,000 Da protein migrated with inter-α mobility in agarose gel electrophoresis and is designated inter-α-trypsin inhibitor (IαI). The 125,000 Da protein migrated with pre-α mobility, appearing to be a newly-recognized protein, and we therefore designate it pre-α-trypsin inhibitor (PαI). Analysis of the proteins, the separated chains, and proteolytic derivatives thereof, revealed that each protein contained a single, identical, trypsin-inhibitory chain of 30,000 Da. IαI contains non-inhibitory heavy chains of 65,000 Da and 70,000 Da, whereas PαI contains a heavy chain of 90,000 Da. Our data allow identification of several recently reported cDNA clones and clarify the confusion surrounding the composition of plasma proteins referred to as "inter-α-trypsin inhibitor."

We compare the unusual chain structure of IαI and PαI with other proteins that contain homologous proteinase-inhibitory domains. These domains, designated "kunins", are often associated with protein material that may target them to selected sites in the body.

ACKNOWLEDGEMENTS

We thank Jan Potempa, James Travis, William Wagner, and John Mort for helpful discussions, Wolfgang Gebhard and Sukanto Sinha for communicating results before publication and Pat Burks for typing of this manuscript. This work was supported by National Institutes of Health grant HL-24066 and American Cancer Society grant IN-1588.

REFERENCES

1. W. Gebhard, and K. Hochstrasser, Inter-α-trypsin inhibitor and its close relatives. In: **Proteinase Inhibitors** (A.J. Barrett, and G. Salvesen, eds), pp. 389-401, Elsevier, Amsterdam (1986).

2. M. Laskowski, Jr. and I. Kato, Protein inhibition of proteinases, **Ann. Rev. Biochem.** 49, 593-626 (1980).

3. W. Gebhard, T. Schreitmüller, K. Hochstrasser, and E. Wachter, Two out of three kinds of subunits of inter-α-trypsin inhibitor are structurally related. **Eur. J. Biochem.**, 181,571-576 (1989).

4. R.E. Tanzi, A.I. McClatchey, E.D. Lamperti, L. Villa-Komaroff, J.F. Gusella, and R.L. Neve, Protease inhibitor domain encoded by an amyloid protein precursor mRNA associated with Alzheimer's disease, **Nature** 331,528-530 (1988).

5. T.-C. Wun, K.K. Kretzmer, T.J. Girard, J.P. Miletech, and G.T. Broze, Cloning and characterization of a cDNA coding for the lipoprotein-associated coagulation inhibitor shows that it consists of three tandem Kunitz-type inhibitory domains. **J. Biol. Chem.** 263, 6001-6004 (1988).

6. J.F. Kaumeyer, J.O. Polazzi, and M.P. Kotick, The mRNA for a proteinase inhibitor related to the HI-30 domain of inter-α-trypsin inhibitor also encodes α_1-microglobulin (protein HC). **Nucleic Acids Res.** 14,7839-7850 (1986).

7. T. Schreitmüller, K. Hochstrasser, P.W.M. Reisinger, E. Wachter, and W. Gebhard, cDNA cloning of human inter-α-trypsin inhibitor discloses three different proteins. **Biol. Chem. Hoppe-Seyler** 368,963-970 (1987).

8. J.P. Salier, K. Kurachi, and J.-P. Martin, Isolation and characterization of cDNAs encoding the heavy chain of human inter-α-trypsin inhibitor (IαTI): unambiguous evidence for multipolypeptide chain structure of IαTI. **Proc. Natl. Acad. Sci. USA** 84,8272-8276 (1987).

9. M. Diarra-Mehpour, J. Bourguignon, R. Sesboüe, M.-G. Mattei, E. Passage, J.P. Salier, and J.-P.Martin, Human plasma inter-α-trypsin inhibitor is encoded by four genes on three chromosomes. **Eur. J. Biochem.** 179,147-154 (1989).

10. P. Reisinger, K. Hochstrasser, K. Albrecht, G.J. Lempart, and J.-P. Salier, Human inter-α-trypsin inhibitor: Localization of the Kunitz-type domains in the N-terminal part of the molecule and their release by a trypsin-like proteinase. **Biol. Chem. Hoppe-Seyler** 366,479-483 (1985).

11. M. Morii, and J. Travis, The reactive site of human inter-α-trypsin inhibitor is in the amino-terminal half of the protein. **Biol. Chem. Hoppe-Seyler's** 366,19-21 (1985).

12. T.E. Jessen, K.L. Faarvang, and M. Ploug, Carbohydrate as covalent crosslink in human inter-α-trypsin inhibitor: A novel plasma protein structure. **FEBS Lett** 230,195-200 (1988).

13. A.F. Bury, Analysis of protein and peptide mixtures. Evaluation of three different sodium dodecyl sulphate polyacrylamide gel electrophoresis buffer systems. **J. Chromatogr.** 213,491-500 (1981).

14. J. Uriel, and J. Berges, Characterization of natural inhibitors of trypsin and chymotrypsin by electrophoresis in acrylamide-agarose gels. **Nature** 218,578-580 (1968).

15. P. Matsudaira, Sequence from picomole quantities of protein electroblotted onto polyvinylidene difluoride membranes. **J. Biol. Chem.** 262,10035-10038 (1987).

16. A.W. Brauer, C.L. Oman, and M.N. Margolies, Use of O-phthalaldehyde to reduce background during automated Edman degradation. **Anal. Biochem.** 137,134-142 (1984).

17. H.T. Sojar, and O.P. Bahl, Chemical deglycosylation of glycoproteins. **Meth. Enzymol.** 138,341-350 (1987).

18. G.S. Salvesen, and H. Nagase, Inhibition of proteolytic enzymes. In: **Proteolytic Enzymes: A Practical Approach** (R. Beynon, and J. Bond, eds), pp. 83-104 IRL Press, Oxford (1989).

19. L. Ødum, and S. Ingwersen, Electrophoretic investigations of acid-stable proteinase-inhibitory activity in human serum. **Hoppe-Seyler's Z. Physiol. Chem.** 364,1671-1677 (1983).

20. J.P. Salier, J.P. Martin, P. Lambin, H. McPhee, and K. Hochstrasser, Purification of the human serum inter-α-trypsin inhibitor by zinc chelate and hydrophobic interaction chromatographies. **Anal. Biochem.** 109,273-283 (1980).

21. J.J. Enghild, I.B. Thogersen, S.V. Pizzo, and G. Salvesen, Analysis of inter-α-trypsin inhibitor and a novel trypsin inhibitor, pre-α-trypsin inhibitor, from human plasma. Polypeptide chain stoichiometry and assembly by glycan. **J. Biol. Chem.**, 264, 15975-15981 (1989).

22. W. Gebhard, T. Schreitmüller, K. Hochstrasser, and E. Wachter, Complementary DNA and derived amino acid sequence of the precursor of one of the three protein components of the inter-α-trypsin inhibitor complex. **FEBS Lett** 229,63-67 (1988).

23. G.S. Salvesen, and A.J. Barrett, Covalent binding of proteinases in their reaction with α_2-macroglobulin. **Biochem. J.** 187,695-701 (1980).

24. H. Kido, Y. Yokogoshi, and N. Katanuma, Kunitz-type protease inhibitor found in rat mast cells. Purification, properties, and amino acid sequence. **J. Biochem.** 91,1519-1530 (1988).

25. K. Kondo, H. Toda, K. Narito, and C.Y. Lee, Amino acid sequence of β_2-bungarotoxin from bungarus multicinctus serum. The amino acid substitutions in the B chains. **J. Biochem.** 91,1519-1530 (1982).

26. T. Nakamura, T. Hirai, F. Tohunaga, S. Kawabata, and S. Iwanaga, Purification and amino acid sequence of Kunitz-type protease inhibitor found in the hemocytes of horseshoe crab (Tachypleus tridentatus). **J. Biochem.** 101,1297-1306 (1987).

27. T. Sasaki, Amino acid sequence of a novel Kunitz-type chymotrypsin inhibitor from hemolymph of silkworm larvae, Bombyx mori. **FEBS Lett.** 168,227-270 (1984).

28. T.E. Creighton, and I.G. Charles, Sequence of the genes and polypeptide precursors for two bovine protease inhibitors. **J. Mol. Biol.** 194,11-22 (1987).

29. T. Oltersdorf, L.C. Fitz, D.B. Schenk, I. Lieberburg, K.L. Johnson-Wood, E.C. Reattie, P.J. Ward, R.W. Blacher, H.F. Dovey, and S. Sinha, The secreted form of the Alzheimer's amyloid precursor protein with the Kunitz domain is protease nexin II. **Nature,** 341, 144-147 (1989).

30. A. Weidemann, G. König, D. Bunke, P. Fischer, J.M. Salbaum, C.L. Masters, and K. Beyreuther, Identification, biogenesis and localization of precursors of Alzheimer's disease A4 amyloid protein. **Cell** 57,115-126 (1989).

31. S. Huber, K.H. Winterhalter, and L. Vaughan, Isolation and sequence analysis of the glycosamino-glycan attachment site of type IV collagen. **J. Biol. Chem.** 263,752-756 (1988).

32. K. Hochstrasser, O.L. Schönberger, I. Rossmanith, and E. Wachter, E., Kunitz-type proteinase inhibitors derived from limited proteolysis of the inter-α-trypsin inhibitor. V. Attachments of carbohydrates in the human urinary trypsin inhibitor isolated by affinity chromatography. **Hoppe-Seyler's Z. Physiol. Chem.** 362,1357-1362 (1981).

33. W. Balduyck, C. Mizon, H. Loutfi, C. Richet, P. Roussel, and J. Mizon, The major human urinary trypsin inhibitor is a proteoglycan. **Eur. J. Biochem.** 158,417-422 (1986).

34. M.W. Swaim, J.J. Enghild, E.A. Auerswald, S.V. Pizzo, and G. Salvesen, Biosynthesis of inter-α-inhibitor. **J. Cell Biol.** 107, p 834a (1988)

35. G. P. Frenette, R.W. Ruddon, R.F. Krzesicki, J.A. Naser, and B.P. Peters, Biosynthesis and deposition of a noncovalent laminin-heparan sulfate proteoglycan complex and other basal lamina components by a human malignant cell line. **J. Biol. Chem.** 264,3078-3088 (1989).

36. A. Ullrich, L. Cousens, J.S. Hayflick, T.J. Dull, A. Gray, A.W. Tam, J. Lee, Y. Yarden, T.A. Liebermann, J. Schlessinger, J. Downward, E.L.V. Mayes, N. Whittle, M.J. Waterfield, and P. H. Seeburg, Human epidermal growth factor receptor cDNA sequence and aberrant expression of the amplified gene in A431 epidermal carcinoma cells. **Nature** 309,418-425 (1984).

37. C.W. Pratt, and S.V. Pizzo, Mechanism of action of inter-α-trypsin inhibitor. **Biochemistry** 26, 2855-2863 (1987).

38. J. Potempa, K. Kwon, R. Chawla, and J. Travis, Inter-α-trypsin inhibitor. Inhibition spectrum of native and derived forms. **J. Biol. Chem.**, in press (1989).

REGULATION OF PROTEASE NEXIN-1 ACTIVITY AND TARGET PROTEASE SPECIFICITY BY THE EXTRACELLULAR MATRIX

DENNIS D. CUNNINGHAM, DAVID H. FARRELL* AND
STEVEN L. WAGNER

Department of Microbiology and Molecular Genetics
College of Medicine
University of California
Irvine, CA 92717

INTRODUCTION

PN-1 is a 44 kDa protein protease inhibitor that is synthesized and secreted by a variety of cultured cells including fibroblasts, smooth muscle cells and astrocytes. [1-4] It rapidly inhibits thrombin, urokinase and plasmin by forming SDS-stable complexes with the catalytic site serine of the protease.[1-5] The protease-PN-1 complexes bind back to the cells, via the PN-1 moiety of the complex, and are rapidly internalized and degraded.[2] This provides a mechanism to regulate and clear certain serine proteases in the extracellular environment.

PN-1 recently was shown to be identical to the glial-derived neurite promoting factor/glial-derived nexin that has been identified and studied by Monard and colleagues.[6,7] It stimulates neurite outgrowth in neuroblastoma cells[8,9] and primary sympathetic neurons[10] and is present in rat[11] and human[12] brain. The neurite outgrowth activity of PN-1/glial-derived nexin on neuroblastoma cells is dependent on inhibition of thrombin. This conclusion is derived from studies which have shown that hirudin, a potent and highly specific thrombin inhibitor from leeches, stimulates neurite outgrowth from neuroblastoma cells to the same extent and with the same kinetics as PN-1.[8,9] Also, thrombin not only blocks but also reverses the neurite outgrowth activity of both PN-1 and hirudin.[9] The thrombin-mediated retraction of neurites is not due to a general proteolytic effect since it does not occur with much higher concentrations of urokinase, plasmin or trypsin.[9] With neuroblastoma cells, PN-1 stimulates neurite outgrowth only if thrombin is present.[13] Recent studies have shown that PN-1 and thrombin can also reciprocally regulate the stellation of cultured early passage astrocytes.[14] This effect of PN-1 also depends on inhibition of thrombin.

The above studies have shown that PN-1 can regulate the morphology/ differentiation of certain cultured neural cells, and that under the experimental conditions employed, this requires thrombin inhibition. This paper will review recent

---

*Present address: Department of Biochemistry
University of Washington, Seattle, WA 98195

Serine Proteases and Their Serpin Inhibitors in the Nervous System
Edited by B. W. Festoff
Plenum Press, New York, 1990

93

studies which have shown that cells, in turn, can regulate PN-1. It will summarize experiments which have shown that the surface/ECM regulates the activity and target protease specificity of PN-1. PN-1 binds to and is localized to the ECM.[15] This interaction accelerates its inactivation of thrombin[16,17] and blocks its inactivation of urokinase and plasmin.[18] Thus, PN-1 *in vivo* is likely to be primarily a thrombin inhibitor since much of it would be bound to the ECM.

RESULTS

Fibroblasts Accelerate the Inactivation of Thrombin by PN-1

The first indication that cells could regulate the activity of PN-1 came from the experiment shown in Figure 1. Here, ^{125}I-thrombin was incubated with PN-1 or with plasma antithrombin III (ATIII) in the presence or absence of normal human fibroblasts in serum-free medium. The cells had been fixed with 2% paraformaldehyde to prevent endocytosis or exocytosis during the course of the experiment. Figure 1 shows that the fixed cells accelerated the formation of [125]I- thrombin-PN-1 complexes but that they did not accelerate the formation of [125]I- thrombin-ATIII complexes.[16] It should be emphasized that the paraformaldehyde treatment inactivated surface-bound PN-1 so that the amount of active PN-1 in the reaction mixtures was precisely known. Similar results were obtained with membranes prepared from unfixed cells, indicating that the acceleration was not due to an artifact of para-formaldehyde fixation.[16] Previous studies had shown that heparin accelerates thrombin inhibition by both PN-1[1-5] and ATIII.[19] To determine if the inability of the fibroblasts to stimulate thrombin-ATIII complex formation was due to loss of the heparin cofactor activity of ATIII, it was incubated with thrombin under the same experimental conditions in the presence or absence of heparin. The ATIII retained heparin cofactor activity under these conditions.[16] These results indicated that the ability of cells to accelerate the reactions of certain protease inhibitors might be cell-type specific. Studies with several protease inhibitors have now shown cell-type specificities in accelerating reactions between these inhibitors and their target proteases.[20,21]

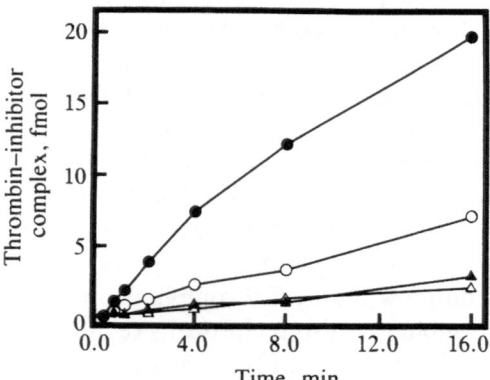

Figure 1. *Fixed human fibroblasts accelerate the rate of thrombin-PN-1 complex formation.* Confluent early passage human foreskin fibroblasts were incubated in serum-free medium for 2 days and then fixed in 2% paraformaldehyde. Medium containing 13.7 nM PN1 (•) or ATIII (▲) was incubated with the cells at 37°C; 13.7 nM PN-1 (○) or ATIII (△) was also incubated with plates without cells. ^{125}I-thrombin was added at 1.37 nM to initiate the reaction. Electrophoresis sample buffer was added at the times indicated by the data points to stop the reaction. Aliquots were electrophoresed, and radioactivity in the bands corresponding to thrombin-PN-1 and thrombin-ATIII complexes was quantitated. Reprinted with permission from reference 16.

In view of the ability of purified heparin to accelerate thrombin inactivation by PN-1,[1-5] we evaluated the hypothesis that the acceleration of this reaction by fibroblasts might be due to heparan sulfate or other sulfated glycosaminoglycans contained in their ECM.[16,17] The first step in this analysis was to determine if ECM preparations of fibroblasts accelerated the inactivation of thrombin by PN-1. Figure 2 shows an experiment in which PN-1 and ^{125}I thrombin were incubated with no cells (open circles), with fibroblasts that had been fixed in 2% paraformaldehyde (closed triangles) or with ECM prepared from an equivalent number of cells (closed circles). These results showed that virtually all of the accelerative activity resided with the ECM.[16] It should be emphasized that the 2% paraformaldehyde treatment to fix the cells and the 0.25M NH4OH treatment to prepare ECM both inactivate PN-1. Thus, the active PN-1 in each reaction mixture was totally due to added purified PN-1.

The next step in our analysis was to determine what molecules in the ECM were responsible for accelerating the inactivation of thrombin by PN-1. As noted above, the ability of purified heparin to accelerate this inactivation suggested that it might be ECM heparan sulfate or some other sulfated glycosaminoglycan. Experiments which employed a number of purified glycosaminoglycans showed that heparan sulfate and chondroitin sulfate significantly accelerated the inactivation of thrombin by PN-1.[17] We next employed highly purified specific glycosidases, which selectively hydrolyze these glycosaminoglycans, to determine if they might remove the accelerative activity from plasma membranes. Figure 3 shows that heparitinase, a glycosidase which selectively hydrolyzes heparan sulfate, removed about 80% of the accelerative activity of plasma membranes prepared from human fibroblasts. Chondroitinase ABC removed the remaining 20% of this activity. Thus, most of the acceleration of the reaction between PN-1 and thrombin is due to heparan sulfate present in the ECM.[17]

Fibroblasts Block the Ability of PN-1 to Inactivate Urokinase and Plasmin.

Studies conducted by Scott et al. on the rates of inactivation of various proteases by PN-1 showed that it rapidly inactivated thrombin, urokinase and plasmin.[5] The

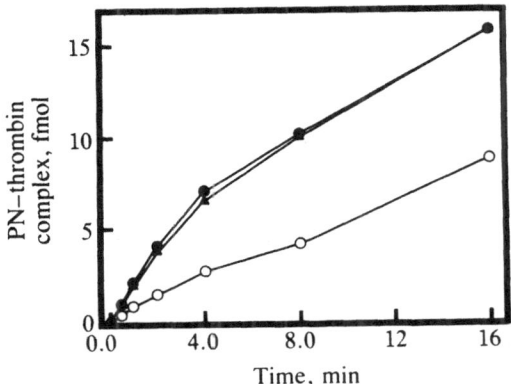

Figure 2. *Human fibroblast ECM accelerates the rate of thrombin-PN-1 complex formation to the same extent as fixed cells.* Parallel cultures were either fixed (▲) or else treated with 0.25 M NH4OH to prepare ECM (●). Medium containing 13.7 nM PN-1 was added to these plates or to plates without cells (○) and incubated at 37°C. ^{125}I-thrombin (1.37 nM) was added to initiate the reaction. At the times indicated by the data points, electrophoresis sample buffer was added to stop the reaction. Aliquots were electrophoresed, and the ^{125}I-thrombin-PN-1 complexes were quantitated. Reprinted with permission from reference 16.

second order rate constants were $6.0 \pm 1.3 \times 10^5$, $1.5 \pm 0.1 \times 10^5$ and $1.3 \pm 0.1 \times 10^5$ $M^{-1}s^{-1}$ for thrombin, urokinase and plasmin, respectively.[5] In view of our studies which showed that PN-1 binds to the ECM of human fibroblasts,[15] and the studies described in the preceding section, we determined if the interaction of PN-1 with cells or their ECM regulated its inactivation of urokinase or plasmin. The studies summarized below showed that it completely blocked the ability of PN-1 to form complexes with these two proteases.[18]

The autoradiogram of Figure 4 shows an important control experiment conducted with ^{125}I-thrombin.[18] Lane A shows the ^{125}I-thrombin used in this experiment, and lane B shows ^{125}I-thrombin-PN-1 complexes that resulted from incubating it with PN-1 from serum-free medium from the fibroblasts. For lanes 0 through 5, fibroblasts were incubated with an anti-PN-1 monoclonal antibody that blocks its ability to form complexes with its target proteases; the antibody-containing solution was then removed. This step was necessary, because active PN-1 is bound to these cells. For lanes 1 through 5, purified PN-1 was added in increasing amounts to the cells and the PN-1 containing solution was then removed. Then, ^{125}I-thrombin was added to the cells represented in lanes 0 through 5, and incubation was continued for 30 min at 37°C. The cells were then rinsed and solubilized and ^{125}I-containing proteins were visualized by autoradiography following SDS-PAGE. These results showed that PN-1 bound to the surface of the cells in a dose-dependent manner, and that the bound PN-1 formed complexes with ^{125}I-thrombin.[18]

The autoradiogram of Figure 5 shows a parallel experiment in which the cells were incubated with ^{125}I-urokinase instead of ^{125}I-thrombin.[18] As shown, the PN-1 which was bound to the cell surface did not form detectable complexes with ^{125}I-urokinase, even though it readily formed complexes with ^{125}I-urokinase in solution. Figure 5 shows that the surface of the human fibroblasts contained a component that formed complexes

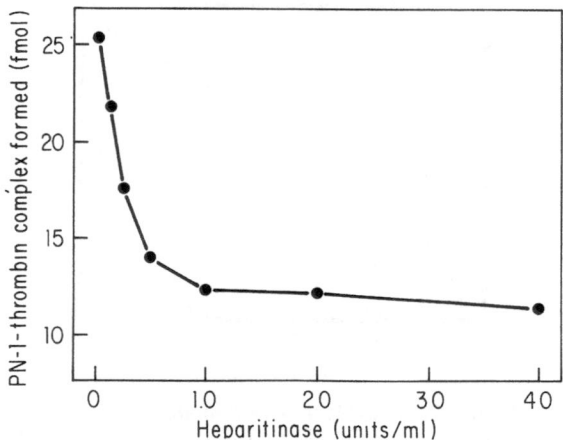

Figure 3. *Heparitinase removes ability of membranes to accelerate thrombin-PN-1 complex formation.* Plasma membranes were incubated for 30 min at 37°C with the indicated concentrations of heparinitase. Then, ^{125}I-thrombin and PN-1 were added and incubation was continued for 8 min at 37°C. The reaction was stopped by addition of electrophoresis sample buffer. Aliquots were electrophoresed and ^{125}I-thrombin-PN-1 complexes were quantitated. Reprinted with permission from reference 17.

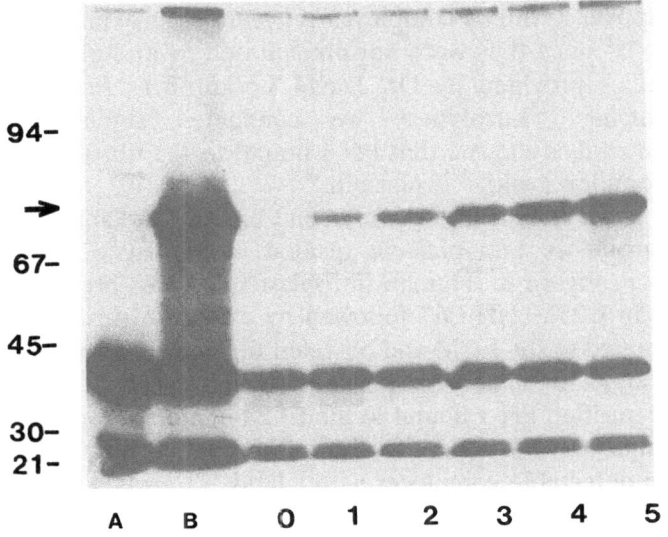

Figure 4. *PN-1 bound to the surface of human fibroblasts forms complexes with* ^{125}*I-thrombin.* For lanes 0 to 5, rinsed confluent fibroblasts were incubated for 60 min at 37°C with 100 μg/ml of an anti-PN-1 monoclonal antibody that blocks its ability to form complexes with its target proteases. After removal of the antibody solution, the cells were incubated for 60 min at 37°C with either no PN-1 (lane 0), or PN-1 at 250 ng/ml (lane 1), 500 ng/ml (lane 2), 750 ng/ml (lane 3) 1 μg/ml (lane 4) or 2 μg/ml (lane 5). After removal of the PN-1 solution, the cells were rinsed and then incubated for 30 min at 37°C with 150 ng/ml of ^{125}I-thrombin. After removing the ^{125}I-thrombin, the cells were rinsed and solubilized and cell-associated proteins were visualized by autoradiography following SDS-PAGE. Lane A, ^{125}I-thrombin; lane B, ^{125}I-thrombin plus PN-1. The arrow denotes thrombin-PN-1 complexes. Reprinted with permission from reference 18.

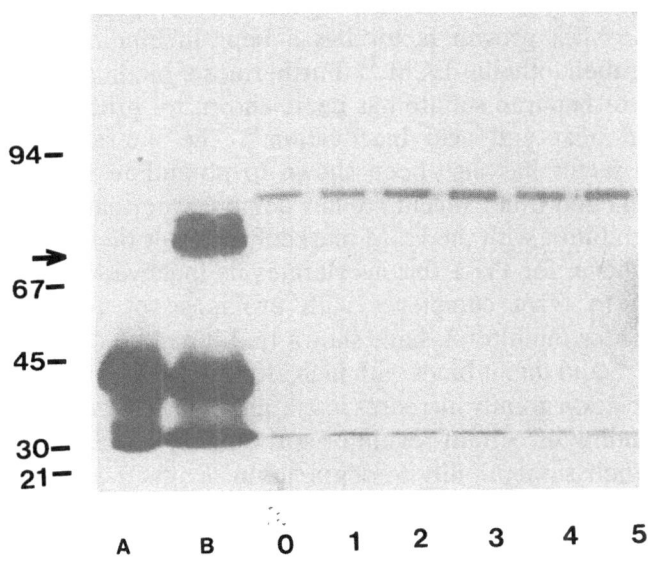

Figure 5. *PN-1 bound to the cell surface does not form complexes with* ^{125}*I-uro-kinase.* The experiment was conducted as described in the legend to Figure 4 for ^{125}I-thrombin except that after removal of the PN-1 solution the cells were incubated for 30 min at 37°C with 500 ng/ml ^{125}I-urokinase. Lane A, ^{125}I-uro-kinase; lane B, ^{125}I-urokinase plus PN-1. The arrow denotes urokinase-PN-1 complexes. Reprinted with permission from reference 18.

with ^{125}I-urokinase that were larger than ^{125}I-urokinase-PN-1 complexes. Importantly, the formation of these complexes was not inhibited by the anti-PN-1 monoclonal antibody. (They appear not to be complexes between ^{125}I-urokinase and plasminogen activator inhibitor-1[22] since they were not precipitated by anti-plasminogen activator inhibitor-1 antibodies provided by Dr. David Loskutoff.) In parallel with these experiments involving ^{125}I-urokinase we conducted similar experiments with ^{125}I-plasmin. These studies showed that PN-1 bound to the fibroblast surface did not form detectable complexes with ^{125}I-plasmin.[18]

To determine if the cellular components that blocked the ability of PN-1 to form complexes with urokinase and plasmin resided in the ECM, we conducted the experiment shown in Figure 6. Human fibroblast ECM was prepared by incubating the cultured cells in 0.25M NH_4OH followed by extensive rinsing. This treatment inactivated PN-1 bound to the ECM and obviated the need to treat with the anti-PN-1 monoclonal antibody as in the above experiments with intact cells. The data in Figure 6 show that added purified PN-1 bound to the ECM in a dose-dependent manner and that it formed complexes with added ^{125}I-thrombin (panel A). However, ECM-bound PN-1 did not form detectable complexes with added ^{125}I-urokinase (panel B) or with added ^{125}I-plasmin (panel C).[18] The ECM components that are responsible for its ability to block the inactivation of urokinase and plasmin by PN-1 are currently unknown. Heparin slightly enhanced the inactivation of ^{125}I-urokinase and ^{125}I-plasmin by PN-1 in solution.[18] Thus, it appears that ECM components other than heparan sulfate are responsible for blocking the ability of PN-1 to inhibit urokinase and plasmin.

DISCUSSION

Recent studies have shown that several regulatory polypeptides and proteins bind to the ECM and that these interactions regulate key properties of these molecules. For example, fibroblast growth factor has a heparin binding site[23] and has been identified in the subendothelial ECM.[24] Furthermore, binding of fibroblast growth factor to heparin or heparan sulfate has been shown to protect it from proteolytic degradation[25] and heat and acid inactivation.[26] The interaction of heparin with fibroblast growth factor has also been shown to potentiate its mitogenic activity.[27] The present results and other recent studies demonstrate that association of certain serine protease inhibitors with the ECM markedly changes their functional properties. As summarized above, for PN-1 this accelerates its inactivation of thrombin[16,17] and blocks its ability to form complexes with urokinase or plasmin.[18] Studies on plasminogen activator inhibitor-1, have shown that it can be detected in the ECM of endothelial cells[22,28] and that it binds with high affinity and specificity to ECM preparations.[29] This association greatly increases its stability.[22,28] Taken together, these results imply that the binding of certain regulatory proteins to the ECM can localize these molecules, alter their susceptibility to degradation or inactivation and regulate their activities.

The cell-conferred thrombin specificity of PN-1 that is summarized in this paper is consistent with the biological activities of PN-1 on cultured cells that depend on its ability to inhibit thrombin. Thrombin is mitogenic for a variety of cultured cells,[30] and early studies showed that added PN-1 can modulate this mitogenic response.[31] This result is consistent with findings that the mitogenic activity of thrombin depends on its proteolytic activity.[30] Moreover, the neurite outgrowth activity of PN-1 on cultured mouse neuroblastoma cells depends on thrombin inhibition. This conclusion is derived from studies which have shown that: (a) hirudin, a specific thrombin inhibitor from leeches, stimulates neurite outgrowth in neuroblastoma cells with the same kinetics and to the same extent as PN-1;[8,9] (b) thrombin brings about retraction of neurites under serum-free conditions;[9] (c) the stimulation of neurite outgrowth in a thrombin-containing serum-free medium by PN-1 and by ATIII plus heparin is quantitatively related to a corresponding decrease in active thrombin;[13] and (d) an anti-PN-1 monoclonal antibody that blocks the ability of PN-1 to inhibit target proteases also blocks its stimulation of neurite outgrowth.[13] Recent studies have shown that PN-1 and thrombin can also regulate the stellation of cultured early passage rat astrocytes.[14] Thrombin treatment of stellate astrocytes causes them to assume a flattened nonstellate morphology. Addition of excess PN-1 to the flattened cells leads to a stellate morphology. This effect of PN-1 results from inhibition of thrombin.

Although the studies summarized in this paper were conducted with cultured human fibroblasts, our studies have shown that a similar regulation of PN-1 activity and target protease specificity occurs with cultured neuroblastoma and glioma cells.[32] Thus, binding of PN-1 to these cells accelerates its inhibition of thrombin and blocks its ability to inhibit urokinase or plasmin.

It should be emphasized that in solution PN-1 inhibits urokinase or plasmin almost as rapidly as it inhibits thrombin.[5] Thus, PN-1 not bound to the ECM is a highly effective urokinase or plasmin inhibitor. Indeed, Bergman et al.[33] showed that human fibrosarcoma cells rapidly degrade the ECM synthesized by cultured rat aortic smooth muscle cells and that this degradation was inhibited by added PN-1. Antibodies against urokinase partially protected the ECM from degradation by the fibrosarcoma cells, indicating that urokinase may have been a target of PN-1.[33] In these experiments, it seems likely that the inhibition by PN-1 occurred in solution. This type of inhibition could also occur *in vivo* under certain conditions. For example, degradation or turnover of certain components of the ECM could release PN-1 in a form that inactivates urokinase or plasmin.

The ECM component that is primarily responsible for accelerating the inactivation of thrombin by PN-1 is heparan sulfate.[16,17] This conclusion came from studies which showed that treatment of membranes with heparitinase, a glycosidase that specifically hydrolyzes heparan sulfate, removed 80% of their ability to accelerate thrombin inhibition by PN-1. This conclusion is consistent with previous findings that PN-1 has a heparin binding site[1] and that both heparin[1,5] and heparan sulfate[17] accelerate inhibition of thrombin by PN-1. It should be emphasized, however, that the effects of heparin in solution and heparan sulfate in the ECM can have differential effects on protease inhibitors that are not well understood. For example, heparin in solution accelerates the inactivation of thrombin by both ATIII[19] and PN-1.[1,5] On the other hand, the fibroblast surface, which contains heparan sulfate, accelerates the inactivation of thrombin by PN-1 but not by ATIII.[16]

Virtually nothing is currently known about the ECM components that block the ability of PN-1 to inhibit urokinase or plasmin. Studies with heparin in solution have shown that it slightly accelerates the formation of complexes between PN-1 and urokinase or plasmin.[18] This suggests that heparan sulfate is probably not the molecule that blocks this activity of PN-1, although heparin in solution might have different effects from heparan sulfate in the ECM as noted in the preceding paragraph. The component(s) that block the ability of PN-1 to inhibit urokinase or plasmin may also be present on platelets. This suggestion is based on findings that a platelet-bound form of PN-1 inhibits thrombin but not urokinase or plasmin whereas

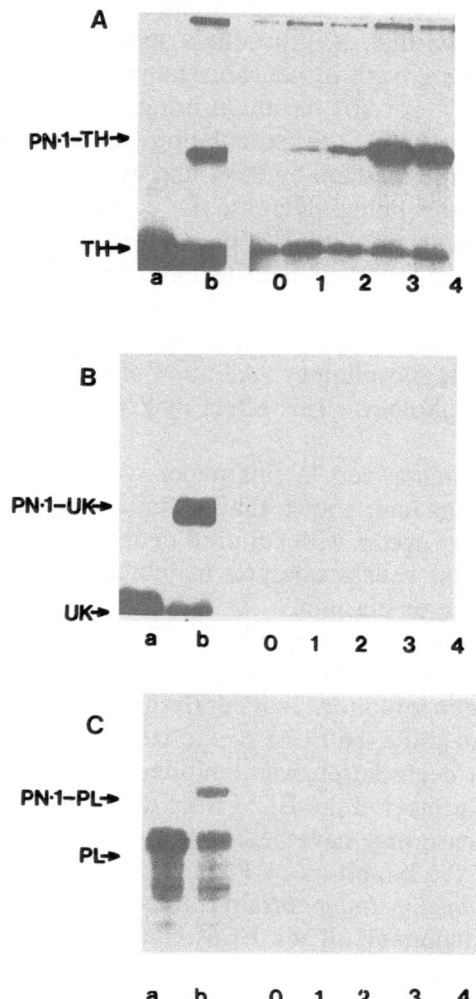

Figure 6. *PN-1 bound to ECM forms complexes with ^{125}I-thrombin but not ^{125}I-urokinase or ^{125}I-plasmin.* ECM was prepared from confluent human fibroblasts by treatment with 0.25 M NH$_4$OH. For lanes 0 to 4, the dishes containing ECM were incubated for 60 min at 37°C with either no PN-1 (lane 0) or PN-1 at at 250 ng/ml (lane 1), 500 ng/ml (lane 2), 750 ng/ml (lane 3), or 1 μg/ml (lane 4). After removal of the PN-1 solution, the dishes were rinsed and then incubated for 30 min with 150 ng/ml ^{125}I-thrombin (panel A), 500 ng/ml ^{125}I-urokinase (panel B), or 1.0 μg/ml ^{125}I-plasmin (panel C). After removal of the ^{125}I-proteases, the ECM preparations were rinsed and solubilized in electrophoresis sample buffer. The solubilized proteins were visualized by autoradiography following SDS-PAGE. Lane a, ^{125}I-protease; lane b, ^{125}I-protease plus purified PN-1. The arrows denote positions of the ^{125}I-protease and the corresponding ^{125}I-protease-PN-1 complexes. Reprinted with permission from reference 18.

an apparently similar form of PN-1 extracted from platelets inhibits thrombin, urokinase and plasmin.[34] It will be important in future studies to identify these molecules that regulate PN-1 target protease specificity and examine their interactions with PN-1 and its target proteases.

SUMMARY

Protease Nexin-1 (PN-1) is a protease inhibitor that is secreted by a variety of cultured extravascular cells. It rapidly inhibits thrombin, urokinase and plasmin in solution. Here we summarize studies which have shown that PN-1 binds to the cell surface and extracellular matrix (ECM) and that this interaction accelerates its inactivation of thrombin and blocks its inhibition of urokinase and plasmin. This cell-conferred thrombin specificity of PN-1 is consistent with studies which have shown that the ability of PN-1 to stimulate neurite outgrowth in neuroblastoma cells and its ability to promote stellation of astrocytes depends on thrombin inhibition.

ACKNOWLEDGMENTS

We thank Alice L. Lau for excellent technical assistance. This work was supported by NIH Research Grant GM 31609 and American Cancer Society Research Grant BC-602.

REFERENCES

1. J. B. Baker, D. A. Low, R. L. Simmer, and D. D. Cunningham, Protease nexin: a cellular component that links thrombin and plasminogen activator and mediates their binding to cells. **Cell** 21:37 (1980).
2. D. A. Low, J. B. Baker, W. C. Koonce, and D. D. Cunningham, Released protease nexin regulates cellular binding, internalization and degradation of serine proteases. **Proc. Natl Acad.Sci. USA** 78:2340 (1981).
3. D. E. Rosenblatt, C. W. Cotman, M. Nieto-Sampedro, J. W. Rowe, and D.J. Knauer, Identification of a protease inhibitor produced by astrocytes that is structurally and functionally homologous to human protease nexin I. **Brain Res.** 415:40 (1987).
4. W. E. Laug, R. Aebersold, A. Jong, W. Rideout, B. L. Bergman, and J. B. Baker, Isolation of multiple types of plasminogen activator inhibitors from vascular smooth muscle cells. **Thrombosis & Hemostasis**, 61:517-521 (1989).
5. R. W. Scott, B. L. Bergman, A. Bajpai, R. T. Hersh, H. Rodriguez, B. N. Jones, C. Barreda, S. Watts, and J. B. Baker, Protease nexin: properties and a modified purification procedure. **J. Biol. Chem.** 260:7029 (1985).
6. M. McGrogan, J. Kennedy, M. Li, C. Hsu, R. Scott, C. Simonsen, and J. Baker, Molecular cloning and expression of two forms of human protease nexin 1. **Bio/Technology** 6:172 (1988).
7. S. Gloor, K. Odink, J. Guenther, N. Hanspeter, and D. Monard, A glia-derived neurite promoting factor with protease inhibitory activity belongs to the protease nexins. **Cell**, 47:687 (1986).
8. D. Monard, E. Niday, A. Limat, and F. Solomonson, Inhibition of protease activity can lead to neurite extension in neuroblastoma cells. **Prog. Brain Res.** 56:359 (1983).
9. D. Gurwitz and D. D. Cunningham, Thrombin modulates and reverses neuroblastoma neurite outgrowth. **Proc. Natl. Acad. Sci. USA** 85:3440 (1988).
10. A. D. Zurn, H. Nick, and D. Monard, A glia-derived nexin promotes outgrowth in cultured chick sympathetic neurons. **Dev. Neurosci.** 10:17 (1988).
11. E. Reinhard, R. Meier, W. Halfter, G. Rovelli, and D. Monard, Detection of glia-derived nexin in the olfactory system of the rat. **Neuron** 1:387(1988).
12. S. L. Wagner, J. W. Geddes, C. W. Cotman, A. L. Lau, D. Gurwitz, P. J. Isackson, and D. D. Cunningham, Protease nexin-1, an antithrombin with neurite outgrowth activity, is reduced Alzheimer's disease. **Proc. Natl. Acad. Sci. USA**, 86:8284-8288 (1989).
13. D. Gurwitz and D. D. Cunningham, Neurite outgrowth activity of protease nexin-1 on neuroblastoma cells requires thrombin inhibition. **J. Cell. Physiol.**, 142:155-168 (1990).
14. K. Cavanaugh, D. Gurwitz, D. D. Cunningham, and R. A. Bradshaw, Reciprocal modulation of astrocyte stellation by thrombin and protease nexin-1. **J. Neurochem.**, 54:1735-1743 (1990).
15. D. H. Farrell, S. L. Wagner, R. H. Yuan, and D. D. Cunningham, Localization of protease nexin-1 on the fibroblast extracellular matrix. **J. Cell. Physiol.** 134:179 (1988).

16. D. H. Farrell and D. D. Cunningham, Human fibroblasts accelerate the inhibition of thrombin by protease nexin. **Proc. Natl. Acad. Sci. USA** 83:6858 (1986).

17. D. H. Farrell and D. D. Cunningham, Glycosaminoglycans on fibroblasts accelerate thrombin inhibition by protease nexin-1. **Biochem. J.** 245:543 (1987).

18. S. L. Wagner, A. L. Lau, and D. D. Cunningham. Binding if protease nexin-1 to the fibroblast surface alters its target proteinase specificity. **J. Biol. Chem.** 264:611 (1989).

19. R. D. Rosenberg and P. S. Damus, The purification and mechanism of action of human antithrombin-heparin cofactor. **J. Biol. Chem.** 248:6490 (1973).

20. E. A. McGuire and D. M. Tollefsen, Activation of heparin cofactor II by fibroblasts and vascular smooth muscle cells. **J. Biol. Chem.** 262:169 (1987).

21. S. A. Hiramoto and D. D. Cunningham, Effects of fibroblasts and endothelial cells on inactivation of target proteases by protease nexin-1, heparin cofactor II and C-1 inhibitor. **J. Cell Biochem.** 36:199 (1988).

22. J. Mimuro, R. R. Schleef, and D. Loskutoff, Extracellular matrix of cultured bovine aortic endothelial cells contains functionally active type 1 plasminogen activator inhibitor. **Blood** 70:721 (1987).

23. Y. Shing, J. Folkman, R. Sullivan, C. Butterfield, J. Murray, and M. Klagsbrun, Heparin affinity: purification of a tumor-derived capillary endothelial cell growth factor. **Science** 223:1296 (1984).

24. A. Baird and A. Ling, Fibroblast growth factors are present in the extracellular matrix produced by endothelial cells *in vitro*: implication for a role of heparinase-like enzymes in the neovascular response. **Biochem. Biophys. Res. Commun.** 142:428 (1987).

25. O. Saksela, D. Moscatelli, A. Sommer, and D. B. Rifkin, Endothelial cell-derived heparan sulfate binds basic fibroblast growth factor and protects it from proteolytic degradation. **J. Cell. Biol.** 107:743 (1988).

26. D. Gospodarowicz and J. Cheng, Heparin protects basic and acidic FGF from inactivation. **J. Cell. Physiol.** 128:475 (1986).

27. A. B. Schreiber, J. Kenney, W. J. Kowalsky, R. Friesel, T. Mehlman, and T. Maciag, Interaction of endothelial cell growth factor with heparin. Characterization by receptor and antibody recognition. **Proc. Natl. Acad. Sci. USA** 82:6138 (1985).

28. E. G. Levin and L. Santell, Association of a plasminogen activator inhibitor (PAI-1) with the growth substratum and membrane of human endothelial cells. **J. Cell Biol.** 105:2543 (1987).

29. J. Mimuro and D. Loskutoff, Binding of type 1 plasminogen activator inhibitor to the extracellular matrix of cultured bovine endothelial cells. **J. Biol. Chem.** 264:5058 (1989).

30. K. C. Glenn, D. H. Carney, J. W. Fenton II, and D. D. Cunningham, Thrombin active site regions required for fibroblast receptor binding site and initiation by cell division. **J. Biol. Chem.** 255:6609 (1980).

31. D. A. Low, R. W. Scott, J. B. Baker, and D. D. Cunningham, Cells regulate their mitogenic response to thrombin through release of protease nexin. **Nature** 298:476 (1982).

32. S. L. Wagner, A. L. Lau, and D. D. Cunningham, Regulation of protease nexin-1 activity by glioma and neuroblastoma cells. Submitted for publication.

33. B. L. Bergman, R. W. Scott, A. Bajpai, S. Watts, and J. B. Baker, Inhibition of tumor cell-mediated extracellular matrix destruction by a fibroblast proteinase inhibitor protease nexin-1. **Proc. Natl. Acad. Sci. USA** 83:996 (1986).

34. R. S. Gronke, B. L. Bergman, and J. B. Baker, Thrombin interaction with platelets. **J. Biol. Chem.** 262:3030 (1987).

SECTION II

Molecular biology of serine proteases and serpins

INDUCTION OF THE UROKINASE-TYPE PLASMINOGEN ACTIVATOR GENE BY CYTOSKELETON-DISRUPTING AGENTS

FLORENCE M. BOTTERI, HERMAN VAN DER PUTTEN[1], BHANU RAJPUT, KURT BALLMER-HOFER AND YOSHIKUNI NAGAMINE

Friedrich Miescher-Institut
Postfach 2543
CH-4002 Basel, Switzerland;
[1]*Biotechnology, Ciba-Geigy AG*
Postfach 2543
CH-4002 Basel, Switzerland

INTRODUCTION

The interaction of a cell with specific components of the extracellular matrix can result in alterations of cell-shape and morphology.[1,2] To a large extent such structural changes are the consequence of modifications of the cytoskeletal network. Dynamic cytoskeletal changes take place in migrating cells as well as in transformed cells.[3] Most likely, migration of normal or metastatic tumor cells requires the expression of specific endogenous genes whose products assist in reshaping the intracellular cytoskeleton and the extracellular matrix. How and whether changes in cell morphology and cytoskeletal components may cause alterations in the expression of certain genes has not yet been investigated extensively.

Plasminogen activators (PA) are implicated in various aspects of cellular function involving localized extracellular proteolysis. For instance, high levels of urokinase-type PA (uPA) expression were observed in migrating cells during early development[4,5] and in transformed cells.[6] We therefore initiated a series of experiments aimed at delineating a possible causal relationship between alterations in specific components of the cytoskeleton and expression of the uPA gene. We conclude that disruption of cytoskeletal structures by various drugs correlates with an induction in uPA mRNA and with increased rates of uPA gene transcription. The role of cAMP-dependent protein kinase, of protein kinase C, and of cis-acting DNA elements in the uPA gene, as mediators in such signal transduction pathway was also investigated. Finally, to initiate experimental designs that permit the dissection of the role of uPA *in vivo* we have generated transgenic mice.

Serine Proteases and Their Serpin Inhibitors in the Nervous System
Edited by B. W. Festoff
Plenum Press, New York, 1990

MATERIALS AND METHODS

Materials

Colchicine was obtained from Fluka. Cytochalasin B, nocodazole, cycloheximide, and phorbol myristate acetate were from Sigma. 8-bromo-cAMP was from Boehringer Mannheim. [α-^{32}P]dCTP and [α-^{32}P]GTP were obtained from Amersham and New England Nuclear, respectively. Synthetic salmon calcitonin was a gift of Dr. S. Guttman (Sandoz AG, Basel, Switzerland). Kemptide (Leu-Arg-Arg-Ala-Ser-Ala-Gly), a synthetic substrate for protein kinase A, was purchased from Bachem.

Cell Culture

Porcine kidney derived epithelial cells, LLC-PK1, were cultured as described.[7] In most experiments about 5×10^5 cells were plated on 35 mm plastic dishes. Two days later, the cells were treated with reagents as indicated in the figure legends.

cDNA Probes

A two kilobases XbaI-XbaI DNA fragment of the pig uPA cDNA, pYN 15,[8] was isolated and labeled with [α-^{32}P]dCTP using a standard random-oligo primed reaction.[9]

RNA Analysis

Total RNA was isolated according to Chomczynski and Sacchi.[10] 10 μg of total RNA were resolved by electrophoresis under denaturing conditions and transferred to nitrocellulose filters as described.[7] Prehybridization and hybridization were done as described,[11] and filters were exposed to Kodak XAR films with intensifying screens at -70 °C.

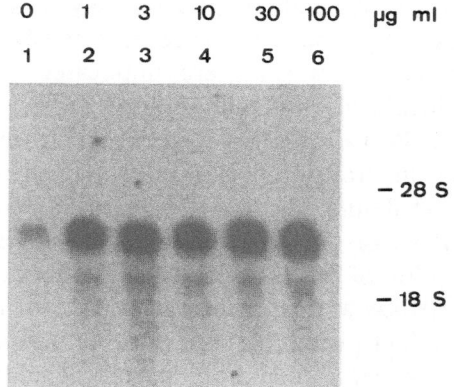

Figure 1 . *Effects of various concentrations of colchicine on uPA mRNA levels (dose response).* Each lane represents 10 μg of total RNA hybridised to the uPA cDNA probe. Lane 1 shows uPA mRNA isolated from untreated LLC-PK$_1$ cells. The rest of the lanes represent RNA isolated from LLC-PK$_1$ cells exposed for 4 hours to varying amounts of colchicine: 1 μg/ml (lane 2), 3 μg/ml (lane 3), 10 μg/ml (lane 4), 30 μg/ml (lane 5), or 100 μg/ml (lane 6). The 18S and 28S ribosomal RNAs are indicated for reference.

Nuclear Run-on Transcription

The isolation of nuclei, nuclear run-on transcription, and quantitation of specific transcripts by hybridization were done as described.[11]

cAMP-Dependent Protein Kinase Assay

LLC-PK$_1$ cells (5×10^5) were plated on 35-mm plastic dishes with 2 ml of Dulbecco's modified Eagle's medium (DMEM) containing 10% fetal calf serum. After 48 hours, cells were exposed to the appropriate drug for 1 hour. Then, cell extracts were prepared and protein kinase activities were measured using the substrate kemptide in absence or presence of 10 μM cAMP.[12] The activity ratio [(-cAMP)/(+cAMP)] is a measure of the amount of *in vivo*-activated cAMP-dependent protein kinase.

DNA Transfection and CAT Assay

Plasmids containing various deletions of the pig uPA promoter linked to a bacterial gene encoding chloramphenicol acetyltransferase (CAT) were cotransfected with pSV2neo into LLC-PK1 cells by the calcium phosphate-DNA precipitation method.[13] Several hundred individual clones growing in the medium containing G418 were pooled for each chimeric gene. Cells were plated at 5×10^5/50 mm plastic dish with 5 ml DMEM containing 10% fetal calf serum. Two days later cells were treated with 5×10^{-7} M colchicine or 10^{-5} M cytochalasin B. After 24 h cell extracts were prepared and chloramphenicol transferase activities were measured as described.[14]

Transgenic Mice

Transgenic mice were produced and analyzed according to Hogan et al.[15] ß-galactosidase activity in the whole embryo was revealed according to Sanes et al.[16]

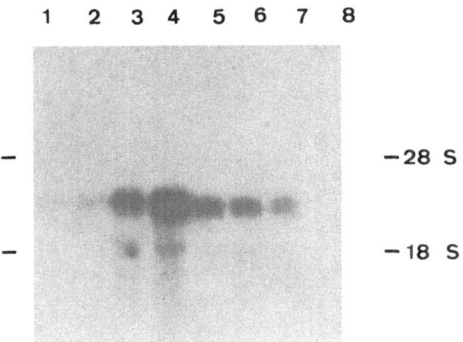

Figure 2. *Time course of colchicine treatment on uPA mRNA levels.* The lanes show uPA mRNA from untreated LLC-PK$_1$ cells (lane 8) or cells exposed to 30 μg/ml of colchicine for variable lenghts of time: 30 minutes (lane 1), 1 hour (lane 2), 2 hours (lane 3), 4 hours (lane 4), 8 hours (lane 5), 16 hours (lane 6), 24 hours (lane 7). The 18S and 28S rRNA are shown.

RESULTS

Colchicine Induces uPA mRNA

To assess a possible relationship between the organization of the cytoskeleton and the expression of the uPA gene, we treated the LLC-PK$_1$ cells with the plant alkaloid colchicine, a drug known to bind to alpha-beta tubulin dimers thereby disrupting microtubules.[17,18] An increase in uPA mRNA level was observed using colchicine concentrations as low as 1 μg/ml (Figure 1). Higher concentrations of colchicine (3 to 100 μg) did not result in further increase in the uPA mRNA level. Induction of uPA mRNA was also obtained with another alkaloid, nocodazole which disrupts microtubules in a reversible manner (data not shown), suggesting that the induction of uPA mRNA by colchicine is not an effect of colchicine *per se*. The data suggest that the disruption of the microtubule network might correlate with an increase in uPA mRNA.

The kinetics of induction of uPA mRNA after colchicine treatment showed that uPA mRNA accumulated as early as 1 hour after exposure to the drug. Maximal levels were reached after approximately 4 hours and remained at this level for at least 14 hours (Figure 2).

Colchicine Induces uPA Gene Transcription

The basal level of uPA mRNA in LLC-PK$_1$ cells is very low and reflects the very low transcriptional activity of the uninduced uPA gene.[7,18,19] Therefore, the increased uPA mRNA level observed in colchicine treated cells is most likely due to an increase in the transcription rate of the uPA gene. As expected, results shown in Figure 3 indicated that uPA gene transcription was increased 11-fold after cells were exposed to colchicine, while transcription of the ß-actin gene was increased only 3-fold. A 10-fold increase in the steady state level of uPA mRNA (from Northern analysis)

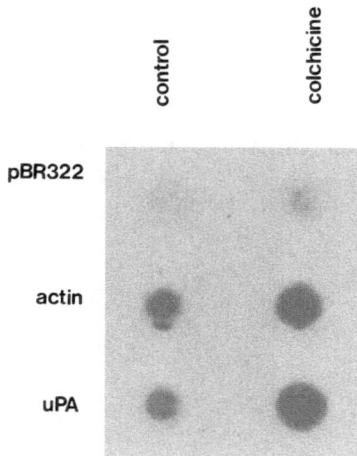

Figure 3 . *Nuclear transcription.* Nuclei were prepared from untreated LLC-PK$_1$ cells (control) or LLC-PK$_1$ cells treated with 100 μg/ml colchicine for 1 hour (colchicine). Transcription in the isolated nuclei was analysed by hybridisation of ^{32}P-labelled transcripts (1x10^7 cpm) to 1 μg each of pBR322, ß-actin, and uPA plasmid DNA immobilised on individual nitrocellulose filters. Hybridisation of each transcript to the three different filters was done in the same tube.

is therefore due to a similar fold increase in transcriptional activity of the uPA gene. Hence, the overall increase in uPA mRNA level after colchicine treatment can be ascribed mainly if not exclusively to increased transcription of the uPA gene.

Cytochalasin B also Induces uPA Gene Expression

A second major component of cytoskeletal structures are the microfilaments. In contrast to the action of colchicine, cytochalasin B binds to the ends of growing microfilaments thereby inhibiting actin polymerization,[20] and actin filament-filament interactions.[21,22] We investigated the effect of cytochalasin B on the level of uPA mRNA. Like colchicine, cytochalasin B treatment induced the uPA mRNA within 1 hour and reached a maximum at 4 hours. This level was maintained for at least 8 hours (Figure 4). The rate of the uPA gene transcription was also increased by cytochalasin B treatment (data not shown). Therefore, alterations of the cytoskeleton by the disruption of either microtubules or microfilaments caused an increase in the uPA gene transcription.

Induction of uPA mRNA by Colchicine or Cytochalasin B Treatment is not Mediated through cAMP-dependent Protein Kinase but Involves Protein Kinase C

In LLC-PK$_1$ cells the uPA gene can be activated by cAMP and phorbol esters, which act through cAMP-dependent protein kinase (cAMP-PK) and protein kinase C, respectively. It has been shown that the integrity of the cytoskeleton plays an important role in determining signal transduction via membrane-bound adenylate cyclase,[23-26] and that subunits of G-proteins associate with cytoskeletal components.[27] Also, there is evidence that phorbol ester treatment causes the disorganization of cytoskeletal structures.[28-31] Therefore, the induction of uPA mRNA after disruption of the cytoskeleton could be regulated by one or both of these kinase pathways.

The results summarized in Table 1 indicate that neither colchicine nor cytochalasin B treatment activates cAMP-PK.

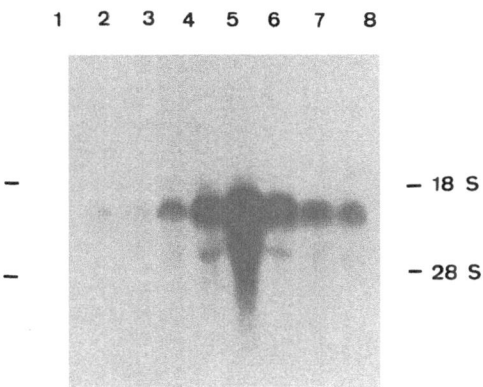

Figure 4 . *Time course of cytochalasin B treatment on uPA mRNA levels.* The lanes show uPA mRNA from untreated LLC-PK$_1$ cells (lane 1) or cells exposed to 30 μg/m. of cytochalasin B for variable lengths of time: 30 minutes (lane 2), 1 hour (lane 3), 2 hours (lane 4), 4 hours (lane 5), 8 hours (lane 6), 16 hours (lane 7), 24 hours (lane 8). The 18S and 28S rRNA are shown.

Table 1 . *cAMP-dependent protein kinase activity under different conditions*

| Treatment | Activity[a] | | Activity ratio |
	-cAMP	+cAMP	[(-cAMP)/(+cAMP)]
control	0.46	19.1	0.024
colchicine	1.24	25.9	0.047
cytochalasin B	1.06	23.5	0.045
8-bromo-cAMP	27.2	28.0	0.97

[a]nmoles phosphate incorporation per mg protein per 10 min

Prolonged exposure (24 hours) of LLC-PK$_1$ cells to phorbol myristate acetate (PMA) causes desensitization of the cells to PMA.[19] This leads to down-regulation of protein kinase C. As shown in Figure 5, the uPA mRNA reached maximal levels after 3 hours of PMA treatment (lane 2) and subsequently decreased to basal level after 24 hours (lane 3).[19] When cells, at this point, were restimulated with PMA for an additional 3 hours, the level of uPA mRNA remained low, but increased slightly above the basal level (lane 4). In contrast, 3 hours of colchicine (lanes 6) or cytochalasin B (lanes 9) treatment after exposure of the cells to PMA for 24 hours, resulted in an appreciable induction of the uPA mRNA. A similar induction of the uPA mRNA level was detected after 3 hours of colchicine (lane 7) or cytochalasin B (lane 10) treatment after exposure of the cells to PMA for 24 hours followed by an additional 3 hours of PMA treatment. These levels, however, were lower when compared to those in cells treated with colchicine (lane 5) or cytochalasin B (lane 8) alone. The prolonged exposure of LLC-PK$_1$ cells to PMA did affect the ability of both drugs to induce uPA mRNA. These results suggest that the induction of uPA mRNA in LLC-PK$_1$ cells as a consequence of exposure to colchicine or cytochalasin B may be regulated by a pathway involving protein kinase C (Figure 5).

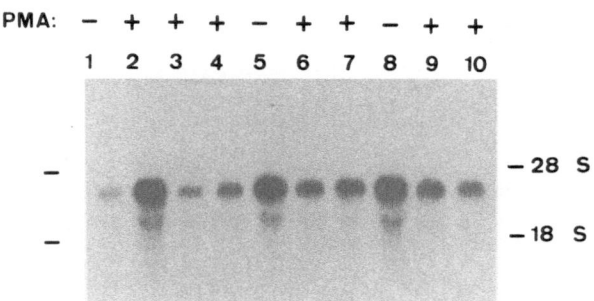

Figure 5 . *Effect of PMA treatment on uPA mRNA levels.* uPA mRNA level from untreated LLC-PK$_1$ cells (lane 1) or treated with PMA for 3 hours (lane 2), or 24 hours (lane 3), or 24 hours followed by an additional 3 hours treatment (lane 4). uPA mRNA levels of LLC-PK$_1$ cells exposed to colchicine for 4 hours (lane 5), or to colchicine for 4 hours after exposure to PMA for 24 hours (lane 6), or to colchicine for 4 hours after exposure to PMA for 24 hours followed by an additional 3 hours treatment (lane 7). uPA mRNA levels of LLC-PK$_1$ cells treated with cytochalasin B for 4 hours (lane 8), or with cytochalasin B for 4 hours after exposure to PMA for 24 hours (lane 9), or with cytochalasin B for 4 hours after exposure to PMA for 24 hours followed by an additional 3 hours treatment (lane 10).

Induction of uPA Gene Expression is Mediated by Cis-acting Element

To test whether the induction of uPA mRNA as a result of breakdown of cytoskeletal components is mediated by a specific cis-acting element, LLC-PK$_1$ cells were stably transfected with various uPA-CAT hybrid genes containing different lengths of the 5' flanking region of the pig uPA gene linked to the reporter gene, chloramphenicol acetyltransferase (CAT). CAT assays were performed on cell extracts prepared from transfected cells treated with colchicine or cytochalasin B. Both drugs induced CAT activity when 4.6 kb of the pig uPA promoter were present in the hybrid gene, whereas neither of them showed inductive effects on DNA templates containing only 148 or 53 bp of the 5' flanking region (Figure 6). Transcription from all three DNA templates started from the correct initiation site (data not shown). The results suggest that there is a DNA sequence localized between 4660 bp and 148 bp 5' of the transcription initiation site that is required for the regulation governed by alterations in the cytoskeleton.

Expression of uPA-LacZ Chimeric Gene in Transgenic Mice

In vivo cytoskeletal changes are expected to take place in many populations of cells during embryogenesis such as migrating cells (germ cells, neural crest cells) and cells undergoing extensive elongation (neurons). Therefore, we have attempted to develop a system in which the expression of the uPA gene can be studied *in vivo* in all cell types at all stages of embryogenesis. To this end, we generated transgenic mice which harbor a uPA-LacZ chimeric gene. The chimeric gene consisted of the uPA promoter containing 4.66 kb of 5' flanking sequences linked to a gene encoding a modified bacterial ß-galactosidase. The amino terminus of this ß-galactosidase includes a strech of amino acids derived from the nuclear targeting signal of SV40 large

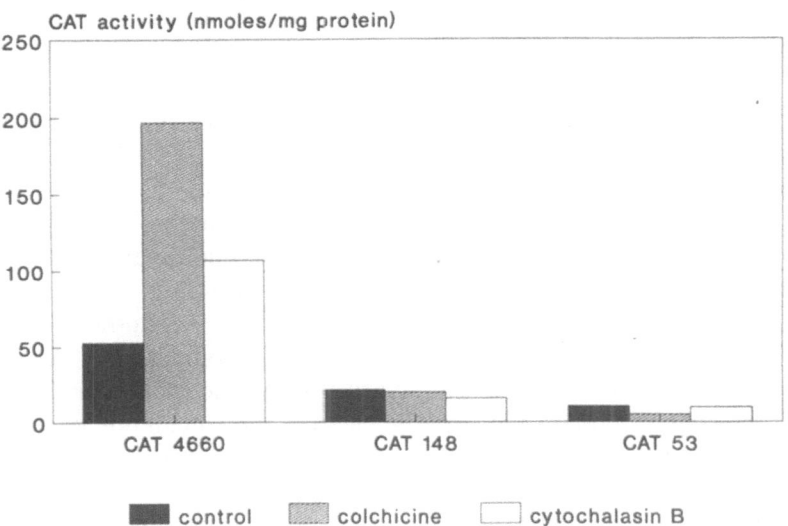

Figure 6 . *Effects of colchicine and cytochalasin B on transfected uPA-CAT hybrid genes.* Schematic representation of CAT activities of LLC-PK$_1$ cells stably transfected with various uPA-CAT hybrid genes containing 4660, 148 or 53 bp of the pig uPA promoter linked to the reporter gene CAT. CAT assays were performed on cells extracts prepared from transfected cells treated with colchicine or cytochalasin B or untreated (control).

T-antigen.[30] Therefore, transgenic ß-galactosidase activity will be localized to the nucleus. This allows its expression to be assessed at the single cell level. As shown in Figure 7 the expression of the pig uPA-LacZ transgene was detected mainly in neurons of the dorsal root and trigeminal ganglia of the peripheral nervous system in mouse embryos at day 12.5 of development. Histochemical analysis confirmed that ß-galactosidase expression was confined to the ganglia and not seen in adjacent somites nor neural cord (data not shown). No ß-galactosidase activity was detected in non-transgenic embryos (data not shown).

DISCUSSION

In this report we have shown that drugs like colchicine or cytochalasin B which disrupt the cytoskeleton induce uPA mRNA in LLC-PK$_1$ cells. We also show that this induction of uPA mRNA was in part regulated by a signal-transduction pathway involving protein kinase C. This conclusion is based on the observation that prolonged exposure of LLC-PK$_1$ cells to PMA partially abolished the ability of colchicine and cytochalasin B to induce uPA mRNA. It seems likely that a signal-transduction pathway involving protein kinase C and pathways involving cytoskeleton-mediated gene regulation share a common transcriptional factor(s). Alternatively, prolonged activation of protein kinase C could result in the inactivation or suppression of a transcription factor which is involved in the regulation imposed by changes in the cytoskeleton. At the moment we cannot distinguish between these two possibilities. In contrast, it seems unlikely that the induction of uPA mRNA by colchicine and cytochalasin B is mediated through cAMP-dependent protein kinase.

It is conceivable that the perturbation of the cytoskeletal network alters the configuration of signal transduction systems across the plasma membrane. For example, this could result in an agonist-independent activation of certain signal transduction pathways. From our results it is, however, not possible to say whether such a mechanism could be operational in the signal transduction between the cytoskeletal structures, protein kinase C, and the uPA gene. An alternative explanation focusing on the organization of the nuclear matrix could be postulated as well. It has been suggested that transcriptional activity of certain genes is dependent on the association of chromatin with the nuclear matrix.[32-34] Proteins of the high-mobility-group (HMG 14 and HMG 17)[35] as well as topoisomerase I[36] may be involved in such contacts and play a role in gene expression as well. Perturbation of cytoskeletal structures possibly alters the integrity of the nuclear matrix, which in turn changes its interaction with chromatin resulting in changes in gene activity.

The induction of the uPA gene as the result of the breakdown of cytoskeletal structures seems to be mediated by specific cis-acting element(s) in the 5' flanking sequences of the uPA gene. Further characterization of the cis-acting element(s) and interacting trans-acting factor(s) will hopefully lead to the elucidation of the molecular mechanism underlying the regulation of uPA gene expression by alterations of the cytoskeleton. In this context the expression of the uPA-LacZ transgene in the developing peripheral nervous system in the mouse is very interesting. When cis-acting element(s) mediating regulation via the cytoskeleton will be characterized in tissue culture, then its physiological significance can be examined by generating several strains of transgenic mice that harbor the same chimeric uPA-LacZ gene lacking the cytoskeleton-dependent cis-acting element(s).

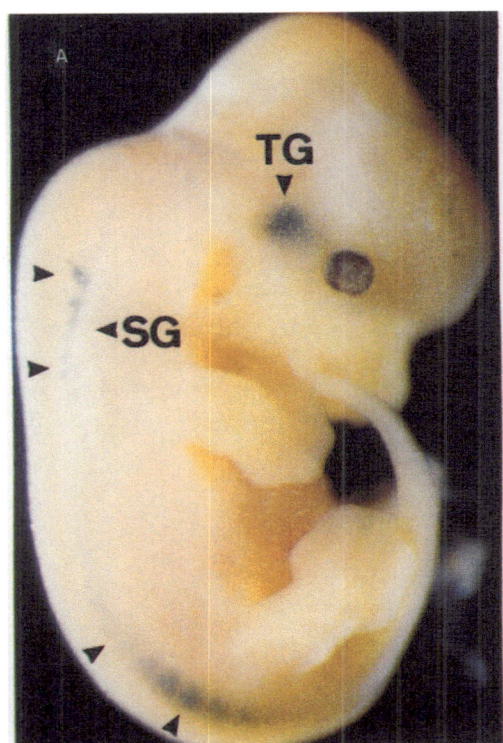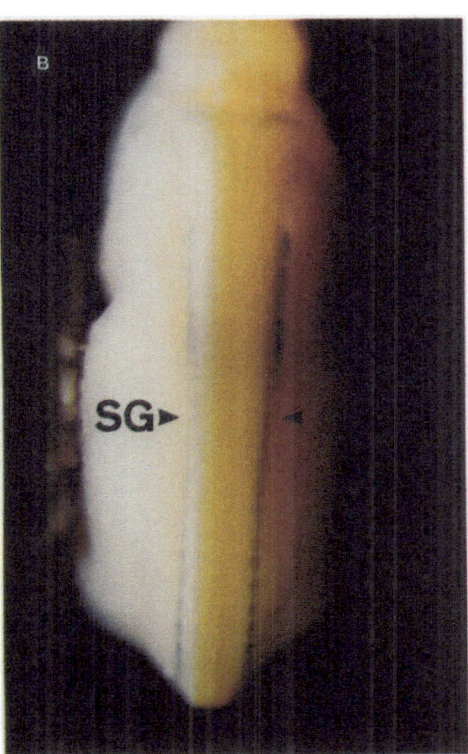

Figure 7 : *Expression of uPA-LacZ gene in transgenic mice.* Transgenic uPA-LacZ embryo at day 12.5 of development was fixed then assayed *in situ* for ß galactosidase activity as described.[16] The lateral view (A) reveals transgene expression in the dorsal root (DRG) and trigeminal ganglia (TG). The dorsal view (B) reveals two columns of blue cells in the dorsal root ganglia (DRG).

It has long been a puzzling question why transformed cells produce higher amounts of plasminogen activator compared to their non-transformed counterparts.[6] As transformed cells in general have a less organized cytoskeletal structure,[3] the results presented above might offer a clue to this question. Raz and Ben-Ze'ev[37] have shown previously that the metastatic potential of a malignant cell (B16 melanoma) can be modulated by altering its cell shape. These same B16 melanoma cells produce high levels of uPA, and when treated with specific antibodies against uPA their metastatic potential significantly decreases as compared to the metastatic capabilities of untreated B16 melanoma cells.[38] These results suggest a certain correlation between the modulation of the cytoskeleton, the development of the metastatic capacity and the expression of the uPA gene.

Finally, it seems conceivable that other genes responsible for the transformed phenotype, such as those involved in anchorage-independent growth and loss of contact inhibition, might be activated in a similar way. It remains to be seen whether an altered cytoskeleton of neoplastic cells is in part the cause or merely a consequence of the onset of the transformation process.

REFERENCES

1. F. Ungar, B. Geiger, and A. Ben-Ze'ev, Cell contact- and shape-dependent regulation of vinculin synthesis in cultured fibroblasts. **Nature** 319: 787 (1986).
2. A. Ben-Ze'ev, G.S. Robinson, N.L.R. Bucher, and S.R. Farmer, Cell-cell and cell-matrix interactions differentially regulate the expression of hepatic and cytoskeletal genes in primary cultures of rat hepatocytes. **Proc. Natl. Acad. Sci. USA** 85: 2161 (1988).
3. A. Ben-Ze'ev, The cytoskeleton in cancer cells. **Biochem. Biophys. Acta** 780: 197 (1985).
4. J.E. Valinsky, E. Reich, and N.M. Le Douarin, Plasminogen activator in the bursa of Fabricius: correlations with morphogenetic remodeling and cell migrations. **Cell** 25: 471 (1981).
5. J.E. Valinsky, and N.M. Le Douarin, Production of plasminogen activator by migrating cephalic neural crest cells. **EMBO J.** 4: 1403 (1985).
6. K. Danø, P.A. Andreasen, J. Grøndahl-Hansen, P. Kristensen, L. S. Nelson, and L. Skriver, Plasminogen activators, tissue degradation and cancer. **Adv. Cancer Res.** 44: 139 (1985).
7. Y. Nagamine, M. Sudol, and E. Reich, Hormonal regulation of plasminogen activator mRNA production in porcine kidney cells. **Cell** 32: 1181 (1983).
8. Y. Nagamine, D. Pearson, M.S. Altus, and E. Reich, cDNA and gene nucleotide sequence of porcine plasminogen activator. **Nucleic Acids Res.** 12: 9525 (1984).
9. A.P. Feinberg, and B. Vogelstein, A technique for radiolabeling DNA restriction endonuclease fragments to high specific activity. **Anal. Biochem.** 132: 6 (1983).
10. P. Chomczynski, and N. Sacchi, Single-step method of RNA isolation by acid guanidinium thyocyanate-phenol-chloroform extraction. **Anal. Biochem.** 162: 156 (1987).
11. L. Andrus, M.S. Altus, D. Pearson, M. Grattan, and Y. Nagamine, hsp70 mRNA accumulates in LLC-PK$_1$ pig kidney cells treated with calcitonin but not with 8-bromo-cyclic AMP. **J. Biol. Chem.** 263: 6183 (1988).
12. B.A. Hemmings, cAMP mediated proteolysis of the catalitic subunit of cAMP-dependent protein kinase. **FEBS Lett.** 196: 126 (1986).
13. F.L. Graham, and A.J. van der Eb, A new technique for the assay of infectivity of human adenovirus 5 DNA. **Virology** 52: 456 (1973).
14. C.M. Gorman, L.F. Moffat, and B.H. Howard, Recombinant genomes which express choramphenicol acetyltransferase in mammalian cells. **Mol. Cell. Biol.** 2: 1044 (1982).
15. B. Hogan, F. Costantini, and E. Lacy, in: "Manipulating the Mouse Embryo: A laboratory manual" Cold Spring Harbor, New York: Cold Spring Habor Laboratory (1986).
16. J.R. Sanes, J.L.R. Rubenstein, and J.-L. Nicolas, Use of recombinant retrovirus to study post-implantation cell lineage in mouse embryos. **EMBO J.** 5: 3133 (1986).
17. E.W. Taylor, The mechanism of colchicine inhibition of mitosis. **J. Cell. Biol.** 25: 145 (1965).
18. G.G. Borisy, and E.W. Taylor, The mechanism of action of colchicine. **J. Cell. Biol.** 34: 525 (1967).

19. J.L. Degen, R.D. Estensen, Y. Nagamine, and E. Reich, Induction and desensitization of plasminogen activator gene expression by tumor promoters. **J. Biol. Chem.** 260: 12426 (1985).

20. M.D. Flanagan, and S. Lin, Cytochalasins block actin filament elongation by binding to high affinity sites associated with F-actin. **J. Biol. Chem.** 255: 835 (1980).

21. J.H. Hartwig, and T.P. Stossel, Cytochalasin B and the structure of actin gels. **J. Mol. Biol.** 134: 539 (1979).

22. S. MacLean-Fletcher, and T.D. Pollard, Mechanism of action of cytochalasin B on actin. **Cell** 20: 329 (1980).

23. P.A. Insel, and M.S. Kennedy, Colchicine potentiates beta-adrenoreceptor-stimulated cyclic AMP in lymphoma cells by an action distal to the receptor. **Nature** 273: 471 (1978).

24. B.D. Cherksey, J.A. Zadunaisky, and R.B. Murphy, Cytoskeletal constraint of the beta-adrenergic receptor in frog erythrocyte membranes. **Proc. Natl. Acad. Sci. USA** 77: 6401 (1980).

25. N.E. Sahyoun, H. LeVine III, J. Davis, G.M. Hebdon, and P. Cuatrecasas, Molecular complexes involved in the regulation of adenylate cyclase. **Proc. Natl. Acad. Sci. USA** 78: 6158 (1981).

26. M.M. Rasenick, P.J. Stein, and M.W. Bitensky, The regulatory subunit of adenylate cyclase interacts with cytoskeletal components. **Nature** 294: 560 (1981).

27. K.E. Carlson, M.J. Woolkalis, M.G. Newhouse, and D.R. Manning, Fractionation of the beta subunit common to guanine nucleotide-binding regulatory proteins with the cytoskeleton. **Molec. Pharmacol.** 30: 463 (1986).

28. M. Schliwa, T. Nakamura, K.R. Porter, and U. Euteneuer, A tumor promoter induces rapid and coordinated reorganisation of actin and vinculin in cultured cells. **J. Cell Biol.** 99: 1045 (1984).

29. B. Herman, M.A. Harrington, N.E. Olashaw, and W.J. Pledger, Identification of the cellular mechanisms responsible for platelet-derived growth factor induced alterations in cytoplasmic vinculin distribution. **J. Cell. Physiology** 126: 115 (1986).

30. D. Kalderon, B.L. Roberts, W.D. Richardson, and A.E. Smith, A short amino acid sequence able to specify nuclear location. **Cell** 39: 499 (1984).

31. M. Järvinen, J. Ylänne, T. Vartio, and I. Virtanen, Tumor promoter and fibronectin induce actin stress fibers and focal adhesion sites in spreading human erythroleukemia (HEL) cells. **Europ. J. Cell Biol.** 44: 238 (1987).

32. N.M. Mironov, V.V. Lobanenkov, and G.H. Goodwin, The distribution of nuclear proteins and transcriptionally-active sequences in rat liver chromatin fractions. **Exp. Cell Res.** 167: 391 (1986).

33. R. Abulafia, A. Ben-Ze'ev, N. Hay, and Y. Aloni, Control of late simian virus 40 transcription by the attenuation mechanism and transcriptionally active ternary complexes are associated with the nuclear matrix. **J. Mol. Biol.** 172: 467 (1984).

34. S.M. Rose, and W.T. Garrard, Differentiation-dependent chromatin alterations precede and accompany transcription of immunoglobulin light chain genes. **J. Biol. Chem.** 259: 8534 (1984).

35. R. Reeves, and D. Chang, Investigations of the possible functions for glycosylation in the high mobility group proteins. **J. Biol. Chem.** 258: 679 (1983).

36. G. Fleischmann, G. Pflugfelder, E.K. Steiner, K. Javaherian, G.C. Howard, J.C. Wand, and S.C.R. Elgin, *Drosophila* DNA topoisomerase I is associated with transcriptionally active regions of the genome. **Proc. Natl. Acad. Sci. USA** 81: 6958 (1984).

37. A. Raz, and A. Ben-Ze'ev, Modulation of the metastatic capability in B16 melanoma by cell shape. **Science** 221: 1307 (1983).

38. V.J. Hearing, L.W. Law, A. Corti, E. Appella, and F. Blasi, Modulation of metastatic potential by cell surface urokinase of murine melanoma cells. **Cancer Res.** 48: 1270 (1988).

USE OF PROTEIN CHEMISTRY AND MOLECULAR BIOLOGY TO DETERMINE INTERACTION AREAS BETWEEN PROTEASES AND THEIR INHIBITORS: THE THROMBIN-HIRUDIN INTERACTION AS AN EXAMPLE

STUART R. STONE, STANLEY DENNIS, ANDREW WALLACE
AND JAN HOFSTEENGE

Friedrich Miescher-Institut
P. 0. Box 2543
CH-4002 Basel, Switzerland

INTRODUCTION

Thrombin is a serine protease that exhibits the same primary specificity as trypsin, i.e., it cleaves peptide bonds on the C-terminal side of basic amino acids (preferably arginine). Thrombin, however, exhibits a narrower specificity than trypsin with respect to the peptide bonds that it will cleave. This narrower specificity of thrombin is partly due to the presence on thrombin of secondary binding sites and this article will concentrate on the contribution of these sites to the formation of the complex between thrombin and the inhibitor hirudin. After a brief introduction on thrombin, hirudin and the kinetics of the formation of their complex, studies aimed at identifying secondary binding sites for hirudin on thrombin will be considered. These studies have used mainly the techniques of protein chemistry. The last part of the article will consider the use of site-directed mutagenesis to elucidate areas of hirudin important for its inhibitory activity.

PROPERTIES OF THROMBIN AND HIRUDIN

Thrombin

Thrombin is formed in blood from its zymogen prothrombin through the action of factor Xa in the presence of factor Va, calcium ions and phospholipid. Cleavage of prothrombin by factor Xa results in the release of two activation fragments and the formation of the A- and B-chains of thrombin. The B-chain of thrombin is homologous to other serine proteases and contains the catalytic triad (histidine, aspartate and serine) that is characteristic of these enzymes. For a review of the activation process, the reader is referred to the review by Mann.[1] Once formed, thrombin is able to cleave fibrinogen to yield fibrin monomers which subsequently

Serine Proteases and Their Serpin Inhibitors in the Nervous System
Edited by B. W. Festoff
Plenum Press, New York, 1990

polymerize to form the basis of a blood clot. Thrombin also interacts with numerous other components of the blood coagulation system.[2] Besides its activity within the blood coagulation system, thrombin also has effects on a number of cell types. The effect of thrombin on different cell types varies; for example, it elicits a chemotactic response from monocytes[3,4] and is mitogenic with fibroblasts.[5] The properties of thrombin have been extensively reviewed by Fenton.[2,6,7]

Hirudin

Hirudin is a polypeptide inhibitor of thrombin that can be isolated from the salivary gland of the medicinal leech *Hirudo medicinalis*. The anticoagulant activity in the saliva of leech was first described by Haycraft in 1884.[8] A polypeptide with antithrombin activity was first isolated by Markwardt in 1957[9] and named hirudin. The amino acid sequence (65 residues) of the major form of hirudin isolated from leeches was determined by Bagdy et al.[10] and Dodt et al.[11] (see also ref. 12) and is given in Figure 1. In the leech, Tyr-63 is post-translationally modified to tyrosine-sulphate. Hirudin contains six cysteines that form three disulfide bridges.[13] The primary structures of two other forms of hirudin have been published[14-16] and are also given in Figure 1. The three forms are highly homologous (about 85 % identical) and all appear to have antithrombin activity. The most striking feature of the primary structure of hirudin is the number of acidic residues (Figure 1). The C-terminal region is particularly acidic in all three hirudin molecules.

The tertiary structure of hirudin has been determined in solution by N.M.R. studies.[17-21] These studies indicate that hirudin is composed of a compact N-terminal

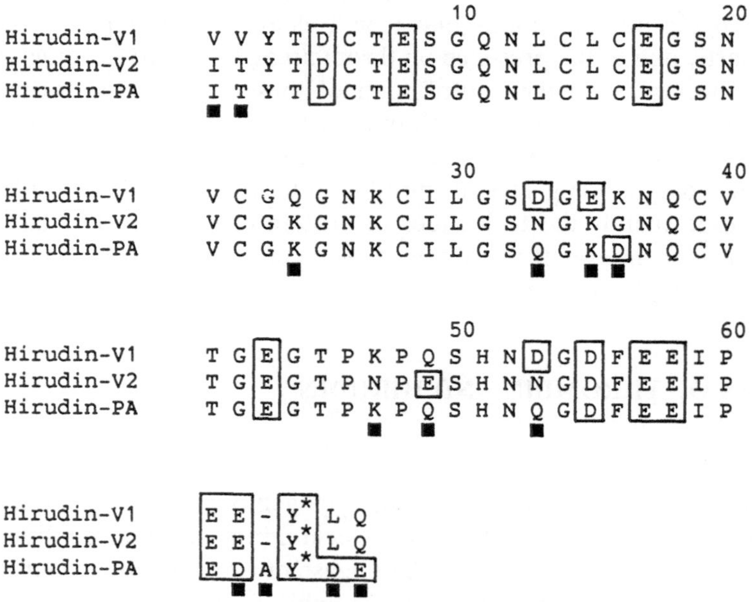

Figure 1 . *Sequences of hirudin variants.* The sequences are given using the single letter code and Y* indicates a sulphated tyrosyl residue. The sequence of hirudin-V1 was determined by Bagdy et al.[10] and Dodt et al.[11] and those of hirudin-V2 and hirudin-PA were determined by Harvey et al.[16] and Dodt et al.,[14] respectively. The acidic residues have been enclosed in blocks and positions where differences occur are indicated with a black rectangle.

domain (residues 1-49) held together by the three disulfide bonds and a disordered C-terminal tail (residues 50-65). The N-terminal domain is composed of a hydrophobic core with several well defined turns and a protruding "finger" domain (residues 31-36) whose orientation with respect to the core could not be precisely determined.[19,20]

Hirudin is uniquely specific for thrombin. It forms a tight complex with human α-thrombin; the dissociation constant (K_d) for the complex is 20 fM (2×10^{-14} M) which corresponds to a binding energy of 81 kJ mol^{-1}.[22] Tyr-63 is not sulphated in recombinant hirudin and this molecule exhibits a slightly higher K_d value of 200 fM.[23] The unique specificity of hirudin is illustrated by the fact that factor Xa, kallikrein, plasmin, tissue plasminogen activator, trypsin, chymotrypsin and the complement proteases are not inhibited by micromolar concentrations of recombinant hirudin.[24]

KINETIC MECHANISM OF THE INHIBITION OF THROMBIN BY HIRUDIN

Hirudin was shown to be a slow, tight binding inhibitor of thrombin, forming a complex that involved the active site and other regions of thrombin.[22] Hirudin combined with thrombin in a two-step mechanism. The first step was rate-limiting and was not influenced by the binding of the substrate to the active site. The rate of this step was also dependent on ionic strength. In a second, faster step, hirudin combined with the active site of thrombin to form a tighter complex. The association rate of hirudin with thrombin appears to be diffusion-limited; values of 4.7×10^8 and 1.4×10^8 M^{-1}s^{-1} were obtained for native and recombinant hirudin, respectively, at an ionic strength of 0.125.[23]

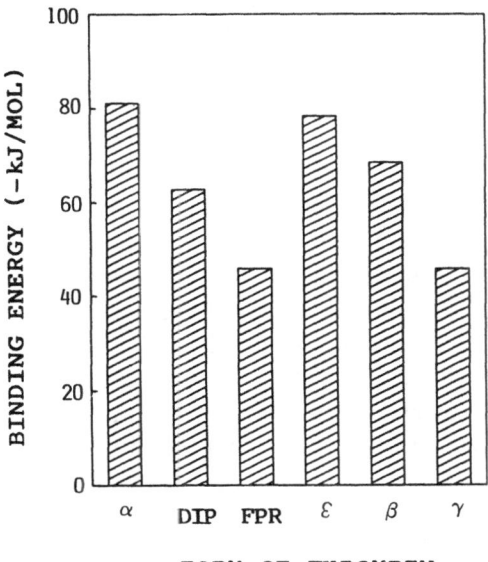

Figure 2 . *Affinities of modified thrombins for hirudin.* Affinities are expressed in terms of the standard free energies for the complex formation (binding energies); these were calculated from data presented in Stone et al.[31] The abbreviation FPR is used for D-Phe-Pro-ArgCH$_2$-thrombin; all other forms of thrombin are as defined in the text.

Values obtained for the dissociation constant of the thrombin-hirudin complex vary from an estimate of 20 fM[22] to estimates over three orders of magnitude larger.[25,26] There are at least three possible reasons for this variation:

1. Differences in the Reaction Conditions. The dissociation constant for hirudin is strongly dependent on the ionic strength;[22,27] higher values for the dissociation constant will be observed in buffers of higher ionic strength. The K_d value for hirudin also appears to be temperature dependent.[28]

2. Differences in the Experimental Protocols. Hirudin is a slow, tight-binding inhibitor and appropriate protocols must be used in order to obtain meaningful data and the appropriate equations should be used to analyze the data.[22,29] The concentrations of thrombin and hirudin used should be as near as possible to the observed value of K_d; this value can be increased by increasing the substrate concentration. In addition, the concentration of active hirudin molecules in the preparation should be determined by titration with α-thrombin. For the data analysis protocols, the reader is referred to the article by Morrison and Walsh;[30] application of these protocols to the thrombin-hirudin interaction is found in the work from our laboratory[22,23] and Dodt et al.[29]

3. Differences in the Quality of the Thrombin Preparations. α-Thrombin is degraded autolytically to form ß- and γ-thrombin. These forms of thrombin have a much reduced affinity for hirudin[31] and, therefore, their presence in thrombin preparations can lead to higher estimates for the value of K_d.

IDENTIFICATION OF INTERACTION AREAS ON THROMBIN

Two regions, that seem to contribute to the specificity of thrombin, have been identified:[6] (1) an apolar binding site that binds proflavine[32] and has been exploited in the design of low molecular weight tripeptidyl substrates and inhibitors[33,34] and (2) an anion binding region that is thought to be involved in the binding of fibrinogen.[6] The importance of these two regions to the binding of hirudin was established using the experimental approaches outlined below.

Derivatives of α-Thrombin with Modified Active Sites

Diisopropylfluorophosphate covalently modifies the active-site serine of thrombin to form diisopropylphosphoryl(DIP)-thrombin. Access to the primary specificity pocket also appears to be blocked in DIP-thrombin.[31] Thrombin can also be inactivated by peptidyl chloromethanes through alkylation of the active-site histidine.[35] D-Phe-Pro-ArgCH$_2$Cl is a specific inactivator of α-thrombin and appears to interact with an apolar binding site as well as the primary specificity pocket.[36,37] The affinities of DIP-thrombin and D-Phe-Pro-ArgCH$_2$-thrombin for hirudin have been determined by the ability of these modified thrombins to compete for the binding of hirudin with α-thrombin.[31] It had been previously reported that DIP-thrombin was still able to bind hirudin[38] and it was shown that the affinity of this form of thrombin for hirudin was reduced about 10^3-fold with respect to α-thrombin; this corresponds to a decrease in binding energy of 18 kJ mol^{-1} (see Figure 2). The affinity of D-Phe-Pro-ArgCH$_2$-thrombin for hirudin was further reduced; its affinity was 10^6-fold lower than that of α-thrombin which corresponds to a decrease in binding energy of 35 kJ mol^{-1} (Figure 2). From the results obtained with DIP-thrombin, a maximum contribution of 18 kJ mol^{-1} to the binding energy can be attributed to the interactions of hirudin with the active-site serine and the primary specificity pocket. This value, however, should be regarded as a maximum contribution since the blocking of active-site regions may force hirudin to bind to thrombin in such a way that other interactions with regions distant from the active site are also disrupted. The binding sites for D-Phe-Pro-Arg and proflavine on thrombin overlap and, therefore, it has been assumed that the D-Phe-Pro portion of the peptide occupies the apolar binding site that binds proflavine.[36] This site is unlikely to be blocked in DIP-thrombin and, thus, the difference in binding energy between DIP-thrombin and D-Phe-Pro-ArgCH$_2$-thrombin can be used to assess the importance of the apolar binding site. The observed difference of 17 kJ mol^{-1} suggests a significant contribution for this site. Once again, this value should be considered a maximum contribution because of the possibility that other interactions have also been altered.

Proteolysed Forms of Thrombin and Peptide-specific Antibodies

Proteolytically cleaved derivatives of thrombin have been widely used in the study of structure-function relationships with this enzyme.[39] The most widely used derivative is γ-thrombin which can be produced by autolysis or by cleavage with trypsin. It contains cleavages in at least two surface regions of the B-chain at the following positions: Arg-62, Arg-73, Lys-154.[40-43] The cleavages at Arg-62 and Arg-73 result in the loss of a surface loop; since this loop contains the cleavage site for the production of ß-thrombin (see below), it is conveniently termed the ß-loop. γ-Thrombin retains an activity similar to that of α-thrombin with tripeptidyl p-nitroanilide substrates[44-46] but retains little activity with fibrinogen.[40,47]

Recently, two derivatives of thrombin have been made that are cleaved in only one of the two regions altered in γ-thrombin. Treatment of α-thrombin with pancreatic elastase results in the formation of ε-thrombin that contains a single cleavage at Ala-150.[48] ß$_T$-thrombin can be isolated after limited trypsinolysis of α-thrombin and contains a single cleavage at Arg-73.[43]

The affinity of each of these forms of thrombin for hirudin is presented in Figure 2.[31] ß$_T$-Thrombin displayed a 100-fold decrease in affinity which indicates that hirudin probably interacts with the ß-loop of thrombin. The results of other experiments are also consistent with this hypothesis (see below). ε-Thrombin had only a slightly reduced affinity for hirudin compared with α-thrombin and these results were taken to indicate that the region around this cleavage site was not involved in an interaction with hirudin.[31] A marked decrease in affinity was observed with γ-thrombin; the affinity of this thrombin derivative for hirudin was 10^6-fold lower than that of α-thrombin which represents a 35 kJ mol^{-1} decrease in binding energy (Figure 2). The large decrease in binding energy from ß$_T$- to γ-thrombin could be due to the removal of the ß-loop or to more extensive structural alterations in γ-thrombin caused by the additional cleavages.

The importance of the ß-loop for the binding of hirudin to thrombin was demonstrated by making antibodies to a synthetic peptide whose sequence corresponded to this loop. These antibodies were found to compete with hirudin for a binding site on thrombin.[49]

Protection against Chemical Modification and Proteolysis

Recently, Chang[50] has used chemical modification to identify that hirudin protected lysyl residues at positions 52, 65, 106, 107 and 154 from modification by a lysine-specific reagent. Lys-65 is located within the ß-loop and Lys-154 is a cleavage site in γ-thrombin.

Hirudin also protected α-thrombin from proteolysis at Arg-62, Arg-73 and Lys-154 by trypsin and from cleavage at Ala-150 by elastase.[51] Thus, results obtained from four different types of experiments indicate that hirudin is bound to the ß-loop (residues 62-73) of thrombin. The results from both chemical modification and proteolysis protection experiments indicate that a region of α-thrombin in the vicinity of residues 150-154 is also involved in an interaction with hirudin. However, the affinity of ε-thrombin for hirudin is not markedly different from that of α-thrombin. Thus, it appears that the effect of cleavage at Ala-150 on the structure of the surrounding region was relatively small such that the binding of hirudin to this region was not affected. Alternatively, this region may not make a great contribution to the overall binding energy.

Fragments of hirudin have been produced by protein engineering to investigate which regions of thrombin and hirudin interact with each other.[51] A C-terminal fragment of hirudin consisting of the last 10 residues is able to block the cleavage of fibrinogen by α-thrombin.[52-54] A similar fragment comprising hirudin residues 53-65 was able to protect α-thrombin from cleavage by trypsin. The N-terminal fragment (residues 1-52) was unable to protect against trypsinolysis. In contrast, the N-terminal fragment protected against elastase cleavage at Ala-150 whereas residues 53-65 afforded only partial protection.[51] These results suggest that (1) the C-terminal region of hirudin is bound to the ß-loop of α-thrombin; and (2) a region of hirudin located towards the N-terminus of residue 52 interacts with the loop in thrombin containing Ala-150. It is interesting to note that the K_d value for the 52-residue N-terminal fragment of hirudin, that does not interact with the ß-loop, is 24 nM. This value can be compared with the K_d value of 19 nM for native hirudin with γ-thrombin where the interaction with the ß-loop also does not take place.

IDENTIFICATION OF INTERACTION AREAS ON HIRUDIN

The production of recombinant hirudins in *Escherichia coli* and *Saccharomyces cerevisiae* have been described[15,16,55-57] and, thus, it is now possible to study binding areas on hirudin by site-directed mutagenesis.[23,29,58] The results of studies concerning three different interaction areas are presented below.

Importance of Basic Amino Acid Residues

Since thrombin's primary specificity is for basic amino acids, it was assumed that one of the basic amino acids in hirudin would be bound to the primary specificity pocket. On the basis of this assumption, three groups have performed site-directed mutagenesis experiments in which basic amino acid residues in hirudin were exchanged.[23,29,58] In the study of Braun et al.,[23] each of the three lysines (Lys-27, 36 and 47) and His-51 in hirudin were mutated to glutamine. With the exception of the mutation Lys-47 to Gln-47, the mutations were without effect on the binding energy. Even the effect of replacement of Lys-47 was small, a decrease in binding energy of 6 kJ mol^{-1} was observed.[23] Similar results have been obtained by Dodt et al.[29] The substitution of Lys-47 for Glu-47 by these workers resulted in a similar decrease in binding energy as that observed for the Gln-47 substitution. Replacement of Lys-47 by isoleucine actually resulted in a better inhibitor and replacement of the other basic residues was without effect.[29] The effect observed on the replacement of Lys-47 by glutamate or glutamine could be due to a change in the tertiary structure of hirudin.

The structure of the Lys-47→Glu mutant has been determined by N.M.R. and it was found to differ slightly from the non-mutant structure.[19] Thus, none of the basic amino acid residues in hirudin is of crucial importance for its interaction with thrombin. Hirudin does not seem to require a specific interaction with the primary specificity pocket of thrombin in order to form a tight complex with the enzyme. In this respect, comparison of hirudin with the cysteine protease inhibitors called cystatins is interesting. Cystatins also do not require interactions with the active site of their target enzymes in order to form a complex.[59]

Hirudin is very different from most other naturally occurring inhibitors of serine proteases. The most common structural feature of these inhibitors is the so-called reactive-site loop. The conformation of this loop is complementary to the active site region of the target enzyme. The reactive-site residue is found within this loop and is bound to the primary specificity pocket of the protease.[60] Changes in the reactive site residue can lead to marked alterations in the specificity of the inhibitor.[61,62] For instance, the importance of the reactive site residue in determining the specificity of α_1-proteinase inhibitor has been demonstrated by the existence of natural variants[62] and by site-directed mutagenesis.[63] The native molecule rapidly inactivates neutrophil elastase but is without effect on thrombin activity. On mutation of the reactive site methionine of α_1-proteinase inhibitor to an arginine, its ability to inhibit neutrophil elastase is lost and it becomes an excellent inactivator of thrombin.

It should be stressed, however, that although the binding of a basic residue of hirudin in the primary specificity pocket of thrombin is not of utmost importance for the formation of the enzyme-inhibitor complex, hirudin nevertheless blocks access of p-nitroanilide substrates to this pocket.[22] Whether the side chain of an amino acid residue of hirudin is bound in this pocket is not known and will probably not be known until the tertiary structure of the complex is determined by X-ray crystallography.

Ionic Interactions Involving the C-terminal Region

Evidence for the importance of the C-terminal region for inhibitory activity came from the experiments of Chang[64] in which the effect of carboxypeptidase Y digestion on the activity of hirudin was examined. In these experiments, it was found that hirudin progressively lost activity as amino acids were removed from its C-terminus. The C-terminal region of hirudin is acidic in nature (Figure 1) and the strong ionic strength dependence of the thrombin-hirudin interaction suggested that it may be involved in an ionic interaction with thrombin.[22] The ß-loop of α-thrombin contains a large number of basic residues and, considering the reduced affinity of β_T-thrombin for hirudin, it was proposed that the ß-loop interacted with the C-terminal region of hirudin.[6,31] The ability of a C-terminal fragment of hirudin to prevent trypsin cleavage at Arg-73 lends credence to this hypothesis.[51] The importance of the acidic residues in the C-terminal region of hirudin was demonstrated by site-directed mutagenesis. Firstly, it was found that recombinant hirudin, which lacks the negatively charged sulphate group on Tyr-63, had a 10-fold lower affinity for α-thrombin than the native molecule.[23,29] The full affinity was restored by introduction of a phosphate group on to Tyr-63.[65] Sequential removal of four more negative charges one at a time by mutation of Glu-57, 58, 61 and 62 to glutamine resulted in a progressive decrease in the binding energy.[23] It was of some concern to exclude the possibility that the four glutamate to glutamine mutations had caused structural changes in hirudin. In order to investigate this possibility, the contributions to the binding energy of ionic and nonionic interactions for each of the mutants were determined by examining the effect of ionic strength on the complex formation;[27] the results are shown in Figure 3. In removing acidic residues from the C-terminal region of hirudin, it was hoped that only ionic interactions would be affected and the data of Figure 3 show that this was indeed the case. The nonionic contribution to binding energy was essentially the same for all the mutants and for recombinant and native hirudin. In contrast, the ionic contribution decreased in a linear manner as negative charges were removed. It was possible to calculate that each negative charge in the C-terminal region contributed about 4 kJ mol^{-1} to the overall binding energy at zero ionic strength. Ionic interactions accounted for about one-third of the total binding energy at zero ionic strength (Figure 3). This contribution will, of course, decrease as the ionic strength increases.

Interactions with the N-terminus

Attempts to express recombinant hirudin indicated that any extension of the N-terminus was deleterious to its inhibitory activity.[55,56] Addition of a single methionine to the N-terminus of hirudin led to a decrease in binding energy of 26 kJ mol^{-1}.[66] Removal of this methionyl residue by CNBr resulted in the recovery of full inhibitory activity for this form of hirudin. The amount by which the binding energy decreased was dependent on the nature of the amino acid that was used to extend the N-terminus. Extension with a glycyl residue resulted in a 20 kJ mol^{-1} decrease in binding energy. A large part of the decrease in binding energy caused by the additional amino acid was due to the displacement of the positively charged α-amino group as shown by experiments in which this group was specifically modified.[66] In order to conduct these experiments, a mutant was constructed in which all the other primary amino groups were removed by site-directed mutagenesis; this allowed specific chemical

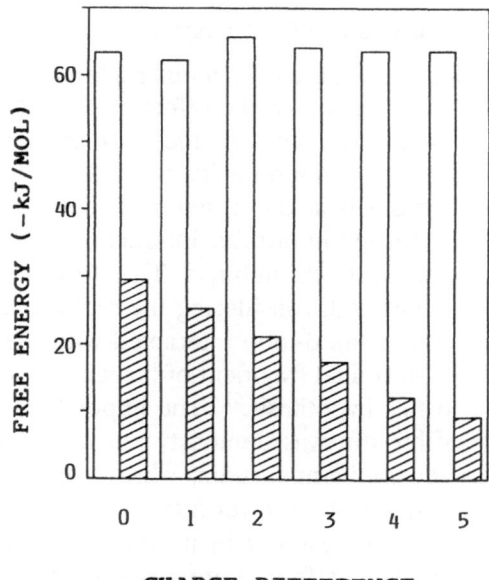

CHARGE DIFFERENCE

Figure 3 . *Contribution of ionic and nonionic interactions to the binding energies of different forms of hirudin.* Ionic (striped bars) and nonionic interactions energies (open bars) for the different forms were taken from Stone et al.[27] The charge differences were calculated with respect to native hirudin and the free energy values for charge differences from 1 to 5 represent those obtained for recombinant hirudin, E62Q, E57,58Q, E57,58,62Q and E57,57,61,62Q, respectively.

modification of the α-amino group. Removal of the positive charge of the α-amino group by acetylation resulted in a reduction of binding energy by 27 kJ mol[-1]. In contrast, reaction of the α-amino group with ethyl acetimidate caused only a 12 kJ mol[-1] decrease in binding energy.[66] This modification adds a group about the size of the acetyl moiety, while retaining the positive charge of the α-amino group. Thus, it appears that a positive charge immediately adjacent to the N-terminal valyl residue is required for optimal binding to thrombin. The positively charged α-amino group is presumably involved in an ionic interaction with a carboxyl group on thrombin. The results obtained by Chang[50] also indicate that the α-amino group is involved in an interaction with thrombin. This group is readily modified in free hirudin but is protected from chemical modification in the complex with thrombin.

All naturally occurring hirudins, whose primary structures have been determined, have an N-terminal hydrophobic amino acid with branched side chain (Figure 1). The importance of the hydrophobic nature of the N-terminus to the strength of the hirudin-thrombin interaction was demonstrated by the observation that replacement of the two N-terminal valyl residues by polar amino acids caused a marked decrease in the binding energy.[66] In contrast, conservative replacements by other hydrophobic amino acids resulted in only moderate changes of the strength of inhibition (Figure 4). By far the largest decrease in binding energy was observed when Val-1 and 2 were replaced by glutamate. In contrast, the effect of substitution by the lysine was much smaller (Figure 4). The differential effect observed between the glutamyl and lysyl mutants is presumably due to the effect of the glutamate on the proposed interaction between the positively charged α-amino group of hirudin and a negatively charged carboxyl group on thrombin. The presence of a negatively charged glutamate residue in the first and second positions of hirudin would weaken such an interaction. On the other hand, a positively charged lysyl residue would also be able to participate in this ionic interaction.

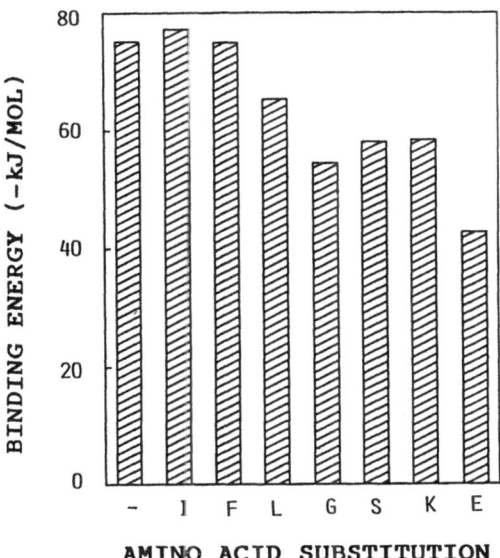

AMINO ACID SUBSTITUTION

Figure 4 . *Effect of N-terminal substitutions on the binding energy of hirudin.* The N-terminal valyl residues at positions 1 and 2 of recombinant hirudin-V1 were replaced with the indicated amino acids (one letter code) by site-directed mutagenesis; the data were taken from Wallace et al.[66]

CONCLUSIONS

The results of studies on thrombin and hirudin using protein chemistry and site-directed mutagenesis indicate that the inhibition of thrombin by hirudin is unique among the interactions of inhibitors with serine proteases in that both molecules contain a number of different areas that are important for complex formation. The areas on thrombin that were found to contribute to the binding energy were two surface loops, an apolar binding site and the active site. The hirudin molecule interacts *via* the acidic C-terminus, the positively charged α-amino group, and the N-terminal hydrophobic amino acid residues.

REFERENCES

1. K.G. Mann, The assembly of blood clotting complexes on membranes. **Trends Biochem. Sci.** 12: 229 (1987).
2. J.W. Fenton, Structural regions and bioregulatory functions of thrombin, *in:* "Cell proliferation: Recent advances", Vol. II, Boynton, A. L., and Leffert, H. L., eds., Academic Press, New York (1987).
3. R. Bar-Shavit, A. Kahn, J.W. Fenton, II, and G.D. Wilner, Chemotactic response of monocytes to thrombin. **J. Cell Biol.** 96: 282 (1983).
4. R. Bar-Shavit, A. Kahn, G.D. Wilner, and J.W. Fenton, II, Monocyte chemotaxis: stimulation by specific exosite region in thrombin. **Science** 220: 728 (1983).
5. L.B. Chen, and J.M. Buchanan, Mitogenic activity of blood components. I. Thrombin and prothrombin. **Proc. Natl. Acad. Sci. U.S.A.** 72: 131 (1975).
6. J.W. Fenton, II, Thrombin specificity. **Ann. N. Y. Acad. Sci.** 370: 468 (1981).
7. J.W. Fenton, II, and D.H. Bing, Thrombin active-site regions. **Semin. Thromb. Hemostasis** 12: 200 (1986).
8. J.B. Haycraft, Secretion obtained from the medicinal leech. **Proc. R. Soc. London** 36B: 478 (1884).
9. F. Markwardt, Die Isolierung und chemische Charakterisierung des Hirudins. **Hoppe-Seylers Z. Physiol. Chem.** 308: 147 (1957).
10. D. Bagdy, E. Barabas, L. Graf, T.E. Peterson, and S. Magnusson, Hirudin. **Methods Enzymol.** 45: 669 (1976).
11. J. Dodt, H. Müller, U. Seemüller, and J.-Y. Chang, The complete amino acid sequence of hirudin, a thrombin specific inhibitor. **FEBS Lett.** 165: 180 (1984).
12. S.J.T. Mao, M.T. Yates, D.T. Blankenship, A.D. Cardin, J.L. Krstenansky, W. Lovenberg, and R.L. Jackson, Rapid purification and revised N-terminal sequence of hirudin: a specific thrombin inhibitor of the bloodsucking leech. **Anal. Biochem.** 161: 514 (1987).
13. J. Dodt, U. Seemüller, R. Maschler, and H. Fritz, The complete covalent structure of hirudin. Localisation of the disulfide bonds. **Biol. Chem. Hoppe-Seyler** 366: 379 (1985).
14. J. Dodt, N. Machleidt, U. Seemüller, R. Maschler, and H. Fritz, Isolation and characterization of hirudin isoinhibitors and sequence analysis of hirudin PA. **Biol. Chem. Hoppe-Seyler** 367: 803 (1986).
15. J. Dodt, T. Schmitz, T. Schäfer, and C. Bergmann, Expression, secretion and processing of hirudin in *E. coli* using the alkaline phosphatase signal sequence. **FEBS Lett.** 202: 373 (1986).
16. R.P. Harvey, E. Degryse, L. Stefani, F. Schamber, J.-P. Cazenave, M. Courtney, P. Tolstoshev, and J.-P. Lecocq, Cloning and expression of a cDNA coding for the anticoagulant hirudin from the bloodsucking leech, *Hirudo medicinalis*. **Proc. Natl. Acad. Sci. USA** 83: 1084 (1986).
17. G.M. Clore, D.K. Sukumaran, M. Nilges, J. Zarbock, and A.M. Gronenborn, The conformations of hirudin in solution: a study using nuclear magnetic resonance, distance geometry and restrained molecular dynamics. **EMBO J.** 6: 529 (1987).
18. D.K. Sukumaran, G.M. Clore, A. Preuss, J. Zarbock, and A.M. Gronenborn, Proton nuclear magnetic resonance study of hirudin: resonance assignment and secondary structure. **Biochemistry** 26: 333 (1987).
19. P.J.M. Folkers, G.M. Clore, P.C. Driscoll, J. Dodt, S. Köhler, and A.M. Gronenborn, Solution structure of recombinant hirudin and the Lys-47 - Glu mutant: a nuclear magnetic resonance and hybrid distance geometry-dynamical simulated annealing study. **Biochemistry** 28: 2601 (1989).

20. H. Haruyama, and K. Wüthrich, The conformation of recombinant desulfatohirudin in aqueous solution determined by nuclear magnetic resonance. **Biochemistry** 28: 4301 (1989).

21. H. Haruyama, Y.-Q. Qian, and K. Wüthrich, Static and transient hydrogen bonding interactions in recombinant desulfatohirudin studied by ^{1}H nuclear magnetic resonance measurements of amide proton exchange rates and pH dependent chemical shifts. **Biochemistry** 28: 4312 (1989).

22. S.R. Stone, and J. Hofsteenge, Kinetics of the inhibition of thrombin by hirudin. **Biochemistry** 25: 4622 (1986).

23. P.J. Braun, S. Dennis, J. Hofsteenge, and S.R. Stone, Use of site-directed mutagenesis to investigate the basis for the specificity of hirudin. **Biochemistry** 27: 6517 (1988).

24. R.B. Wallis, Hirudins and the role of thrombin: lessons from leeches. **Trends Pharmac. Sci.** 9: 425 (1988).

25. F. Markwardt, and P. Walsmann, Die Reaktion zwischen Hirudin und Thrombin. **Hoppe-Seylers Z. Physiol. Chem.** 312: 85 (1958).

26. B.H. Landis, M.P. Zabinski, G.J.M. Lafleur, D.H. Bing, and J.W. Fenton, II, Human α-thrombin and γ-thrombin differential inhibition with hirudin. **Fed. Proc.** 37: 1445 (1978).

27. S.R. Stone, S. Dennis, and J. Hofsteenge, Quantitative evaluation of the contribution of ionic interactions to the formation of the thrombin-hirudin complex. **Biochemistry** in press (1989).

28. W.G. Landis, and D.F. Waugh, Interactions of bovine fibrinogen and thrombin. Early events in the development of clot structure. **Arch. Biochem. Biophys.** 168: 498 (1975).

29. J. Dodt, S. Köhler, and A. Baici, Interaction of site specific hirudin variants with α-thrombin. **FEBS Lett.** 229: 87 (1988).

30. J.F. Morrison, and C.T. Walsh, The behavior and significance of slow-binding enzyme inhibitors. **Adv. Enzymol. and Rel. Areas Mol. Biol.** 61: 201 (1987).

31. S.R. Stone, P.J. Braun, and J. Hofsteenge, Identification of regions of α-thrombin involved in its interaction with hirudin. **Biochemistry** 26: 4617 (1987).

32. L.J. Berliner, and Y.Y.L. Shen, Physical evidence for an apolar binding site near the catalytic center of human α-thrombin. **Biochemistry** 16: 4622 (1977).

33. G. Claeson, L. Aurell, G. Karlsson, and P. Friberger, Substrate structure and activity relationship, *in:* "New methods for analysis of coagulation using chromogenic substrates", Witt, I., ed., de Gruyter, Berlin (1977).

34. C. Kettner, and E. Shaw, Inactivation of trypsin-like enzymes with peptides of arginine chloromethyl ketone. **Methods Enzymol.** 80: 826 (1981).

35. G. Glover, and E. Shaw, The purification of thrombin and isolation of a peptide containing the active center histidine. **J. Biol. Chem.** 246: 4594 (1971).

36. S.A. Sonder, and J.W. Fenton, II, Proflavin binding within the fibrinopeptide groove adjacent to the catalytic site of human α-thrombin. **Biochemistry** 23: 1818 (1984).

37. B. Walker, P. Wikström, and E. Shaw, Evaluation of inhibitor constants and alkylation rates for a series of thrombin affinity labels. **Biochem. J.** 230: 645 (1985).

38. J.W. Fenton, II, B.H. Landis, D.A. Walz, D.H. Bing, R.D. Feinman, M.P. Zabinski, S.A. Sonder, L.J. Berliner, and J.S. Finlayson, Human thrombin: preparative evaluation, structural properties, and enzymic specificity, *in:* "The chemistry and physiology of human plasma proteins", Bing, D. H., ed., Pergamon Press, New York (1979).

39. L.J. Berliner, Structure-functions relationships in human α- and γ-thrombins. **Mol. Cell. Biochem.** 61: 159 (1984).

40. J.W. Fenton, II, B.H. Landis, D.A. Walz, and J.S. Finlayson, Human thrombins, *in:* "Chemistry and biology of thrombin", Lundblad, R. L., Fenton, J. W., II, and Mann, K. G., eds., Ann Arbor Science, Ann Arbor (1977).

41. J.-P. Boissel, B. Le Bonniec, M.-J. Rabiet, D. Labie, and J. Elion, Covalent structure of β- and γ autolytic derivatives of human α-thrombin. **J. Biol. Chem.** 259: 5691 (1984).

42. J.-Y. Chang, The structures and proteolytic specificities of autolyzed human thrombin. **Biochem. J.** 240: 797 (1986).

43. P.J. Braun, J. Hofsteenge, J.-Y. Chang, and S.R. Stone, Preparation and characterization of proteolyzed forms of human α-thrombin. **Thromb. Res.** 50: 273 (1988).

44. R. Lottenberg, J.A. Hall, J.W. Fenton, II, and C.M. Jackson, The action of thrombin on peptide p-nitroanilide substrates; hydrolysis of Tos-Gly-Pro-Arg-pNA and D-Phe-Pip-Arg-pNA by human α and γ and bovine α and β-thrombins. **Thromb. Res.** 28: 313 (1982).

45. S.A. Sonder, and J.W. Fenton, II, Thrombin specificity with tripeptide chromogenic substrates: comparison of human and bovine thrombins with and without fibrinogen clotting activities. **Clin. Chem.** 32: 934 (1986).

46. J. Hofsteenge, P.J. Braun, and S.R. Stone, Enzymatic properties of proteolytic derivatives of human α-thrombin. **Biochemistry** 27: 2144 (1988).

47. S.D. Lewis, L. Lorand, J.W. Fenton, II, and J.A. Shafer, Catalytic competence of human α- and γ-thrombin in the activation of fibrinogen and factor XIII. **Biochemistry** 26: 7597 (1987).

48. S. Kawabata, T. Morita, S. Iwanaga, and H. Igarashi, Staphylocoagulase-binding region in human prothrombin. **J. Biochem. (Tokyo)** 97: 325 (1985).

49. G. Noé, J. Hofsteenge, G. Rovelli, and S.R. Stone, The use of sequence specific antibodies to identify a secondary binding site in thrombin. **J. Biol. Chem.** 263: 11729 (1988).

50. J.-Y. Chang, The hirudin-binding site of human α-thrombin. **J. Biol. Chem.** 264: 7141 (1989).

51. S. Dennis, A. Wallace, J. Hofsteenge, and S.R. Stone, Use of fragments of hirudin to investigate the thrombin-hirudin interaction. **Eur. J. Biochem.** in press (1989).

52. J.L. Krstenansky, and S.J.T. Mao, Antithrombin properties of C terminus of hirudin using synthetic unsulfated N^α-acetyl-hirudin$_{45-65}$. **FEBS Lett.** 211: 10 (1987).

53. J.L. Krstenansky, T.J. Owen, M.T. Yates, and S.J.T. Mao, Anticoagulant peptides: Nature of the interaction of the C-terminal region of hirudin with a noncatalytic binding site on thrombin. **J. Med. Chem.** 30: 1688 (1987).

54. S.J.T. Mao, M.T. Yates, T.J. Owen, and J.L. Krstenansky, Interaction of hirudin with thrombin: identification of a minimal binding domain of hirudin that inhibits clotting activity. **Biochemistry** 27: 8170 (1988).

55. C. Bergmann, J. Dodt, S. Köhler, E. Fink, and H.G. Gassen, Chemical synthesis and expression of a gene coding for hirudin, the thrombin-specific inhibitor from the leech *Hirudo medicinalis*. **Biol. Chem. Hoppe-Seyler** 367: 731 (1986).

56. B. Meyhack, J. Heim, H. Rink, W. Zimmermann, and W. Märki, Desulfatohirudin, a specific thrombin inhibitor: expression and secretion in yeast. **Thromb. Res. Suppl.** 7: 3 (1987).

57. G. Loison, A. Findeli, S. Bernard, M. Nguyen-Juilleret, M. Marquet, N. Riehl-Bellon, D. Cavallo, L. Guerra-Santos, S.W. Brown, M. Courtney, C. Roitsch, and Y. Lemoine, Expression and secretion in *S. cerevisiae* of biologically active leech hirudin. **Biotechnology** 6: 72 (1988).

58. E. Degryse, M. Acker, G. Defreyn, A. Bernat, J.P. Maffrand, C. Roitsch, and M. Courtney, Point mutations modifying the thrombin inhibition kinetics and antithrombin activity *in vivo* of recombinant hirudin. **Protein Eng.** 2: 459 (1989).

59. A.J. Barret, The cystatins: a new class of peptidase inhibitors. **Trends Biochem. Sci.** 12: 193 (1986).

60. R.J. Read, and M.N.G. James, Introduction to the proteinase inhibitors: X-ray crystallography, *in:* "Proteinase Inhibitors" Barret, A. J., and Salvesen, G, eds., Elsevier, Amsterdam (1986).

61. M. Laskowski, Jr., I. Kato, W. Ardelt, J. Cook, A. Denton, M.W. Empie, W.J. Kohr, S.J. Park, K. Parks, B.L. Schatzley, O.L. Schoenberger, M. Tashiro, G. Vichot, H.E. Whatley, A. Wieczorek, and M. Wieczorek, Ovomucoid third domains from 100 avian species: Isolation, sequences, and hypervariability of enzyme-inhibitor contact residues. **Biochemistry** 26: 202 (1987).

62. M.C. Owen, S.O. Brennan, J.H. Lewis, and R.W. Carrell, Mutation of antitrypsin to antithrombin. α_1-Antitrypsin Pittsburh (358 Met - Arg), a fatal bleeding disorder. **N. Engl. J. Med.** 309: 694 (1983).

63. S. Jallat, D. Carvallo, L.H. Tessier, D. Roecklin, C. Roitsch, F. Ogushi, R.G. Crystal, and M. Courtney, Altered specificities of genetically engineered α_1-antitrypsin variants. **Protein Eng.** 1: 29 (1986).

64. J.-Y. Chang, The functional domain of hirudin, a thrombin-specific inhibitor. **FEBS Lett.** 164: 307 (1983).

65. J. Hofsteenge, S.R. Stone, A. Donella-Deana, and L.A. Pinna, The effect of substituting phosphotyrosine for sulphotyrosine on the activity of hirudin. **Eur. J. Biochem.** in press (1989).

66. A. Wallace, S. Dennis, J. Hofsteenge, and S.R. Stone, Contribution of the N-terminal region of hirudin to its interaction with thrombin. **Biochemistry** in press (1989).

SIGNAL TRANSDUCTION CHAINS INVOLVED IN THE CONTROL OF THE FIBRINOLYTIC ENZYME CASCADE

WOLF-DIETER SCHLEUNING* AND ROBERT L. MEDCALF

Central Hematology Laboratory
University of Lausanne Medical School (CHUV)
CH-Lausanne, Switzerland

INTRODUCTION

The fibrinolytic enzyme cascade is a summary term for several regulatory serine proteases and serine protease inhibitors, which cooperate in the digestion of extracellular matrix protein in processes of tissue repair, growth, and remodelling. Tissue-type plasminogen activator (t-PA) and urinary-type plasminogen activator (u-PA) activate the proenzyme plasminogen by the cleavage of a single peptide bond, converting it into plasmin, a proteolytic enzyme with a specificity similar to the pancreatic digestive enzyme trypsin. Plasminogen is synthesized in the liver and secreted into the bloodstream where it circulates in relatively large amounts (80-160 mg/l). Blood plasma plasminogen provides a reservoir of proteolytic activity which is recruited for the removal of fibrin deposits or for the digestion of extracellular matrix proteins during morphogenesis, wound healing or malignant growth. t-PA and u-PA are structurally and enzymatically related. Both proteins display a characteristic mosaic molecular architecture: they consist of a series of structural motifs homologous to other proteins. Thus, starting from the amino terminus, t-PA is composed of a "finger" domain, which is found tandemly arranged in fibronectin; an epidermal growth factor (EGF) like motif, found likewise in various blood clotting factors, receptor proteins and developmentally regulatory proteins; two "kringle" regions, present in u-PA, plasminogen, prothrombin, clotting factor XII and apolipoprotein (a) and finally a sequence homologous to the pancreatic proteases trypsin, chymotrypsin, elastase and kallikrein. The "finger"- and one of the two "kringle"-domains are absent in u-PA.

t-PA differs from u-PA enzymatically because it attains its optimal activity only in the presence of fibrin (reviewed in ref. 1 and 2). *In vivo* PA activity is tightly controlled by the local presence of plasminogen activator inhibitors 1 and 2 (PAI-1 and PAI-2). PAs and PAIs are produced by most - if not all - cells and their biosynthesis

---

*Present address: Schering AG, Institute of Biochemistry, Müllerstraße 170-178, Postfach 65 03 11, D-1000 Berlin 65, Germany

Serine Proteases and Their Serpin Inhibitors in the Nervous System
Edited by B. W. Festoff
Plenum Press, New York, 1990

127

is under developmental, tissue specific and hormonal control.[3-6] A deregulation of the biosynthesis of PAs and PAIs may occur under pathological circumstances in which regulatory molecules e.g.inflammatory cytokines are produced untimely or in inappropriate quantity. An overproduction of PAIs and an inhibition of PA biosynthesis may be one of the contributing factors of occlusive vascular disease.

HORMONAL MODULATION OF THE FIBRINOLYTIC SYSTEM

Protein Kinase C Dependent Pathway

Protein kinase C dependent pathways (Figure 1) play a role in a wide variety of signaltransduction chains often related to the regulation of growth and differentiation.[7,8] The distal end of the pathway consists of a hormone receptor which is probably coupled to a G-Protein.[9-11] The G-Protein regulates the activity of phospholipase C, an enzyme associated with the inner face of the plasma membrane. Phospholipase C cleaves phosphoinositol diphosphate (a metabolite of the phosphoinositol-cycle), yielding inositol triphosphate (IP_3) and diacylglycerol (DAG). IP_3 liberates Ca^{++} from storage pools within the endoplasmic reticulum which act in various ways as second messengers.[12] DAG activates protein kinase C, an enzyme that phosphorylates a number of substrate proteins at serine and threonine residues. The pathway that leads to protein kinase C activation is cut short by a class of plant derived diterpene esters which directly activate protein kinase C and which were originally recognized as potent tumor promoters, their prototype being phorbol 12-myristate 13-acetate (PMA). Activation of protein kinase C results in the rapid (15 min) induction of *c-fos* gene transcription[13] and a somewhat delayed (2-4 h) activation (or sometimes a shut off) of the *c-myc* gene. These nuclear protooncogene products appear to act like a class of "third messengers" that cooperate with other nuclear proteins in the regulation of gene transcription. The hallmark of PMA regulated gene

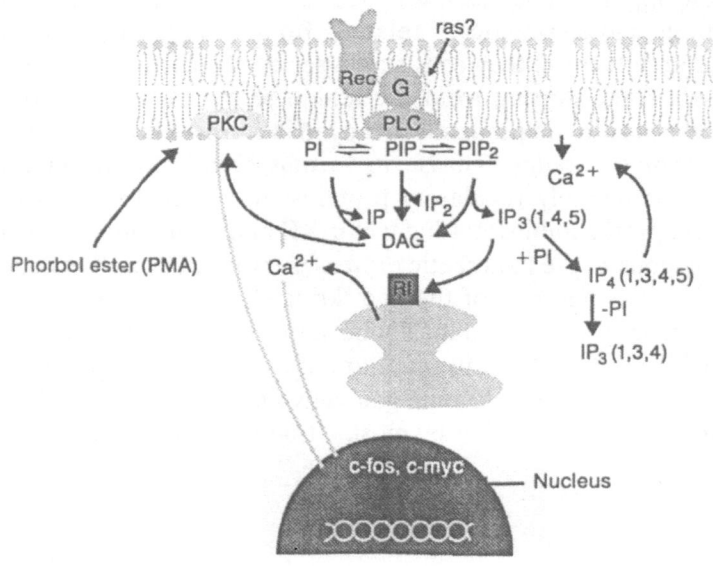

Figure 1. *Protein kinase C dependent pathway.*

promoters is a consensus sequence (ATGACTCA) which acts as a binding site for the transcription factor activator protein-1 (AP-1).[14] AP-1 is identical with the protooncogene product *c-jun* and interacts with *c-fos* via structural elements consisting of interdigitating leucine residues named the "leucine zipper".[15] The induction of PA-activity in HeLa cells by PMA was already observed by Wigler and Weinstein.[16] This activity was subsequently purified and characterized as t-PA.[17] These authors have also shown that the induction of t-PA activity and antigen is associated with an increase of t-PA mRNA, that this effect is late (maximum after 24 hours) and requires ongoing protein biosynthesis. Employing "run-on" transcription assays, it was recently demonstrated that the PMA effect results in a tenfold increase of t-PA gene template activity, and DNAase I protection (foot print) analysis revealed the presence of an element identical to an element found in the AP-1 gene promoter (ATGACATCA) and probably responsible for the positive feed back loop that controls the activity of this gene (R.L. Medcalf, M. Rüegg, and W.-D. Schleuning, submitted for publication). This *cis*-acting element is also related to the cAMP responsive element[18] ATGACGTCA. Deletion of this element in t-PA promoter - chloramphenicol acetyl transferase (CAT) hybrid genes, which were analysed by DNA-transfection assays in HeLa cells, demonstrates that it is required for PMA mediated induction of t-PA gene transcription and for the cooperative effect of PMA. PMA also increases u-PA biosynthesis in human carcinoma cells,[19] mouse 3T3 cells[20] and LLC-PK 1 porcine kidney cells[21,22] and stimulates PAI-2 transcription in the U-937 monocytic cell line.[23]

Tyrosine Specific Protein Kinase Dependent Pathway (Figure 2)

Tyrosine specific protein kinases are a class of membrane spanning proteins that consist of a ligand binding site located on the outside of the plasma membrane and an enzymatic domain located on its inner side. Interaction of a hormone with the receptor site initiates enzymatic activity directed at a variety of substrates among which are the lipocortins and inositol kinases.[24] Prominent examples of this class of proteins are the receptors for insulin, platelet derived growth factor (PDGF), EGF and macrophage colony stimulating factor (M-CSF). The receptors for EGF and M-CSF have been shown to be related to the oncogenes *v-erb B*, and *v-fms* respectively. Other

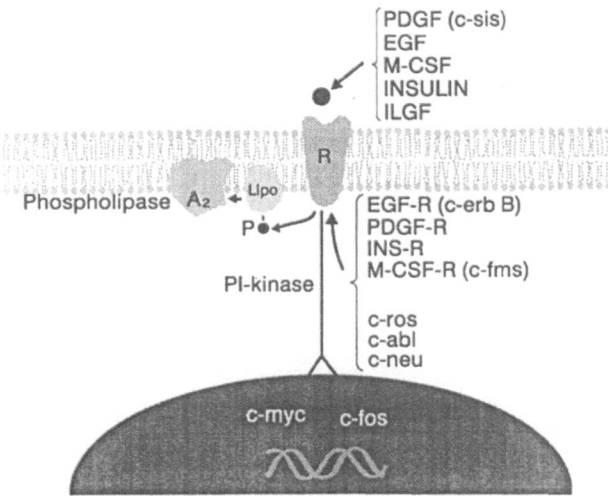

Figure 2. *Tyrosine specific kinase dependent pathway.*

129

proteins of this type, notably *c-neu* and *c-ros* have so far only been recognized as protooncogenes, whereas ligands for these putative receptors so far have not been discovered. The stimulation of this pathway leads - like the protein kinase C dependent signal transduction cascade - to the activation of the nuclear protooncogenes *fos* and *myc*. The homology of elements of the pathway with retroviral oncogene products strongly suggests a role in the regulation of growth and development. Lee and Weinstein have demonstrated that EGF induces PA activity in HeLa cells.[25] Recently we have shown that this activity is (a) due to t-PA, (b) the result of a 2.5 fold increase of t-PA gene template activity and (c) that on the levels of antigen, mRNA but not transcription the effects of EGF and PMA are additive (R.L. Medcalf, M. Rüegg, and W.-D. Schleuning, submitted for publication). It is at present not clear whether our inability to observe an additive effect on the level of transcription is due to limitations of the assay or whether EGF mediated advents also have an influence on the stability of t-PA mRNA. EGF also induces u-PA mRNA in the A-431[26] and PAI-1 in the Hep G2 (hepatoma) cell line.[27]

Regulation by Steroid Hormones (Figure 3)

The mechanism of steroid hormone action is one of the best understood pathways of gene regulation. Steroid hormone receptors are ligand dependent transcription factors which are located in the cytosol. Because of their hydrophobic structure, steroid hormones pass readily through the plasma membrane and bind to the receptor. In a process that involves dissociation from the heat shock protein hsp 90,[28] the hormone receptor complex is subsequently translocated into the nucleus and binds to defined regions of DNA which are usually composed of incomplete palindromes.[29] Steroid hormone receptors belong to a larger family of ligand dependent transcription factors that also include the retinoic acid, the thyroid hormone, and the vitamin D receptor.[30] The notion that plasmin may have an important role in the inflammatory process instigated an investigation of the modulatory effects of antiinflammatory

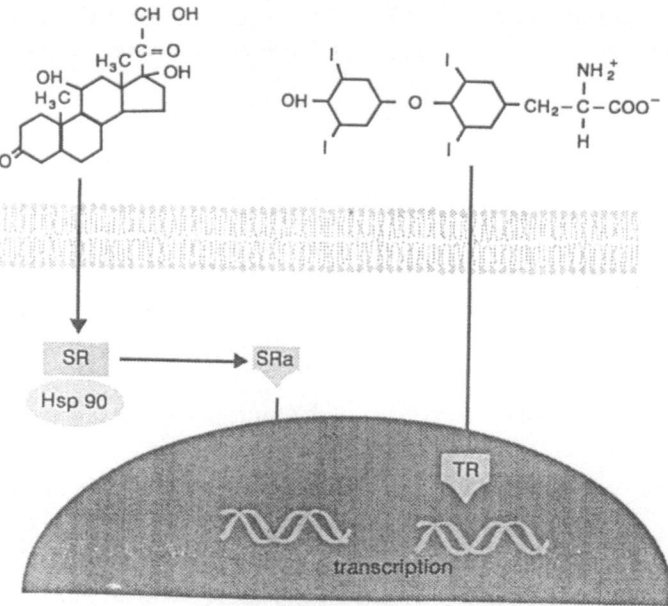

Figure 3. *Steroid hormone and thyroid hormone dependent pathway.*

steroids on plasminogen activator generation. In a study employing activated mouse peritoneal macrophages, the class of steroid with the most potent suppressive effect on the fibrinolytic system were the antiinflammatory glucocorticoids. The inhibition *in vitro* correlated with their antiinflammatory potency *in vivo*.[31] Although these early studies suggested a role for the fibrinolytic pathway in inflammatory processes, they provided no clear indication as to which particular component of the fibrinolytic system was affected. Using tumor cell lines as model systems, the suppression of fibrinolytic activity by glucocorticoids has since been shown to be due to suppression of u-PA biosynthesis[32] in concert with a concomitant increase in PAI-1 biosynthesis and secretion.[33-35] Positive regulation of t-PA mRNA steady state levels by glucocorticoids was shown by Busso et al. in a series of mammary tumor cells.[33,36] Medcalf et al. have shown a positive regulation of t-PA gene transcription by dexamethasone[37] in the human fibrosarcoma cell line HT-1080. These results are surprising in the context of the forementioned effects on u-PA and PAI-1 biosynthesis.

Regulation by Cyclic AMP (Figure 4)

Cyclic AMP dependent metabolic pathways belong to the classical subjects of the study of cellular metabolism.[9,10] There are two types of plasma membrane receptors known, that either stimulate or inhibit adenylcyclase via either stimulatory (G_s) or inhibitory (G_i) G-proteins. Sustained G_s-activation can be achieved by cholera toxin, whereas the action of G_i can be blocked by pertussis toxin. An increase of intracellular cAMP can also be achieved by the plant derived drug forskolin. cAMP activates protein kinase A by binding to its regulatory subunit. Activated protein kinase A modifies through phosphorylation reactions many proteins and thus modulates a variety of parameters of cellular metabolism. Regulatory DNA elements that are responsive to changes in cAMP have recently been identified.[18] They are similar to PMA responsive elements. The protein that binds to this element, cAMP responsive element binding protein (CREB) belongs to a family of immunologically related transcription factors which also include AP-1, Adenovirus E1A product and yeast transcription factor GCN 4.[37,38] The t-PA promoter contains an element (ATGACATCA) which is very similar to the cyclic AMP consensus ATGACGTCA.

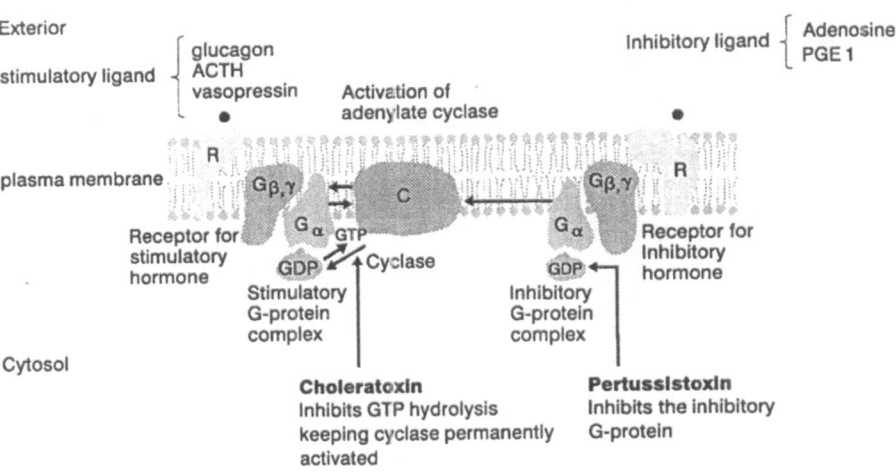

Figure 4. *cAMP dependent pathway.*

Calcitonin and vasopressin which are known to operate via the activation of adenylcyclase are very potent stimulators of u-PA biosynthesis in the porcine kidney cell line LLC PK 1.[22] On the other hand cAMP alone seems to have none or very little effects on t-PA biosynthesis (at least in mouse and human systems). It acts as a potent stimulator however if the cells were primed with PMA[39] (R.L. Medcalf, M. Rüegg, and W.-D. Schleuning, submitted for publication) or - in the case of the F9 mouse teratocarcinoma cells - with retinoic acid.[40]

Regulation via the Hypothalamic-gonadal Axis

A considerable body of evidence suggests a role for the fibrinolytic system in ovulation and early embryonic development. Regulation of t-PA biosynthesis has been of particular interest because of its association with the meiotic maturation of oocytes both *in vitro* and *in vivo*.[41,42] The regulation of t-PA in this system has been studied at the level of gene expression. Bicsak et al.[43] showed that rat oocytes contained t-PA mRNA and synthesized active protease at levels that depended on the maturational state of the oocyte, being low in preovulatory oocytes and increasing at ovulation. After fertilization, t-PA remained high during the first four days of pregnancy and decreased thereafter.

Undifferentiated rat granulosa cells *in vitro* have been shown to express u-PA and t-PA activities that are regulated independently by gonadotropins.[44] The Production of t-PA was shown to require continuous exposure to the gonadotropin, follicle stimulating hormone (FSH), whereas u-PA was unaffected. Low amounts of PAI-1, detected in these cells were shown to be suppressed by physiological concentrations of either FSH or luteinizing hormone (LH). Similar results were obtained by Reich et al.,[45] who studied the distribution of t-PA and u-PA in the granulosa and thecal compartments of the preovulatory follicle, and by Casslén et al.,[46] who investigated regulation of t-PA and PAI-1 by estradiol and progesterone in endometrial tissue.

These observations on t-PA and PAI-1 regulation by gonadotropins and gonadal steroids underpin previous findings that suggested a role of the fibrinolytic system during ovulation and implantation.

REGULATION BY CYTOKINES

Transforming growth factor-ß (TGF-ß) is a regulator of cellular proliferation which induces some cell types to exhibit characteristics[47] of transformation, whereas it inhibits the proliferation of others. It has been shown stimulate the increment of PAI-1 mRNA in WI 38 embryonic lung fibroblasts, and human lung carcinoma cells[48] while simultaneously inhibiting u-PA and t-PA mRNA accumulation.[49,50] On the other hand TGF-ß induces u-PA mRNA and activity in human lung carcinoma cells[48] confirming tissue specific differences of its action. TGF-ß also inhibits collagenase expression in human MRC-5 fibroblasts[51] and stimulates the biosynthesis of components of the extracellular matrix, including fibronectin and procollagen.[52] Investigations of the collagen type 1 gene promoter have identified a protein binding site which mediates the transcriptional activation of this gene by TGF-ß.[53] This element, however is not present in the PAI-1 promoter.

The dynamic equilibrium that characterizes the stability of the extracellular matrix requires a balance between the deposition of structural components, and their degradation by extracellular proteases such as metalloproteinases, cathepsins and plasmin. TGF-ß mediated stimulation of matrix deposition and the inhibition of the biosynthesis of fibrinolytic and collagenolytic factors may be disturbed in pathological conditions associated with aberrant matrix breakdown e.g. in tumor invasion and metastasis and the joint tissue destruction commonly observed in rheumatoid arthritis.

In the human fibrosarcoma cell line HT-1080, that constitutively expresses t-PA, t-PA gene transcription is suppressed by tumor necrosis factor (TNF), a cytokine produced by activated macrophages whereas PAI-1 and PAI-2 are strongly induced.[54] Whether related cytokines like interleukin 1 or TGF-ß have similar effects, remains to be established. Hence, hormones produced by activated macrophages may induce the cells in their neighborhood to decrease the biosynthesis of profibrinolytic factors in favor of antifibrinolytic factors. This mechanism may contribute to the development of fibrin deposits which are frequently found associated with inflamed tissue, or to disseminated intravascular coagulation, a common complication of Gram negative septicemia.

SUMMARY

Known pathways of signal transduction that regulate the biosynthesis of components of the fibrinolytic system are dependent on protein kinase C, tyrosine specific kinases, cAMP, steroid hormones and cytokines. A detailed knowledge of the gene regulation of these compounds may guide us to a better understanding of the underlaying mechanisms that lead to the removal of fibrin and other extracellular matrix proteins in the course of tissue growth, repair and remodelling processes that are central to morphogenesis, neoplasia and the inflammatory response.

REFERENCES

1. F. Bachmann, Fibrinolysis, Thrombosis and Haemostasis, J. Verstraete, H. R. Lijnen, Arnout (eds), Leuven University p 227 (1987).
2. D. Collen, On the regulation and control of fibrinolysis. **Thromb. Haemost.** 43: 77 (1980).
3. D.J. Loskutoff, J.A. van Mourik, L.A. Erickson, and D. Lawrence, Detection of an unusually stable fibrinolytic inhibitor produced by bovine endothelial cells. **Proc. Nat. Acad. Sci. USA** 80: 2956 (1983).
4. R.K.O. Kruithof, J.D. Vassalli, W.-D. Schleuning, R.J. Mattaliano, and P. Bachmann, Purification and characterization of a plasminogen activator inhibitor from the histiocytic lymphoma cell-line U-937. **J. Biol. Chem.** 261: 11207 (1986).
5. O. Saksela, and D.B. Rifkin, Cell associated plasminogen activation: Regulation and physiological functions. **Ann. Rev. Cell Physiol.** 4: 93 (1988).
6. F. Blasi, J.D. Vassalli, and K. Danø, Urokinase type plasminogen activator: proenzyme receptor and inhibitors. **J. Cell Biol.** 104: 801 (1987).
7. Y. Nishizuka, Studies and perspectives of protein kinase C. **Science** 233: 305 (1986).
8. Y. Nishizuka, The molecular heterogeneity of protein kinase C and its implication for cellular regulation. **Nature** 334: 661 (1988).
9. L. Stryer, and H.R. Bourne, G proteins: a family of signal transducers. **Ann. Rev. Cell Biol.** 2: 391 (1986).
10. E. J. Neer, and D.E. Clapham, Roles of G protein subunits in transmembrane signalling. **Nature** 333: 129 (1988).
11. P.W. Majerus, T.M. Connolly, H. Deckmyn, T.S. Ross, T.H. Bross, H. Ishii, V.S. Bansal, and D.B. Wilson, The metabolism of phosphoinositide-derived messenger molecules. **Science** 234: 1519 (1986).
12. H. Rasmussen, The calcium messenger system. **New Engl. J. Med.** 314: 1094 (1986).

13. M.E. Greenberg, and E.B. Ziff, Stimulation of 3T3 cells induces transcription of the *c-fos* proto-oncogene. **Nature** 311: 433 (1984).

14. P. Angel, M. Imagawa, R. Chiu, B. Stein, R.J. Imbra, H.J. Rahmsdorf, C. Jonat, P. Herrlich, and M. Karin, Phorbol ester-inducible genes contain a common cis element recognized by a TPA-modulated trans-acting factor. **Cell** 49: 720 (1987).

15. W.H. Landschulz, P.F. Johnson, and S.L. McKnight, The Leucine Zipper: A hypothetical structure common to a new class of DNA binding proteins. **Science** 240: 1759 (1988).

16. M. Wigler, and I.B. Weinstein, Tumour promoter induces plasminogen activator. **Nature** 259: 232 (1976).

17. E.K. Waller, and W.-D. Schleuning, Induction of fibrinolytic activity in HeLa cells by phorbol myristate acetate. Tissue-type plasminogen activator antigen, and mRNA augmentation require intermediate protein biosynthesis. **J. Biol. Chem.** 260: 6354 (1985).

18. A.R. Montminy, K.A. Sevarino, J.A. Wagner, G. Mandel, and R.H. Goodman, Identification of a cyclic AMP responsive element within the rat somatostatin gene. **Proc. Natl. Acad. Sci. USA** 83: 6682 (1986).

19. M.P. Stoppelli, P. Verde, G. Grimaldi, E.K. Locatelli, and F. Blasi, Increase in urokinase plasminogen activator mRNA synthesis in human carcinoma cells is a primary effect of the potent tumour promoter, phorbol myristate acetate. **J. Cell Biol.** 102: 1235 (1986).

20. D. Belin, F. Godeau, and J.D. Vassalli, Tumor promoter PMA stimulates the synthesis and secretion of mouse prourokinase in MSV-transformed 3T3 cells; this is mediated by an increase in urokinase mRNA content. **EMBO J.** 3: 1901 (1984).

21. J.L. Degen, R.D. Estensen, Y. Nagamine, and E. Reich, Induction and desensitization of plasminogen activator gene expression by tumor promoters. **J. Biol. Chem.** 260: 12426 (1985).

22. Y. Nagamine, M. Sudol, and E. Reich, Hormonal regulation of plasminogen activator mRNA production in porcine kidney cells. **Cell** 32: 1181 (1983).

23. W.-D. Schleuning, R.L. Medcalf, K. Hession, R. Rothenbühler, A. Shaw, and E.K.O. Kruithof, Plasminogen activator inhibitor 2: regulation of gene transcription during phorbol ester mediated differentiation of U-937 human histiocytic lymphoma cells. **Mol. Cell Biol.** 7: 4564 (1987).

24. T. Hunter, and J.A. Cooper, Protein-tyrosine kinases. **Ann. Rev. Biochem.** 54: 897 (1985).

25. L.S. Lee, and I.B. Weinstein, Epidermal growth factor, like phorbol esters, induces plasminogen activator activity in HeLa cells. **Nature** 274: 4406 (1978).

26. G. Grimaldi, P. DiFiore, E.K. Locatelli, J. Falco, and F. Blasi, Modulation of urokinase plasminogen activator gene expression during the transition from quiescent to proliferative state in normal mouse cells. **EMBO J.** 5: 855 (1986).

27. C.L. Lucore, S. Fujii, T.-C. Wun, B.E. Sobel, and J.J. Billadello, Regulation of the expression of type 1 plasminogen activator inhibitor in Hep G2 cells by epidermal growth factor. **J. Biol. Chem.** 263: 15845 (1988).

28. M.G. Catelli, N. Binart, I. Jung-Testas, J.M. Renoir, E.E. Beaulieu, J.R. Feramisco, and W.J. Welch, The common 90 kd protein component of non-transformed 8 S steroid receptors is a heat-shock protein. **EMBO J.** 4: 3131 (1985).

29. G. Schütz, Control of gene expression by steroid hormones. **Biol. Chem. Hoppe-Seyler** 77 (1988).

30. R.M. Evans, The steroid and thyroid hormone receptor superfamily. **Science** 240: 889 (1988).

31. J.D. Vassalli, J. Hamilton, and E. Reich, Macrophage plasminogen activator: modulation of enzyme production by anti-inflammatory steroids, mitotic inhibitors, and cyclic nucleotides. **Cell** 8: 271 (1976).

32. R.L. Medcalf, R.I. Richards, R.J. Crawford, and J. Hamilton, Suppression of urokinase-type plasminogen activator mRNA levels in human fibrosarcoma cells and synovial fibroblasts by anti-inflammatory glucocorticoids. **EMBO J.** 5: 2217 (1986).

33. N. Busso, D. Belin, C. Failly-Crépin, and J.D. Vassalli, Glucocorticoid modulation of plasminogen activators and one of their inhibitors in the human mammary carcinoma cell line MDA-MB-321. **Cancer Res.** 47: 364 (1987).

34. P.A. Andreasen, C. Pyke, A. Riccio, P. Kristensen, L.S. Nielsen, L.R. Lund, F. Blasi, and K. Danø, Plasminogen activator inhibitor type 1 biosynthesis and mRNA level are increased by dexamethasone in human fibrosarcoma cells. **Mol. Cell Biol.** 7: 3021 (1987).

35. R.L. Medcalf, E. Van den Berg, and W.-D. Schleuning, Glucocorticoid modulated gene expression of tissue and urinary-type plasminogen activator and plasminogen activator inhibitor 1 and 2. **J. Cell Biol.** 106: 971 (1988).

36. N. Busso, D. Belin, C. Failly-Crépin, and J.D. Vassalli, Plasminogen activators and their inhibitors in a human mammary cell line (HBL-100). Modulation by glucocorticoids. **J. Biol. Chem.** 261: 9309 (1986).

37. Y.S. Lin, and M.R. Green, Interaction of a common transcription factor, ATF, with regulatory elements in both E1A and cyclic AMP-inducible promoters. **Proc. Natl. Acad. Sci.** 84: 3396 (1988).

38. T. Hai, F. Liu, E.A. Allegretto, M. Karin, and M.R. Green, A family of immunologically related transcription factors that includes multiple forms of ATF and AP-1. **Genes & Development** 2: 1216 (1988).

39. L. Santell, and E.G. Levin, Cyclic AMP potentiates phorbol ester stimulation of tissue plasminogen activator release and inhibits secretion of plasminogen activator inhibitor-1 from human endothelial cells. **J. Biol. Chem.** 263: 16802 (1988).

40. R.J. Rickles, A.L. Darrow, and S. Strickland, Differentiation response elements in the 5' region of the mouse tissue plasminogen activator gene confer two stage regulation by retinoic acid and cyclic AMP in teratocarcinoma cells. **Mol. Cell Biol.** 9: 1691 (1989).

41. J. Huarte, D. Belin, and J.D. Vassalli, Plasminogen activator in mouse and rat oocytes: induction during meiotic maturation. **Cell** 43: 551 (1985).

42. Y.-X. Lui, and A.J.W. Hsueh, Plasminogen activator activity in cumulus-oocyte complexes of gonadotropin-treated rats during the periovulatory period. **Biol. Reprod.** 36: 1055 (1988).

43. T.A. Bicsak, S.B. Cajander, X.-R. Peng, T. Ny, P.S. LaPolt, J.K.H. Lu, P. Kristensen, A. Tsafriri, and A.J.W. Hsueh, Tissue-type plasminogen activator activity in rat oocytes: expression during the periovulatory period, after fertilization, and during follicular atresia. **Endocrinology** 124: 187 (1989).

44. T. Ny, L. Bjersing, A.J.W. Hsueh, and D.J. Loskutoff, Cultured granulosa cells produce two plasminogen activators and an antiactivator, each regulated differently by gonadotropins. **Endocrinology** 116: 1666 (1985).

45. R. Reich, R. Miskin, and A. Tsafriri, Intrafollicular distribution of plasminogen activators and their hormonal regulation *in vivo*. **Endocrinology** 119: 1588 (1986).

46. B. Casslén, A. Andersson, I.M. Nilsson, and B. Åstedt, Hormonal regulation of the release of plasminogen activators and of a specific activator inhibitor from endometrial tissue in culture. **Proc. Soc. Exp. Biol. Med.** 182: 419 (1986).

47. T.J. Todaro, C. Fryling, and J.B. DeLarco, Transforming growth factors produced by certain human tumor cells: polypeptides that interact with epidermal growth factor receptors. **Proc. Natl. Acad. Sci. USA** 77: 5258 (1980).

48. J. Keski-Oja, F. Blasi, E.B. Leof, and H.L. Moses, Regulation of the synthesis and activity of urokinase plasminogen activator in A549 human lung carcinoma cells by transforming growth factor-β. **J. Cell Biol.** 106: 451 (1988).

49. M. Laiho, O. Saksela, P.A. Andreasen, and J. Keski-Oja, Enhanced production and extracellular deposition of the endothelial-type plasminogen activator inhibitor in cultured human lung fibroblasts by transforming growth factor-β. **J. Cell Biol.** 103: 2403 (1986).

50. L.R. Lund, A. Riccio, P.A. Andreasen, L.S. Nielsen, P. Kristensen, M. Laiho, O. Saksela, F. Blasi, and K. Danø, Transforming growth factor β is a strong and fast acting positive regulator of the levels of type-1 plasminogen activator inhibitor mRNA in W1-39 human lung fibroblasts. **EMBO J.** 6: 1281 (1987).

51. D.R. Edwards, G. Murphy, J.J. Reynolds, S.B. Whitham, A.J.P. Doeherty, P. Angel, and J.K. Heath, Transforming growth factor-β modulates the expression of collagenase and metalloproteinase inhibitor. **EMBO J.** 6: 1899 (1987).

52. R.A. Ignotz, and J. Massagué, Transforming growth factor-β stimulates the expression of fibronectin and collagen and their incorporation into the extracellular matrix. **J. Biol. Chem.** 261: 4337 (1986).

53. P. Rossi, G. Karenty, A.B. Roberts, N.S. Roche, M.B. Sporn, and B. de Crombrugghe, A nuclear factor 1 binding site mediates the transcriptional activation of a type 1 collagen promoter by transforming growth factor-β. **Cell** 52: 405 (1988).

54. R.L. Medcalf, E.K.O. Kruithof, and W.-D. Schleuning, Plasminogen activator inhibitor 1 and 2 are tumor necrosis factor/cachectin responsive genes. **J. Exp. Med.** 168: 751 (1988).

RODENT SERPINS :
ACCELERATED EVOLUTION AND NOVEL SPECIFICITIES

JOHN D. INGLIS AND ROBERT E. HILL

Medical Research Council
Human Genetics Unit
Western General Hospital
Crewe Road
Edinburgh EH4 2XU, United Kingdom

INTRODUCTION

In the basic studies of mammalian physiology, development, pharmacology, anatomy, and other disciplines, rodents (rats and inbred mice) have been assumed to be appropriate model systems in the study of human biology and diseases. In many cases, direct comparisons are valid and valuable. However, recently several disappointing results have been reported using the mouse as a model for studying human diseases. The deletion by homologous recombination of the mouse HPRTase gene has not resulted in Lesch-Nyan Disease[1] and the *mdx* mutation of the dystrophin gene shows few phenotypic similarities to muscular dystrophy.[2] In the HPRTase deficiency in mouse, the neurological pathology of Lesch-Nyan is bypassed by a more efficient manner of dealing with toxic metabolites of uric acid.[2] These physiological differences are simply due to the fact that the present day rodents and primates shared their last common ancestor some 80 million years ago (Mya). Therefore, different evolutionary pressures have been operating on these species' ancestors for an appreciable evolutionary time period.

In our studies of rodent serpins we have encountered a gene system which, when compared to human, has diverged considerably over this course of time. Attempts to directly relate the human α_1-antichymotrypsin (α_1-AChy) and α_1-antitrypsin (α_1-AT) genes to rodent counterparts have been problematic and at times confusing. However, as more is learned about the rodent genes it is becoming evident that the serpins provide a window in which to observe some of the events that occur during molecular evolution. In turn understanding the evolutionary events has provided clues as to the biological functions of these proteins.

Serine Proteases and Their Serpin Inhibitors in the Nervous System
Edited by B. W. Festoff
Plenum Press, New York, 1990

MATERIALS AND METHODS

Genetic Analysis

The mouse serpin genes were mapped using recombinant inbred mouse strains.[3] DNA from the BXD strains was digested with Hind III to map the *Spi-2* complex and Bgl II to map the *Spi-1* complex.[4]

Cosmid Clone Analysis

An extensive mouse cosmid library was obtained from L. Stubbs (ICRF, Lincoln's Inn Field, London) cloned into the vector pCos2EMBL.[5] The library was screened using standard procedures.[6] The contiguous arrangement of overlapping cosmids was obtained by establishing the restriction pattern for each cosmid clone using Bam HI, EcoR I, Hind III and Xho I and comparing the pattern to those of other clones. Those clones which contained identical patterns over part of the insert were considered to overlap. The regions of each clone which contained *Spi-2* genes were subcloned into pTZ vectors for further analysis.

Sequence Analysis of Exon 5

Sequence analysis was performed using the dideoxy chain termination reaction of Sanger.[7] Sequence information was obtained directly from the cosmid or from subcloned fragments using oligonucleotides complementary to conserved regions of exon 5.

RESULTS AND DISCUSSION

Serpin Genes and Genetics

In our studies of the rodent serpins it was not immediately apparent that there were genes which encoded the rodent counterparts to human α_1-AChy and α_1-AT. This confusion has arisen from the fundamental differences that exist in the genetic complexity between the rodent and human serpin genes. The simplest situation is found in the human genome in which only a single gene codes for α_1-AChy (Figure 1). In contrast, the gene which encodes α_1-AT has undergone a duplication resulting in a copy which resides just downstream of the α_1-AT gene.[8] The activity of this duplicate has been questioned since in an extensive search no transcripts from this gene have been detected.[9] The α_1-AT and α_1-AChy genes have been placed on the human chromosomal map; both reside on human chromosome 14 linked to the immunoglobulin heavy chain region.[10,11] Methods of determining physical distances have shown that these two serpin genes are within 120 kb of each other.[12]

The mouse counterpart of human α_1-AT and α_1-AChy present a far more complicated picture. In mouse the primordial α_1-AT and α_1-AChy genes have undergone an extensive series of duplication events.[4] We have identified approximately 12 closely related genes which have α_1-AChy as their closest human relative, and, as we show later, each gene is distinctly different from the other. Similarly, the primordial α_1-AT gene has multiplied in mouse; we have estimated the number to be in the region of 6-8 genes. The amplifications have resulted in two gene clusters for these gene families in the mouse genome. The genetics of these two families show similarities to that of α_1-AChy and α_1-AT in human. The 12

α_1-AChy-like genes reside on mouse chromosome 12 closely linked (within one per cent recombination) to the multiple α_1-AT-like genes showing the same close relationship determined in human.[4] These genes are also linked to the immunoglobulin heavy chain region which is genetically 8-10 centiMorgans away. Therefore these genes in mouse and human constitute a conserved linkage group.

Because of the close relationship of these two multigene families both functionally and physically within the genome we have termed this whole region on chromosome 12 the *Spi*-complex; the genes most related to α_1-AT, the *Spi*-1 genes and those most related to α_1-AChy, the *Spi*-2 genes.[4] Each gene in the cluster, until a physiological function can be defined, is given a number as it is elucidated.

We have also studied these genes in the rat,[4] however, the rat is not a well-defined genetic system. The situation in rat most closely reflects that of the mouse as may be expected in this close evolutionary relationship. The rat genome also contains multiple *Spi*-1 and -2 genes although fewer than in mouse. There are approximately 4 or 5 genes at each of the *Spi* loci.

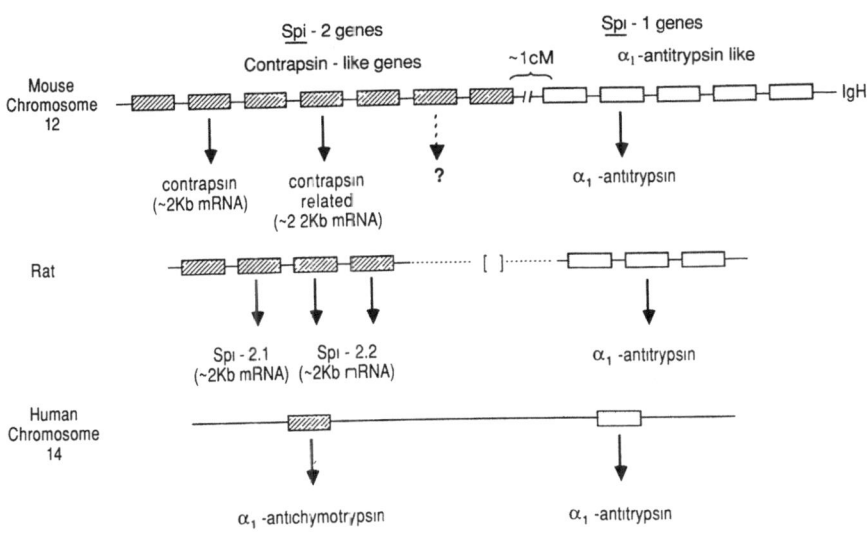

Figure 1. *Spi complex.* Genetics of α_1-AT and α_1-AChy and the counterpart gene complexes in rodents. The boxes represent genes; those that are most similar to and including human α_1-AChy (α_1-antichymotrypsin) are crosshatched. The stippled boxes are those genes most similar to and including human α_1-AT (α_1-antitrypsin). The diagram shows the duplications that have occurred in rodent lineages, and the close genetical relationship of the genes in mouse and proposed in rat. The exact number of genes in the rodent complexes is not shown. The arrows represent several genes that are known to be active by Northern blot analysis of liver mRNA; the size of the mRNA is shown in brackets for the *Spi*-2 genes. Sequence analysis of cDNAs has shown more of these genes to be active (see below). Contrapsin was the first mouse α_1-Achy-like protein identified as being transcribed from an *Spi*-2 gene. The rat *Spi*-2 gene products have been given numbers until their functions have been elucidated. The chromosomal assignments of the genes in mouse and human are listed on the left of the figure.

Structure of the *Spi-2* Locus

Our studies have focused primarily on the mouse *Spi-2* (α_1-AChy-like) cluster of genes. We have undertaken the task of sorting through these genes to determine which are the active genes and how each of the coded proteins function. Much of the *Spi-2* cluster has been isolated on a series of overlapping cosmids. Figure 2 shows the contiguous arrangement of 2 of these genes. This cosmid clone contains the two genes on a DNA insert fragment of 36 kb, showing the tight clustering of the genes of this complex. The whole *Spi-2* complex may constitute about 250-300 kb, or about 0.2% of chromosome 12. The multigene families studied to date, which include the globin genes, the kallikrein genes, and the mouse urinary protein genes, among others, contain genes which are expressed at different times in development, in different tissues, and genes which are not expressed at all (pseudogenes). The answer to the question of why different genes in a family are necessary has not been fully answered. We have attempted to answer this question for the *Spi-2* family by first, studying the structure of the genes to elucidate the primary structure of the protein potentially encoded by each gene and secondly, by analyzing the evolution of the gene.

Evolution of the Reactive Centre Region of the Serpins

The reactive site of a particular serpin can be defined as the amino acids which surround the peptide bond that is cleaved, usually *in vitro*, by the substrate proteinase. The amino acid to the N-terminal side of the cleavage site is termed the P1 amino acid and each adjacent amino acid is numbered sequentially (P2, P3 etc). Those to the C-terminal side are numbered sequentially starting at the P1' amino acid. The amino acid at the P1 position is the most important in determining the specificity of the serpin.[13] Alteration of the chemical nature of this critical residue often leads to changes of inhibitory specificity. The 'classic' example of just such a natural occurrence was found in the mutation α_1-AT Pittsburg.[14] A P1 Met to Arg alteration reduced dramatically native anti-elastase activity, but resulted in a significant gain

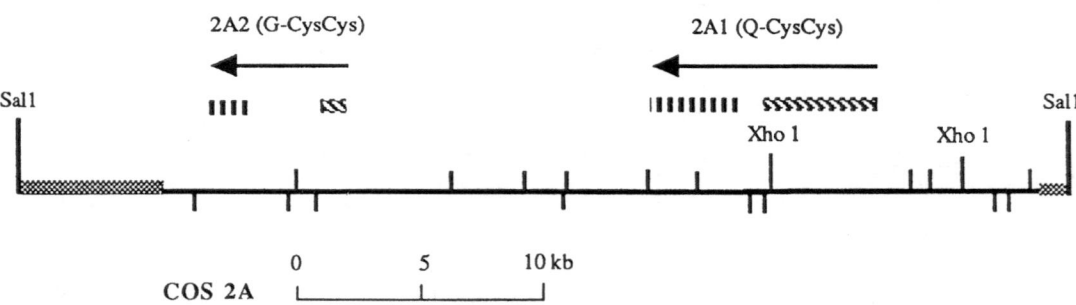

Figure 2. *Arrangement of two Spi-2 genes in a mouse cosmid.* Analysis of this clone shows the close physical relationship of the *Spi-2* genes predicted by the genetic analysis. Restriction sites for the enzymes Hind III and Bam HI are marked by vertical dashes above and below the line, respectively. Sites for the enzymes Sal I and Xho I are also shown. A DNA probe from the 5' end and one from the 3' end (from conserved regions of an *Spi-2* cDNA) were used to place the genes within the cosmid, and to establish the orientation of each gene. The restriction fragments which each probe hybridized to are represented by hatched boxes above the contigs, and the predicted direction of transcription represented by arrows. The amino acid residues at the putative P1-P1' sites are shown above the arrows; the P2 amino acid is also shown, in the one letter code.

(several thousand fold) in antithrombin activity. Clearly, other amino acids that are adjacent to the P1 position affect more subtly the activity of the serpin.[15,15]

The reactive centre is encoded by the last exon (exon 5) of the gene. Primary structural comparison of diverse serpins shows the P1 amino acid to be embedded in a region of hypervariability (of 11-15 amino acids) which in turn is flanked on both sides by highly conserved protein motifs. Mutations in these conserved regions have lead to inactivity of the inhibitor.[17] In our studies, we refer to the P1 containing hypervariable domain as the reactive centre regions since the composition affects inhibitor specificity, and the rate of reactivity. It is this region which has been most valuable in studies of the evolution of the genes.

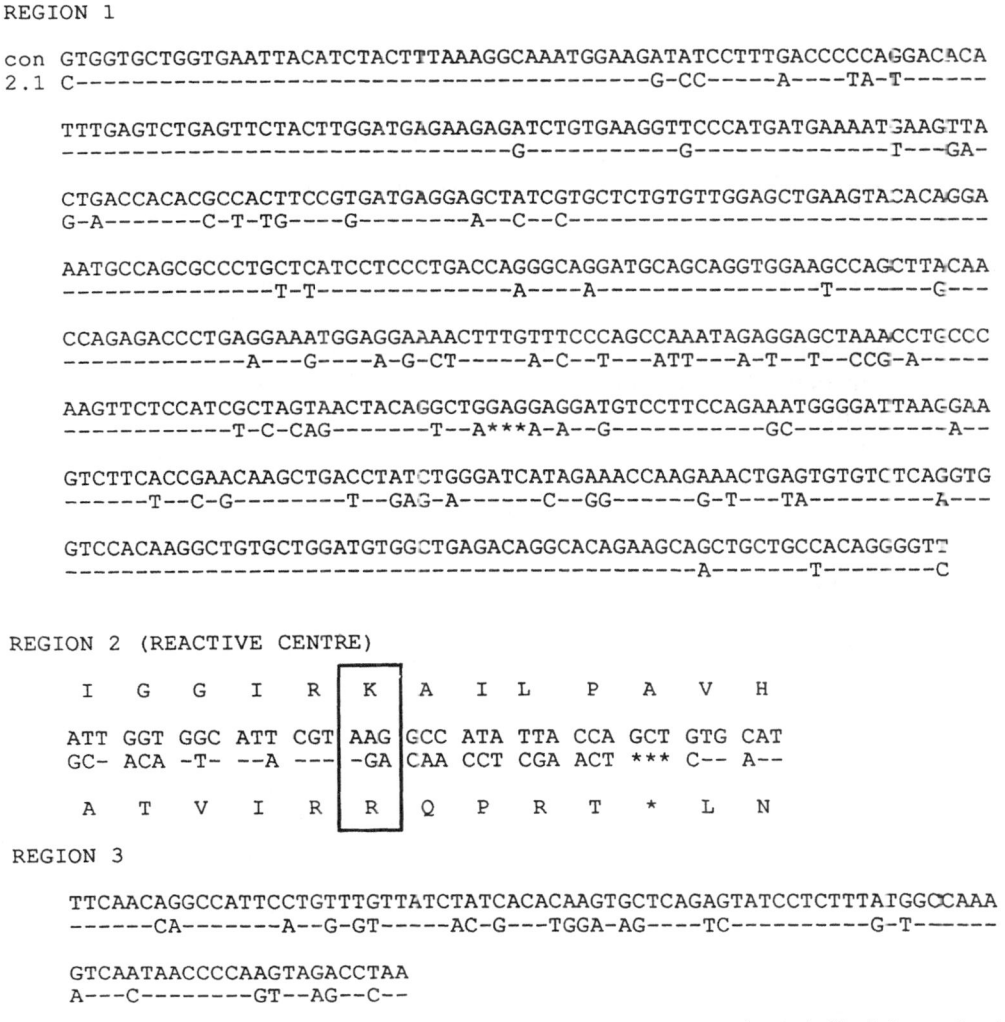

Figure 3. *Sequence comparison of contrapsin (mouse) and rat Spi-2.1.* The 3' half of the nucleotide coding sequence is shown for each gene. Additionally, one letter amino acid sequence is shown for the reactive centre domain, region 2. Sequence identity of rat *Spi*-2.1 with contrapsin is indicated by dashes, deletions by asterisks. The P1 residue for each gene is boxed. con, contrapsin; 2.1, rat *Spi*-2.1.

Region	Nucleotide sequence similarity (%)	Amino acid sequence similarity (%)	Rates of substitution KA	KS
1	84	75	2.8	7.0
2	33	15	28.0	NA
3	75	41	8.7	5.2

For pseudogenes KA - KS

	K
Rodent pseudogenes	4.9
introns	3.7

Rates are expressed as the no. of substitutions x 10^{-9} site/year

KA - rate of substitution at asynonymous sites

KS - rate of substitution at synonymous sites

NA - rate not calculable.

These rates are based on a divergence time for mouse and rat of 25 million years.

Accelerated Evolution in the Reactive Center Domain

The neutral theory of evolution predicts mechanisms of adaptive change of organisms at the molecular level.[18] The basic tenet suggests that regions of the genome which have no function, or parts of a gene that code for unimportant regions of proteins, will accumulate changes most rapidly in the genome. Since there are few or no selective constraints in these regions, the rate of change will reflect the rate of mutation for any particular species. The rate of mutation is assumed to be roughly equivalent between related families and may only be affected by the generation time of the organism.[19] This has led to the idea of the molecular clock, which suggests that the accumulation of molecular changes can be directly correlated to an evolutionary time span. Most of the molecular data, notably that derived from the advent of DNA sequencing, has supported the ideas of proponents of neutral evolution.

We have published an analysis on the rate of change which occurred in members of the *Spi*-2 gene family based on the neutral theory.[20] In this analysis we compared the sequence of a mouse gene which encodes a serpin called contrapsin and a rat gene which encodes a previously unreported serpin which we gave the designation *Spi*-2.1. Nucleotide sequence for these two genes is shown in Figure 3. These two genes are by several criteria orthologous, i.e. direct descendants of the same ancestral gene. These criteria include tissue distribution, mRNA size, and relative high degree of expression. Comparison of these two sequences showed that overall (in particular region 1 [Figure 3]) the genes behaved quite normally in evolution; however, as mentioned above, the predicted reactive center region differed dramatically between these two genes. Using the computer program of Dr W. H. Li[21] we were able to predict the number of nucleotide changes that had occurred in this region. The rate of change is shown in Table 1 compared to the rates for predictably neutral region of DNA. This rate of substitution at the asynonymous positions of codons exceeds the neutral mutation rate as reflected in the changes that have accumulated in synonymous codon positions and in pseudogenes and introns. This contradicts the basic tenet of the neutral theory. Without invoking a mechanism of hypermutability in this region, for which there is no evidence, we have concluded that positive selective forces have acted, resulting in rapid fixation of mutations within a species. This has been termed accelerated evolution. Positive selection acting in evolution necessarily implies novel functions for these proteins. It is unlikely that these inhibitors are acting exclusively, or at all, upon endogenous proteases as this would require rapid co-evolution of the target protease. We feel it is more likely that these inhibitors are acting upon exogenous proteases brought in by infectious agents upon wounding or parasitic infection.

Reactive Centre Regions within the *Spi*-2 Gene Cluster

It follows from the existence of multiple genes in the *Spi*-2 cluster and the variability possible in the reactive centre regions that numerous opportunities exist to establish genes with new functions within this cluster. We have now determined the P1 residue for several genes within the cosmid cluster. Some of these are listed in Figure 4. Within these genes, none of the reactive centre domains are identical and at the P1 amino acid position five different amino acids are present. Most of these genes with the exception of the one designated Cos3E2 (which may be a pseudogene) also contain the conserved protein motifs which border the reactive centre. Therefore, as in the interspecies comparison above these Mus genes contain a hypervariable region defined by distinct conserved boundaries. Contrapsin as previously shown[22] contains at the predicted P1-P1 positions Lys-Ala which we suggested has antitryptic activity due to similarities to bovine pancreatic trypsin inhibitor. The protein encoded by Cos3E46 contains an Arg-Ser at the reactive centre residues which confer specificity. Again, the comparison to other serpins such as protein C inhibitor, antithrombin and α_1-AT-Pittsburg suggests this molecule will be antitryptic. There are indeed several very unexpected amino acids at the P1-P1' positions. Three genes predictably encode a Cys-Cys (Cos6C-28,Cos2Al,Cos2A2) and one an Arg-Cys (sub D). These have not been previously reported as P1-P1' amino acids in active serpins. Only the gene subC containing a Leu-Ser closely resembles the human α_1-ACEy.

We have compared the number of nucleotide changes which have occurred in the reactive centre of four of these genes to that of partial intron sequence. The intron sequences appear remarkably similar, suggesting these genes are derived from quite recent duplication events. We find that the number of changes that have occurred in the reactive centre exceeds those in the intron, again suggesting accelerated evolution as the mode of change. By inference, some or all of these genes have acquired new functions.

Expression of the *Spi*-2 Genes in Mouse

The question arises as to which of these genes are active. We have found at least five are expressed in different tissues. In the liver, we have cloned three different cDNAs. These are contrapsin (Lys-Ala reactive centre), and the genes encoding the Arg-Ser (Cos3E46), and the Met-Ser (EB22/5) reactive centres (Figure 4). We have also cloned two cDNAs from a mouse cell line called EB22. This is a teratocarcinoma cell line which has differentiated towards a chondroblast lineage. These two genes would encode P1-P1' aminoacids of Met-Ser (same gene as found in liver) and, very surprisingly, Cys-Cys (Cos2A2). Contrapsin mRNA is the most abundant of the *Spi*-2RNAs in the liver constituting about 1% of the poly(A)+ fraction. The Arg-Ser and Met-Ser serpins are about 50-100 fold lower. Absolute levels of the two serpins in EB22 cells have not been determined; however the Cys-Cys and Met-Ser mRNA concentration is approximately equal. We have not found the Leu-Ser gene expressed in the liver or EB22 cells.

From the available data we can suggest which of the mouse genes fulfills the physiological role of human α_1-AChy. Two lines of evidence suggest that the Met-Ser serpin is the mouse α_1-AChy equivalent. Firstly, the genes which show the greatest sequence similarity within the last exon to human α_1-AChy are the Leu-Ser gene and the Met-Ser gene. As already stated, however, the Leu-Ser gene has not been found to be expressed in the liver. Secondly in chronic and acute inflammation in human there is a rapid rise in the levels of serum α_1-AChy. Induction of the inflammatory state in mouse using bacterial lipopolysaccharides induces only one of the three mRNAs.[4] Using specific oligonucleotide hybridization we have shown this to be the Met-Ser gene (data not shown). If the human α_1-AChy has an important physiological function, the mouse will have retained this function. We suggest that α_1-AChy in mouse is produced from the Met-Ser encoding gene.

In conclusion, the mouse and human α_1-AChy gene systems are vastly different. There is, within the *Spi*-2 complex of genes, a mouse counterpart to the human α_1-AChy which in expression studies has many of the properties of the human α_1-AChy. In studies of the physiological activity and areas of expression (i.e. such as brain, bone and liver) the mouse Met-Ser encoding gene may be a useful model system. The one unexpected outcome of these studies is that the mouse genome provides a rich new store of serpins with different reactive centre regions. Determining the mechanism of action and the physiological function of these murine proteins is the next challenge.

```
Sequence                P2  P1  P1'  P2'  P3'

cos3E-46                Phe-Arg-Ser-Arg-Arg
subC                    Pro-Leu-Ser-Ala-Lys
subD                    Leu-Arg-Cys-***-Gly
2A1                     Gln-Cys-Cys-Gln-Gly
2A2                     Gly-Cys-Cys-Ala-Val       Mouse Genes
subE                    Phe-Met-Ser-Ala-Lys
3E2                     Phe-Gln-Ser-Ser-Lys
Mouse α1-PI             Pro-Tyr-Ser-Met-Pro
Contrapsin              Gly-Arg-Lys-Ala-Ile


Human α1-PI             Pro-Met-Ser-Ile-Pro
Human PCI               Phe-Arg-Ser-Ala-Arg       Human Genes
Human α1-AChy           Leu-Leu-Ser-Ala-Leu
```

Figure 4. *Partial reactive centres of some inhibitors.* Partial amino acid sequence around the reactive centre residues of mouse *Spi*-2 genes as predicted from cosmid analysis. Sequences were derived from cosmid clones isolated from 129/J inbred mice. Sequences of α_1-PI and contrapsin are from Ref. 22. Position of amino acids with respect to the P1 residue is indicated. The amino acids are designated by the three letter code.

REFERENCES

1. T. Doetschman, R.G. Gregg, N. Maeda, M.L. Hooper, D.W. Melton, S. Thompson, and O. Smithies, Targeted correction of a mutant HPRT gene in mouse embryonic stem cells. **Nature** 330: 576 (1987).

2. R.P. Erickson, Why isn't a mouse more like a man? **Trends Genet.** 5: 1 (1989).

3. K.L. Bennett, P.A. Lalley, R.K. Barth, and N.D. Hastie, Mapping the structural genes coding for the major urinary proteins in the mouse: combined use of recombinant inbred strains and somatic cell hybrids. **Proc. Natl. Acad. Sci. USA** 79: 1220 (1982).

4. R.E. Hill, P.H. Shaw, R.K. Barth, and N.D. Hastie, A genetic locus closely linked to a protease inhibitor gene complex controls the level of multiple mRNA transcripts. **Molec. Cell Biol.** 5: 2114 (1985).

5. E. Ehrich, A. Craig, A. Poustka, A.-M. Frischauf, and H. Lehrach, A family of cosmid vectors with the multi-copy R6K replication origin. **Gene** 57: 229 (1987).

6. T. Maniatis, E.F. Fritsch, and J. Sambrook, Amplification, storage, and screening of cosmid libraries, in: "Molecular Cloning: A Laboratory Manual" CSH, Cold Spring Harbor, NY p 304 (1982).

7. F. Sanger, S. Nicklen, and A.R. Coulson, DNA sequencing with chain terminating inhibitors. **Proc. Natl. Acad. Sci. USA** 74: 5463 (1977).

8. J.-J. Bao, L. Reed-Fourquet, R.N. Sifers, U.J. Kidd, and S.L.C. Woo, Molecular structure and sequence homology of a gene related to α_1-antitrypsin in the human genome. **Genomics** 2: 165 (1988).

9. G.D. Kelsey, M. Parkar, and S. Povey, The human alpha-1-anti-trypsin-related sequence gene: isolation and investigation of its expression. **Ann. Hum. Genet.** 52: 151 (1988).

10. D.W. Cox, V.D. Markovic, and I.E. Teshima, Genes for immunoglobulin heavy chains and for α_1-antitrypsin are localized to specific regions of chromosome 14q. **Nature** 297: 428 (1982).

11. M. Rabin, M. Watson, V. Kidd, S.L.C. Woo, W.R. Beeg, and F.H. Ruddle, Regional Localization of α_1-antichymotrypsin and α_1-antitrypsin genes on human chromosome 14. **Som. Cell Mol. Genet.** 12: 209 (1986).

12. G.D. Kelsey, D. Abeliovich, C.J. McMahon, D. Whitehouse, G. Corney, S. Povey, D.A. Hopkinson, J. Wolfe, G. Mieli Vergani, and A.P. Mowat, Cloning of the human α_1 antichymotrypsin gene and genetic analysis of the gene in relation to α_1 antitrypsin deficiency. **J. Med. Genet.** 25: 361 (1988).

13. R. Carrell, Therapy by instant evolution. **Nature** 312: 14 (1984).

14. M.C. Owen, S.O. Brennan, J.H. Lewis, and R.W. Carrell, Mutation of antitrypsin to antithrombin. **New Engl. J. Med.** 309: 694 (1983).

15. A.W. Stephens, B.S. Thalley, and C.H.W. Hirs, Antithrombin III Denver, a reactive site variant. **J. Biol. Chem.** 262: 1044 (1987).

16. W.E. Holmes, H.R. Lijnen, and D. Collen, Characterization of recombinant human α_2- antiplasmin and of mutants obtained by site directed mutagens is of the reactive site. **Biochemistry** 26: 5133 (1987).

17. W.E. Holmes, H.R. Lijnen, L. Nelles, C. Kluft, H.K. Nieuwenhuis, D.C. Rijken, and D. Collen, α_2-Antiplasmin Enschede: alanine insertion and abolition of plasmin inhibitory activity. **Science** 238: 209 (1987).

18. M. Kimura, The neutral theory of molecular evolution. Cambridge Univ Press, London (1983).

19. W.H. Li, and M. Tanimura, The molecular clock runs more slowly in man than in apes and monkeys. **Nature** 326: 93 (1987).

20. R.E. Hill, and N.D. Hastie, Accelerated evolution in the reactive centre regions of serine protease inhibitors. **Nature** 326: 96 (1987).

21. W.-H. Li, C.-I. Wu, and C.-C. Luo, A new method for estimating synonymous and nonsynonymous rates of nucleotide substitution considering the relative likelihood of nucleotide and codon changes. **Molec. Biol. Evol.** 2: 150 (1985).

22. R.E. Hill, P.H. Shaw, P.A. Boyd, H. Banmann, and N.D. Hastie, Plasma protease inhibitors in mouse and man: divergence within the reactive centre regions. **Nature** 311: 175 (1984).

STRUCTURE OF THE HUMAN PROTEASE NEXIN GENE AND EXPRESSION OF RECOMBINANT FORMS OF PN-I

MICHAEL McGROGAN, JACKIE KENNEDY, FRED GOLINI,
NINA ASHTON, FRANCES DUNN, KIMBERLY BELL,
EMILY TATE, RANDY W. SCOTT, AND
CHRISTIAN C. SIMONSEN

INVITRON Corp.,
301 Penobscott Dr.,
Redwood City, CA 94063

INTRODUCTION

Protease nexin (PN-I) is a member of the serpin family of serine protease inhibitors that are characterized by the formation of an irreversible complex with the catalytic site of their target serine proteases.[1] PN-I is known to inhibit a number of biologically relevant serine proteases such as thrombin, urokinase, plasmin, and plasminogen activators.[2-4] The significance of regulatory inhibitors in controlling the activity of the serine proteases has only begun to be appreciated in the areas of cell movement, blood coagulation, fibrinolysis, extracellular matrix modulation, and mitosis.[5-8] Recently PN-I has been shown to be identical to glial derived nexin, which has been reported to possess neurite extension activity on peripheral nerve cells.[3,9-11] Native PN-I is a glycoprotein of approximately 45,000 daltons that is secreted by various fibroblasts and extravascular cells.[12] Multiple forms of PN-I have been described which differ in their behavior on SDS-PAGE, pH gradient gels, and heparin affinity chromatography.[4,13] Although glycosylation differences can most certainly account for some of these differences, we have identified two species of human PN-I which we have designated αPN-I and βPN-I that differ by a net change of three amino acids.[14]

In order to better understand the physiological role of PN-I, we have undertaken a molecular characterization of the two forms of PN-I. Here, we determine the structure of the PN-I gene providing insight into the mechanism involved in the synthesis of α and β PN-I. We have compared the levels of expression of the PN-I gene and one of its target proteases, tissue plasminogen activator, in a variety of mammalian cells. We have also taken the approach of expressing α and β PN-I as individual recombinant proteins using a mammalian cell culture system[14] to produce quantities of each form of PN-I suitable for comparative activity studies. Here, we show that the α form of recombinant PN-I is produced as a biologically active serpin which retains its target protease specificity and high affinity heparin binding site.

Serine Proteases and Their Serpin Inhibitors in the Nervous System
Edited by B. W. Festoff
Plenum Press, New York, 1990

MATERIALS AND METHODS

Tissue Culture

The human promyelocytic leukemia cell line (HL60), and the acute myelogenous leukemia cell lines (K562, KGla) were obtained from the American Type Culture Collection (Rockville, MD). These lines were maintained at 37°C in suspension culture at 0.5 to 2.0 x 10^6 cells/ml in modified DMEM supplemented with 10% fetal calf serum, penicillin, streptomycin, and glutamine. The human adenocarcinoma cell line SK-HEP-1, Bowes melanoma, human glioblastoma lines (U373, U251) and human colon cell line (HCC18Co) were obtained from American Type Culture Collection. The SV-40 transformed SV-colon fibroblasts were obtained from Dr. Charles Hsu (Invitron Corp.) and the human foreskin fibroblast cells (HFF) were obtained as a primary culture at passage 9 from Dr. Joffre Baker, University of Kansas. These cells were carried as monolayer cultures in complete DMEM supplemented with 10% fetal calf serum.

Preparation of mRNA and Northern Blot Analysis

Cytoplasmic RNA was isolated from cells as previously described.[14] Membrane bound and soluble RNA was isolated as described above from hypotonic cell extracts that were fractionated by centrifugation.[15] Polyadenylated mRNA was purified by oligo-dT chromatography.[16] Poly A-containing cytoplasmic RNA was denatured in formamide/formaldehyde buffer and samples were applied to a 1.2% agarose formaldehyde gel and electrophoresed at 100 volts for 6 hr.[17] The RNA was electrotransferred to a Genescreen membrane according to the manufacturer's recommendation (NEN-Dupont, Wilmington, DE).

Contruction of Human Genomic Library

High molecular weight DNA was prepared from a human colon cell line (HCC18Co) or from the SK hepatoma cell line as previously described.[14] 12-18 kb sized DNA fragments from an incomplete SauIIIA digest were cloned into the λ - phage vector EMBL-3 (Promega Corp., Madison, WI) according to the supplier's recommendations. The resulting library contained greater than 10^6 independent clones having an average insert size of 14 kb.

Hybridization Procedures

DNA fragment probes were hybridized under standard conditions in a buffer containing 50% formamide at 42°C.[18] Oligomer probe hybridization was carried out at 37°C or 42°C in hybridization buffer containing 0.9 M NaCl and processed as described.[14] Filters were exposed to Kodak XAR X-ray film at -80°C. The oligomer probes were labeled by the addition of ^{32}P to the 5' base using polynucleotide kinase and purified over a Bio Gel P-4 spin column.[18] DNA fragment probes were labeled by primer extension using Klenow polymerase in the presence of α^{32}-P-dCTP (Prime-Time C kit, IBI, New Haven, CT) and purified over a Biogel P-60 spin column or Sephadex G-50 column.

DNA Manipulations

Recombinant GT10 DNA[19] was purified from phage plate lysates of the positively hybridizing clones according to the method of Davis et al.[20] Positively hybridizing inserts were subcloned into pUC plasmids. Recombinant EMBL-3 DNA was purified from liquid lysates of positively hybridizing clones by PEG 6000 precipitation followed by CsCl banding.[21] Fragments of interest were subcloned into pUC plasmids for further analyses. Plasmid DNA was purified after alkaline-SDS lysis of saturated cultures of JM101, followed by polyethylene glycol precipitation. Synthetic DNA was prepared using a Pharmacia Gene Assembler (Pharmacia LKB Biotechnology, Inc., Piscataway, NJ) and purified on OPC columns according to the manufacturer's recommendations (Applied BioSystems, Foster City, CA).

DNA Sequencing and Analysis

Restriction fragments encompassing the regions of interest were inserted into phage M13 vectors, mp18 and mp19, or PUC vectors[22,23] and sequenced by the dideoxynucleotide chain termination method[24] using T7 polymerase (U.S. Biochemicals and Pharmacia-LKB) according to the manufacturer's recommendations. Sequence analysis was performed using the PC Gene programs (Intelligenetics Corp., Mountain View, CA) on a Compaq 386 microcomputer.

Expression of PN-I in CHO Cells

The chinese hamster ovary cell line, CHO DUX B-11, lacking dihydrofolate reductase[25] was transfected with SV40-based plasmids containing either the α or β form of PN-I linked to a murine DHFR cDNA.[26,27] Clones arising after growth in F-12 medium lacking glycine, hypoxanthine, and thymidine were isolated and expanded in selective media.[28] The level of PN-I was determined on conditioned media from confluent monolayers grown in serum free media using a thrombin inhibition assay (below). Positive clones were isolated, expanded in selective medium, and subjected to increasing concentrations of the folate analog, methotrexate.[28]

Expression of Recombinant PN-I in Insect Cells

The insect cell line, Sf9 (ATCC, CRL1711), was grown in suspension in low protein media (Excell, JR Scientific) at 27°C.[29] The baculovirus expression plasmid, pAC373,[30] was modified to contain the coding sequence of α PN-I inserted behind the polyhedrin promoter. Virus isolated from recombinant plaques were evaluated for PN expression by infecting cultures of Sf9 cells and assaying the serum free conditioned media 72 hrs later.

Thrombin Inhibition Assay of PN-I

Thrombin (0.25 N.I.H. Units/ml) was incubated with an equimolar amount of αPN-I (determined by titration of the PN-I against thrombin) in 0.1 M Tris, pH 8.1 containing 0.15 M NaCl, 1% polyethylene glycol and 1 μg/ml aprotinin in the presence or absence of 10 μg/ml heparin (170 USP units/mg) as indicated. The reaction was stopped by the addition of substrate (Kabi S-2238) to a final concentration of 0.2 mM and residual thrombin activity determined at 405 nM on a kinetic microtiter plate reader (Molecular Devices, Menlo Park, Ca).

PN-I/Protease Complex Formation

Recombinant αPN-I (25 μg) was incubated with urokinase (22 μg) or thrombin (8 μg) in phosphate buffered saline in volume of 0.2 ml for 60 minutes at 37°C. The reaction was stopped by the addition of an equal volume of 20% trichloroacetic acid and the protein precipitated at 4°C overnight. Samples were then solubilized in

149

reducing SDS sample buffer and loaded onto a 10% polyacrylamide gel. SDS-PAGE was performed by the method of Laemmli.[31]

RESULTS

Identification of Two Different Forms of PN-I

We have previously described two different types of PN-I cDNA clones that were isolated from human foreskin fibroblasts (HFF) cDNA library using mixed oligomer probes designed from the amino terminal sequence of protease nexin I.[14] Two of the presumably full length clones, approximately 2000 bp long, were sequenced and found to contain an open reading frame that encoded PN-I. The DNA sequence of the first clone predicted a protein containing a 19 amino acid signal peptide that preceded the 378 residues characteristic of secreted PN-I (Figure 1).

Upon careful comparison of the DNA sequence, the cDNA clones were found to fall into two classes, differing only by an insertion of three nucleotides within the coding sequence. A total of six clones were sequenced across this region and two of the clones were found to contain the additional codon. The sequence of the region that defines the two forms of PN-I, designated alpha PN-I (αPN-I) and beta PN-I (βPN-I), is shown in Figure 1. αPN-I has an Arg at amino acid residue 310 which was found to be interrupted by a CAG codon in the βPN-I clone. This alteration maintains the open reading frame and predicts that βPN-I would be 379 amino acids (Figure 1), containing one more amino acid than the α form.

In order to determine the relative levels of expression of αPN-I versus βPN-I in HFF cells, specific oligomer probes were designed and synthesized: a 14 base α probe and a 17 base β probe spanning the altered codon. We have previously reported the results of analyzing 26 independent PN cDNA clones, originally isolated from the HFF cDNA library,[14] where we found that two thirds of the clones hybridized to the αPN-I oligomer and one third specifically to the βPN-I probe. These results suggested that the two forms of PN-I are made at a ratio of approximately 2 αPN to 1 βPN in HFF cells and that native PN-I is a mixture of α and β PN-I. Although the produc-

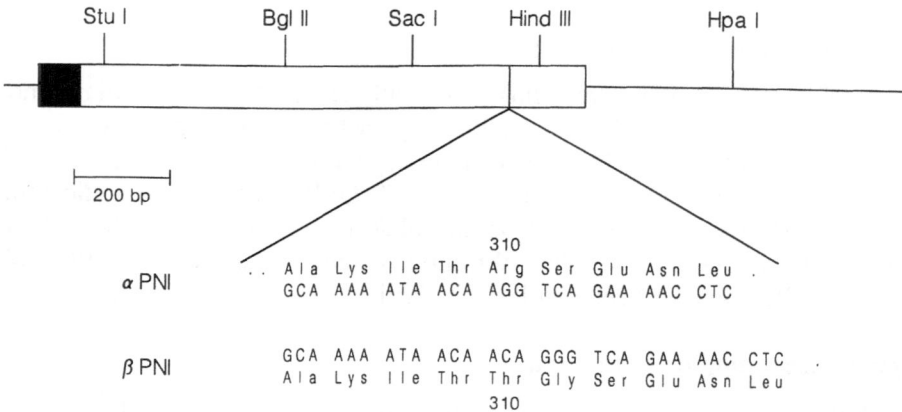

Figure 1. *Location of the α/β splice region of PN-I.* The linear restriction map of the approximately 2000 bp PN-I cDNA clone showing an expanded region encoding α and β forms. The coding sequence is indicated by the boxed region. The portion comprising the 19 amino acid secretory signal peptide is filled in. Relevant unique restriction sites are indicated. The nucleotide and deduced amino acid sequence spanning position 310 is shown for α and β PNI.

tion of αPN-I by HFF cells has been confirmed by amino acid sequencing of a tryptic peptide that is predicted by cleavage at Arg310, the corresponding tryptic peptide from βPN-I, lacking Arg310, has not been characterized.

Analysis of PN-I mRNA Expression

The distribution of PN-I mRNA in HFF cells was examined by Northern hybridization analysis to determine if the PN-I message is present on membrane bound polysomes as predicted for a secreted product. The cytoplasmic extracts prepared from HFF cells were fractionated into membrane bound and soluble components. The RNA isolated from these samples was electrophoresed on a denaturing gel, the resulting filter blot hybridized to the PN-I cDNA probe, and the results are shown in Figure 2. The approximately 2300 nucleotide PN-I RNA can be seen easily in total cytoplasmic RNA (lane 2), and appears to be quantitatively polyadenylated as measured by the intense band in the poly-A enriched sample (lane 6) as opposed the poly-A minus sample (lane 1). The hybridization pattern obtained for equivalent amounts of soluble RNA (lane 3) and membrane bound RNA (lane 4), demonstrate that virtually all of the detectable PN-I RNA is present in the membrane bound sample, as expected for mRNA associated with membrane bound polysomes.

The expression of PN-I in a number of human cell lines was examined by Northern and slot blot hybridization analysis. Poly-A containing RNA isolated from two normal primary cell lines, HFF (foreskin fibroblasts) and HCF (colon fibroblasts), hybridized very strongly to the PN probe (Figure 2, lanes 6 and 10). In comparison,

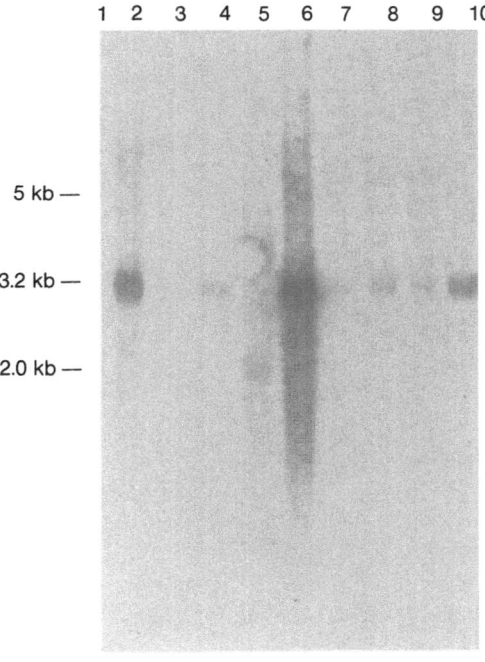

Figure 2. *Distribution of PN-I mRNA and a survey of PN-I expression levels.* Cytoplasmic RNA was fractionated on denaturing agarose gel and electrophoretically transferred to a membrane filter. The filter was hybridized to ^{32}P labeled PN-I cDNA probe and exposed to X-ray film. Samples were loaded in the following lanes: 1) HFF Poly-A minus (20 μg); 2) HFF total soluble (20 μg); 3) HFF total soluble (5 μg); 4) HFF total membrane (5 μg), 5) RNA size marker; 6) HFF Poly-A+ (5 μg); 7) human 293 Poly-A+ (10 μg), 8) SK-HEP Poly-A+ (10 μg); 9) Bowes melanoma Poly-A+ (10 μg); and 10) HCF Poly-A+ (5 μg).

the PN-I RNA levels in the other cell lines ranged from 5 to 20 fold lower. RNA samples from a number of the tumor lines (SK Hepatoma, Bowes melanoma, and 293 cells) gave barely detectable signals and no significant hybridization was found to samples derived from the leukemic lines HL-60, KG-1, or K562. By contrast, tumor lines that were of neural origin (U373 and U251 glioblastoma) were found to hybridize well, although the levels were several fold lower than that seen for the normal fibroblasts. The RNA samples from the HFF, HCF, and U373 cells were found to hybridize to both the α and β PN-I oligomer probes suggesting that both forms of PN-I are expressed. Hybridization of the oligomer probes to the remaining samples was not significantly above background.

Differential Expression of PN-I and Plasminogen Activator

In order to examine the possibility that the expression level of the PN-I gene may effect the regulation of a target serine protease such as tissue plasminogen activator (tPA), the relative levels of PN-I and tPA mRNA was determined in several human cell lines. A Northern blot containing cytoplasmic Poly-A RNA from primary, transformed, and tumor cell lines was first hybridized to a tPA cDNA probe (Figure 3A). Very high levels of tPA RNA were found in the normal colon (lane 1) and the SV40 transformed colon fibroblast cells (lane 4). Interestingly enough, the other primary fibroblast cell, HFF (lane 3), gave low, but detectable levels of hybridization.

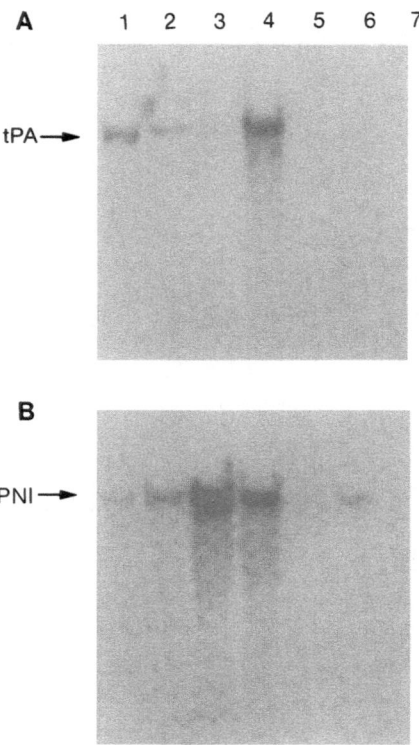

Figure 3. *Expression of tPA and PN-I mRNA in human cells.* Northern blot containing Poly-A cytoplasmic RNA was prepared and hybridized as in Figure 2. Panel A shows auto-radiograph of blot hybridized to ^{32}P labeled tPA cDNA probe. Probe was eluted from filter at 80°C in 80% formamide buffer. Panel B shows autoradiograph of blot probed with PN-I cDNA. Lane 1) HCF (2 μg); 2) U373 (5 μg); 3) HFF (5 μg); 4) SV-colon (5 μg), 5) RNA size marker; 6) SK-HEP (10 μg); 7) HL-60 (10 μg).

The series of transformed cells; U373 glioblastoma (lane 2, SK hepatoma (lane 6) and HL-60 leukemia (lane 7) showed wide variation in the levels of tPA RNA.

In contrast to the results above, a very different pattern is obtained upon rehybridization of this blot to the PN-I cDNA probe (Figure 3A). The HFF cells show a band that is several-fold more intense than any other sample. The SV-40 transformed colon cells contain about five times more PN-I RNA than the parental normal colon cell. Detectable levels of PN-I RNA were found in the U373 (lane 2) and SK hepatoma (lane 6), and as expected the HL-60 cell line was negative. It is interesting to note that the increase in PN-I expression relative to tPA in comparing the primary colon fibroblasts and the SV-40 transformed colon cells, may explain the increase in tPA present as an inactive complex in the SV-colon cells (data not shown). These results indicate that the PN-I gene and the tPA gene are independently regulated in a variety of different normal and transformed cell types. The only apparent correlation that is consistent with the data is that the cells expressing higher levels of PN-I appear to more attachment dependent during cell growth.

Structure of the PN-I Gene

Previous studies performed in our laboratory were designed to resolve the structure of the PN-I gene surrounding the region encoding the α and β forms of PN-I.[14] Based upon preliminary Southern mapping of the human PN-I gene, the most likely mechanism for the synthesis of α and β appeared to be due to the alternate splicing of the precursor RNA immediately preceding the region encoding Arg310. In order to confirm the specific structure which is predicted by this model, the PN-I gene was cloned and characterized. Although the possibility of allelic genes encoding the two forms cannot be totally eliminated, the data presented here supports the mechanism of alternate splicing.

In order to determine the structure of the PN-I gene, human genomic libraries were constructed in the EMBL-3 bacteriophage vector by cloning incomplete SauIIIa-digested DNA. A library made from DNA isolated from the SK hepatoma cell lines was screened using the full length PN-I cDNA as probe. Several overlapping clones were isolated and characterized by restriction mapping. A representative genomic clone, GN-3, was sequenced throughout the coding regions. Clone GN-3 (Figure 4) containing approximately 15,000 bp was found to span the 3' half of the PN-I gene. This clone accounts for about 500 bases of coding sequence that is divided into 5 exons by four intervening regions, each of approximately 2000 bp. Intron VI of GN-3 (Figure 4) was found to map to the sequence encompassing amino acid 310 of PN-I, and contained a second splice acceptor site utilized by the αPN-I mRNA adjacent to the CAG triplet of the βPN-I mRNA.

A second genomic library, constructed from human colon cell DNA, was screened with the 700 bp 5' BglII fragment as probe. Several positive clones were analyzed and found to represent a series of similar clones, none of which overlapped with the GN-3 clone. As predicted by the genomic Southern results, clone GN-11 (Figure 4) contains a 11,000 bp BglII fragment which encodes the 5' half of PN-I localized to three exons. However, upon sequencing of the exons in this clone another intron was found to interrupt the 5' most exon at a position 22 bases upstream from the translational start codon. Rescreening of the 5'-PN positive genomic clones for the missing 5' noncoding region proved negative. These results necessitated a third round of library screening using a 120 bp BamHI fragment comprising the very 5' end of the cDNA. Clone GN-5 was isolated and found to contain a single BamHI site which mapped to the 5' noncoding region and presumptive PN-I promotor. The region

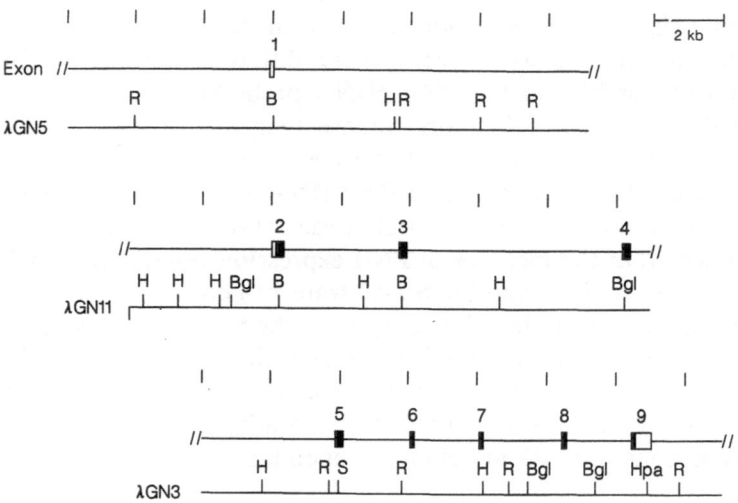

Figure 4. *Restriction map of the human PN-I gene.* The map of three nonoverlapping recombinant phage clones is shown. GN5 and GN11 were isolated from the normal colon cell genomic library. GN3 was isolated from the SK-Hep genomic library. The exons were localized by Southern blotting of restriction digests[32] and by subcloning the appropriate restriction fragments. Positions of the coding sequences are indicated by the boxed areas. Open boxes represent noncoding information; filled in boxes represent the protein coding region. Exons are numbered 1-9 and are arrayed in a 5'-3' direction. Relevant restriction sites used to map each clone are shown: R, EcoRI; B, BamHI; H, HindIII; Bgl, BglII; S, SalI; and Hpa, HpaI.

encompassing exon I was sequenced and found to be extremely G + C rich (over 78%) and was preceded by putative promoter elements upstream from the predicted start of transcription.

Taken together, the three clones shown in Figure 4 describe a composite of the human PN-I gene. The PN-I gene has been mapped to three nonoverlapping genomic clones which span a minimum of 36 kb, a size which is considerably larger than reported for other serpin genes.[33,34] The 2000 bases represented in the PN-I mRNA are interrupted by 8 introns the largest of which is intron I which maps to the 5' promotor region and is estimated to be over 12,000 bp long (Figure 5). Since the genomic clones we have isolated are bounded by intervening sequences and do not overlap, the exact size of the gene and the map of the two boundary introns are given as minimum estimates. The DNA sequence was determined for all of the exons; each of the intron/exon junctions uphold the consensus GT/AG splice sequence with the GT dinucleotide at the 5' border and the AG at the 3' end of the intron.[35] The smallest exon (8) contains but 83 bases and the largest exon (9), containing the 3' end, is 728 bp long of which only 34 bases are of coding sequence.

Comparison of the PN-I Gene and the PAI-1 Gene

The gene encoding the plasminogen activator inhibitor PAI-1 has recently been characterized.[33] Although the PN-I gene is approximately three times the size of the PAI-1 gene, the location and number of introns, as shown in Figure 5, are remarkably similar. Both genes have eight introns the first of which is located in the 5' noncoding region. When the positions of the introns are aligned according to amino acid sequence of the mature protein, it readily becomes apparent that the corresponding introns are located at virtually identical positions and, as a result, the size of the respective exons are the same. Similar to the two forms of PN-I we have described,

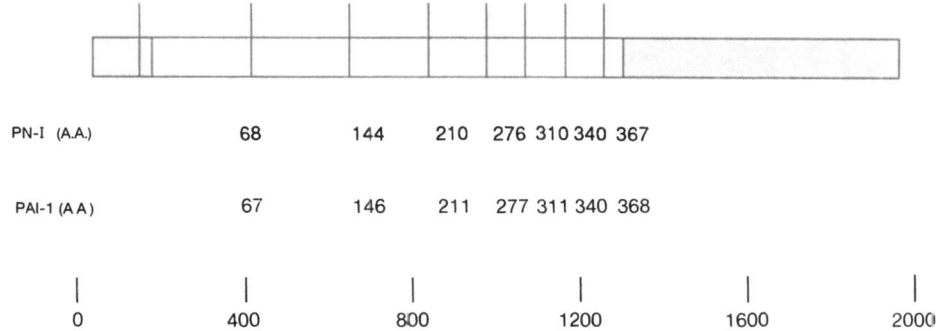

Figure 5. *Comparison of intron location in the PN-I gene and the PAI-1 gene.* Linear map of the 2000 bp PN-I cDNA showing 5' and 3' noncoding region (shaded areas), and the region encoding PN-I protein (open area). The position of the introns are designated by roman numerals at the top. Below the map the amino acid position, based upon the mature protein, is shown for each intron. The positions for the PAI-1 introns are from Loskutoff et al.[33]

an alternate form of PAI-1 has also been reported wherein a 21 base insertion occurs, adding seven amino acids at position 339 in the predicted protein sequence.[36] Interestingly enough, this also maps to an intron/exon junction; but in the case of PAI-1 the region involved in alternate splicing occurs at the 3' border of intron VII, as compared to intron VI for the PN-I gene. The high degree of conservation in maintaining the location of the introns between these genes is even more suprising when one realizes that the amino acid sequence homology is only about 40%.[37,38] The distribution of the introns in the coding sequence appears consistent with separation of the proposed subdomain structure of these two serpins. The conservation of intron position in PN-I and PAI-I suggest that these genes are very closely related in the evolution of the serpin gene superfamily and that intron position may serve a function.

Expression of Recombinant PN-I in Mammalian and Insect Cells

In order to demonstrate functionality of the α and β forms of PN-I, the coding sequences of α and β PN-I cDNAs were cloned into a mammalian expression vector under control of the early SV-40 promoter.[26,27] After transfection of DHFR minus CHO cells, recombinant cells were allowed to grow in selective media. The PN positive clones were identified by assaying conditioned media for thrombin inhibition activity. The recombinant cells resulting from transfection with the α and β expression vectors produce a protein which comigrates with purified PN-I at approximately 48,000 daltons in size, and is specifically recognized by the anti PN monoclonal antibody (Figure 6). Consistent with native PN which is secreted as a processed glycoprotein, it appears that CHO cells similarly modify and secrete the recombinant forms of PN-I.

Since a variety of mammalian proteins have been produced at high levels in insect cells using a baculovirus expression system,[30] we elected to evaluate this system for production of protease nexin. We constructed an expression vector that contained the PN-I coding sequence driven by the polyhedron promotor to allow us to express PN-I in insect cells. The αPN-I produced by Sf9 cells infected with PN-I recombinant baculovirus was subjected to immunoblot analysis (Figure 6). In contrast to the CHO

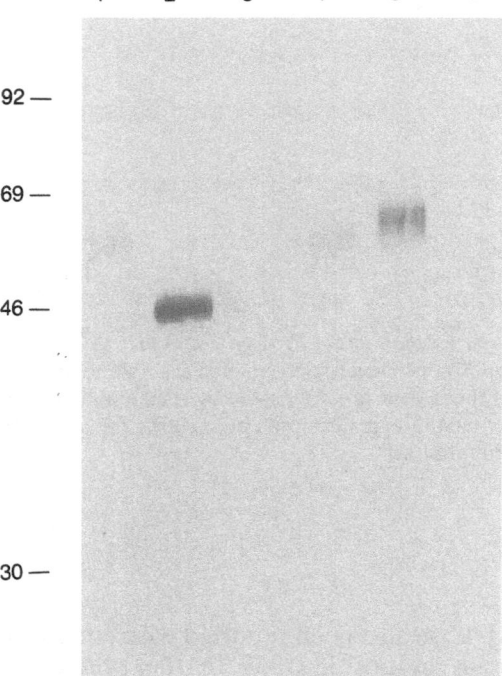

Figure 6. *Western blot of recombinant PN-I expressed in insect and mammalian cells.* Sf9 cells were infected with wild type baculovirus or αPN recombinant baculovirus and media was harvested 72 hrs post-infection. Semi-confluent cultures of recombinant CHO cells were fed serum free media was processed after 24 hrs. Concentrated samples were resolved on a reducing SDS polyacrylamide gel and electrophoretically transferred to a nitrocellulose membrane.[21] The blot was developed using an anti PN-I monoclonal antibody (16.352.1) and alkaline phosphatase-conjugated goat anti-mouse IgG. Lanes: 1) Purified αPN-I (50 ng); 2) Baculovirus αPN-I; 3 Baculovirus control; 4) CHO βPN-I; 5) CHO αPN-I; 6) CHO control.

derived PN-I, the PN-I produced in insect cells by the baculovirus vector migrates at approximately 44,000 daltons, significantly smaller than mammalian PN-I. This result suggests that insect produced PN-I is processed and modified in a different manner than that produced by mammalian cells, possibly reflecting a deficient glycosylation pattern. It is noteworthy that despite the apparent altered glycosylation, insect αPN-I is both secreted by infected cells, and maintains its activity as a potent thrombin inhibitor.

Characterization of CHO Recombinant αPN-I

Recombinant αPN-I was purified from CHO cells for further study and comparison to human fibroblast PN-I. Purified αPN-I was found to be virtually identical to fibroblast PN-I as judged by SDS-PAGE, amino acid composition analysis and amino terminal sequence analysis (data not shown). Fibroblast PN-I is known to inhibit both thrombin and urokinase, two serine proteases with seemingly opposite roles in blood clotting and clot dissolution respectively. Recombinant PN-I was added to thrombin or urokinase and the mixtures analyzed by SDS-PAGE (Figure 7). Both proteases formed SDS stable high molecular weight complexes with αPN-I as would

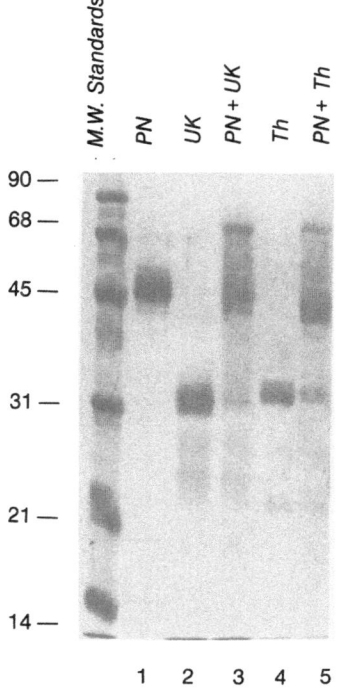

Figure 7. *Recombinant αPN-I complex formation with thrombin and urokinase.* αPN-I was purified from conditioned serum free media harvested from recombinant CHO cells. Coomassie stained SDS polyacrylamide gel of αPN-I-protease complexes formed at 37°C and heated in SDS reducing buffer. Lanes 1) PN-I; 2) urokinase; 3) PN-I + urokinase; 4) thrombin; 5) PN-I + thrombin.

be predicted for a member of the serpin family. Thus recombinant αPN-I retains the broad protease specificity identified for fibroblast PN-I.

In addition to its broad protease specificity, PN-I has been characterized by its similarity to antithrombin III (ATIII). Both PN-I and ATIII contain high affinity heparin-binding domains and react with thrombin at a rate which is greatly accelerated in the presence of heparin. To test whether αPN-I maintained this important regulatory domain we reacted PN-I with thrombin in the presence or absence of heparin. The results (shown in Figure 8) indicate that heparin does indeed accelerate the rate at which αPN-I inhibits thrombin by at least 10 fold, with the reaction being essentially complete within the first minute in the presence of heparin as compared to 30 minutes in the absence of heparin. Thus, CHO cells express αPN-I as an active, functional molecule with the expected protease specificity and having an intact heparin binding domain.

DISCUSSION

We have described here the structure of the PN-I gene and have found that it is consistent with the model of alternate splicing as the mechanism for the synthesis of α and β PN-I. The alternate splicing of the PN-I RNA illustrates one of the smallest

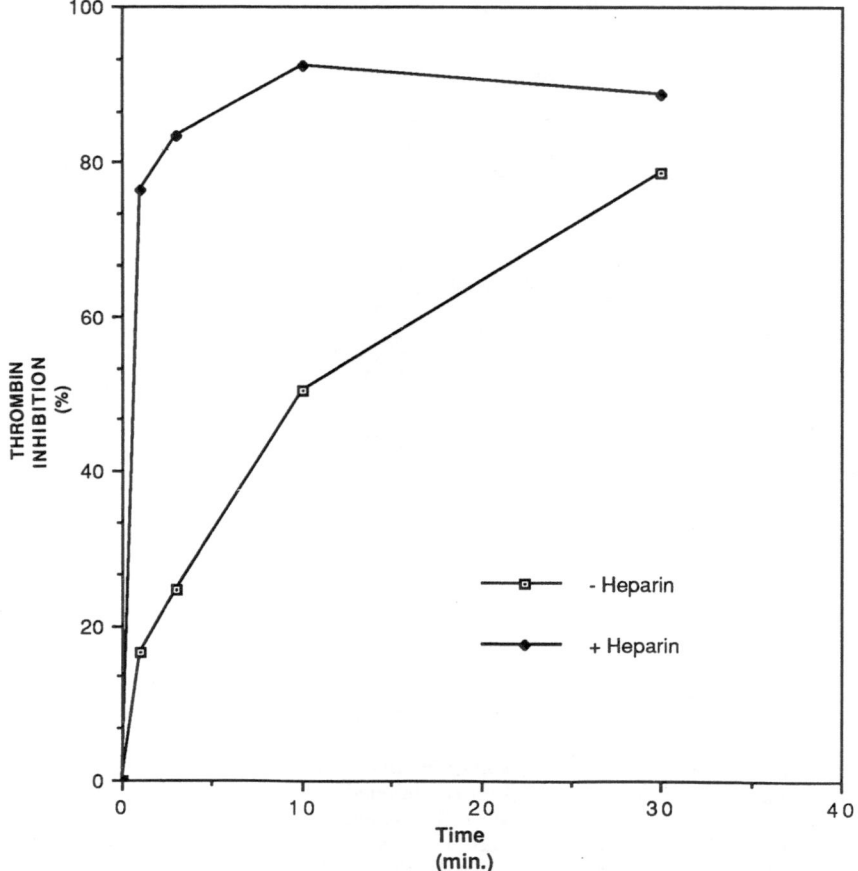

Figure 8. *Heparin accelerates the inhibition of thrombin by αPN-I.* Thrombin was incubated with an equimolar amount of recombinant αPN-I purified from CHO cells (determined by titration of the PN-I) in the presence (closed diamonds) or absence (open squares) of heparin.

alterations which has been characterized; the transfer of a triplet codon, the minimum genetic unit that can maintain the same open reading frame in the mRNA. Since most examples of alternate splicing involve larger units of information which encode functional domains or entire exons, the significance of inserting three bases into a coding sequence does not have an immediate impact. But in the context of the 40,000 bp that we estimate span the PN-I gene, the precise processing of three nucleotides in the precursor mRNA resulting in α and β PN-I is amazing. For the three cell types where we have detected both forms of PN-I, the β form, containing the three base insert, represents a significant fraction of the PN-I synthesized and we estimate that it represents about one-third of the PN-I transcripts. Although the process of alternate splicing of gene transcripts has been shown in a number of systems to result in functionally distinct products; such as immunoglobulins, myosin, tropomyosin, fibrinogen, and many viral gene products,[39] the functional differences between α and β PN-I remain to be demonstrated.

The survey of PN-I expression in a number of normal and transformed human cell lines in culture indicated that very few of the cell lines were making significant levels of PN-I RNA. The leukemic lines which exhibit a pronounced attachment-in-dependence for growth did not contain detectable levels of PN-I RNA. Of the transformed cells analyzed, the glioblastoma lines had the highest levels of PN-I RNA, with the exception of the SV-40 transformed colon line which was derived from normal colon cell. The normal primary foreskin and colon fibroblasts were clearly expressing PN-I RNA at levels many fold higher than any other cell examined. By contrast, the results of a collaborative study (B.A. White, University of Kansas) to examine RNA levels in human tissue found that fresh foreskin did not express detectable levels of PN-I. These data suggest that the expression of PN-I may be induced upon establishing primary cells in culture. The high levels of PN made in primary cultures may reflect involvement of protease inhibitors in protecting the established focal adhesions that attachment dependent cells require for growth. In support of the idea that PN-I may protect attachment sites and maintain the extracellular matrix, we have made the observation that in recombinant CHO cells overproducing PN-I there is a tremendous build up of extracellular material around the cells.

In order to address the question as to functional differences between the two forms of PN-I, we have developed recombinant cell lines that express α or β PN-I. The strategy of overproducing the two cloned forms of PN-I in CHO cells has given us a source of purified α and β PN-I to study. Preliminary evaluation of both forms of PN-I have shown that they are similarly processed and secreted into the media. α and β PN-I made in CHO cells have the expected activities of heparin binding and thrombin inhibition. The PN-I produced by insect cells is secreted into the media and behaves as a lower molecular weight form which probably is under-glycosylated, but surprisingly, the insect PN-I retains high levels of activity.

The significance of the PN-I gene expressing two different forms of PN-I product is not clear. The subtle change in the amino acid sequence, from Arg-Ser in αPN-I to Thr-Gly-Ser in βPN-I would not seem to dramatically change the structure of the protein. This region is located approximately 36 residues N-terminal of the active site of PN-I and would appear, as deduced from the structure of α_1-antitrypsin to be a flex region connecting β-pleated sheets on the exterior of PN-I.[38] Although the question of functional differences remains to be addressed, it is interesting to note that if Arg 310 is exposed to αPN, it could be susceptible to cleavage by a serine protease such as thrombin. The substitution of Arg310 as we have found occurs for βPN-I could result in an inhibitor that would be more stable in the presence of thrombin and perhaps complex thrombin more efficiently. To date the recombinant forms of PN-I appear to have all of the expected activities of natural PN-I, experiments are underway to more critically compare the reaction rates, substrate specificity, receptor binding, and stability of complex formation of the α and β forms of recombinant PN-I.

SUMMARY

The analysis of PN-I cDNA clones isolated from a human fibroblast cDNA library has shown that two forms of PN-I mRNA are expressed in these cells. The two species, designated αPN-I and βPN-I, differ by the alternate splicing of a single codon into the mRNA that results in the change from Arg at position 310 for αPN-I to Thr-Gly for βPN-I. A survey of PN-I RNA expression levels in a variety of normal and transformed human cell lines found that the highest levels were present in cells that were attachment dependent. The human PN-I gene was cloned and found to span over 40 kb and contain 8 introns, one of which is present in the 5' non-coding region. The position of the introns in the coding sequence shows remarkable similarities in comparison with the PAI-gene, suggesting a close evolutionary relationship. Examination of the coding regions of the gene show that the codon for Arg310 in αPN-I is interrupted by intron VI, and that the 3' donor site of βPN-I (CAG) is itself preceded by an AG splice acceptor site. The structure of the PN-I gene thus confirms alternate splicing as a mechanism for the synthesis of the two forms of PN-I.

Recombinant α and βPN-I have been expressed in both mammalian and insect cell cultures. Purified αPN-I from recombinant CHO cells migrated as a broad band on SDS-PAGE of approximately 45,000 molecular weight and formed SDS-stable complexes with thrombin and urokinase. The rate of inhibition of thrombin by αPN-I was accelerated by heparin in a dose dependent manner. The results of preliminary comparison of CHO produced recombinant PN-I and natural HFF PN-I are consistent with both forms having comparable activities *in vitro*.

REFERENCES

1. R.W. Carrell, and D.R. Boswell, Serpins: The super family of serine protease inhibitors. p. 403-418. In: Protease Inhibitors. A. Barret and G. Salvesen (eds.). Elsevior, Amsterdam (1987).
2. J.B. Baker, D.A. Low, R.L. Simmer, and D.D. Cunningham, Protease nexin: a cellular component that links thrombin plasminogen activator and mediates their binding to cells. Cell 21:37-45 (1980).
3. J.B. Baker, M. McGrogan, C.C. Simonsen, R.W. Scott, R.S. Gronke, and A. Honeyman, Protease nexin I. Structure and potential functions. In: The Pharmacology and Toxicology of Proteins. Holrenberg and Winklehalee, eds., U.C.L.A. Symposium (1987).
4. R.W. Scott, B.L. Bergman, A. Bajpai, R.T. Hersh, H. Rodrigues, B.N. Jones, S. Watts, and J.B. Baker, Protease nexin. Properties and a modified purification procedure. J. Biol. Chem. 260:7029-7034 (1985).
5. D.A. Low, R.W. Scott, J.B. Baker, and D.D. Cunningham, Cells regulate their mitogenic response to thrombin through secretion of protease nexin. Nature 298:476-478 (1982).
6. J.B. Baker, D.A. Low, D.L. Eaton, and D.D. Cunningham, Thrombin-mediated mitogenesis. The role of secreted protease nexin. J. Cell. Physiol. 112:291-297 (1982).
7. B.L. Bergman, R.W. Scott, A. Bajpai, S. Watts, and J.B. Baker, Protease nexin I inhibits destruction of extracellular matrix by human tumor cells. Proc. Natl. Acad. Sci. USA 83:996-1000 (1986).
8. R.S. Gronke, B.L. Bergman, and J.B. Baker, Thrombin interaction with platelets: influence of platelet protease nexin. J. Biol. Chem. 262:3030-3036 (1987).
9. S.K. Gloor, K. Odink, J. Guenther, H. Nick, and D. Monard, A glial-derived neurite promoting factor with protease inhibitory activity belongs to the protease nexins. Cell 47:687-593 (1986).
10. D.J. Knauer, R.A. Orlando, and D. Rosenblatt, The glioma cell-derived neurite promoting activity protein is functionally and immunologically related to human protease nexin-1. J. Cell Physiol. 132:318-324 (1987).
11. D. Gurwitz, and D.D. Cunningham, Thrombin modulates and reverses neuroblastoma neurite outgrowth. Proc. Natl. Acad. Sci. USA 85:3440-3444 (1988).
12. J.B. Baker, D.J. Knauer, and D.D. Cunningham, Protease nexins: secreted protease inhibitors that regulate protease actions at and near the cell surface, 153-172. In: The Receptors, Vol. 3. P.M. Conn, (ed.) Academic Press, N.Y. (1986).
13. R.W. Scott, and J.B. Baker, Purification of human protease nexin. J. Biol. Chem. 258:10439-10444 (1983).
14. M.J. McGrogan, J. Kennedy, M.P. Li, C. Hsu, R.W. Scott, C.C. Simonsen, and J.B. Baker, Molecular cloning and expression of two forms of human protease nexin I. Bio/Tech. 6:172-177 (1988).
15. S.L. Berger, Isolation of cytoplasmic RNA: Ribonucleoside-vanadyl complexes. In: Methods of Enzymology. S.L. Berger, and A.R. Kimmel, (eds.) 152:227-234, Academic Press, N.Y. (1987).
16. J. Aviv, and P. Leder, Purification of biologically active globin messenger RNA by chromatography on oligothymidylic acid-cellulose. Proc. Natl. Acad. Sci. USA 69:1408-1412 (1972).
17. P. Thomas, Hybridization of denatured RNA and small DNA fragments to nitrocellulose. Proc. Natl. Acad. Sci. USA 77:5201-5205 (1980).
18. T. Maniatis, E.F. Fritsch, and J. Sambrook, Molecular cloning. Cold Spring Harbor Laboratory Press. Cold Spring Harbor, N.Y. (1982).
19. T.V. Hunyh, R.A. Young, and R.W. Davis, In: DNA Cloning: A practical approach. D.M. Glover (ed)., Vol. 1 (1985).
20. R.W. Davis, D. Botstein, and J.R. Roth, Advanced bacterial genetics. Cold Spring Harbor

Laboratory Press. Cold Spring Harbor, N.Y. (1980).

21. F.M. Ausubel, R. Brent, R.E. Kingston, D.D. Moore, J.A. Smith, J.G. Seidman, and K. Struhl (ed), Current Protocols in Molecular Biology. Wiley Interscience, John Wiley and Sons. New York, N.Y. (1987).

22. J. Messing, and J. Viera, A new pair of M13 vectors for selecting either DNA strand of double-digest restriction fragments. Gene 19:269-276 (1982).

23. J. Messing, New M13 vectors for cloning. Methods Enzymol. 101:20-78 (1983).

24. F. Sanger, S. Nicklen, and R. Coulson, DNA sequencing with chain termination inhibitors. Proc. Natl. Acad. Sci. USA 74:5463-5467 (1977).

25. G. Urlaub, and Chasin, Isolation of Chinese hamster cell mutants deficient in dihydrofolate reductase activity. Proc. Natl. Acad. Sci. USA 77:4216-4220 (1980).

26. C.C. Simonsen, and A.D. Levinson, Isolation and expression of an altered mouse dihydrofolate reductase cDNA. Proc. Natl. Acad. Sci. USA 80:2495-2499 (1983a).

27. C.C. Simonsen, and A.D. Levinson, Analysis of processing and polyadenylation signals of the hepatitis B virus surface antigen using SV40-HBV chimeric plasmids. Molec. and Cell. Biol. 3:2250-2258 (1983b).

28. G.S. Gasser, C.C. Simonsen, J.W. Schilling, and R.T. Schimke, Expression of abbreviated mouse dihydrofolate reductase genes in cultured hamster cells. Proc. Natl. Acad. Sci. USA 79:6522-6528 (1982).

29. M.D. Summers, and G.E. Smith, A manual of methods for Baculovirus Vectors and Insect Cell Culture Procedures. Texas Agricultural Experiment Station Bulletin No. 1555 (1987).

30. V.A. Luckow, and M.D. Summers, Trends in the development of baculovirus expression vectors. Bio/Tech. 6:47-55 (1988).

31. U.K. Laemmli, Cleavage of structural proteins during the assembly of the head of bacteriophage T4. Nature 227:680-685 (1970).

32. E. Southern, Detection of specific sequences among DNA fragments separated by gel electrophoresis. J. Mol. Bio. 98:503-517 (1975).

33. D.J. Loskutoff, M. Linders, J. Keijer, H. Veerman, H. van Heerikhuizen, and H. Pannekoek, Structuree of the human plasminogen activator inhibitor 1 gene: nonrandom distribution of introns. Biochem. 26:3763-3768 (1987).

34. L. Strandberg, D. Lawrence, and T. Ny, The organization of the human plasminogen-activator-inhibitor-1 gene. Implications on the evolution of the serine-protease inhibitor family. Eur. J. Biochem. 176:609-616 (1988).

35. R. Breathnach, and P. Chamdon, Organization and expression of eukaryotic split genes coding for proteins. Ann. Rev. Biochem. 50:349-383 (1981).

36. H. Pannekoek, H. Veerman, H. Lambers, P. Diergaarde, C.L. Verwiej, A.-J. van Zonneveld, and J.A. van Mourik, Endothelial plasminogen activator inhibitor (PAI): a new member of the serpin gene family. EMBO J. 5:2539-2544 (1986).

37. J. Sommer, S. Gloor, G.F. Rovelli, J. Hofsteenge, H. Nick, R. Meier, and D. Monard, cDNA sequence coding for a rat glial derived nexin and its homology to members of the serpin family. Biochem. 26:6407-6410 (1987).

38. R. Hubert, and R. Carrell, Implications of the three-dimensional structure of α_1-antitrypsin for structure and function of serpins. Biochem. 28:8951-8966 (1989).

39. S.E. Leff, M.G. Rosenfeld, and R.M. Evans, Complex transcriptional units: Diversity in gene expression by alternative DNA processing. Ann. Rev. Biochem. 55:1091-1117 (1986).

EVOLUTIONARY ADAPTATIONS OF SERPINS AND THEIR USE IN DESIGNING NEW PROTEINASE INHIBITORS

JAN POTEMPA, BIE-HUE SHIEH, AMELIA GUZDEK, ADAM DUBIN,
AMY COLQUHOUN AND JAMES TRAVIS

Dept. of Biochemistry
University of Georgia
Athens, Georgia 30602

INTRODUCTION

Human plasma contains a myriad of proteins with specialized functions. Of these, proteinase inhibitors represent 10% of the total protein and can be classified as the third major group in this fluid, after albumin and the immunoglobulins. Their functions are to regulate proteolytic enzymes which are normally utilized in coagulation, fibrinolysis, complement activation, phagocytosis, etc. Thus, since these processes involve a variety of proteinases with vastly different specificities it is only logical that specific inhibitors must also be utilized to regulated each system on an individual basis.

Examination of the numerous proteinase inhibitors in serum has resulted in the observation that several are structurally related. Indeed, these inhibitors are now classified as a superfamily of proteins referred to as Serpins (serine proteinase inhibitors).[1] Their common features may be summarized as follows:

1. They are of high molecular weight (40 kD to 50 kD)
2. They form NaDodSO$_4$ stable complexes with serine proteinases.
3. They are single headed, forming only 1:1 complexes.
4. They usually have a single, primary, target enzyme.
5. Complexes have an extremely low Ki.
6. Most, but not all members, are proteinase inhibitors.

The above features of Serpins have resulted in the posing of a number of questions with regard to how this family of inhibitors function. In the next several sections of this review we will attempt to answer some of the more perplexing ones.

HOW DO SERPINS FUNCTION AS INHIBITORS?

The mechanism by which Serpins form complexes with proteinases is only partially understood. It seems clear that these inhibitors have a reactive site loop which is

Serine Proteases and Their Serpin Inhibitors in the Nervous System
Edited by B. W. Festoff
Plenum Press, New York, 1990

exposed on the protein surface and which offers a sequence of amino acids readily attacked by proteinases. During complex formation the proteinase attempts to hydrolyze a specific peptide bond at the reactive site of the inhibitor with the specificity for proteinase interactions being primarily decided by the amino acid residue (P_1) donating the carboxyl group to the peptide bond being attacked.[2] However, as will be discussed later this is not the only criteria for inhibitor specificity.

IS SERPIN COMPLEX FORMATION REVERSIBLE?

Complexes which are formed between Serpins and specific target proteinases only slowly hydrolyze the reactive site peptide bond so that the proteinase remains tightly bound to their controlling inhibitor. Such complexes are unusual in that they cannot be dissociated by boiling in NaDodSO$_4$, thus suggesting the formation of a covalent linkage between proteinase and inhibitor. However, recent studies indicate that this effect may be artificial.

Experiments performed in our laboratory have shown that one inhibitor, α-2-antiplasmin (α-$_2$-AP), can form complexes at one of two overlapping but independent sites.[3] The first position, with an Arg-Met P_1-P_1' reactive site, is specific for complex formation with plasmin or trypsin, while in the second site, where the Met residue becomes P_1 and the reactive site is a Met-Ser linkage, chymotrypsin is specifically complexed. As shown in the schematic given below, the addition of α_2-macroglobulin (α-$_2$-M) to either complex could result in the trapping of any free proteinase obtained from complex dissociation. Since dissociation can, theoretically, occur in two directions to give either active inhibitor (I) or cleaved, inactive inhibitor (I*), inhibition of a new proteinase added to this mixture would indicate the presence of functional inhibitor and that complex formation is reversible. However, it should be noted that k_{-1} must be greater than k_2 for I to be accumulated.

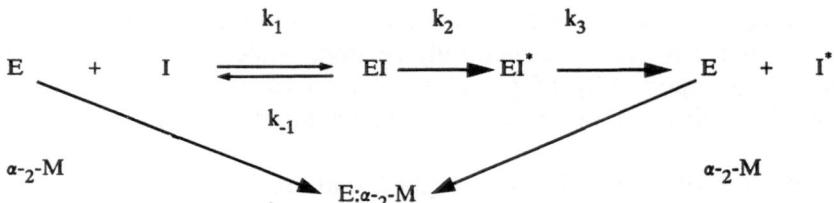

We chose α-$_2$-AP to demonstrate complex reversibility because by using two very different proteinases we could clearly detect new inhibitory activity. In this case we formed α-$_2$-AP:chymotrypsin complexes, allowed them to dissociate in the presence of α-$_2$-M, and then removed the latter from the system. Residual inhibitor (I and/or I*) was then mixed with trypsin to determine if inhibitory activity had been re-generated. A similar experiment was performed using α-$_2$-AP:trypsin complexes and later adding chymotrypsin.

In both experiments we found that α-$_2$-M accelerated complex dissociation, resulting in the production of active inhibitor.[4] Thus, with α-$_2$-AP it is possible that both reactive sites could be utilized, alternately, if α-$_2$-M were present to capture dissociating proteinases. Our results strongly suggest that α-$_2$-AP, and probably other Serpins with inhibitory activity, forms Michaelis and/or tetrahedral complexes with target proteinases but does not normally form acyl intermediates since this would involve irreversible peptide bond cleavage, major conformational changes,[5] and inhibitor inactivation.

HOW ARE SERPINS REGULATED?

Three possible mechanisms would appear to be utilized to control Serpin activity and to insure that some proteolytic events are continually maintained. The first method, oxidative inactivation, only seems to involve inhibitors with Met in the P_1 position, such as α_1-proteinase inhibitor (α-$_1$-PI). As shown previously,[6] oxidation of this inhibitor by either chemical or biologically derived oxidants results not only in the conversion of the P_1 Met to Met(SO) but also a 2000-fold reduction in the k_{ass} for neutrophil elastase. As a result the modified inhibitor cannot compete with substrates such as elastin for elastase binding sites and some proteolysis of this connective tissue protein occurs.[7]

A second mechanism for Serpin regulation is through proteinase inactivation. In this case advantage is taken of the fact that all Serpins with inhibitory activity have an exposed reactive site loop which is readily attacked by a multitude of proteinases. Thus, those proteinases which are not complexed by a Serpin may use the latter as a substrate and cleave peptide bonds at a variety of positions within the reactive site loop. These proteinases may be from prokaryotes or eukaryotes, from pathogenic organisms or endogenous systems, and from all four classes of proteinases. Invariably, the result is a major conformational change in the clipped protein and inhibitor inactivation.[8-10] In some cases enzyme to Serpin ratios of 1:2000 can result in complete inhibitor inactivation in a matter of minutes and, thereby, a significant reduction in local active inhibitor levels.

Finally, examination of k_{ass} values of a variety of proteinases with specific Serpins indicates that many of those which are not oxidation sensitive react more slowly with their target enzyme. This may allow some competition by substrates so that conversion to products may still be maintained at a slow but regular rate.

IS CARBOHYDRATE REQUIRED FOR INHIBITORY ACTIVITY?

All Serpins present in human plasma are glycoproteins. However, the role of the carbohydrate moiety in inhibitor function has not been delineated. In the case of α-$_1$-PI, it has been suggested that glycosylation of the Z variant is probably slower than that for the normal M form of the protein, thus accounting for its lower concentration in plasma.[11] This inability to correctly glycosylate and secrete the Z inhibitor is apparently due to a conformational change in protein structure brought about by a double charge mutation (Glu_M to Lys_Z).[12]

In our laboratory we have also been interested in the role of carbohydrate in Serpin function. We treated human HEPG-2 cells, which make and secrete several Serpins, with a variety of glycosylation inhibitors and measured both secretion and inhibitory activity. As shown in Table 1 tunicamycin reduced secretion of α-$_1$-PI dramatically, while 1-deoxynojirimycin was less effective, and swainsonine and 1-deoxymannojirimycin had no effect. Similar results were obtained with α-$_2$-M but, surprisingly, α_1-antichymotrypsin (α-$_1$-achy) synthesis and secretion seemed unaffected. These results indicate that the absence of carbohydrate (tunicamycin treatment) or inhibition of carbohydrate trimming prior to complex carbohydrate formation (1-deoxynojirimycin) affects inhibitor secretion.

In contrast, carbohydrate plays no apparent role in inhibitor function, at least with regard to α-$_1$-PI function, since all forms of this inhibitor were equally effective as inhibitors of either porcine trypsin or human neutrophil elastase (Table 2). Significantly, we found that both the plasma and liver forms of α-$_1$-PI obtained from

Table 1. Effect of Inhibitors of Glycosylation on Serpin Secretion in Human HEPG-2 Cells.

Sample	Inhibitor[a]		
	α-$_1$-PI	α-$_1$-Achy	α-$_2$-M
Control	3.31	0.98	1.39
+ Tunicamycin	1.69	n.d.	0.81
+ Swainsonine	3.86	1.04	1.52
+ 1-deoxynojirimycin	2.25	0.87	1.01
+ 1-deoxymannojirimycin	3.43	0.93	1.16

[a] micrograms/10^6 cells

a homozygous ZZ phenotype complexed neutrophil elastase at less than one-half the rate of normal α-$_1$ PI (MM phenotype). This confirms the results of others[13] and indicates that the role of the carbohydrate moiety in α-$_1$-PI, and presumably other Serpins, is to aid in inhibitor secretion (and probably complex uptake) and not at all in complex formation.

ARE ALL ELASTASE SPECIFIC SERPINS OXIDATION SENSITIVE?

In humans it has already been noted that the two major inhibitors of neutrophil elastase, α-$_1$-PI and the bronchial mucous inhibitor, each have a Met residue in P_1 which, when oxidized, causes inhibitor inactivation.[14] However, elastase can hydrolyze peptide bonds not only after Met residues[15] but also Leu, Val, and Ala[16] so that replacement of Met in the P_1 residue of α-$_1$-PI by any of these amino acids should give an oxidation resistant inhibitor. Indeed, this is exactly what was found in a mutant inhibitor obtained by recombinant DNA technology,[17] the protein being fully functional in regulating elastase activity but oxidation insensitive.[18] Unfortunately, such inhibitors have been produced in bacteria or yeast systems and thus have no carbohydrate side chains. As a result they have very short half-lives[18] and limited value in long term replacement therapy for α-$_1$-PI deficient individuals. For short term therapy, however, they would appear to be perfectly adequate.

We have been interested in whether oxidation sensitive Serpins with elastase inhibitory activity were common in the animal kingdom. To test this we examined sera from 62 different species and determined their elastase and trypsin inhibitory

Table 2. Effect of Carbohydrate Side Chain Alteration on α-$_1$-Proteinase Inhibitor Function

Inhibitor Variant	Association Rate Constant[a]	
	Trypsin	Elastase
MM Plasma	2.2×10^5	2.4×10^7
ZZ Plasma		1.1×10^7
HEPG-2 Cells		
Control		2.7×10^7
+ Swainsonine	2.2×10^5	2.8×10^7
+ Tunicamycin	2.4×10^5	2.8×10^7
+ 1-deoxynojirimycin	2.2×10^5	2.2×10^7
+ 1-deoxymannojirimycin	2.1×10^5	2.3×10^7

[a] $M^{-1}sec^{-1}$

Table 3. Levels of Elastase Inhibitory Activity In Animal Sera

1. Less than 25% of Human	20%
2. 25% - 50% of Human	34%
3. 50% - 80% of Human	26%
4. Greater than 80% of Human	20%

capacity both before and after oxidation against both porcine pancreatic and human neutrophil elastases, as well as porcine pancreatic trypsin. All samples were first treated with methylamine to inactivate any α_{-2}-M which might interfere with the assays.

As shown in Table 3 we found a wide variation in total elastase inhibitory activity in the samples tested, with an equal percentage of animals having either very low or very high levels of elastase inhibitors relative to humans. After oxidation and re-assay we could categorize the animal sera in terms of containing either a) completely oxidation sensitive elastase inhibitory activity (66%), b) completely oxidation resistant activity (12%), or c) a mixture of sensitive and resistant inhibitors (44%). This is further correlated in Table 4 which indicates that those animals with low to moderately low activity had primarily oxidation sensitive inhibitors while those with much higher levels had much more oxidation resistant activity.

Analysis of individual species was surprisingly consistent in that virtually all carnivores had oxidation sensitive activity in low concentrations, while herbivores had high levels of oxidation resistant activity. This is summarized in Table 5. In contrast, all animals were consistent in having moderately high to higher than normal trypsin inhibitory activity relative to humans.

WHAT ARE THE P_1 RESIDUES IN NATURALLY OCCURING OXIDATION RESISTANT ELASTASE INHIBITORS?

To answer this question we examined two species, one which appeared to contain elastase inhibitors which were totally insensitive to oxidation (goat) and one which was partially sensitive (horse). The inhibitors were isolated and their reactive sites determined by methods which have been previously described for other inhibitors.[19,20]

Two elastase inhibitors from horse and one from goat were obtained. In the case of horse, one inhibitor was oxidation sensitive and had Met in the P_1 position while the other was insensitive and had Ala at P_1. These results are in agreement with the partial sensitivity of horse serum to oxidation but do not explain why the horse has two elastase inhibitors. The goat serum yielded a single, oxidation resistant elastase inhibitor with Leu in the P_1 position. Significantly, neither oxidation resistant inhibitor could inactivate trypsin. In contrast, the oxidation sensitive inhibitor from the horse was a good inactivator of trypsin in agreement with results originally obtained with human α_{-1}-PI which could also inactivate this enzyme. Clearly, Met in the P_1 position

Table 4. Correlation Between Elastase Inhibitory Activity and Oxidation Sensitivity in Animal Sera

Activity			Sensitivity	
		Total	Partial	Resistant
0 - 25%	(12)	12	0	0
25 - 50%	(21)	18	2	1
50 - 80%	(16)	8	7	1
Over 80%	(13)	3	5	5

Table 5. Correlation of Elastase Inhibitory Type with Species

1. Low Inhibitory Activity
   - Carnivores (e.g., bears, cats, dogs, aquatic mammals)
2. High Inhibitory Activity
   - Herbivores (e.g., primates, zebra, goat)
3. Oxidation Sensitive Activity
   - Primates
   - Carnivores
4. Oxidation Resistant Activity
   - Most Herbivores (except primates)

somewhat reduces inhibitor specificity while non-oxidizable hydrophobic amino acids have the opposite effect.

HOW HAVE ELASTASE INHIBITORS EVOLVED?

Structural studies on several elastase and trypsin inhibitors from different species indicate that both α_1-PI and α_1-Achy like molecules serve this function. As shown in Table 6, either type of inhibitor can have homology at the amino terminus or reactive site with either or both human proteins. Thus, human α_1-Achy with a Leu in the P_1 position should, conceivably, have been an elastase inhibitor. However, treatment of this protein with neutrophil elastase, while causing rapid cleavage after this residue, does not result in enzyme inactivation indicating that the P_1 residue is certainly not the only criteria for inhibition of this enzyme. Indeed, if one looks at reactive site P_1 residues from a number of Serpins (Table 7), one can readily see that a single P_1 residue can be involved in the regulation of a number of very different proteinases so that this amino acid only defines a potential binding site and not necessarily an inactivation site.

Obviously, homology to either α_1-PI or α_1-Achy, while being of interest and implying that a combination of the two may have represented a primitive Serpin, tells us nothing about the importance of specific residues in inhibitor function. This can only come from a determination of the structure of an entire series of elastase inhibitors in the Serpin family, similar to that already derived with ovomucoids.[20]

CAN WE CHANGE INHIBITOR SPECIFICITY?

From the information presented so far it is clear that substitutions in the reactive site loop, especially in the P_1 position, will not necessarily result in the inactivation of a proteinase with a specificity towards the new residue. In the case of α_1-PI,

Table 6. Structural Relationships Between Animal and Major Human Serpin Inhibitors

Species	P_1	Specificity	Relationship Amino Terminus	Reactive Site
Goat	Leu	Elastase	α_1-Achy	α_1-PI
Horse	Met	Elastase	α_1-PI	α_1-PI
	Ala	Elastase	α_1-PI	α_1-PI
	Arg	Trypsin	α_1-PI	α_1-Achy
Bovine	Lys	Trypsin	α_1-Achy	α_1-Achy
Sheep	-	Trypsin	α_1-Achy	-
Mouse	Arg	Trypsin	-	α_1-Achy

Table 7. Reactive Site P_1 Residues of Serpins

Residue	Target Enzyme	Inhibitor
Met	Elastase	α-$_1$-PI (man)
	Chymotrypsin	"
Leu	Elastase	α-$_1$-PI (goat)
	Cathepsin G	α-$_1$-Achy (man)
	Thrombin	HC II (man)
Ala	Elastase	α-$_1$-PI (horse)
Arg	Trypsin	α-$_1$-PI (man)*
	Thrombin	AT III (man)
		Nexin I (man)
	Urokinase	Nexin I (man)
	Plasmin	α-$_2$-AP (man)
	t-PA	PAI-1 (man)
	Protein C	Protein C Inh (man)
	Kallikrein	C1-Inh (man)
Lys	Unknown	Spi II (rat)
Tyr	Elastase	α-$_1$-PI (mouse)

* Pittsburgh Mutant [22]

however, conversion to non-oxidizable elastase inhibitors with Val, Leu, or Ala in the P_1 position by recombinant DNA technology is without a problem[18,21] and the natural alteration to an Arg residue (Pittsburgh mutant) results in trypsin and thrombin inhibition.[22] Therefore, in at least this one inhibitor binding and inhibition of elastase or trypsin seem to occur at similar if not identical sites and to be primarily directed by the P_1 residue. However, with α-$_1$-Achy this fails completely since a Leu residue in the P_1 position should promote elastase inhibition but, in fact, does not. Similarly, with heparin cofactor II (HC II) inhibition of thrombin is its primary function. However, there is a Leu in the P_1 postion, a residue not at all expected to be cleaved by a trypsin-like enzyme. Again, this strongly supports the role of other residues in Serpins in delineating inhibitor specificity.

SUMMARY

Serpins are an intriguing family of macromolecular proteinase inhibitors which function by forming tight but reversible complexes with target proteinases. Inhibitor specificity, while primarily being designated by the amino acid residue in the P_1 position, is also dependent on other contact residues. Indeed, some variance in the P_1 residue is even allowed as shown by reactive loop sequences present in elastase inhibitors from several animal species. It remains to be determined exactly which amino acids in Serpins other than the P_1 residue play major roles in conferring specificity so that a given proteinase is regulated by a single inhibitor. The availability of methods for rapidly obtaining site specific mutations should soon make such information available so that designed Serpins may be developed which are singularly targeted with a high reaction rate towards a given proteinase.

REFERENCES

1. R. Carrell and J. Travis, α_1-antitrypsin and the serpins: Variation and countervariation. Trends In Biochemical Sciences, 10:20, (1985).
2. J. Travis and G. Salvesen, Human plasma proteinase inhibitors. Ann. Review Biochem., 83:655, (1983).

3. J. Potempa, B.-H. Shieh, and J. Travis, α_2-antiplasmin: A serpin with two separate but overlapping reactive sites. Science 241:699 (1988).

4. B.-H. Shieh, J. Potempa, and J. Travis, The use of α_2-antiplasmin as a model for the demonstration of complex reversibility in serpins. J. Biol. Chem. 264:13420 (1989).

5. H. Loebermann, R. Tokuoka, J. Deisenhofer, and R. Huber, Human α_1-proteinase inhibitor. Crystal structure analysis of two crystal modifications, molecular model and preliminary analysis of the implications for functions. J. Mol. Biol., 177:531 (1984).

6. K. Beatty, J. Bieth, and J. Travis, Kinetics of association of serine proteinase inhibitors with native and oxidized α_1-proteinase inhibitor and α_1-antichymotrypsin. J. Biol. Chem., 255:3931, (1980).

7. K. Beatty, N. Matheson, and J. Travis, Kinetic and chemical evidence for the inability of oxidized α_1-proteinase inhibitor to protect lung elastin from elastolytic degradation. Hoppe-Seyler's Z. für Physiol. Chemie, 365:1131 (1984).

8. D. Johnson and J. Travis, Inactivation of human α_1-proteinase inhibitor by thiol proteinases. Biochem. J., 163:639 (1977).

9. J. Potempa, W. Watorek, and J. Travis, Inactivation of human α_1-proteinase inhibitor by S. aureus proteinases. J. Biol. Chem., 261:14330 (1986).

10. M. Banda, S. Sinha, and J. Travis, Inactivation of human α_1-proteinase inhibitor by macrophage elastase. J. Clin. Invest., 79:1314 (1987).

11. C.B. Laurell and S. Eriksson, The electrophoretic α_1-globulin pattern of serum in α_1-antitrypsin deficiency. Scand. J. Clin. Lab. Invest., 15:132 (1963).

12. A. Yoshida, J. Lieberman, L. Gaidulis, and C. Ewing, Molecular abnormality of human α_1-antitrypsin variant (Pi ZZ) associated with plasma activity deficiency. Proc. Natl. Acad. Sci. USA, 73:1324 (1976).

13. H. Takahashi, T. Nukiwa, K. Satoh, F. Ogushi, M. Brantly, G. Fells, L. Stier, M. Courtney, and R. Crystal, Characterization of the gene and protein of the α_1-antitrypsin "deficiency" allele. J. Biol. Chem., 263:15528 (1988).

14. D. Johnson and J. Travis, Structural evidence for methionine at the reactive site of human α_1-proteinase inhibitor. J. Biol. Chem., 253:7142 (1978).

15. W. Watorek and J. Travis, The action of neutrophil elastase on intact and oxidized (Met)-enkephalin-Arg-6-Gly-7-Leu-8 peptide. Biochem. Biophys. Res. Commun., 147:416 (1987).

16. A. Blow, The action of human neutrophil elastase on the oxidized insulin B chain. Biochem. J., 161:13 (1977).

17. S. Rosenberg, P. Barr, R. Najarian, and R. Hallewell, Synthesis in yeast of a functional oxidation-resistant mutant of human α_1-antitrypsin. Nature, 312:77 (1984).

18. J. Travis, M. Owen, P. George, R. Carrell, S. Rosenberg, R. Hallewell, and P. Barr, Isolation and properties of recombinant DNA produced variants of human α_1-proteinase inhibitor. J. Biol. Chem., 260:4384 (1985).

19. J. Potempa, A. Dubin, W. Watorek, and J. Travis, An elastase inhibitor from equine leukocyte cytosol belongs to the serpin superfamily. Further characterization and amino acid sequence of the reactive site. J. Biol. Chem., 263:7364 (1988).

20. I. Kato, W. Ardelt, J. Cook, A. Denton, M. Entie, W. Kohr, F. Park, K. Parks, B. Schatzley, O. Schoenberger, M. Tashiro, G. Vechot, H. Whatley, A. Wiecorek, M. Wiecorek, and M. Laskowski, Jr., Ovomucoid third domains from one hundred avian species. Isolation, sequence, and hypervariability of enzyme-inhibitor contact residues. Biochemistry 26:202 (1987).

21. N. Matheson, R. Hallewell, H. Gibson, P. Barr, and J. Travis, Kinetics of association of recombinant DNA produced variants of human α_1-proteinase inhibitor. Studies on Met-358 to Ala-358 and Met-358 to Cys-358. J. Biol. Chem., 261:10404 (1986).

22. R. Carrell and D. Boswell, Serpins: The Superfamily of Plasma Serine Proteinase Inhibitors, in: "Proteinase Inhibitors," A.J. Barrett and G. Salvesen, ed., Elsevier, Amsterdam (1986).

SECTION III

Serine proteases in the nervous system

PLASMINOGEN ACTIVATOR IN THE DEVELOPING NERVOUS SYSTEM

NICHOLAS W. SEEDS, SHAHLA VERRALL, PAUL McGUIRE AND GLENN FRIEDMAN

Department of Biochemistry, Biophysics and Genetics
University of Colorado Medical School
Denver, CO 80262

INTRODUCTION

Extensive cell migration and elaborate axonal growth are characteristic features of the developing nervous system. By the time many of the late generated neurons, such as cerebellar granule neurons, leave their site of origin in proliferating germinal zones and migrate to their permanent positions in the stratified nervous system, these migrating neurons are confronted by a morass of nerve fibers and extracellular matrix components. Although radial glial fibers[1] and Purkinje cell dendrites[2] may provide guidance toward the internal granule cell layer for migrating granule cells, the leading growth cone has been proposed to "push" its way through the tissue.[3] Ramon Cajal[3] likened these growth cones to battering rams which possessed an enormous power of penetration. More recently the possibility that these neuronal movements may be facilitated by the cell's ability to "cut" its way through the tissue has been explored.[4-10] A likely candidate for the cutting activity is cell secreted plasminogen activator which is capable of generating the broad acting protease plasmin from the extracellular proenzyme plasminogen.

METHODS

Cell Cultures

Cerebella were removed from five to eight day old C57Bl/6J mice and finely diced. The tissue was softened enzymatically in 0.2% trypsin in Saline 1 (0.138 M NaCl, 5.4 mM KCl, 1.1 mM Na_2HPO_4, 1.1 mM KH_2PO_4) containing 0.4% glucose, 0.01% $CaCl_2$ and 3 μg/ml DNase and incubated at 37°C with rotation for 12 min. The tissue was allowed to settle and the medium was discarded. Fresh basal Eagle's medium (BEM) containing 10% heated (30 min. at 56°C, which inactivates residual tPA) fetal calf serum (FCS) was added to the flask and the tissue dispersed by gentle pipetting and sieving through a nylon screen to yield single cells. A granule cell enriched population was produced by adsorption of non-neuronal cells on glass beads during a one hour incubation at 37°C and neurons collected by centrifugation. The

Serine Proteases and Their Serpin Inhibitors in the Nervous System
Edited by B. W. Festoff
Plenum Press, New York, 1990

cells were resuspended in BEM with 10% FCS and plated on poly-D-lysine coated rectangular plastic coverslips (10 x 22 mm) or 25 mm glass circles.

Sensory neurons were isolated from dorsal root ganglia of one to three day old mice. The ganglion capsule was digested with 0.1% collagenase at 37°C for 40 min. The cells were dispersed mechanically by pipeting and layered on a cushion of 35% Percoll and centrifuged at 200 x g for 15 min.. The pellet was rinsed twice with BEM and plated at 10^5 cells/25 mm fibronectin coated coverslip in BEM with 10 mM Hepes, 50 ng/ml 2.5S NGF and ITS+ additives: insulin, transferrin, BSA, selenious acid and linoleic acid (Collaborative Research).

Plasminogen Activator Assays

The fibrin overlay procedure of Todd[11] as modified by Krystosek and Seeds[5] was used to visualize the plasminogen activator (PA) activity at the cellular level. Cerebellar cell cultures were washed in Saline 1 and coated with a drop of thrombin (50 U/ml), drained and followed by a drop of fibrinogen (ICN, 75% clottable), drained and the thin fibrin clot was incubated in a moist chamber for 90 min. The clot was fixed and stained with Coomassie brilliant blue, mounted for microscopic observation of fibrinolytic zones.

Plasminogen activator activity in tissue extracts and cell conditioned medium was assessed by the release of ^{125}I-peptides from ^{125}I-fibrin coated on 35-mm plastic dishes in the presence and absence of plasminogen as described by Unkeless et al.[12] and Krystosek and Seeds.[6]

A sensitive amidolytic assay was used to quantify the PA activity by coupling plasmin formation to the conversion of the chromogenic substrate S2251 (Kabi), as described by McGuire and Seeds.[13]

^{125}I-tPA Binding Assay

The binding of tissue-type plasminogen activator (tPA) to the surface of dissociated cerebellar cell cultures was assessed with both iodinated murine tPA[14] and human recombinant tPA (Genentech). After six days in culture cerebellar cells were washed five times in serum-free BEM and placed in fresh BEM for 3 hrs. to facilitate the displacement of endogenously bound tPA. The coverslips were rinsed 3 times in Saline 1 with 1 mM $MgCl_2$ and 0.1 mM $CaCl_2$, then covered with 0.1 ml of Binding Buffer (BEM containing 15 mM Hepes (pH 7.5), 1 mg/ml BSA, 10 KIU aprotinin/ml) and varying amounts of ^{125}I-tPA as indicated, and incubated at room temp. for 30 min. The coverslips were rapidly washed five times in Saline 1 (Mg,Ca) drained and placed in a vial for scintillation counting. Specific binding was defined as the difference between cell-bound radioactivity in the absence and presence of a 100-fold molar excess of non-labeled tPA. Competitive binding studies were performed by incubating the cell cultures with ^{125}I-tPA in the presence of varying concentrations of the unlabeled competing protein.

Proteolysis of Fibronectin

Coverslips were coated with either human plasma fibronectin (10 μg/ml) or FITC-fibronectin for 1 hr., rinsed and dissociated sensory neurons added. After an overnight incubation at 37°C the FITC-fibronectin coverslips were viewed directly with an inverted epi-fluorescence microscope. Fibronectin coated coverslips were either rinsed with calcium- magnesium-free Hank's balanced salt solution at 4°C for 1hr to dislodge the cells prior to antibody binding, or coverslips were directly incubated with antibody to fibronectin and a fluorescently-labeled second antibody to visualize substratum associated fibronectin.

Degradation of the fibronectin was demonstrated with ^{125}I-fibronectin coated culture dishes on which 10^5 sensory neurons were cultured overnight in the serum-free BEM + ITS medium in the presence or absence of plasminogen. Conditioned medium was collected and aliquots counted, or pooled, dialyzed and concentrated prior to SDS-PAGE under nonreducing conditions. Various protease inhibitors: protease nexin a gift of In Vitron Corp., TIMP a gift of Synergen Corp., Hirudin and aminocaporic acid (Sigma), aprotinin and leupeptin (Boehringer) and PAI-1 (American Diagnostica) were tested for their effect on the release of ^{125}I-fibronectin fragments into the culture medium.

tPA mRNA

A synthetic oligonucleotide (30-mer) complementary to a DNA coding sequence near tPA's active-site[15] was end-labeled with ^{32}P-ATP and T4 polynucleotide kinase.[16] This 30-mer was used in Northern hybridization with blots of agarose gels containing poly-A+ RNA, which had been isolated from mouse cerebella by a guanidine isothiocyanate extraction and oligo-dT cellulose chromatography.

RESULTS

The plasminogen activator - plasmin system has been previously implicated in both neoplastic cell movements and tissue remodeling.[17-19] Therefore, it was not too surprising to find that tissue extracts of developing (8d.) mouse cerebellum showed higher (8-fold) levels of plasminogen activator activity than did adult cerebellum or extracts of brain regions, such as cerebral cortex, where cell migration was already complete by this same time.[4,6]

These elevated levels of PA in the developing cerebellum are in agreement with our recent studies quantifying the level of tPA mRNA in cerebella from young as compared to more mature animals (Figure 1). When equal amounts (2 μg) of poly A$^+$ RNA from 10 day and 40 day mouse cerebella are electrophoresed in agarose gels and blotted, RNA from the 10 day cerebella shows 10-fold more binding of the ^{32}P-end-labelled 30-mer DNA complementary to the 2.7 kb tPA mRNA than does RNA from the older animal.

Dissociated cell cultures prepared from neonatal mouse cerebellum at a time of active cell migration secrete plasminogen activator into the culture medium.[4,6] Zymographic analysis, using SDS-PAGE containing casein or fibrinogen gels,[20] of the secreted PA in the culture medium indicated that greater than 90% of the PA activity co-migrated with tPA at a Mr = 65,000 and less than 10% migrated with urokinase (uPA).[14]

A fibrin clot overlay technique[11] was used to localize the PA active cells in these mixed cell cultures. Zones of fibrinolytic activity were found around small neurons tenatively identified as granule neurons.[6] Only a subpopulation of the granule neurons showed PA activity. 3H-thymidine "birthdating" studies (not shown) indicated that postmigratory granule neurons lacked activity; however, migratory and possibly mitotic granule cells were associated with the fibrinolytic zones. Since the mitotic poison cytosine arabinoside failed to influence the number of fibrinolytically active cells in the cultures, the migratory granule cells probably represent the PA active cells

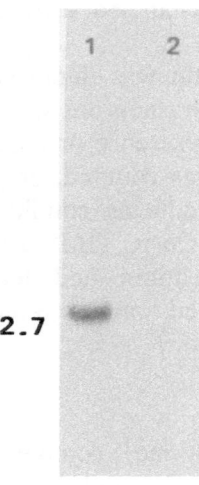

2.7

Figure 1. *Northern blot hybridization of mRNA from 10 day and 40 day mouse cerebella.* A ^{32}P-end-labeled 30-mer complementary to mouse tPA mRNA was used to probe a blot of an agarose gel electrophoresis that contained 2 μg of mRNA per lane.

in the population. Furthermore, glia and other non-neuronal cells in these cultures lacked any appreciable fibrinolytic activity.

Although the fibrinolytic zones are seen around single cells it is more common to see these zones around clusters of granule cells (Figure 2), especially when most of non-neuronal cerebellar cells are removed prior to culture. While PA activity in the conditioned medium from these cells is highest at 3 days of culture the number of fibrinolytic zones doesn't reach its maximum until day 6.[21] This finding suggested that the fibrinolysis may reflect cell-associated PA activity rather than cell secretion during the overlay assay. If these 6 day cultures are subjected to a 5 min. wash at low pH (<4), they lose their fibrinolytic activity, and tPA activity is recovered in the wash fluid.[21] Similarly, if the cultures are rinsed several times and exposed to a large volume of fresh BEM medium they also lose their fibrinolytic activity concomitant with the appearance of tPA activity in the medium.[14] These findings suggest the presence of a cell surface receptor for PA on these granule neurons.

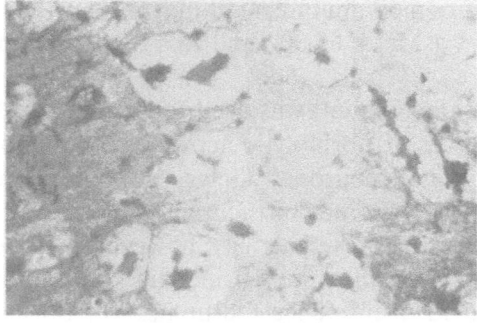

Figure 2. *Fibrin overlay of cerebellar granule cell culture shows fibrinolytic zones around clusters of granule neurons.*

The presence of a cell surface receptor for tPA was indicated by the finding that cerebellar cell clusters stripped of their endogenously bound tPA can bind exogenously added murine tPA, thus restoring about 70% of the original number of fibrinolytic zones.[21] Using ^{125}I-tPA of murine origin we have been able to characterize some of the properties of this binding site.[14] The binding is rapid, reaching completion within 20 min. The binding is specific and saturable as shown in Figure 3A. The binding is of high affinity with a Kd of interaction between 45 - 100 pM and approximately 30 - 50,000 binding sites per cell (Figure 3B). The binding is rapidly reversible with >67% of the tPA released within 5 min. (Figure 4); the residual ^{125}I-tPA is bound non-specifically and not released by low pH. All the specifically bound tPA can be displaced from the cells by simple dilution or pH 3, thus indicating that it is not being internalized by an uptake site on the cell surface. Furthermore, cell bound tPA displaced after several hours is still intact single chain tPA when examined by zymography.

Autoradiography shows that the specifically bound ^{125}I-tPA is associated with granule neurons, and only non-specific binding is associated with glia and other non-neuronal cells in the cultures. Specific binding did not coincide with cells possessing glial fibrillary acidic protein sites, nor with those few regions containing the extracellular matrix protein fibronectin.[14] Although fibronectin has been shown to bind plasminogen activator,[22] the affinity is lower than that demonstrated here (Figure 3). Furthermore, an antibody to fibronectin which blocks its interaction with PA failed to inhibit the binding of ^{125}I-tPA to the cerebellar cell cultures. Under similar conditions an antibody to murine tPA blocked the binding of tPA to the cerebellar cells.[23]

The specificity of the ^{125}I-tPA binding was studied in the presence of an excess of related and other proteases (Table 1). Although our initial studies[21] using American Diagnostica's human melanoma tPA indicated tPA binding was specific for only murine tPA, both human melanoma tPA (Nat. Instit. Biol. Stds. and Control, London) and Genentech's recombinant human tPA compete for the binding site. Furthermore, our more recent studies use ^{125}I-recombinant human tPA, which has only a slightly lower affinity for binding than the murine tPA. Neither murine nor human urokinase compete for the binding. Similarly, two other serine proteases, thrombin and plasminogen, known to bind to cells failed to compete. The binding of tPA to the cell surface does not appear to involve the carbohydrate moieties of the tPA molecule, since tPA treated with Endoglycosidase H and excluded from a Concanavalin A resin actively competes with native tPA for binding. The catalytic site of tPA is not involved in the binding, since its inactivation with diisopropylfluoro-

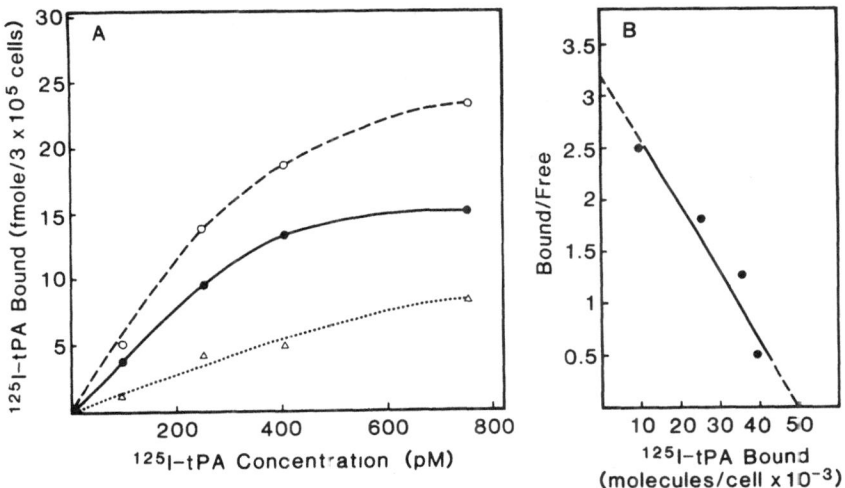

Figure 3. *Characteristics of* [125]*I-tPA Binding to cerebellar neurons.* The saturation of the specific binding is shown in panel A. A Scatchard analysis of the binding is shown in panel B.

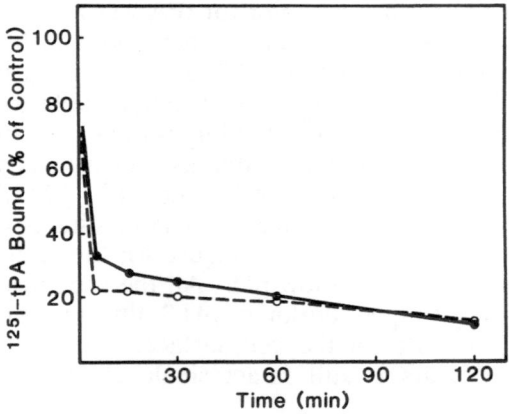

Figure 4. *Reversibility of the ^{125}I-tPA bound to cerebellar cells is shown in BEM (solid line) and at pH 3 (dashed line).*

phosphate has little effect on its ability to compete with native tPA. Furthermore, as predicted by the fibrin overlay assays, DFP-^{125}I-tPA shows specific binding to cerebellar cells indicating that tPA's interaction with the cell surface is not mediated by a serpin protease inhibitor.[23]

Although both the intact single chain tPA and a small amount of the protease generated catalytic carboxy-terminal portion of tPA are seen in the conditioned medium from cerebellar cells, only the single chain tPA binds to the cells and is displaced from the cell surface. This finding suggests that the binding of tPA to the cell surface is mediated by a domain in the amino-terminal portion of the molecule. The amino-terminal portion of tPA contains structural domains shared in common with other proteins and several serine proteases. These stuctures include a fibronectin finger domain, an epidermal growth factor domain followed by two kringle structures, and have probably arisen by exon shuffling.[24,25] Preliminary studies done in collaboration with Drs. M. Gething and J. Sambrook[26] using site-directed mutants of human tPA missing the finger, growth factor, finger + growth factor, kringle 1, or kringle 2, indicate that tPA missing the growth factor domain fails to compete with native tPA for binding to neurons.

An antibody to murine tPA[23] has been used to localize the cell bound tPA at the electron microscope level. Although tPA is bound over the entire surface of cerebellar granule neurons and their growth cones, there is an apparent concentration of bound tPA at the trailing edge of those granule neurons migrating on neurite cables in cell

Table 1. Specificity of ^{125}I-tPA binding to cerebellar neurons. Cerebellar cells stripped of endogenous PA by pH 3 were washed and incubated for 30 min. at 20°C with murine ^{125}I-tPA in the absence or presence of a 50-fold molar excess of the competing proteins. Percent specific binding was calculated relative to inhibition by a 50-fold molar excess of active non-labeled tPA.

Competing Protein	Molar Ratio	Binding %
None	0:1	100%
Murine tPA	50:1	0
Human tPA	50:1	0
Murine urokinase	50:1	106
Human urokinase	50:1	110
Human thrombin	50:1	94
Human plasminogen	50:1	78
Murine DFP-tPA	50:1	5
Deglycosylated tPA	50:1	0

culture.[27] These findings suggested that plasminogen activator may play a role in granule cell migration.

Using video time-lapse recording to track granule cell migration in cell culture, we found that granule cells migrated along neurite cables at an average rate of 220 um/d, similar to the projected rates for their migration *in vivo*.[28] The addition of several low molecular weight protease inhibitors, leupeptin and D-phe-pro-arg-CH_2Cl that block plasminogen activator activity inhibited this migration. Furthermore, the inhibition by leupeptin was readily reversible, as indicated in Table 2. *In vivo* studies, where these protease inhibitors were injected over the external granule cell layer and beneath the pial membrane of 8d. mice, showed a significant retardation of granule cell migration through the molecular layer when compared to sites more than 400um distant from the injection or in adjacent folia.[27] These findings suggest that extracellular proteases play a role in neuronal migration.

To explore the possibility that plasminogen activator activity may be important in axonal growth cone movement, we focused on the peripheral nervous system. Dissociated sensory neurons in culture have relatively large growth cones that display a very dynamic and robust growth. Fibrin overlays of sensory neurons from 2d. mice showed that plasminogen activator activity was especially pronounced around the growth cones.[29] Zymography of the conditioned medium from these sensory neurons indicated that uPA was the major PA secreted by the cells.

The influence of proteolytic activity on neurite outgrowth from sensory ganglia was assessed by the addition of protease inhibitors to the culture medium.[30] Interestingly, several protease inhibitors promoted outgrowth of neurites on tissue culture plastic (a relatively low adhesive substrate); however, at high concentrations these protease inhibitors began to retard neurite growth. Similarly, addition of exogenous proteases such as thrombin and urokinase inhibited growth, and they could reverse the effects of these protease inhibitors on neurite growth. Furthermore, when these PA inhibitors were covalently coupled to the substratum they oriented the direction of neurite outgrowth towards the inhibitor coated surface, while coupled urokinase and thrombin directed growth away from their zones.[31] These findings suggest that a delicate balance of neuronal secreted protease and extracellular inhibitors are necessary in mediating interactions between the growth cone and the extracellular matrix.

Sensory neurons not only secrete PA they deposit PA on the substratum underneath the advancing growth cone where it may be strategically placed to participate in the local degradation of matrix components.[32] Axonal growth cones of sensory neurons growing on a fibronectin substratum can degrade this matrix molecule as they migrate over the substratum[33] as shown in Figure 5. The cleavage of fibronectin is indicated by the release of large 200-220,000 Mr fragments into the culture medium. This cleavage is independent of plasminogen in the medium and insensitive to very high concentrations of plasmin inhibitors. Interestingly, only the

Table 2. Granule cell migration in cell culture. Granule neurons migrating along neurite fascicles in a Dorvak-Stolter chamber were viewed by Nomarski differential interference contrast optics and recorded by video time-lapse for 2-3 hrs. between perfusions for each treatment.

Conditions	Average Rate (um/d)
Granule cells on neurite cables	220
plus 10^{-5}M D-phe-pro-arg-CH_2Cl	20
plus 10^{-4}M Leupeptin	12
3 hr. after washout of leupeptin	230

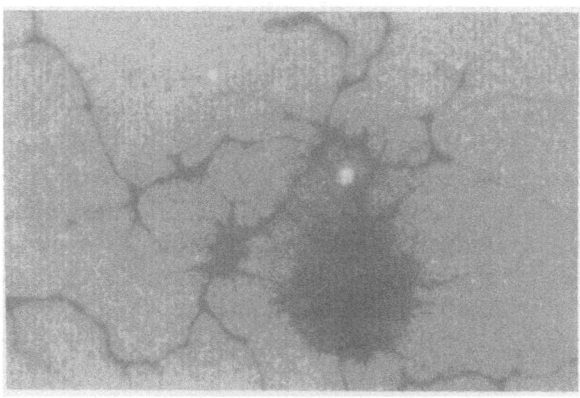

Figure 5. *Displacement of FITC-fibronectin from the substrata under a growing sensory neuron as visualized by epi-immunofluorescence.*

serpin protease nexin was found to inhibit this plasminogen independent degradation of fibronectin as shown in Table 3.

DISCUSSION

These studies have demonstrated that plasminogen activator activity is elevated in the mouse brain at the time of active cell migration, and that cells from the developing nervous system when placed in culture actively secrete PA into the culture medium. Studies by Soreq and Miskin[7,34] using rat brain and a different assay method for PA activity have confirmed these findings. This elevation of PA activity in the developing nervous system and its dramatic decrease in the mature brain coincides with a similar transition in tPA mRNA levels (Figure 1). The 2.7 kb mRNA detected in cerebellar tissue is similar in size to the reported 2.8 kb tPA message seen in mouse teratocarcinoma cells.[35] Our findings suggest that tPA expression in the developing cerebellum is regulated at the transcriptional level, although detailed studies of tPA mRNA turnover are still ongoing. The role of synaptic target tissues and cell interaction in this regulation, as well as, the molecular nature of potential regulators of tPA gene expression are important avenues for future investigation.

The specific, rapid, saturable, reversible and high affinity binding of tPA to granule cell neurons in the developing cerebellum suggests the presence of an unique tPA receptor on these neurons. Only a few previous reports of tPA binding to cells have appeared, and they all differ significantly from the binding reported here. In some cases the tPA was bound through its catalytic site,[36,37] in others tPA was bound irreversibly,[38] or bound with low affinity ($Kd = 0.2uM$).[37] Although a similar high affinity binding of tPA was seen with human endothelial cells,[39] uPA competed with tPA for this binding site in contrast to the binding by granule neurons (Table 2). However, there have been several reports of high affinity ($Kd = 0.4nM$) binding of uPA to human monocytes and fibroblasts.[40-42]

Interestingly, uPA binding to cells is mediated by its EGF-like domain (as reported at this workshop,[43]), thus it should not be too surprising that tPA may use this same EGF-like region for binding to granule neurons. Although the EGF-like

Table 3. Inhibitor sensitivity of sensory neuron protease that cleaves fibronectin. Percentage of ^{125}I-fibronectin CPM released from coated coverslips by growing sensory neurons in 16 hr. is indicated. Cells were cultured in serum-free BEM + ITS and 2.5S NGF with or without plasminogen, and the various protease inhibitors as indicated. ND = not determined.

Treatment	% of Total CPM released	
	- Plasminogen	+ Plasminogen
None	12.92	38.61
Aprotinin (3000 U/ml)	11.35	12.36
Aminocaproic acid (3 mM)	12.63	14.34
Leupeptin (1 mM)	14.30	11.02
Hirudin (10 U/ml)	ND	36.60
TIMP (10 μg/ml)	12.50	ND
Protease Nexin (10 μg/ml)	3.20	8.37

domains in mouse uPA and tPA show large differences in amino acid sequence (possibly explaining why uPA doesn't compete for neuron binding), the center (10 amino acids) portion of the EGF-like domain in both mouse and human tPA, which compete for binding, is identical; thus supporting our proposal of the EGF-like domain's role in tPA binding to neurons.

The presence of tPA receptors on the neuronal cell surface adds an important new level of regulation to the PA-serpin system in neural development.[6-9,44-46] Neuronal receptors for tPA may provide several important functions, including a mechanism to prevent tPA diffusion away from the cell surface. Receptor bound tPA may be protected from inactivation by protease inhibitors in the extracellular environment. In addition, the possibility exists that tPA's interaction with the receptor functions in an autocrine mechanism to initiate a second messenger response within the neuron. This cell surface receptor permits the developing neurons to "arm themselves" with active protease; furthermore, temporal and spatial regulation of the receptors may, during specific stages of development, lead to highly localized and concentrated regions of extracellular protease activity at sites of cell-cell and cell-matrix interaction. The potential ability of the protease to disrupt these interactions may play a role in neuron and axon movement.

Proteolytic activity appears important in neuron and axon movements, since PA inhibitors have been shown to block or retard granule neuron migration in explant cultures,[8] along neurite fascicles in culture (Table 2) and *in vivo*.[27] Similarly, protease inhibitors have a significant effect on neurite outgrowth by cultured neurons from the peripheral nervous system.[30,31,44-46]

Although PA activity in most tissues is primarily directed at converting plasminogen to the broad acting plasmin, this limited specificity has not been established in the nervous system. In fact, there is evidence from Quigley et al.[47] that under some circumstances uPA may act directly to cleave other substrates, such as fibronectin. Since uPA is the major PA released by sensory neurons in culture, the protease nexin sensitive and plasminogen-independent cleavage of fibronectin seen in Figure 5 and Table 3 may be mediated by cell associated PA.

In summary, a hypothetical schematic of PA activity at the growth cone is shown in Figure 6. An excess of cell secreted PA would be neutralized by glial derived PA inhibitors, thus permitting the attachment of filopodia to the substratum via extracellular matrix molecules, and extending the forward edge of the growth cone. The presence of PA receptors on the under surface of the growth cone and concentrated at its trailing edge may bring about the limited cleavage of substrate

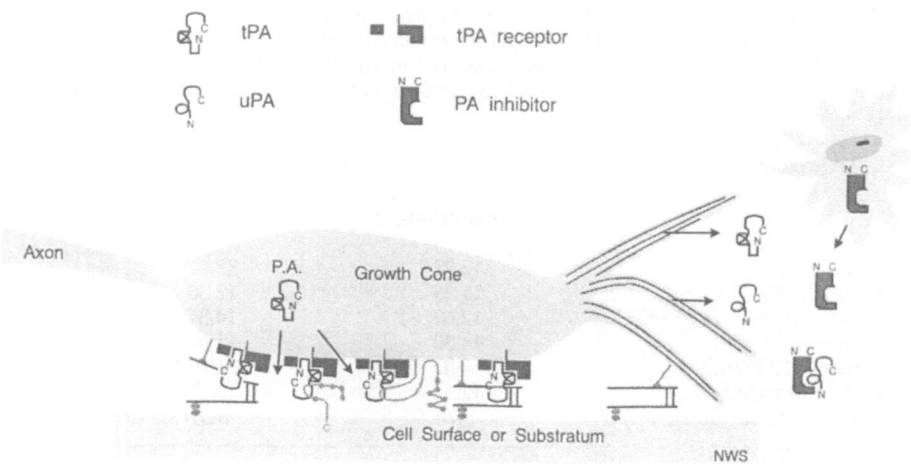

Figure 6. Schematic diagram showing plasminogen activator release from the growth cone, binding to PA-receptors, interaction with PA-inhibitors, interaction of receptor-bound PA with extracellular matrix (fibronectin) and cell adhesion molecules that mediate growth cone attachment to other cells or substrata, as well as the cleavage of fibronectin and the growth cone's detachment from the substratum at its trailing edge.

adhesion molecules or possibly cell adhesion molecules. This cleavage would promote the detachment of the trailing edge, an event apparently necessary for the continued forward movement of the growth cone.

ACKNOWLEDGEMENTS

The authors would like to thank Ms. K. Christensen and Ms. S. Haffke for their excellent technical assistance in the studies reported here.

These studies were supported in part by grants from the National Institutes of Health (NS-09818) Javits Investigator Award, the National Science Foundation (BNS-86-07719) and the Muscular Dystrophy Association.

REFERENCES

1. P. Rakic, Neuron-glia relationship during cell migration in developing cerebellar cortex, **J. Comp. Neurol.** 141:283 (1971).
2. C. Sotelo and J. P. Changeux, Bergmann fibers and granule cell migration in the cerebellum of homozygous Weaver mutant mouse, **Brain Res.** 77:484 (1974).
3. R.S. Cajal, "Studies on Vertebrate Neurogenesis", L. Guth, trans., Thomas Publ., Springfield, IL (1960).
4. A. Krystosek and N. W. Seeds, Plasminogen activator production by cultures of developing cerebellum, **Fed. Proc.** 37:1702 (1978).
5. A. Krystosek and N. W. Seeds, Plasminogen activator release at the neuronal growth cone, **Science** 213:1532 (1981).
6. A. Krystosek and N. W. Seeds, Plasminogen activator secretion by granule neurons in cultures of developing cerebellum, **Proc. Nat. Acad. Sci.** 78:7810 (1981).
7. H. Soreq and R. Miskin, Plasminogen activators in the rodent brain, **Brain Res.** 216:361 (1981).
8. G. Moonen, M. Grau-Wagemans and I. Selak, Plasminogen activator-plasmin system and neuronal migration, **Nature** 298:753 (1982).
9. R.N. Pittman, Release of plasminogen activator and a calcium-dependent metalloprotease from cultured sympathetic and sensory neurons, **Dev. Biol.** 110:91 (1985).
10. R.N. Pittman and A. G. Williams, Neurite penetration into collagen gels requires Ca^{++}-dependent metalloprotease activity, **Dev. Neurosci** 11:41 (1988).
11. A.S. Todd, The histological localization of fibrinolysin activator, **J.Pathol. Bacteriol.** 78:281 (1959).

12. J.C. Unkeless, S. Gordon and E. Reich, Secretion of plasminogen activator by stimulated macrophages, **J. Exp. Med.** 139:834 (1974).
13. P.G. McGuire and N. W. Seeds, The interaction of plasminogen activator with a reconstituted basement membrane matrix and extracellular macromolecules produced by cultured epithelial cells, **J. Cell. Biochem.** 40:215 (1989).
14. S. Verrall and N. W. Seeds, Characterization of ^{125}I-tissue plasminogen activator binding to cerebellar granule neurons, **J. Cell Biol.** 109:265 (1989).
15. R.J. Rickles, A. L. Darrow and S. Strickland, Molecular cloning of complementary DNA to mouse tissue plasminogen activator mRNA and its expression duringteratocarcinoma cell differentiation, **J. Biol. Chem.** 263;1563 (1988).
16. L. G. Davis, M. Dibner and J. Battey, "Basic Methods in Molecular Biology", Elsevier, New York (1986).
17. R. R. Burk, A factor from a transformed cell line that affects cell migration, **Proc. Nat. Acad. Sci.** 70:369 (1973).
18. E. Reich, D. Rifkin and E. Shaw, "Proteases and Biological Control", Cold Spring Harbor Lab, New York (1975).
19. K. Danø, P. Andreasen, J. Grondahl-Hansen, P. Kristensen, L. Nielsen and L. Skriver, Plasminogen activators, tissue degradation and cancer, **Adv. Cancer Res.** 44:139 (1985).
20. C. Heussen and E. B. Dowdle, Electrophoretic analysis of plasminogen activators in polyacrylamide gels containing SDS and copolymerized substrates, **Anal. Biochem.** 102:196 (1980).
21. S. Verrall and N. W. Seeds, Tissue plasminogen activator binding to mouse cerebellar granule neurons, **J. Neurosci. Res.** 21:420 (1988).
22. E. Salonen, O. Saksela, T. Vartio, A. Vaheri, L. Nielsen and J. Zeuthen, Plasminogen and tissue-type plasminogen activator bind to immobilized fibronectin, **J. Biol. Chem.** 260:12302 (1985).
23. S. Verrall, "Characterization and Purification of Tissue Plasminogen Activator and Its Binding to the Surface of Cerebellar Neurons", Ph.D. Thesis Univ. Colorado HSC, Denver (1989).
24. T. Ny, F. Elgh and B. Lund, The structure of the human tissue-type plasminogen activator gene: correlation of intron and exon structures to functional and structural domains, **Proc. Nat. Acad. Sci.** 81:5355 (1984).
25. A. J. van Zonnevald, H. Veerman and H. Pannekoek, Automonous functions of structural domains on human tissue-type plasminogen activator. **Proc. Nat. Acad. Sci.** 83:4670 (1986).
26. M. J. Gething, B. Adler, J. Boose, R. Gerald, E. Madison, D. McGooKey, R. Meidell, L. Roman and J. Sambrook, Variants of human tissue plasminogen activator that lack specific structural domains of the heavy chains, **EMBO J.** 7:2731 (1988).
27. N. W. Seeds, K. Christensen, S. Haffke and J. Schoonmaker, Granule neuron migration involves proteolysis, in: "Molecular Aspects of Development and Aging", J. Lauder (ed.), Allan R. Liss Inc., New York (1989).
28. J. Altman, Autoradiographic and histological studies of postnatal neurogenesis II, **J. Compar. Neurol.** 128:431 (1966).
29. A. Krystosek and N. W. Seeds, Peripheral neurons and Schwann cells secrete plasminogen activator, **J. Cell Biol.** 98:773 (1984).
30. R. L. Hawkins and N. W. Seeds, Effect of proteases and their inhibitors on neurite outgrowth from neonatal sensory ganglia in culture, **Brain Res.** 398:63 (1986).
31. R. L. Hawkins and N. W. Seeds, Protease inhibitors influence the direction of neurite outgrowth, **Dev. Brain Res.** 45:203 (1989).
32. A. Krystosek and N. W. Seeds, Normal and malignant cells, including neurons deposit plasminogen activator on the growth substrata, **Exp. Cell Res.** 166:31 (1986).
33. P. G. McGuire and N. W. Seeds, Regenerating sensory neurons degrade fibronectin during the process of neurite outgrowth, **J. Cell Biol.** 107:374a (1988).
34. H. Soreq and R. Miskin, Plasminogen activator in the developing rat cerebellum, **Dev. Brain Res.** 11:149 (1983).
35. R. J. Rickles and S. Strickland, Tissue plasminogen activator mRNA in murine tissues, **FEBS Lett.** 229:100 (1988).
36. D. P. Beebe, Binding of tissue plasminogen activators to human umbilical vein endothelial cells. **Thrombosis Res.** 46:241 (1987).
37. E. S. Barnathan, A. Kuo, H. van der Keyl, K. McCrae, G. Larsen and D. Cines, Tissue type plasminogen activator binding to human endothelial cells, **J. Biol. Chem.** 263:7792 (1988).
38. E. G. Hoal, E. Wilson and E. Dowdle, The regulation of tissue plasminogen activator activity by human fibroblasts, **Cell** 34:273 (1983).
39. K. A. Hajjar, N. Hamel, P. Harpel and R. Nachman, Binding of tissue plasminogen activator to cultured human endothelial cells, **J. Clin. Invest.** 80:1712 (1987).
40. A. Bajpai and J. B. Baker, Cryptic urokinase binding sites on human foreskin fibroblasts, **Biochem. Biophys. Res. Comm.** 133:475 (1985).
41. M. Stoppelli, A. Corti, A. Soffientini, G. Cassani, F. Blasi and R. Assoian, Differentiation enhanced binding of the amino-terminal fragment of human urokinase plasminogen activator to a specific receptor on U937 monocytes, **Proc. Nat. Acad. Sci.** 82:4939 (1985).

42. J. D. Vassalli, D. Baccino and D. Belin, A cellular binding site for the Mr 55,000 form of the human plasminogen activator urokinase, **J. Cell Biol.** 100:86 (1985).

43. F. Blasi, Structure and regulation of the surface receptor for human urokinase plasminogen activator, in: **NATO-ARW "Serine Proteases and Serpins in the Nervous System"**, B. Festoff (ed), Plenum Press, New York (1989).

44. J. Geunther, H. Nick and D. Monard, A glial derived neurite promoting factor with protease inhibitory activity, **EMBO J.** 4:1963 (1985).

45. D. Monard, Cel-derived proteases and protease inhibitors as regulators of neurite outgrowth, **Trends In NeuroSci.** 11:541 (1988).

46. D. Gurwitz and D. Cunningham, Thrombin mediates and reverses neuroblastoma neurite outgrowth, **Proc. Nat. Acad. Sci.** 85:3440 (1988).

47. J. P. Quigley, L. Gold, R. Schwimmer and L. Sullivan, Limited cleavage of cellular fibronectin by plasminogen activator purified from transformed cells, **Proc. Nat. Acad. Sci.** 84:2776 (1987).

MULTIPLE ROLES FOR PLASMINOGEN ACTIVATOR SYSTEM IN NERVOUS SYSTEM DEVELOPMENT

PIERRE LEPRINCE, BERNARD ROGISTER, PAUL DELRÉE,
PHILIPPE P. LEFEBVRE, JEAN-MICHEL RIGO AND GUSTAVE MOONEN

Laboratory of Human Physiology
University of Liège
17 Place Delcour
B-4020 Liège, Belgium

INTRODUCTION

Brain development involves several steps where extracellular proteolytic activities are required, either in intercellular signalling or in tissue remodeling. Those proteolytic activities are in part due to plasminogen activators (PA) which are released by cells of the nervous system and have been implicated in several aspects of neuronal and glial biology. Both forms of PA, the urokinase type (uPA) and the tissue type (tPA) are produced by various cell populations of the mammalian nervous system. Their production appears to be developmentally regulated, being most prominent at various stages of neuronal or glial cells ontogenesis in the embryo or soon after birth.

PAs have the potential to induce intense and rather non-specific proteolytic activities through the activation of plasminogen into plasmin and subsequent activation of other proteases such as collagenases. The reality of such processes, occurring for example in the blood fibrinolytic system, has not yet been demonstrated in the nervous tissue. Still, efficient control mechanisms of PA activity have recently be found to be associated with glial cells which have been shown to produce highly active plasminogen activator inhibitors (PAI).[1] The wide occurrence and high levels of expression of PAs during many specific developmental stages of the nervous system ontogenesis suggest that these proteases (and their inhibitors) may play multiple roles in brain development. In this chapter, we will describe the involvement of PAs and PAIs in four different developmental processes interesting both glial and neuronal cell types of the central and peripheral nervous system. These processes: neuronal migration, glial proliferation, axonal fasciculation and neuritogenesis have been studied using *in vitro* culture techniques of nervous tissue explants, or of normal or transformed neural cells. They exemplify how, depending on the cell type, the tissue location and the developmental stage, similar molecules can be used in the modulation of highly specific and varied biological responses.

Serine Proteases and Their Serpin Inhibitors in the Nervous System
Edited by B. W. Festoff
Plenum Press, New York, 1990

NEURONAL MIGRATION

The migration of neurons from the germinative layers to their final destination in cellular layers is a prominent feature of cerebral and cerebellar cortical development.[2] This migration involves a close relationship between the radial glia and the migrating neurons.[3] The molecular mechanism underlying such interactions is largely unknown. We have previously reported that PA activity is involved in this process since synthetic proteases inhibitors can block neuronal migration in suspension cultures of cerebellar paraflocullus.[4] Furthermore, glia-derived Nexin (GDN), a PAI produced by glioma cells and normal astrocytes[5] significantly reduces the extent of granule cell migration in a dose-dependent fashion in a different assay system.[6]

Various cell types, at the time of their normal or artificially-induced migration, have been shown to release PA (identified in some cases as tPA[7] and in some cases

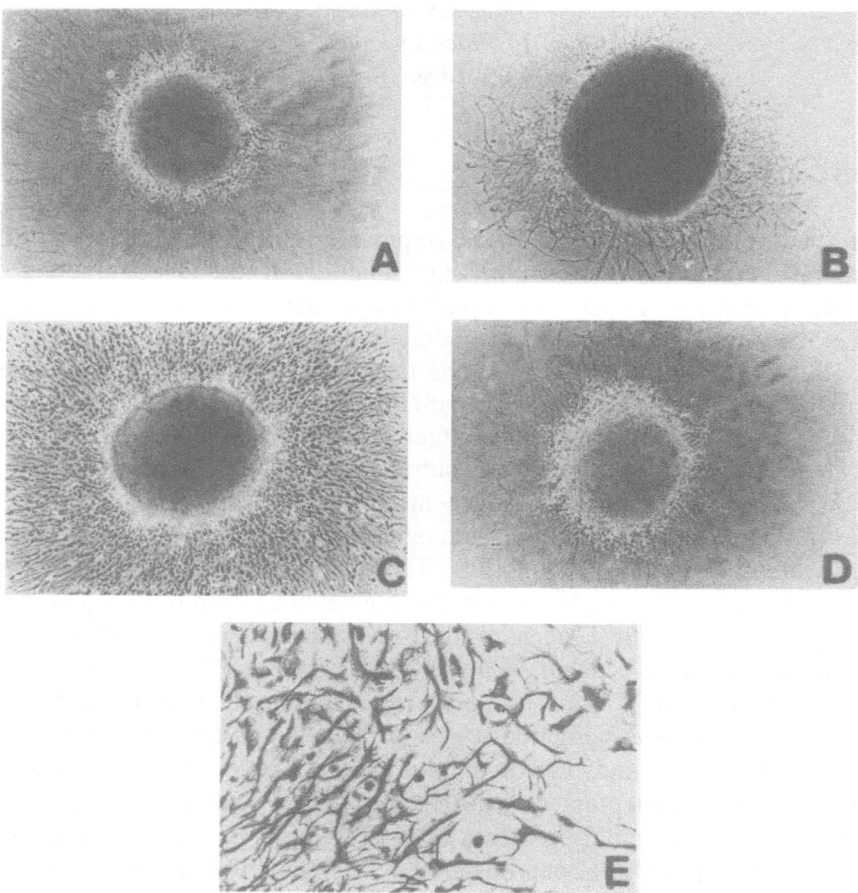

Figure 1. *Mitogenic effect of uPA on astrocytic cells in newborn rat cerebellar microexplants.* Microexplants were grown for 48 hr on polyornithine-coated dishes under different conditions. During that time interval, a neuritic outgrowth as well as a proliferation and subsequent outward migration of astrocytes can be observed (A: MEM medium + Insulin, 5 μg/ml). Upon addition of a serine protease inhibitor (PMSF, 10^{-3}M) only the neuritic outgrowth is observed while cell proliferation is inhibited (B). The addition of purified human uPA (12 units/ml) to MEM medium + Insulin increases very much the proliferation of astrocytes (C). This effect is blocked when PMSF is added together with uPA (D). Immunostaining with anti-GFAP antibody of cells in the neuritic array indicates the astroglial nature of these cells (E).

Table 1. ^{3}H-Thymidine incorporation in cultured
newborn rat cerebellar microexplants

Medium	% of ^{3}H-Thymidine Incorporation
MEM-Insulin	100 ± 10
MEM-HS (10 %)	130 ± 9
MEM-Insulin + uPA (12 U/ml)	244 ± 6
MEM-Insulin + PMSF (10^{-3} M)	11 ± 3
MEM-Insulin + uPA (12 U/ml) + PMSF (10^{-3} M)	19 ± 7

as uPA[8-10]). Cerebellar microexplants in which granule cell migration can be observed in vitro release both tPA and uPA,[11] yet the respective role of these two proteases in neuronal migration remains to be elucidated.

The mechanism by which PAs control neuronal migration is equally unknown. The involvement of PAI activity secreted by the migrating neuron or the guiding glia has not been investigated but should be considered in view of the proposed involvement of PAI in mechanisms modulating neuritic outgrowth (see next sections). Another important question relates to the substrate of PA during neuronal migration. Extracellular matrix components such as laminin and fibronectin are able to promote neuronal migration[12,13] and are substrate of,[14,15] or present binding sites for PAs.[16,17] Since PAs can be deposited on the growth substratum while retaining their proteolytic activity[18] it can be suggested that proteolytic alteration of extracellular matrix components may regulate substrate-neuron adhesiveness during neuronal migration.[15]

GLIAL PROLIFERATION

We have recently reported that uPA but not tPA was a mitogenic signal for astrocytes in cerebellar microexplants and in Schwann cells purified from sciatic nerves[11,19] (Figure 1). In both cases it was found that the proteolytic activity of uPA was required for its mitogenic activity. The mechanism of action of uPA in inducing glial proliferation remains unknown although different sets of data are relevant to that question. This effect of uPA is additive with that of other known mitogens (Insulin for astrocytes and Cholera Toxin for Schwann cells) suggesting that the mechanism of action of uPA is different from that of other glial mitogens. Secondly, the uPA mitogenic effect seems to involve a binding of urokinase to the cells, since the 33 kDa form of uPA, which lacks the cell binding site but not the proteolytic site is uneffective (see also the effect of PMSF, Table 1). Finally, the PA release is a consequence of a mitogenic stimulation in both systems : bFGF treatment of cerebral astrocytes induces an enhanced release of PA in the conditioned medium of these cells[20] while a similar effect can be observed upon addition of Cholera Toxin to Schwann cell cultures[19] (Figure 2).

Mitogenic effects of uPA have been found for other types of cells.[21] Similarly, thrombin has also been reported to induce proliferation of various cell types,[22] including astrocytes.[23] A modulation of the mitogenic effect of thrombin on fibroblasts by secreted protease nexin has been proposed.[24] Such a regulation is likely to constitute a model for the control of the mitogenic activity of uPA on glial cells in view of the demonstration that an astrocytic PAI, the glia-derived nexin, is homologous to protease nexin-I.[25]

The demonstration that uPA is effectively a mitogenic signal acting *in vivo* on astrocytes remains to be established but is strongly suggested by several observations:
- in a serum-free *in vitro* assay, a mitogenic effect for astrocytes of media conditioned by cerebral cortex or cerebellum microexplants can be demonstrated. That mitogenic effect can be suppressed by inhibitors of PA and is mimicked by a pure preparation of human uPA.[11]
- in the peripheral nervous system (dorsal root ganglia [DRG]) where PA-secreting cell populations are limited to one neuronal type and one non-neuronal type (Schwann cells) we found that uPA was secreted by neurons while tPA was secreted by Schwann cells[19] (Figure 3). Thus in DRG, conditions are found for a possible regulation of Schwann cell proliferation through release of uPA by the neuronal cell population.

In the central nervous system where glial and neuronal cells can be recognized as many different subsets, a similarly simple neurono-glial interaction does not seem to be present since both neuronal and glial cells are found to release uPA.[8] This release of uPA by glial cells suggest that an autocrine regulation of astrocytic proliferation through release of mitogenic uPA might be considered.

A glial modulation of glial proliferation has recently been found to occur between members of the glial cell lineage in the developing rat optic nerve and has been shown to involve the activity of several well-known growth factors such as CNTF and PDGF.[26]

In an attempt to determine the involvement of proteases and protease inhibitors activities in this glial developmental process we have studied the PA and PAI activities released by differentiated glial cells : the type-1 astrocyte, type-2 astrocyte and oligodendrocyte, and by proliferating O_2A progenitor cells (Figure 4). This analysis

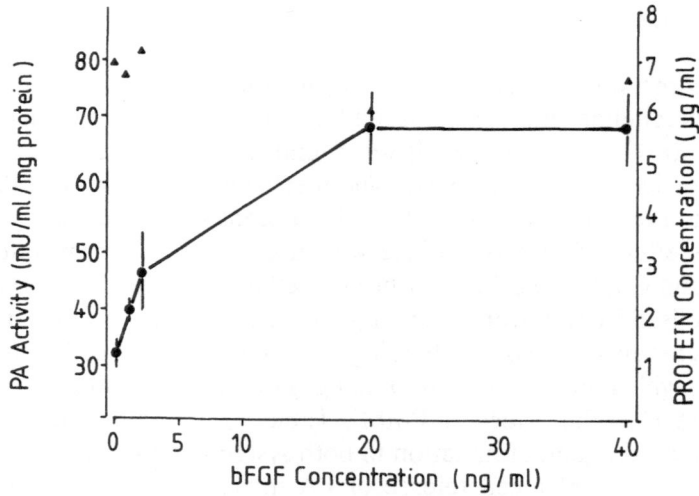

Figure 2. *Stimulation by FGF of the PA activity released by newborn rat cortical astrocytes in culture.* 6-day cultures were incubated for 24 h in serum-free medium containing the indicated concentration of bFGF. PA activity was determined by the ^{125}I-fibrin plate assay. Zymographic analysis indicated a similar effect on both tPA and uPA released by these astrocytes.

reveals the most complex protease profile seen in the course of the nervous system study. Indeed if mature type-2 astrocytes or oligodendrocytes release only uPA, type-1 astrocytes release both uPA and tPA, a result that could be expected from studies of cortical astrocytes in culture. O_2A progenitor cells were found to release a complex mixture of uPA, tPA, PAI and at least one protease of low MW which was later found to be plasminogen-independent. When stimulated to proliferate by addition of type-1 astrocyte conditioned medium (a mitogenic activation which is now believed to be mediated by the release of PDGF), O_2A progenitor cells increased very much their release of tPA and produced other unidentified proteases of low molecular weight. Subsequent analysis indicated the presence of a PA form of about 35 kDa and of a non-PA serine protease of 28-30 kDa. The roles of these proteases in O2A cells proliferation and differentiation remains to be elucidated. It is interesting to note that until they stop dividing and differentiate into oligodendrocytes or type 2 astrocytes, the O2A cells are actively migrating cells both *in vivo* and *in vitro*.[27]

NEURITIC FASCICULATION

When perinatal rat dorsal root ganglia (DRG) or cerebellar microexplants are grown *in vitro*, they are able to extend large neuritic arrays whose behavior can be modulated by modifications of the substratum or by addition of non-neuronal cells conditioned media. For example on polyornithine-coated surfaces, a highly adhering substratum, ganglia and cerebellar microexplants grow strongly twisted and branched neurites presenting in many cases multiple growth cones and some flattening of processes appearing as points of increased substratum adherence (Figure 5). When treated with Schwann cell conditioned medium, those neuritic processes rapidly

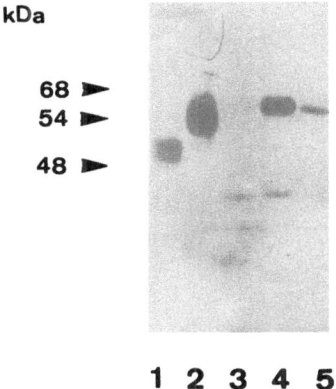

Figure 3. *Zymographic analysis of media conditioned by newborn rat DRG cultures.* The ganglia were dissociated by trypsin and collagenase digestion and the resulting cell suspension was in some cases subjected to selective attachment to produce enriched populations of neuronal and non-neuronal cells. Serum-free media where generated for 24 hr after 1 or 2 days in culture. Lane 1: 5 mU human uPA; lane 2: 5 mU human tPA; lane 3: 24 hr cultures of preplated DRG neurons in the presence of 10 U/ml 7 S NGF (enriched neuronal cultures); lane 4: 48 hr cultures of non-preplated DRG neurons in the presence of 10 U/ml 7 S NGF (mixed neuronal and non-neuronal cultures); lane 5: 48 hr cultures of non-preplated DRG neurons in the absence of 7 S NGF (neuron-free cultures).

detached from the culture dishes and straightened up, forming rectilinear bundles of neurites anchored only at their distal extremities. Based on our knowledge of the presence of tPA in those Schwann cell conditioned media, we tried to reproduce similar morphological alterations by direct addition of purified human tPA. Those experiments were negative, a result which now appears logical in view of the demonstrated species specificity of the interaction of tPA with neuronal cells.[7]

To further substantiate the hypothesis that tPA was modulating the interaction between neuritic processes and the substratum, we grew Schwann cell-containing DRGs onto laminin, a substratum with lower adhesiveness. In this case, neuritic extensions where in general rectilinear and associated into thick bundles in a fashion which is reminiscent of the fasciculation of axonal processes into developing nerve (Figure 6). With this model we found that blockade of serine proteases by PMSF, or neutralization of tPA activity by a tPA active-site blocking preparation of anti-human tPA IgG, resulted in the appearance of individually twisting and branching neuritic extensions displaying an increased adherence for the substratum. These treatments did not inhibit further neuritic outgrowth and, interestingly, were without effect on the already fasciculated neurites that were allowed to grow before addition of anti-tPA IgG.

Those results suggest a possible role for tPA activity in the control of neuritic fasciculation *in vitro*. The available informations do not allow at this stage to distinguish between models involving either a tPA activity-dependent increase in adhesion between neurites or an inhibition by tPA of the interaction of the neuronal membranes with substratum. The results obtained after use of substrata presenting either a low or a high affinity for neuritic processes can be interpreted as being two

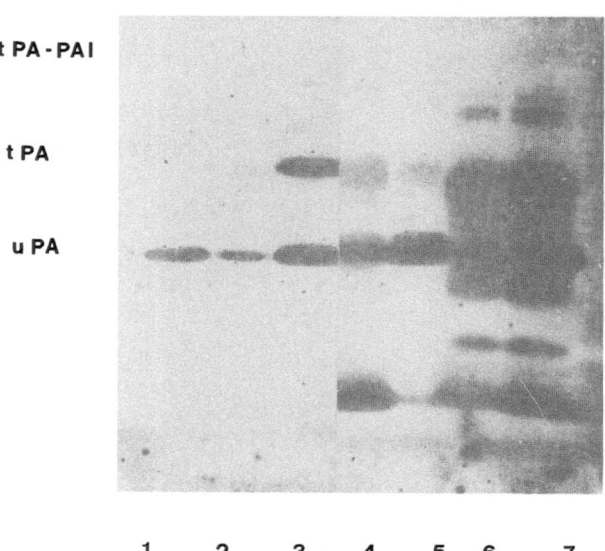

Figure 4. *Zymographic analysis of media conditioned by 7 day-old rat optic nerve glial cells.* Purified cell populations were used for generation of serum-free conditioned media for 24 hr or 48 hr, either as single cell type or in coculture. Lane 1: purified mature oligodendrocytes; lane 2: purified mature type 2 astrocytes; lane 3: purified mature type 1 astrocytes; lane 4 and 5: purified O_2A progenitor cells after 24 hr in culture (lane 4) and 48 hr in culture (lane 5); lane 6 and 7: proliferating O_2A progenitor cells in the presence of type 1 astrocytes conditioned medium (media were provided by Dr. R. Small).

aspects of a same phenomenon occurring at the two extremities of a spectrum of neuron-substrate adhesiveness. One possible mechanism to explain these results involves the deposition of tPA on laminin substratum (for which an affinity cf tPA has been demonstrated) occurring in the vicinity of newly-formed neurites and resulting in a decrease of the adherence between those neurites and laminin. Such an effect would then be absent or controlled by the release of specific inhibitors (e.g. PAI) at the growth cone and along stabilized and already fasciculated neurites. As Schwann cells release tPA, the possibility that this protease is also involved in the control of axon-Schwann cell interactions, e.g. during myelination, should also be considered.

NEURITOGENESIS

Neuritogenesis is a developmental process that has received early attention since the discovery of Nerve Growth Factor (NGF) and because of the potential importance of controlling neurite regeneration after nervous system injury. Neuritogenesis is a phenomenon where the involvement of PA activities in neuronal development has been most documented especially when it was realized that a glia-derived neurite promoting factor was a serine protease inhibitor.[28]

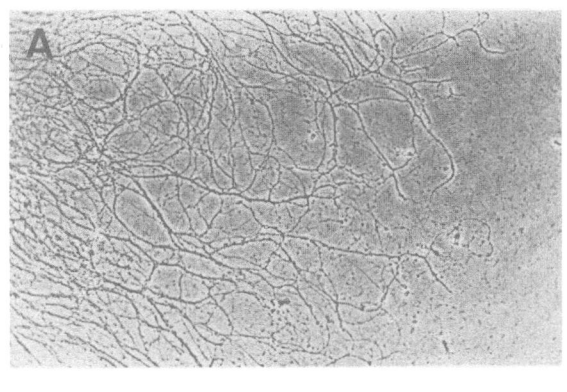

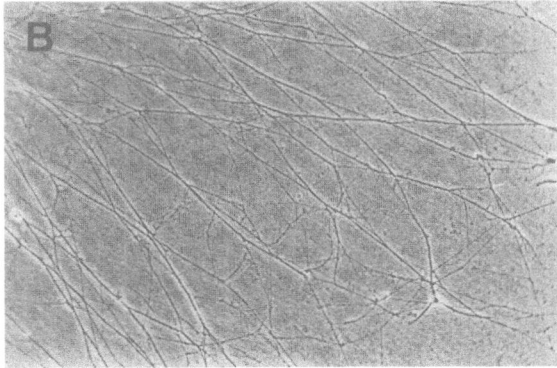

Figure 5. *Newborn rat Dorsal Root Ganglion grown in culture on polylysine-coated dishes.* The cultures were grown for 72 hr in DMEM medium supplemented with NGF either alone (A) or in separate coculture with purified sciatic nerve Schwann cells (B).

Numerous observations on PA and PAI involvement in neuritogenesis and putative mechanisms explaining these observations are described in other chapters of this volume. We have focused our attention on a cloned cell model of neuritogenesis, assuming that if a highly specific signal (in this case NGF) was able to induce in those cells the overall phenotypic alteration of neuronal differentiation, one should expect to find in the responsive cells a complete mechanism for induction and control of the growth of neuritic processes. This cell line, the pheochromocytoma PC12,[29] is a proliferating clone that can be induced by NGF to differentiate as neuron and to grow neurite. This effect of NGF can be mimicked partly by activation of adenylate cyclase or by diButyryl-cyclic AMP (diButcAMP) with, however, a resulting production of neurites presenting a less extensive development as well as differences in stability and cytoskelettal protein content.[30-33] A key feature of neuritogenesis in PC12 cells is that NGF and diButcAMP act synergistically to initially enhance the rate and extension of neuritic outgrowth.[34,35]

PA and PAI production and release by PC12 cells was thus studied with the expectation that neuritogenesis induced by different agents and resulting into different morphologies, would be accompanied by modulation of the PC12 cell-derived proteolytic activity.

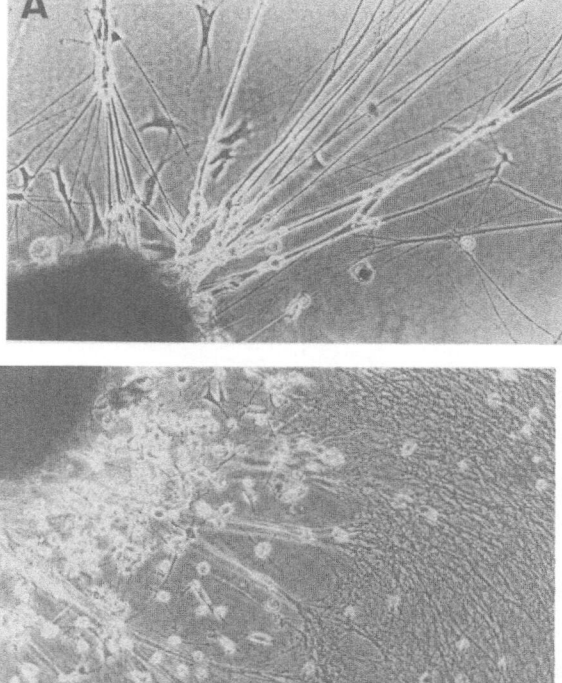

Figure 6. *Newborn rat Dorsal Root Ganglion grown in culture on laminin-coated dishes.* The cultures were treated for 3 days with control medium (DMEM-NGF) (A) or were treated for 1 day in control medium followed by two days in medium supplemented with 10 µg/ml purified anti-human tPA IgG (gift of Dr. D. Collen) (B).

PC12 cells were found to contain and release mostly the tPA form of PA together with lesser amounts (less than 5 % of total PA) of uPA. Direct, as well as reverse, zymographies indicate that those cells release also a PAI of 54-58 kDa which forms SDS-stable complexes with released tPA. Short term treatment (2-4 days) with NGF, diButcAMP or a combination of these two agents resulted in dramatic changes in PA profile in conditioned media (Figure 7) which can be summarized as follows :

- diButcAMP treatment increases more than 20-fold the release of tPA activity, but has little or no effect on PAI production
- NGF has only a limited effect on tPA activity but increases slightly the release of PAI in culture medium
- the increased tPA levels after diButcAMP treatment result into enhanced formation of tPA-PAI complexes with available PAI
- simultaneous addition of NGF and diButcAMP seemingly induce additive modifications of proteolytic activity profiles, with resulting formation of tPA-PAI complexes with all available tPA.

Concomitantly, characteristic patterns of neuritogenesis were observed after treatment with diButcAMP and NGF (Figure 8).

It is thus demonstrated that a neuronal cell line can release in its culture medium both a tPA and a PAI activity. This PAI is in the process of being further characterized and appears by several criteria to be different form protease-nexin I or the glia-derived nexin. Current models of proteases involvement in neuritogenesis require that a finely-tuned control of proteolytic activity is necessary to allow for the mobility of the growth cone, for positive increases in neuritic length and for the subsequent stabilization of the newly-formed neurite. Certainly an interesting point along this line is the demonstration of separate control mechanisms for the release of tPA and PAI in a single cell. Many NGF-induced alterations of PC12 properties have

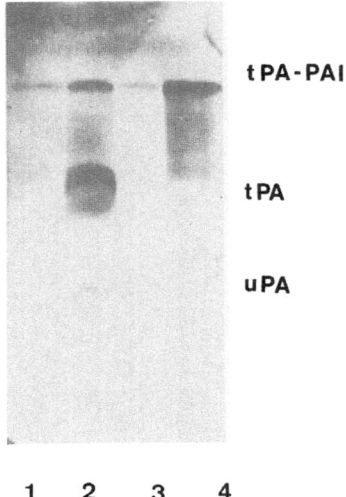

tPA-PAI

tPA

uPA

1 2 3 4

Figure 7. *Zymographic analysis of media conditioned by differentiating PC12 neuronal cells.* Cells were grown for 48 hr in DMEM medium supplemented with N1-derived nutrient formulation. Conditioned media were collected and analyzed using plasminogen-containing casein-agar gels. Lane 1: control culture; lane 2: culture treated with 10^{-5}M diButcAMP; lane 3: culture treated with 40 U/ml NGF; lane 4: culture treated with diButcAMP and NGF.

been shown to be under control of protein kinase C-dependent pathways. The involvement of similar modulatory mechanisms in control of PAI activity in PC12 cells should thus be considered.

Neurite outgrowth is a complex phenomenon requiring changes in membrane-substratum adhesiveness but also many intracellular alterations, such as the reorganization of the cytoskeleton. The involvement of an extracellular proteolytic system during neuritogenesis must thus be relocated in a more complex system as an element that regulates the effect of other intracellular mechanisms rather than as a driving force. It would thus be premature to draw conclusions that would directly link the extent of neurite outgrowth induced by different agents and the appearance in extracellular medium of one or another form of PA or PAI. One can only conclude that a physiological stimulation of neuritogenesis result into an enhanced release of PAI activity by PC12 cells and that a more intense neuritic outgrowth is correlated with the presence of higher amounts of tPA -PAI complexes in the growth medium.

The PC12 model of neuritogenesis constitute thus an interesting model for the study of the involvement of proteases activity in neuritic outgrowth. It is expected that the use of this highly modulable and well-characterized cell line will provide answers to the many questions that remain pending and deal with the nature of substratum for PA, the interaction of PA and PAI with the neuritic membrane and extracellular

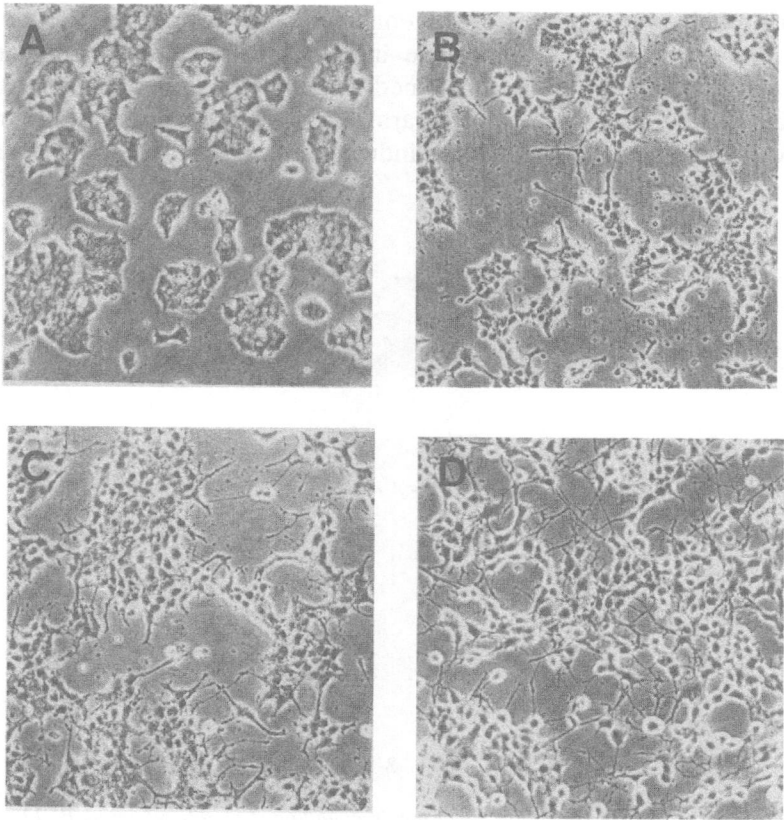

Figure 8. *Phase contrast micrography of the PC12 cell cultures used for the generation of conditioned media analyzed in Figure 7.* Treatments were: A: control; B: diButcAMP (10^{-3}M); C: NGF (40 U/ml); D: diButcAMP (10^{-3}M) + NGF (40 U/ml).

matrix and the mechanism of control of adhesion between cell membranes and between cellular membranes and their environment.

CONCLUSIONS

Plasminogen activators and plasminogen activator inhibitors are suspected to be endowed with multiple roles in nervous system development. In several cases, the data suggest that the proteolytic activity of PAs or the presence of a free active-site able to form covalent complexes with PAI, are necessary for their biological action. Although uPA and tPA show similar affinity for their physiological substrate plasminogen, they clearly constitute different signals in nervous system ontogenesis. Thus this suggest that plasminogen activation may not be a general or even a necessary step in PAs action in the nervous system (data related to the presence of plasminogen in the nervous system are scarce) but rather that other substrates for PA must exist that display different sensitivities to uPA or tPA.

Both forms of PA differ by several structural and functional properties such as their affinity for different membrane receptor or for extracellular matrix components, the presence of "growth-factor domain" and of fibrin-binding sites, which may confer to them their observed biological specificity. Furthermore, the production of PAs and of PAI appears to be unevenly distributed amongst the neural cell populations and to be highly regulated, affording an efficient control of each PA activity in time and space.

As did others,[36] we find that other types of proteases are also released by cells of neural origin. The identity and role of these proteases in neural function remain to be investigated. Together with PAs, they may contribute to the highly organized and complex cellular interactions that are needed for the adequate ontogenesis of the nervous system.

Finally, the molecular mechanism of PA and PAI actions in the nervous system should constitute an exciting field of study since present knowledge suggests their involvement in general and highly-specific proteolytic processes as well as in less characterized actions, dependent on proteases-inhibitors complex formation and interactions with cellular and extracellular receptors.

This work was supported by grants from Fonds de la Recherche Scientifique Médicale and Fonds Médical Reine Elisabeth, Belgium. P.L. is Chargé de Recherches, B.R. is Aspirant F.N.R.S. We thank P. Ernst-Gengoux and A. Brose for their expert technical assistance.

REFERENCES

1. D. Monard, Cell-derived proteases and proteases inhibitors as regulators of neurite outgrowth. **Trends Neurosci.** 11: 541 (1988).
2. M.W. Cowan, The development of the vertebrate central nervous system: an overview, *in:* "Development of the nervous system" D.R. Garrod, ed. Cambridge University Press, Cambridge,UK p 3 (1982).
3. P. Rakic, Mechanisms of neuronal migration in developing cerebellar cortex, *in:* "Molecular basis of neural development" G.M. Edelman, W.E. Gall, and W.M. Cowan, eds. Wiley, New York p 139 (1985).
4. G. Moonen, M.-P. Grau-Wagemans, and I. Selak, Plasminogen activator-plasmin system and neuronal migration. **Nature** 298: 753 (1982).

5. J. Guenther, H. Nick, and D. Monard, A glia derived neurite promoting factor with protease inhibitor activity. **EMBO J.** 4: 1963 (1985).

6. J. Lindner, J. Guenther, H. Nick, G. Zinser, H. Antonicek, M. Schachner, and D. Monard, Modulation of granule cell migration by a glia-derived protein. **Proc. Natl. Acad. Sci. USA** 83: 4568 (1986).

7. S. Verrall, and N.W. Seeds, Tissue plasminogen activator binding to mouse cerebellar granule neurons. **J. Neurosci. Res.** 21: 420 (1988).

8. P. Leprince, B. Rogister, and G. Moonen, Identification and distribution of plasminogen activator and plasminogen-activator inhibitors in the mammalian nervous system. **Arch. Int. Physiol. Biochim.** 94: 145 (1986).

9. J.E. Valinsky, and N.M. Le Douarin, Production of plasminogen activator by migrating cephalic neural crest cells. **EMBO J.** 4: 1403 (1985).

10. M.S. Pepper, J.D. Vassalli, R. Montesano, and L. Orci, Urokinase-type plasminogen activator is induced in migrating capillary endothelial cells. **J. Cell Biol.** 105: 2535 (1987).

11. G. Moonen, M.-P. Grau-Wagemans, I. Selak, P.P. Lefebvre, B. Rogister, J.D. Vassali, and D. Belin, Plasminogen activator is a mitogen for astrocytes in developing cerebellum. **Devel. Brain Res.** 20: 41 (1985).

12. P. Liesi, Do neurons in the vertebrate CNS migrate on laminin. **EMBO J.** 4: 1163 (1985).

13. I. Selak, J.-M. Foidart, and G. Moonen, Laminin promotes cerebellar granule cells migration *in vitro* and is synthesized by cultured astrocytes. **Devel. Neurosci.** 7: 278 (1985).

14. J.P. Quigley, L.I. Gold, R. Schwimmer, and L.M. Sullivan, Limited cleavage of cellular fibronectin by plasminogen activator purified from transformed cells. **Proc. Natl. Acad. Sci. USA** 84: 2776 (1987).

15. P.G. McGuire, and N.W. Seeds, Regenerating sensory neurons degrade fibronectin during the process of neurite outgrowth. **J. Cell Biol.** 107: 374a (1988).

16. E.M. Salonen, A. Zitting, and A. Vaheri, Laminin interacts with plasminogen and its tissue-type activator. **FEBS Lett.** 172: 29 (1984).

17. E.M. Salonen, O. Saksela, T. Vartio, A. Vaheri, S. Nielsen, and J. Zeuthen, Plasminogen and tissue-type plasminogen activator bind to immobilized fibronectin. **J. Cell Biol.** 22: 12302 (1985).

18. A. Krystosek, and N.W. Seeds, Normal and malignant cells, including neurons, deposit plasminogen activator on the growth substratum. **Exp. Cell Res.** 166: 31 (1986).

19. A. Baron van Evercooren, P. Leprince, B. Rogister, P.P. Lefebvre, P. Delrée, I. Selak, and G. Moonen, Plasminogen activators in developing peripheral nervous system. Cellular origin and mitogenic effect. **Devel. Brain Res.** 36: 101 (1987).

20. B. Rogister, P. Leprince, B. Pettman, G. Labourdette, M. Sensenbrenner, and G. Moonen, Brain basic fibroblast growth factor stimulates the release of plasminogen activators by cultured astroglial cells. **Neurosci. Lett.** 91: 321 (1988).

21. J.C. Kirchheimer, J. Wojta, G. Hienert, G. Christ, M.E. Heger, H. Pflüger, and B.R. Binder, Effect of urokinase on the proliferation of primary cultures of human prostatic cells. **Thromb. Res.** 48: 291 (1987).

22. L.B. Chen, and J.M. Buchanan, Mitogenic activity of blood components. I. Thrombin and prothrombin. **Proc. Natl. Acad. Sci. USA** 72: 131 (1975).

23. F. Perraud, F. Besnard, M. Sensenbrenner, and G. Labourdette, Thrombin is a potent mitogen for rat astroblasts but not for oligodendroblasts and neuroblasts in primary culture. **Int. J. Devel. Neurosci.** 5: 181 (1987).

24. D.A. Low, R.W. Scott, J.B. Baker, and D.D. Cunningham, Cells regulate their mitogenic response to thrombin through release of protease nexin. **Nature** 298: 476 (1982).

25. D.E. Rosenblatt, C.W. Cotman, M. Nieto-Sampedro, J.W. Rome, and D.J. Knauer, Identification of a protease inhibitor produced by astrocytes that is structurally and functionally homologous to human protein nexin 1. **Brain Res.** 415: 40 (1987).

26. M.C. Raff, Glial cell diversification in the rat optic nerve. **Science** 243: 1450 (1989).

27. R.K. Small, P. Riddle, M. and Noble, Evidence for migration of oligodendrocyte-type-2 astrocyte progenitor cells into developing rat optic nerve. **Nature** 328: 155 (1987).

28. D. Monard, E. Niday, A. Limat, and F. Solomon, Inhibition of protease activity can lead to neurite extension in neuroblastoma cells. **Prog. Brain Res.** 58: 359 (1983).

29. L.A. Greene, and A.S. Tischler, Establishment of a noradrenergic clonal line of rat adrenal pheochromocytoma cells which respond to nerve growth factor. **Proc. Natl. Acad. Sci. USA** 73: 2424 (1976).

30. M. Black, and L.A. Greene, Changes in the colchicine susceptibility of microtubules associated with neurite outgrowth: studies with nerve growth factor-responsive PC12 pheochromocytoma cells. **J. Cell Biol.** 95: 379 (1982).

31. D. Drubin, S. Kobayashi, D. Kellogg, and M. Kirschner, Regulation of microtubule protein levels during cellular morphogenesis in nerve growth factor-treated PC12 cells. **J. Cell Biol.** 106: 1583 (1988).

32. P. Doherty, D.A. Mann, and F.S. Walsh, Cholera toxin and dibutyryl cyclic AMP inhibit the expression of neurofilament protein induced by nerve growth factor in cultures of naive and primed PC12 cells. **J. Neurochem.** 49: 1676 (1987).

33. J.M. Aletta, S.A. Lewis, N.J. Cowan, and L.A. Greene, Nerve Growth Factor regulates both the phosphorylation and steady-state levels of microtubule-associated protein 1.2 (MAP1.2). **J. Cell Biol.** 106: 1573 (1988).

34. P.W. Gunning, G.E. Landreth, M.A. Bothwell, and E.M. Shooter, Differential and synergistic actions of nerve growth factor and cyclic AMP in PC12 cells. **J. Cell Biol.** 89: 240 (1981).

35. S.R. Heidemann, H.C. Joshi, A. Schechter, J.R. Fletcher, and M.A. Bothwell, Synergistic effects of cyclic AMP and nerve growth factor on neurite outgrowth and microtubule stability of PC12 cells. **J. Cell Biol.** 100: 916 (1985).

36. R.N. Pittman, Release of plasminogen activator and a calcium-dependent metalloprotease from cultured sympathetic and sensory neurons. **Devel. Biol.** 110: 91 (1985).

INTERACTION OF PLASMINOGEN ACTIVATORS WITH THE NEURONAL SURFACE

RANDALL N. PITTMAN, ANN REPKA, JEFF H. WARE,

AND ANN MARIE LaROSA

Department of Pharmacology 6084 University of Pennsylvania,
School of Medicine, Philadelphia, PA 19104

INTRODUCTION

Extensive structural and cellular changes are conspicuous features of the developing organism. The mechanisms underlying these changes are in general unknown and few of the molecules responsible for dynamic events such as morphogenesis and differentiation have been identified. Proteolytic activity is an ideal mechanism for producing many of the changes characteristic of the developing organism. Proteases have been implicated in a number of prominent developmental events including trophoblast implantation,[1,2] pattern formation,[3] migration,[4] cell proliferation,[5-7] and differentiation.[8] Proteases present in the nervous system have been implicated in a number of developmental events including glial proliferation,[9,10] neural crest and granule cell migration,[11-13] and neurite outgrowth.[14-20]

Both serine proteases and metalloproteinases are released from distal processes and growth cones of developing neurons.[21-23] Their release from growth cones is consistent with proteases being involved in functions associated with growth cones such as neurite outgrowth, pathfinding, target recognition, and synaptogenesis. One obvious function for proteases released by growth cones would be to degrade components of the extracellular matrix and to promote neurite outgrowth toward target tissues. This appears to be the function of a Ca^{2+}-dependent metalloproteinase released by sensory and sympathetic neurons of the peripheral nervous system.[23,24] The metalloproteinase degrades native and denatured Type I collagen and fibronectin. Inhibition of the neuronal metalloproteinase activity blocks neurite penetration and outgrowth in 3-dimensional collagen gels but does not affect outgrowth on 2-dimensional collagen films.[24] Functions of neuronal serine proteases are less clearly defined than that of the Ca^{2+}-dependent metalloproteinase. A thrombin-like serine protease released by (or present on) growing neurites has been implicated in neurite outgrowth from neuroblastoma cells[14,19] and from primary cultures of mouse sensory ganglia.[16] Inhibition of urokinase plasminogen activator (uPA) activity increases neurite outgrowth from rat sympathetic neurons,[24] suggesting that a uPA-like serine protease may also be involved in neurite outgrowth.

Serine Proteases and Their Serpin Inhibitors in the Nervous System
Edited by B. W. Festoff
Plenum Press, New York, 1990

Proteases serve a number of important cellular functions; however, their activities must be controlled in order to protect the organism from the potentially lethal effects of nondiscriminate protein degradation. Inhibitors are an effective means of controlling the destructive actions of proteases. Recent studies have resulted in purification and characterization of a number of cellular protease inhibitors. These inhibitors are likely to be important regulators of proteolysis in both neural and non-neural systems.[15,17,25-29]

NEURONAL uPA

The physiological functions of uPA are currently being investigated using a variety of cellular assays. A number of studies are consistent with a role of uPA in changes in cell shape during mitosis and in motility.[30,31] The involvement of uPA in these cellular functions is postulated to result from changes in cell adhesion following the binding of uPA to specific receptors located on focal contacts between the bottom surface of cells and the substratum.[32-34] The mechanism responsible for changes in cell adhesion has not been identified, although activation of plasminogen with subsequent degradation of proteins or limited cleavage of cell surface components or components of the extracellular matrix by uPA are the most likely possibilities. uPA also appears to be the initiating factor in a proteolytic cascade consisting of uPA, plasminogen, and procollagenase that degrades components of the extracellular matrix.[35,36] This cascade of proteolytic activity may serve important functions during morphogenesis, tissue repair, and movement of migrating and malignant cells.[30,31]

Rat sympathetic and sensory neurons in culture release uPA.[23] Most of the uPA appears to be released soon after synthesis, such that there is little stored or cell associated uPA (Pittman, 1985, unpublished observations). Because neurons are highly polarized, it is possible to determine the amount of uPA released from different regions of the neuron. Sympathetic neurons release about 50% of the uPA from the cell body and dendrites, 10% from proximal portions of the axons and 40% of the uPA from distal portions of axons and growth cones.[23] Distal axons and growth cones account for < 0.1% of the total neuronal tissue; therefore, the specific activity of uPA in the area of the growth cone is 3 orders of magnitude higher than at the cell body or along the proximal axon. Because of the fundamental importance of the growth cone in developmental events such as guidance, neurite outgrowth, target recognition, and synaptogenesis, the release of high concentrations of uPA by this cellular specialization is potentially significant.

Possible functions of neuronal uPA have been investigated by exposing sympathetic neurons to a variety of serine protease inhibitors and determining the effects on growth cone function.[20] Inhibition of serine protease activity increases neurite outgrowth about 2 fold. Inhibition of PA activity but not other serine protease activity correlates with the increase in neurite outgrowth (r = 0.89 for uPA, r = 0.86 for tPA, r < 0.2 for plasmin, trypsin, chymotrypsin, and thrombin). Since inhibitors do not effectively distinguish tPA from uPA, antibodies have been used to differentiate between them. Monoclonal antibodies that we have generated against rat tPA do not inhibit activity and preliminary experiments indicate that these antibodies do not affect neurite outgrowth. Therefore, the effect of selectively inhibiting tPA is unknown. An affinity purified antiserum against uPA, however, inhibits activity and increases neurite outgrowth (Figure 1). Antibodies that bind at or near the active site and inhibit uPA activity, increase neurite outgrowth (Figure 1; fraction #2) while antibodies that bind to other regions of uPA and do not inhibit activity do not

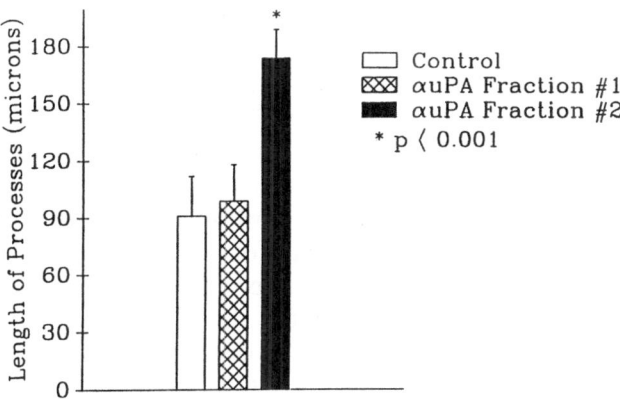

Figure 1. *Increase in neurite outgrowth in the presence of antibodies against uPA.* Sympathetic neurons were grown for 14 hrs in the presence of 5 μg/ml pre-immune IgG (control), or affinity purified anti-uPA antibodies that bind at or near the active site of uPA (Fraction #2), or that bind to other parts of the molecule (Fraction #1). Fraction #2 was 100x more effective at inhibiting uPA activity than Fraction #1. Bars represent the mean ± SEM of neurite lengths.

increase neurite outgrowth (Figure 1; fraction #1). These data indicate that at least part of the increase in neurite outgrowth following treatment of sympathetic neurons with inhibitors of serine proteases probably results from inhibition of uPA activity. Inhibitors of PA increase neurite outgrowth 120-130%, while antibodies that completely inhibit uPA activity increase neurite outgrowth only 75%. This may indicate that part of the increase in neurite outgrowth following treatment of neurons with inhibitors of serine proteases is due to inhibition of other PA-like serine proteases, possibly tPA.

In the presence of inhibitors of PA, neurites not only grow faster, but they are also much straighter than control neurites.[20] This is reflected in a 4-fold increase in the persistence time in the presence of inhibitors. Persistence time is a mathematical measure of tortuosity and in the case of growing neurites indicates the amount of time a growth cone spends "wandering around" the culture dish in contrast to growing straight. The higher the persistence time, the straighter the neurite. The mechanism by which inhibitors of PA increase the straightness of neurites is unknown, although changes in adhesion between the neuronal surface and substrate could alter the persistence time.

Time lapse videomicroscopy studies of neurite outgrowth in the presence of inhibitors of uPA indicates that the major change at the growth cone is an increase in membrane ruffling and lamellipodial activity at the leading edge of the growth cone and less motility on the sides and base of the growth cone.[20] The decrease in lamellipodial activity on the sides of the growth cone along with the increase in activity at the leading edge may be partially responsible for decreasing lateral movement of growth cones and producing straighter and longer neurites. Approximately 85% of the growth cones increase their rate of outgrowth (40-250% increases in rates) in the presence of inhibitors of uPA. Routinely, increases in rates of outgrowth are obvious within 1-2 minutes following addition of the inhibitor.

A small amount of the uPA released by sympathetic neurons and the neuronal cell line, PC12, binds to the cell surface. It is this cellular bound form of uPA that is most likely involved in neurite outgrowth. Surface bound uPA is highly localized in patches on the bottom surface (next to the substrate) of PC12 cells,[20] which places it in an ideal position for being involved in motile events such as neurite outgrowth. This localization is very reminiscent of the binding of uPA to focal contacts on the bottom surface of fibroscarcoma cells and fibroblasts.[32-34'] A variety of potential mechanisms could account for the increase in outgrowth following inhibition of uPA. The simplest hypothesis is that inhibitors of uPA increase neurite outgrowth by increasing the adhesion of the growth cone to the substrate. This increased adhesion of the growth cone to the substrate may decrease the amount of backward "slippage" or retraction of the growth cone[37-39] and produce an increased rate of forward outgrowth. A variety of other hypotheses could account for the increase in neurite outgrowth following inhibition of uPA including differential distribution of uPA receptors on the leading edge compared to other parts of the growth cone and the involvement of the uPA-receptor complex in various adhesion and/or deadhesion events.

NEURONAL INHIBITOR OF uPA

Proteolytic activity is an efficient mechanism for degrading cellular and extracellular molecules during tissue remodeling, migration and neurite outgrowth. Uncontrolled proteolytic activity, however, represents a potentially lethal threat to the organism. Proteolytic activity is controlled by the availability of activators and substrates and by serum and cellular inhibitors. Many different types of cells produce serine protease inhibitors;[15,17,26,27] therefore, it is not surprising that neurons also have an inhibitor associated with them.[20] An inhibitor present on the surface of sympathetic neurons binds to the active site of uPA (as well as plasmin and to a lesser extent tPA) and inhibits proteolytic activity. The neuronal inhibitor forms a 68 kDa complex with ^{125}I-uPA (MW 33 kDa) that is stable in boiling SDS/mercaptoethanol. An inhibitor with similar properties is present on the neuronal cell line, PC12. PC12 cells are derived from a tumor of the rat adrenal medulla;[40] cells with the same developmental lineage as sympathetic neurons. When treated with nerve growth factor (NGF), PC12 cells acquire many of the morphological and biochemical characteristics of sympathetic neurons. Interestingly, the complex formed between ^{125}I-uPA and the inhibitor(s) on PC12 cells is dependent on the differentiation state of the PC12 cells. The inhibitor on undifferentiated PC12 cells forms an 80 kDa complex with ^{125}I-uPA, whereas the inhibitor present on PC12 exposed to NGF for 1 week forms a 68 kDa complex with ^{125}I-uPA (Figure 2). Whether the two different MWs of the complexes with ^{125}I-uPA represent two different inhibitors or two forms of the same inhibitor is unknown. The neuronal inhibitor forming the 68 kDa complex with ^{125}I-uPA has a MW of 50 kDa.[20] It is released from the surface of PC12 cells by heparinase, indicating that it is not an integral membrane protein but rather appears to be bound to the cell surface by binding to cell associated heparan sulfate proteoglycan. The properties of this neuronal/PC12 inhibitor are very similar to the properties of glial derived nexin,[15] protease nexin-I,[25] and an inhibitor of serine proteases released by heart cells[17] (see Table 1).

A likely physiological role of cellular inhibitors is to protect cells during normal or pathological events in which large amounts of serine proteases are present. A specific example of this general role of inhibiting serine proteases is that the neuronal inhibitor may be present on areas of the neuron involved in stable adhesion to the substrate or stable cell/cell contacts. The inhibitor would inactivate uPA or other serine protease in the local area and thereby help maintain stable contacts.

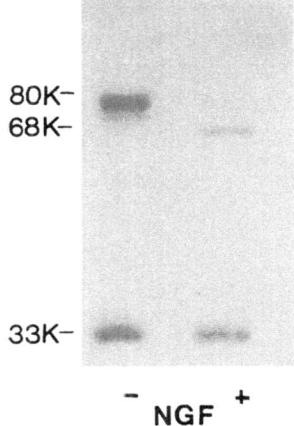

Figure 2. *Complexes between* ^{125}*I-uPA and surface inhibitor(s) on PC12 cells as a function of differentiction.* PC12 cells were grown in serum-free medium (-), or for 4 days in serum-free medium containing 50 ng/ml 2.5 S NGF (+). Cells were then incubated with 5 x 10^6 cpm ^{125}I-uPA for 20 min, washed extensively and prepared for electrophoresis and autoradiography. The bands at 68 kDa and 80 kDa were not formed in the absence of PC12 cells and represent complexes between ^{125}I-uPA and PC12 proteins. The band at 33 kDa is ^{125}I-uPA.

Cardiac myocytes and cells present in peripheral nerve sheaths also release inhibitors that may interact with neuronal uPA (Table 1).[17,41] Cardiac myocytes, a normal target of sympathetic neurons release an inhibitor(s) that forms complexes of 68 kDa and 80 kDa with ^{125}I-uPA (MW 33 kDa). It appears that in presence of heparan sulfate proteoglycan, the 80 kDa complex forms and in the absence of proteoglycan the 68 kDa complex forms. The purified inhibitor increases neurite outgrowth from sympathetic neurons approximately 2-fold,[17] which is consistent with observations that small molecular weight inhibitors and cellular inhibitors of PA increase neurite outgrowth.[15,16,18-20] Cells present in peripheral nerves (superior cervical ganglion, dorsal root ganglion, and sciatic) but not cells present in a central nerve (optic) also release an inhibitor of uPA.[41] The inhibitor present in peripheral nerves may be partially responsible for the increased ability of nerves in the peripheral nervous system to regenerate relative to nerves in the central nervous system. The presence of an inhibitor in peripheral nerves, however, may represent a protective mechanism for cells in a tissue with high levels of endogenous proteolytic activity, which may be required for neurite outgrowth and regeneration.

NEURONAL tPA

tPA is released from endothelial cells, binds to fibrin clots and activates plasminogen which then degrades fibrin. No other clear function has been attributed to tPA; however, based on the large number of cell types containing tPA,[42] it is likely to be involved in non cardiovascular functions as well. The nervous system and neuroendocrine tissues in particular have high concentrations of tPA.[42-44] Rat sympathetic and sensory neurons in serum-free cultures contain significant amounts of intracellular tPA (Figure 3), but release only small amounts of tPA.[20] It may be that tPA has an intracellular function, or that its release is regulated by factors that are not present in our cultures. In contrast to sympathetic neurons, glia appear to release a large amount of tPA (Pittman, unpublished observations).[9,45] tPA

Table 1. Cellular Inhibitors Likely to Interact with Neuronal Plasminogen Activators.

	Neuronal	Heart	GDN/Nexin-I
MW in SDS (kDa)	50	50	43/49-53 42 actual MW
Complex with ^{125}I-uPA (kDa)	68 predominantly 80 occasionally	68 and 80	72
Heparin/heparan sulfate binding	Yes	Yes	Yes
Cell surface attachment	Yes, heparinase sensitive	?	Yes
Free in culture medium	small amounts only	Yes	Yes
Effect on neurite outgrowth	?	Increases	Increases

that is released by neurons and/or glia appears to bind to the surface of neurons. Neurons of both the central and peripheral nervous system have binding sites for tPA on their surface (Seeds et al., this volume).[20,46,47] The binding site on cerebellar granule neurons appears to be a higher affinity site (see Seeds et al., this issue) than the binding site on sympathetic neurons and PC12 cells.[20] There are about 350,000 binding sites/cell on sympathetic neurons with a K_d = 20 nM for human ^{125}I-tPA.[20,47] The binding is not affected by inhibitors that bind to the active site of tPA indicating that the binding is not to a cellular inhibitor. At least one of the binding sites for tPA or a component of the binding site on sympathetic neurons and PC12 cells appears to be the glycoprotein, Thy-1. This tentative identification is based on the following observations: 1) a monoclonal antibody made against a membrane preparation of sympathetic neurons inhibits the binding of ^{125}I-tPA to neurons, this mAb recognizes Thy-1, 2) ^{125}I-tPA is crosslinked to a single protein on the surface of PC12 cells which has a MW of 25-30 kDa (Thy-1 has an apparent MW of 25 kDa on SDS PAGE), 3) treatment of PC12 cells with phosphoinositol specific phospholipase C removes a significant fraction of the binding sites (Thy-1 is bound to the surface through a PI linkage), and 4) ^{125}I-tPA interacts with purified Thy-1 with the same kinetic and equilibrium binding characteristics as its interactions with the neuronal binding site. Preliminary studies however, suggest that interaction of ^{125}I-tPA with the neuronal surface may be considerably more complex than just binding to Thy-1. The reasons for this are several fold: 1) antibodies recognizing Thy-1 only partially inhibit binding of ^{125}I-tPA to neurons, 2) cells such as N18 neuroblastoma that do not have endogenous Thy-1, have a tPA binding site with properties similar to the binding site on sympathetic neurons and PC12 cells, 3) N18 cells transfected with Thy-1.2 (sympathetic neurons and PC12 cells have Thy-1.1) do not show additional binding to 125I-tPA, and 4) the carbohydrate groups of Thy-1 (and probably other surface molecules) may be involved in the binding reaction. It is clear that the binding of tPA

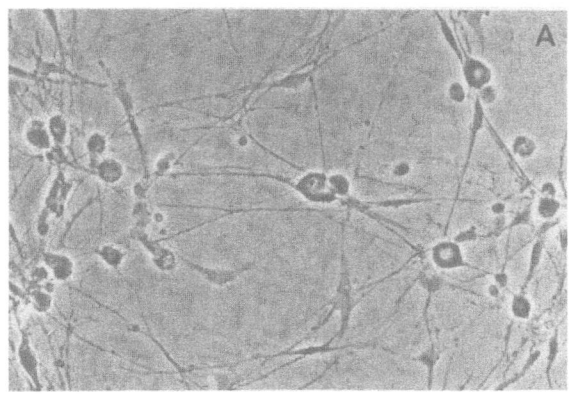

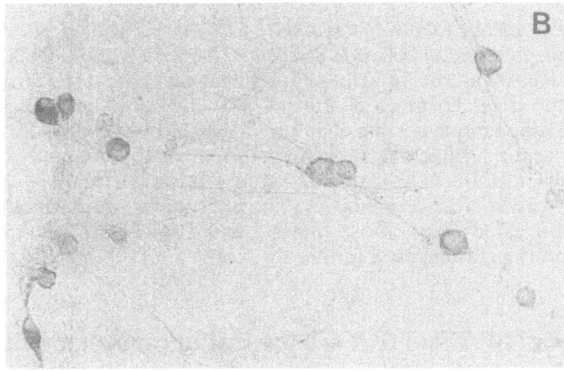

Figure 3. *Immunohistochemical localization of intracellular tPA in trigeminal sensory neurons grown in serum-free culture for 23 days.* Phase contrast micrograph (A) and micrograph of HRP reaction product (B) in neurons incubated with monoclonal antibody 2-1.2 recognizing rat tPA. Biotinylated rabbit anti-mouse, strepavidin-HRP and hydrogen peroxide incubations were used to produce the specific reaction product. Incubation with monoclonal antibody 6-17 (recognizes human tPA only) showed no reaction product (not shown). Note that nonneuronal cells visible with phase contrast are not positive for intracellular tPA-like immunoreactivity.

to the neuronal surface is not as simple as a ligand binding to a single receptor protein.

The function of neuronal tPA is unknown; however, one potential role for tPA is that it is involved in early stages of neuronal differentiation. P19S1801A1 embryonal carcinoma cells differentiate into neurons in the presence of retinoic acid; the earliest differentiation marker found for the neuronal phenotype in these cells is tPA.[48] Treatment of PC12 cells with NGF results in cells acquiring morphological, biochemical, and molecular properties of differentiated neurons.[49] The acquisition of the neuronal phenotype is also characterized by a 50-100 fold increase in tPA mRNA (Figure 4). Other cell types including progenitors for granulocytes and macrophages express tPA early in differentiation, but either lose tPA or greatly decrease its activity later in development.[50] These studies are intriguing because they correlate the presence of tPA with differentiation in a variety of cell types, however, they do not indicate the role of tPA in differentiation or the mechanism of its actions.

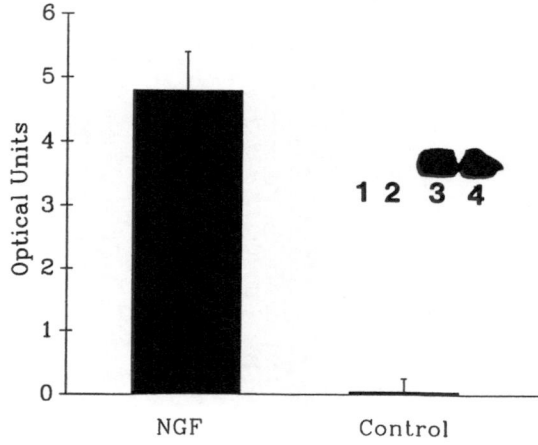

Figure 4. *Increase in tPA mRNA following treatment of PC12 cells with NGF.* PC12 cells were grown in the absence (control) or presence of 200 ng/ml 7S NGF for 7 days and poly A RNA isolated using oligo dT cellulose. Poly A RNA was run in agarose-formaldehyde gels, transferred onto nitrocellulose and probed with a 32P-labeled 1.9 kb piece of the rat tPA cDNA. Autoradiographs were scanned, and background subtracted (non sample portion of autoradiograph) from control and NGF samples (expressed in optical units of the scan). Values have been corrected for the amount of poly A RNA loaded per sample using cyclophilin mRNA as a standard. Values represent mean $\pm$ SEM of 4 different RNA preparations from control and NGF-treated cells. The inset shows Northern blots from two control (lanes 1 and 2) and two NGF-treated preparations (lanes 3 and 4) of PC12 cells. 5 ng of poly A RNA were loaded in each lane and the gels exposed to film for 5 hrs.

In summary, both uPA and tPA are present in rat sympathetic neurons in culture. A significant fraction of the uPA is released from the distal axons and growth cones. Some of the released uPA binds back to the bottom surface of neurons where it is in an ideal location for modulating motile activity. Antibodies that bind to the active site of uPA and inhibit its activity, increase the rate of neurite outgrowth while antibodies that do not inhibit activity do not alter neurite outgrowth. Time lapse videomicroscopy analysis of growth cone movement indicates that inhibition of uPA activity increases lamellipodial activity at the leading edge and decreases lamellipodial activity on the sides of growth cones. Therefore, at least one function of the neuronal uPA appears to be in motile activity during neurite outgrowth. An inhibitor of uPA is present on the surface of sympathetic neurons and PC12 cells. This inhibitor has a MW of about 50 kDa, forms a 68 kDa or 80 kDa complex with ^{125}I-uPA and binds to the cell surface through a heparinase sensitive site (presumably heparan sulfate proteoglycan). Most of the tPA associated with sympathetic neurons in culture is intracellular; however, a small amount is released. There are approximately 350,000 binding sites with a K_D = 20 nM for tPA on the surface of sympathetic neurons and PC12 cells. Based on crosslinking studies, immunological experiments, and protein purification, at least part of the binding site has been identified as the glycoprotein Thy-1. Other unidentified binding sites are also present on the neuronal surface.

REFERENCES

1. S. Strickland, E. Reich, and M.I. Sherman, Plasminogen activator in early embryogenesis:enzyme production by trophoblast and parietal endoderm. **Cell** 9:231-240 (1976).

2. R.H. Glass, J. Aggeler, A. Spindle, R.A. Pendersen, and Z. Werb, Degradation of extracellular matrix by mouse trophoblast outgrowth: a model for implantation. **J. Cell Biol.** 96:1108-1116 (1983).

3. R. DeLotto, and P. Spierer, A gene required for the specification of dorsal-ventral pattern in Drosophila appears to encode a serine protease. **Nature** 323:688-692 (1986).

4. J.E. Valinsky, E. Reich, and N.M. Le Douarin, Plasminogen activator in the bursa of Fabricius. Correlations with morphogenetic remodeling and cell migrations. **Cell** 25:471-476 (1981).

5. D.H. Carney, Perspectives on the cellular and biochemical effects of thrombin interaction with surface receptors and substrate molecules. In Proteases in Biological Control and Biotechnology, D.D. Cunningham and G.L. Long, eds., pp. 277-282, Alan R. Liss, New York (1987).

6. D.H. Carney, and D.D. Cunningham, Initiation of chick cell division by trypsin action at the cell surface. **Nature** 268:602-606 (1977).

7. L.B. Chen, and J.M. Buchanan, Mitogenic activity of blood components. I. Thrombin and prothrombin. **Proc. Natl. Acad. Sci. USA** 72:131-135 (1975).

8. G. Ortolani, E. Patricolo, and C. Mansueto, Trypsin-induced cell surface changes in ascidan embryonic cells. **Exp. Cell Res.** 122:137-147 (1979).

9. N. Kalderon, Schwann cell proliferation and localized proteolysis: Expression of plasminogen-activator activity predominates in the proliferating cell populations. **Proc. Natl. Acad. Sci., USA** 81: 7216-7220 (1984).

10. G. Moonen, M.-P. Grau-Wagemans, I. Selak, Ph.P. Lefebvre, B. Rogister, J.D. Vassalli, and D. Belin, Plasminogen activator is a mitogen for astrocytes in developing cerebellum. **Dev. Br. Res.** 20:41-48 (1985).

11. G. Moonen, M.P. Grau-Wagemans, and I. Selak, Plasminogen activator-plasmin system and neuronal migration. **Nature** 298:753-755 (1982).

12. J.E. Valinsky, and N.M. LeDouarin, The production of plasminogen activator by migrating cephalic neural crest cells. **EMBO J.** 4:1403-1406 (1985).

13. G. Grossman, J.P. Quigley, and J.E. Valinsky, An antibody with anti-catylitic activity against chicken plasminogen activator inhibits neural crest cell migration *in vitro*. **J. Cell Biol. Abstr.** 105:1256 (1987).

14. D. Monard, E. Miday, A. Limat, and F. Solomon, Inhibition of protease activity can lead to neurite extension in neuroblastoma cells. **Prog. Brain Res.** 58:359-64 (1983).

15. J. Guenther, N. Hanspeter, and D. Monard, A glia-derived neurite promoting factor with protease inhibitory activity. **EMBO J.** 4: 1963-1966 (1985).

16. R.L. Hawkins, and N.W. Seeds, Effect of proteases and their inhibitors on neurite outgrowth from neonatal mouse sensory ganglia in culture. **Br. Res.** 398:63-70 (1986).

17. R.N. Pittman, and P.H. Patterson, Characterization of an inhibitor of neuronal plasminogen activator released by heart cells. **J. Neurosci.** 7:2664-2673 (1987).

18. A.D. Zurn, H. Nick, and D. Monard, A glia-derived nexin promotes neurite outgrowth in cultured chick sympathetic neurons. **Dev. Neurosci.** 10:17-24 (1988).

19. D. Gurwitz, and D.D. Cunningham, Thrombin modulates and reverses neuroblastoma neurite outgrowth. **Proc. Natl. Acad. Sci. U.S.A.** 85:3440-3444 (1988).

20. R.N. Pittman, J.K. Ivins, and H.M. Buettner, Neuronal plasminogen activators:Cell surface binding sites and involvement in neurite outgrowth. **J. Neurosci.** 9: (1989a).

21. A. Krystosek, and N.W. Seeds, Plasminogen activator release at the neuronal growth cone. **Science** 213:1532-1534 (1981).

22. A. Krystosek, and N.W. Seeds, Peripheral neurons and Schwann cells secrete plasminogen activator. **J. Cell Biol.** 98:773-776 (1984).

23. R.N. Pittman, Release of plasminogen activator and a calcium- dependent metalloprotease from cultured sympathetic and sensory neurons. **Dev. Biol.** 110:91-101 (1985).

24. R.N. Pittman, and A. G. Williams, Neurite penetration into collagen gels requires Ca^{2-}-dependent metalloproteinase activity. **Dev. Neurosci.** 11:41-51 (1989).

25. R.W. Scott, and J.B. Baker, Purification of human protease nexin. **J. Biol. Chem.** 258:10439-10444 (1983).

26. D.L. Eaton, and J.B. Baker, Evidence that a variety of cultured cells secrete protease nexin and produce a distinct cytoplasmic serine protease-binding factor. **J. Cell. Physiol.** 117:175-182 (1983).

27. L.A. Erickson, C.M. Kekman, and D.J. Loskutoff, The primary plasminogen-activator inhibitors in endothelial cells, platelets, serum, and plasma are immunologically related. **Proc. Natl. Acad. Sci. USA** 82:8710-8714 (1985).

28. S. Gloor, K. Odink, J. Guenther, H. Nick, and D. Monard, A glia-derived neurite promoting factor with protease inhibitory activity belongs to the protease nexins. **Cell** 47:687-693 (1986).

29. D. Monard, Cell-derived proteases and protease inhibitors as regulators of neurite outgrowth. **Trends Neurosci.** 11:541-544 (1988).

30. K. Danø, P.A. Andreasen, J. Grøndahl-Hansen, B. Kristensen, L.S. Nielsen, and L. Skriver, Plasminogen activators, tissue degradation and cancer. **Adv. Cancer Res.** 44:146-239 (1985).

31. O. Saksela, and D.B. Rifkin, Cell-associated plasminogen activation: Regulation and physiological functions. **Ann. Rev. Cell Biol.** 4:93-126 (1988).

32. J. Pöllänen, O. Saksela, E.-M. Salonen, P. Andreasen, L. Nielsen, K. Danø, and A. Vaheri, Distinct localizations of urokinase-type plasminogen activator and its type 1 inhibitor under cultured human fibroblasts and sarcoma cells. **J. Cell Biol.** 104:1085-1096 (1987).

33. J. Pöllänen, K. Hedman, L.S. Nielsen, K. Danø, and A. Vaheri, Ultrastructural localization of plasma membrane-associated urokinase-type plasminogen activator at focal contacts. **J. Cell Biol.** 106:87-95 (1988).

34. C.A. Hebert, and J.B. Baker, Linkage of extracellular plasminogen activator to the fibroblast cytoskeleton: Colocalization of cell surface urokinase with vinculin. **J. Cell Biol.** 106:1241-1247 (1988).

35. Z. Werb, C. Mainardi, C.A. Vater, and E.D. Harris, Endogenous activation of latent collagenase by rheumatoid synovial cells. Evidence for a role of plasminogen activator. **New Engl. J. Med.** 296:1017-1023 (1977).

36. P. Mignatti, E. Robbins, and D.B. Rifkin, Tumor invasion through the human amniotic membrane: Requirement for a proteinase cascade. **Cell** 47:487-498 (1986).

37. M.J. Katz, How straight do axons grow? **J. Neurosci.** 5:589-595 (1985).

38. V. Argiro, M.B. Bunge, and M.I. Johnson, Correlation between growth form and movement and their dependence on neuronal age. **J. Neurosci.** 4:3051-3062 (1984).

39. D. Bray, and P.J. Hollenbeck, Growth cone motility and guidance. **Ann. Rev. Cell Biol.** 4:43-62 (1988).

40. L.A. Greene, and A. S. Tischler, Establishment of a noradrenergic clonal cell line of rat adrenal pheochromocytoma cells which respond to nerve growth factor. **Proc. Natl. Acad. Sci. USA** 73:2424-2428 (1976).

41. R.N. Pittman, P. Vos, J.K. Ivins, H.M. Buettner, and A. Repka, Proteases and inhibitors in the developing nervous system. In: *Assembly of the Nervous System*, L. Landmesser, ed. Alan R. Liss, Inc. New York, 109-128 (1989b).

42. G. Danglot, D. Vinson, and F. Chapeville, Qualitative and quantitative distribution of plasminogen activators in organs from healthy adult mice. **FEBS Lett.** 194:96-100 (1986).

43. P. Kristensen, J.S. Nielsen, J. Grøndahl-Hansen, P.B. Andresen, L.-I. Larsson, and K. Danø, Immunocytochemical demonstration of tissue-type plasminogen activator in endocrine cells of the rat pituitary gland. **J. Cell Biol.** 101:305-311 (1985).

44. P. Kristensen, J.S. Nielsen, L.-I. Larsson, and K. Danø, Tissue-type plasminogen activator in somatostatin cells of rat pancreas and hypothalamus. **Endocrinology** 121:2238-2244 (1987).

45. M.B. Clarke, and L.D. Snellinger, Schwann cells release both plasminogen activator and a plasminogen activator inhibitor. **J. Cell Biol. Abstr.** 4143 (1988).

46. S. Verrall, and N.W. Seeds, Tissue plasminogen activator binding to mouse cerebellar granule neurons. **J. Neurosci. Res.** 21:420-425 (1988).

47. R.N. Pittman, Cell surface binding sites for plasminogen activators. **Soc. Neurosi. Abstr.** 14:240.9 (1988).

48. T. Whitford, and J.M. Levine, Secretion of tissue type plasminogen activator during neuronal differentiation of an embryonal carcinoma cell line. **Soc. Neurosci. Abstr.** 13:892 (1987).

49. L.A. Greene, and A.S. Tischler, PC12 pheochromocytoma cultures in neurobiological research. **Adv. Cell. Neurobiol.** 3:373-414 (1982).

50. E.L. Wilson, and G.E. Francis, Differentiation-linked secretion of urokinase and tissue plasminogen activator by normal human hemopoietic cells. **J. Exp. Med.** 165:1609-1623 (1987).

GLIAL PLASMINOGEN ACTIVATORS IN DEVELOPING AND REGENERATING NEURAL TISSUE

NURIT KALDERON

The Rockefeller University,
New York, New York 10021

INTRODUCTION

In the following chapter I would like to introduce the postulate that a population of neural cells -- the glia -- serve as construction/remodelling agents in morphogenetic processes during development and regeneration of the nervous system. The capacity of these glial cells to perform their plasticity function effectively is dependent on their potential to generate extracellular proteolytic activities, namely, to produce plasminogen activator (PA) of the urokinase type (u-PA). Previously published and new data are being presented in support of this hypothesis.

The nervous system is composed essentially of two cell types, neurons and glia. The neurons perform the communication function of the tissue, whereas the glia provide numerous support functions to the neurons; their role is maintenance and housekeeping of the nervous tissues. The nervous system is divided into two parts: the central nervous system (CNS), i.e., spinal cord, brain, olfactory bulb, and optic nerve, and the peripheral nervous system (PNS), i.e., those neuronal pathways and centers which are not included in the CNS and are found in the periphery. These two systems differ in their glial cell composition; the astrocytes and oligodendrocytes are the central glial cells, whereas the Schwann cell is the peripheral glial component.[1] Even though the tissue's communication function is conducted exclusively by the neurons, defects in the glial function and/or viability are expressed and reflected in impairment of neuronal function. A few examples are epilepsy, multiple sclerosis, and the mouse mutant weaver[2] in which a genetic deficiency in glial cells results in a secondary neuronal cell death. It is clear that for normal function of the tissue neuronal-glial cell interactions should be maintained at all times. The establishment and the maintenance of these complex cellular interactions requires plasticity.

Recent studies show that glia play a determinant role, not only in the maintenance of the adult nervous system, but also during histogenesis[3,4] and in the fate of regenerative processes evoked by injury to the adult nervous system.[5,6,7] To be able to perform these supportive functions, glia are equipped with plasticity tools. These tools enable the cells, for example, to move in front of and/or accompany the growing neurites during development and regeneration of the nervous system.

Serine Proteases and Their Serpin Inhibitors in the Nervous System
Edited by B. W. Festoff
Plenum Press, New York, 1990

One of the physiological systems controlling cell migration/invasion and plasticity properties of the tissue is the extracellular proteolysis generated by the enzyme plasminogen activator of the urokinase type (u-PA) (e.g., Danø, ibid.).[8] PA is a key enzyme in the cascade of proteolytic activities which are generated and which operate in concerted fashion in tissue remodelling processes. For example, such activities are plasmin and collagenase which participate in tumor metastasis.[9]

The focus of our research is the biochemical machineries which endow the glia with plasticity properties. In this chapter the following issues are being examined: (1) the expression of PA activity and PA molecular forms in CNS and PNS glia, and (2) the possible role of extracellular proteolysis in control of Schwann cell migration during peripheral nerve regeneration *in situ*.

RESULTS AND DISCUSSION

Developing nervous tissues produce PA and their PA activity levels are developmentally regulated.[10,11] It was found in these studies that there is a marked enhancement in PA activity levels during certain time periods which are specific to each of the regions of the developing brain and spinal cord. The enhancement in PA activity levels is suggestive of a remodeling event that is a prerequisite step in histogenesis of the nervous system. On the other hand, the adult neural tissues express very low levels of PA activity. The fact that extracellular proteolysis is involved in regulation of morphogenesis and neuronal-glial cell interactions was demonstrated in differentiating spinal cord cells in culture.[12] The major observation of this study was a marked difference in the spatial organization of the cells on the substratum which was dependent on the presence or absence of plasmin activity in the culture medium. Plasmin activity in the growth medium induced changes in the cytoarchitecture, namely, from a random distribution the cells seemed to reorganize themselves into aggregates of neuronal cells connected with axonal bundles which were accompanied by glial cells. The spatial organization induced by plasmin activity was reminiscent of the typical cellular organization of differentiated neural tissue.

Modulation of Neural Cytoarchitecture by PA/Plasmin *In Vitro*

The effects of plasmin activity described above on the heterotypic cellular organization of differentiating neural cells were monitored in growth media which contained serum.[12] In an attempt to isolate the effect of PA and/or plasmin on the morphogenetic processes, the immature dissociated chick spinal cord cells were maintained in a chemically-defined medium (CDM), and the effect of addition of plasminogen on their cytoarchitecture on the substratum was studied (Figure 1). The effect of plasmin activity was reproduced in studies with the chemically-defined medium, lending support to the presumed primary role of PA in initiating plasticity/remodelling processes. In the absence of plasminogen, the cells were randomly spread on the substratum (Figure 1A), whereas addition of plasminogen (Figure 1B) led to the specific typical neuronal-glial cell organization, e.g., axonal bundles accompanied and enveloped by glia. Since the immature cells secrete PA into their growth medium, the CDM,[13] the selective effects of addition of plasminogen into the medium implicate PA activity in plasticity of the differentiating neural cells *in vitro*.

Studies described above might not represent the actual processes and the physiological role of PAs since they were conducted in artificial conditions, *in vitro*. Our goal is to assess or obtain a clue to the role of glial PA in nervous tissue *in situ*. For this purpose the expression of PA activity by the various glial cells under different

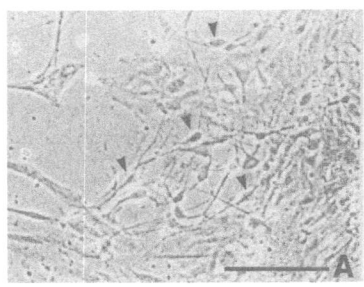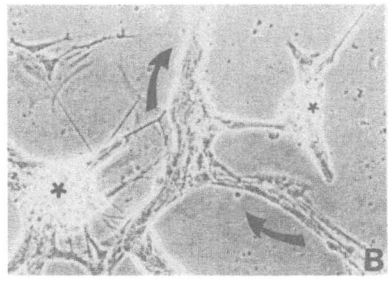

Figure 1. *Plasminogen-dependent spatial organization of embryonic neural cells on the substratum in vitro.* The spatial organization of embryonic spinal cord cells maintained in either: (A) chemically-defined medium (CDM), or (B) CDM containing plasminogen. In the presence of CDM, neuronal cells (arrowheads in A) are randomly spread on a carpet of flat cells; in contrast, in the presence of plasminogen in the same growth medium (B) the cells exhibit a cytoarchitecture typical for differentiated neural tissue (asterisk). Addition of plasminogen induced cell aggregation into rounded clusters (asterisks) that are connected by axonal fascicles (arrows). The fascicles are accompanied by cells, presumably glia. Cells were photographed 5 days after plating. Spinal cord cells from 7-day-old chicken embryos were prepared as previously described[13] except that the tissue dissociation step which was mechanical rather than proteolytic digestion. Small pieces of spinal cord were gently forced through a sterile Nitex^R bag (pore size 75 μm). The composition of the CDM growth medium was as previously described.[3] Bar, 100 μm.

physiological conditions has been examined. Schwann cells and astroglia share many common features during their differentiation process. Differentiation of the two cell types is controlled by the neurons with which they are associated. Immature Schwann cells and astrocytes stop mitotic activity and develop their mature phenotypes.[14,15] The process of differentiation is reversible in Schwann cells but not in the astrocytes, as expressed in response to neuronal injury.[16,17,18] These unique properties of the mature glia parallel the plasticity properties of their corresponding mature tissue.[18,19]

Successful regeneration occurs in the mammalian PNS but not in the mammalian CNS. Injury results in disruption of the various cellular interactions of the tissue, whereby successful regeneration is expressed in the re-establishment of the synaptic connectivity and the specific neuronal-glial cell interactions (e.g., myelin ensheathement). Plasticity implies that the cells are free to move within the tissue, and in biochemical terms that the cells have the tools which allow them to breakdown the pre-existing attachment bonds and to establish new cellular interactions. Data summarized in the following link the capacity of the glial cells to express u-PA activity with the plasticity properties of the tissue. PA activity levels and PA molecular forms were examined in glial cells in culture.

PAs in Schwann Cells

Schwann cells in culture produce and secrete PA into their growth medium.[20,21,22] Significant differences in the capacity to express PA activity were found between proliferating and quiescent Schwann cells.[19] The mitotic rate of Schwann cells in culture is minimal and can be manipulated with a pituitary derived mitogenic factor.[23,24] Addition of this factor to the growth medium leads to a proliferating cell population (presumably undifferentiated). PA activity levels in proliferating Schwann cell cultures are 3-4 fold higher than those of the quiescent cell populations.[20] Analysis of the cellular content of Schwann cells by zymography shows that the proliferating cell populations contain two PA molecular forms (Figure 2). These were identified as the u-PA type (Figure 3) and the tissue PA type (t-PA) (data not shown). The u-PA molecular form appears to predominate in this cell population (Figures 2-

3). No PA activity could be detected by zymography in the quiescent Schwann cell populations, those maintained without the growth factor.

Based on these results (Figures 2-3) and on the presumed physiological role of u-PA, one can speculate that the undifferentiated Schwann cells produce primarily the u-PA enzymatic form thus implicating these cells in migration/invasion processes.

PAs in Differentiating Astroglia

In light of the apparent changing role of the CNS.glia in the developing and mature tissue, at first studies were focused on the transition stages..The capacity of the differentiating glial cells of the CNS, astroglia and oligodendroglia, to express and regulate the two types of plasminogen activator (PA) activities has been examined as a function of cell age in culture.[28] Purified rat astroglia and oligodendroglia were prepared following the procedure of McCarthy and de Vellis.[29] Cellular PA activity levels of differentiating rat astroglia were found to be developmentally regulated. The specific activity of PA reached its highest level in rat astroglia at cell age corresponding to 20-32 postnatal days (P20-P32). Thereafter, the specific activity declined (3-4 fold decrease) to a low value, e.g., in rat astroglia by age P45. At comparable ages (P0-P35), the magnitudes of the PA specific activities of the differentiating rat astroglia and of the developing cerebrum, the tissue from which these cells were purified, were similar. No PA activity was found in the purified oligodendroglia, tested at ages P11-P19.

Differentiating rat astroglia produce two molecular forms of PA as identified by zymography, the u-PA and the t-PA types, and the cellular content of both is deveop-

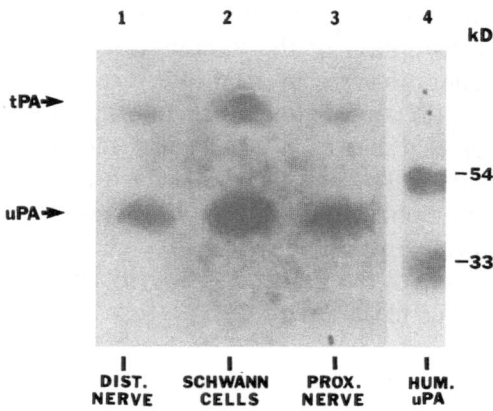

Figure 2. *Zymography of PA molecular forms of proliferating Schwann cells in vitro and of migrating cells from the two sciatic nerve stumps within a silicone chamber in situ.* The following samples were analyzed: proliferating rat Schwann cells (lane 2), the proximal (lane 1), and distal (lane 3) regenerating portions of the rat sciatic nerve within the silicone chamber. For calibration, human uPAs (lane 4) were analyzed. The three cell samples contain two PA forms that were identified as uPA and tPA types (see Figure 3). Note in the three samples the heavy band of uPA in contrast with the faint activity of the tPA molecular form. Purified rat Schwann cells were prepared according to the procedure of Brockes et al[23] and as previously described in detail.[20] The identification of PAs molecular forms was performed by PA zymography.[25] Cell samples (200-300 μg protein/sample) were separated by electrophoresis on SDS polyacrylamide slab gels under non-reducing conditions.[26] Gels were analyzed for PA molecular species by overlaying the gel on indicator agar-gel containing 1.6% casein and 10 μg/ml plasminogen. At the end of the overlay assay (20-24 h) the indicator agar gels were stained with 0.1% amidoblack in 70% methanol-10% acetic acid.

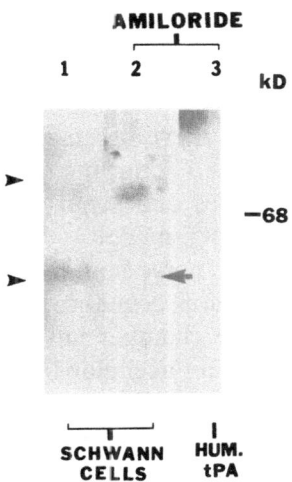

AMILORIDE

1   2   3   kD

−68

SCHWANN   HUM.
CELLS   tPA

Figure 3. *Characterization of the Schwann cells' PA molecular forms.* Zymography of dividing Schwann cells (lanes 1-2), and the single chain human t-PA (lane 3) in the absence (lane 1) and presence (lanes 2-3) of amiloride, the u-PA specific inhibitor. Note that in the presence of amiloride of the two PAs (arrowheads) the low mol wt form (arrow) was abolished, since its activity was inhibited. Equal amounts of Schwann cell samples (~200 μg protein) were applied to the SDS gels. For the identification of the enzymatic types of PA amiloride which selectively inhibits u-PA[27] at final concentration of 1 mM was incorporated into the indicator agar gel.

mentally regulated.[30] u-PA is the predominant form in the immature astrocyte, until age P13. At cell ages P14-P30 the cells manifest both forms and at later stages u-PA disappears while the t-PA type persists as the sole form. In older cells, e.g., P54 astrocytes, neither of the PA types was detected. Along with PA activity, astroglia expressed PA inhibitory activity. The rat astroglial PA inhibitor (PAI), as identified by immunoblots, seemed to be immunologically identical to rat PAI-1.

In many cell types, e.g., Schwann cell, stimulation of cell division is accompanied by an increase in the cells' PA specific activity. Subculturing of the rat cells and stimulation of astroglial proliferation, however, did not lead to an increase; rather, beyond a certain cell age (P13) it resulted in a 3-fold irreversible decline in the PA specific activity of the daughter cells.

It has been established that various biochemical properties of CNS mature glia appear on schedule with cell age in culture, thus defining "mature" glia *in vitro*.[31] According to our study, among differentiating and "mature" CNS glial cells the immature astrocytes appear to be the sole source of the u-PA molecular form and of high levels of PA activity. The astroglia manifest this enhanced activity level and the active u-PA molecular form in a limited developmental period, and upon maturation lose these capacities altogether. The major role attributed to the u-PA type enzyme is in regulating cell migration and tissue plasticity (Danø, ibid.).[8] Based on the findings of this study it is proposed that the astroglia, only at a limited period in their lifetime at the immature stages, are involved in tissue construction and remodelling processes.

Role of Plasmin/PA Activities in PNS Regeneration

Very little is known about the biochemical mechanism which underlies successful regeneration of a damaged nervous system. The peripheral nervous system has the capacity to regenerate and regain functional activity. The Schwann cell plays a crucial role in the capacity of the tissue to regenerate. The role which this cell plays in support of axonal regrowth was demonstrated and established by various grafting and transplantation experiments.[5,32] It seems that the Schwann cell is equipped with appropriate biochemical machineries which are instrumental in the regeneration process.

The temporal process of peripheral nerve regeneration was studied and analyzed in a model system by Williams et al.[33] and according to this study in peripheral nerve regeneration the Schwann cell precedes and seems to be "going" in front of the growing axons. The PNS regenerating model system which was developed by Lundborg et al.,[34] consists of rat sciatic nerve stumps attached to an empty silicone chamber (e.g., Figure 4A). The temporal sequence of events in peripheral nerve regeneration through a 10 mm silicone chamber has been established (Williams et al.[33]): (a) within one week postsurgery an acellular fibrin bridge is formed between the nerve stumps; and (b) in the second week extensive migration within this bridge of Schwann cells and other cells takes place, while the regenerating axons lag by 1-2 days behind the Schwann cells. The formation of blood vessels is the last phase, neovascularization seems to follow axonal growth. The fibrin bridge appears to be the solid substratum which Schwann cells and other cells utilize for migration. The cells migrate into the chamber from both nerve stumps, whereas axonal regeneration occurs from the proximal stump only.

This PNS regenerating model system provides a convenient "test tube" for the study of the biochemical machineries which are involved in nerve regeneration and for our purpose in Schwann cell migration. This model system (Figure 4A) was used by us in the studies described in the following.

Adult sciatic nerves do not express any PA activity, whereas high PA activity is found in the regenerating nerve after Wallerian degeneration.[35] Experiments were performed in a model system consisting of a regenerating sciatic nerve through a 10 mm silicone chamber as described by Lundborg et al.[34] The purpose of this part of the study was to identify the PA molecular forms expressed by the emigrant cells within the chamber during the period when these cells are highly mobile. To capture the cells at their migratory phase the cellular content of the silicone chamber was analyzed at 10-12 days postsurgery. Analysis of the cellular content of the chamber emigrating from the two nerve stumps by zymography, shows that the two contain primarily the u-PA form (Figure 2). Since the distal portion of the chamber does not contain neuronal processes but only Schwann cells and other cells we can conclude that PAs and the u-PA is not dependent or associated to neuronal elements. Since the emigrating cell populations contain cells other than Scwhann cells (e.g., macrophages), even though in small amounts, it is impossible at this stage to conclusively identify the cellular source of the u-PA.

We have utilized the PNS model system as a biochemical test tube to examine the role of extracellular proteolysis in regeneration by studying the effects of protease inhibitors on sciatic nerve regeneration through a 10 mm silicone chamber (Juhasz et al., 1989 in preparation).[36,37] In the design of the experiments we took advantage of the fact that the major cell migration event into the chamber starts at the second week after the surgery. It was assumed that if proteolysis is utilized for cell migration/invasion, protease inhibitors should exert their effects and thus tested when massive cell migration proceeds from both nerve stumps into the chamber. The inhibitors, with different protease specificities, were injected into the chamber at 7 days postsurgery and their effect on nerve regeneration was examined a week later. To facilitate analysis of the process, we have developed a "quick" assay which enables a complete view of the content of the entire chamber, examining in parallel cell migration, Schwann cell migration, and axonal growth utilizing histological proce-

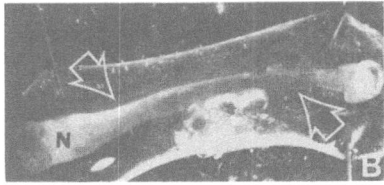
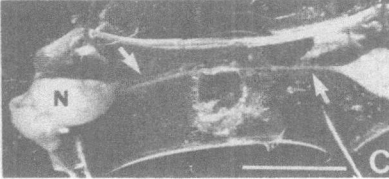

Figure 4. *Inhibition of sciatic nerve regeneration within a silicone chamber by the plasmin inhibitor, ε-ACA.* (A) At the end of the surgery, the stumps of the sciatic nerve (N) were sutured (asterisks) and enclosed within the silicone chamber (arrows), while the gap between the stumps was filled with a physiological solution. (B-C) The content of the chambers two weeks after the surgery of regenerating nerves treated with either physiological solution -- control (B) or 1 mM ε-ACA (C). Note the thick cable (clear arrows in B) extending from the proximal nerve stump (N) through the chamber in contrast with the thin amorphous bridge (arrows in C) linking the proximal stump (N) with its distal counterpart in the protease inhibitor-treated nerve. Sciatic nerve surgery and attachment of the silicone chamber were performed as previously described.[7] Bar, 4 mm.

dures.[7] Various plasmin inhibitors including ε-aminocaproic acid (1 mM) inhibited Schwann cell migration as compared with control samples; accordingly, axonal regrowth and nerve regeneration was impaired (Figure 4B-4C). Results generated thus far show that the activities of plasmin and a metalloprotease, probably of collagenase, are involved in Schwann cell migration (Juhasz et al., 1990 in preparation). None of the tested inhibitors showed any adverse effect on Schwann cells in culture. Our results are similar to the inhibitory effects of protease inhibitors on the invasion of tumor cells through amniotic membrane.[9]

Our results indicate that plasmin activity is elaborated by the Schwann cells for migration/invasion and is instrumental for successful nerve regeneration. Since Schwann cell migration from both nerve stumps was blocked by the inhibitors, we can conclude that their migration is not dependent on neurite outgrowth. The presumed role of u-PA in Schwann cell migration during PNS regeneration remains to be clarified.

The facts that injured PNS can recover from the trauma and that Schwann cells,[5] immature astrocytes,[6] and embryonic CNS[18] support neurite regeneration lead to the conclusion that certain biochemical processes exist in these tissues that are essential for the regenerative process to take place. Furthermore, since regeneration is abortive in the adult mammalian CNS, this implies that this tissue is deficient in some of these biochemical processes. One possible interpretation of these observations is that the capacity to support neural and neuronal regeneration is linked to the high glial PA activity levels. In plasticity terms, the supposition implies that, as long as the glia (i.e., the Schwann cell and the immature astrocyte) are mobile they maintain the capacity to support neural regeneration.

CONCLUDING REMARKS

The data summarized above constitute four independent pieces of evidence which are related to glial PAs and to the processes of development and regeneration of neural tissues. It is evident that neural morphogenetic events can be manipulated both in cell culture and *in situ* by enhancing or inhibiting plasmin activity. In addition, it appears that the capacity to express u-PA activity is maintained in Schwann cells throughout their entire lifetime whereas in astrocytes it is limited to their early developmental period. The notion that glial capacity to express u-PA activity has a pivotal role in regulating plasticity processes in PNS and CNS remains to be resolved in future studies.

ACKNOWLEDGEMENT

The research reported herein was supported by the National Institutes of Health, grants NS 23064 and in part BRSG SO7 RR07065, and by The Spinal Cord Research Foundation.

REFERENCES

1. A. Peters, S.L. Palay, and H. deF. Webster, In: "The fine structure of the nervous system: The neurons and supporting cells," W.B. Saunders Company (1976).
2. R.L. Sidman, and P. Rakic, Neuronal migration, with special reference to developing human brain: A review, Brain Res., 62:1-35 (1973).
3. J. Silver, S.E. Lorenz, D. Wahlsten, and J. Coughlin, Axonal guidance during development of the great cerebral commissures: Descriptive and experimental studies in vivo on the role of preformed glial pathways, J. Comp. Neurol., 210:10-29 (1982).
4. J.R. Jacobs, and C.S. Goodman, Embryonic development of axon pathways in the Drosophila CNS. I. A glial scaffold appears before the first growth cones, J. Neurosci., 9:2402-2411 (1989).
5. P.M. Richardson, A.J. Aguayo, and U.M. McGuinness, Role of sheath cells in axonal regeneration, In: "Spinal Cord Reconstruction," Kao, Bunge, and Reier, eds., Raven Press, New York, pp. 293-304 (1983).
6. G.S. Smith, R.H. Miller, and J. Silver, Astrocyte transplantation induces callosal regeneration in postnatal acallosal mice, Ann. N.Y. Acad. Sci., 485:185-205 (1987).
7. N. Kalderon, Differentiating astroglia in nervous tissue histogenesis/regeneration: studies in a model system of regenerating peripheral nerve, J. Neurosci., 21:501-512 (1988a).
8. F. Blasi, J.-D. Vassalli, and K. Danø, Urokinase- type plasminogen activator: proenzyme, receptor, and inhibitors, J. Cell Biol., 104:801-804 (1987).
9. P. Mignatti, E. Robbins, and D.B. Rifkin, Tumor invasion through the human amniotic membrane: requirement for a proteinase cascade, Cell, 47:487- 498 (1986).
10. H. Soreq, and R. Miskin, Plasminogen activator in the rodent brain, Brain Res., 216:361-374 (1981).
11. N. Kalderon, and C.A. Williams, Extracellular proteolysis: developmentally regulated activity during chick spinal cord histogenesis, Dev. Brain Res., 25:1- 9 (1986).
12. N. Kalderon, Migration of Schwann cells and wrapping of neurites in vitro: A function of protease activity (plasmin) in the growth medium, Proc. Natl. Acad. Sci. USA, 76:5992-5996 (1979).
13. N. Kalderon, Role of the plasmin-generating system in developing nervous tissue: I. Proteolysis as a mitogenic signal for the glial cells. J. Neurosci. Res., 8:509-519 (1982).
14. J.L. Salzer, and R.P. Bunge, Studies of Schwann cell proliferation. I. An analysis in tissue culture of proliferation during development, Wallerian degeneration, and direct injury, J. Cell Biol., 84:739-752 (1980).
15. E.M. Hatten, Neuronal regulation of astroglial morphology and proliferation in vitro, J. Cell Biol., 100:384-396 (1985).
16. H.J. Weinberg, and P.S. Spencer, The fate of Schwann cells isolated from axonal contact, J. Neurocyt., 7:555- 569 (1978).
17. R.G. Pellegrino, and P.S. Spencer, Schwann cell mitosis in response to regenerating peripheral axons in vivo, Brain Res., 341:16-25 (1985).
18. P.J. Reier, Gliosis following CNS injury: The anatomy of astrocytic scars and their influence on axonal elongation, in: "Astrocytes", Fedoroff, S., and Vernadakis, A., eds., Academic Press, New York, Vol. 3, pp. 263-324 (1986).
19. P.J. Reier, and J.D. Houlé, The glial scar: Its bearing on axonal elongation and transplantation approaches to CNS repair, Adv. Neurol., 47:87-136 (1988).

20. N. Kalderon, Schwann cell proliferation and localized proteolysis: expression of plasminogen-activator activity redominates in the proliferating cell populations, Proc. Natl. Acad. Sci. USA, 81:7216-7220 (1984).

21. A. Krystosek, and N.W. Seeds, Peripheral neurons and Schwann cells secrete plasminogen activator, J. Cell Biol., 98:773-776 (1984).

22. A. Alvarez-Buylla, and J.E. Valinsky, Production of plasminogen activator in cultures of superior cervical ganglia and isolated Schwann cells, Proc. Natl. Acad. Sci. USA, 82:3519-3523 (1985).

23. J.P. Brockes, K.L. Fields, and M.C. Raff, Studies on cultured rat Schwann cells. I. Establishment of purified populations from cultures of peripheral nerve, Brain Res., 165:105-118 (1979).

24. J.P. Brockes, G.E. Lemke, and D.R. Balzer, Purification and preliminary characterization of a glial growth factor from the bovine pituitary, J. Biol. Chem., 255:8374-8377 (1980).

25. A. Granelli-Piperno, and E. Reich, A study of proteases and protease-inhibitor complexes in biological fluids, J. Exp. Med., 147:223-234 (1978).

26. U.K. Laemmli, Cleavage of structural proteins during the assembly of the head of bacteriophage T4, Nature, 227:680-685 (1970).

27. J.-D. Vassalli, and D. Belin, Amiloride selectively inhibits the urokinase-type plasminogen activator, FEBS (Fed. Eur. Biochem. Soc.) Lett. 214:187-191 (1987).

28. N. Kalderon, K. Ahonen, and S. Fedoroff, The immature astrocyte as the predominant source of plasminogen-activator activity and of the urokinase type: Studies in differentiating astroglial cell cultures. Submitted for publication (1990).

29. McCarthy and de Vellis, Preparation of separate astroglial and aliogodendroglia cell cultures from rat cerebral tissue. J. Cell Biol., 85:890-902 (1980).

30. N. Kalderon, The molecular forms of plasminogen activator in differentiating astroglia are developmentally regulated, Soc. Neurosci. Abstr., 14:1056 (1988b).

31. E.R. Abney, P.P. Bartlett, M.C. Raff, Astrocytes, ependymal cells, and oligodendrocytes develop on schedule in dissociated cell cultures of embryonic rat brain, Dev. Biol., 83:301-310 (1981).

32. A.J. Aguayo, M. Vidal-Sanz, M.P. Villegas-Perez, G. M. Bray, Growth and connectivity of axotomized retinal neurons in adult rats with optic nerves substituted by PNS grafts linking the eye and the midbrain, Ann. N.Y. Acad. Sci., 495:1-9 (1987).

33. L.R. Williams, F.M. Longo, H.C. Powell, G. Lundborg, and S. Varon, Spatial-temporal progress of peripheral nerve regeneration within a silicone chamber: parameters for a bioassay, J. Comp. Neurol., 18:460-470 (1983).

34. G. Lundborg, R.H. Gelberman, F.M. Longo, C.H. Powell, and S. Varon, In vivo regeneration of cut nerves encased in silicone tubes. Growth across a six- millimeter gap, J. Neuropathology Exp. Neurol., 41:412-422 (1982).

35. A. Bignami, G. Cella, N.H. Chi, Plasminogen activators in rat neural tissues during development and in Wallerian degeneration, Acta Neuropathol. (Berl), 58:224-228 (1982).

36. N. Kalderon, J.P. Kirk, and A. Juhasz, Impairment of sciatic nerve regeneration by protease inhibitor treatment: inhibition of Schwann cell migration, Soc. Neurosci. Abstr., 13:1208 (1987).

37. N. Kalderon, K. Ahonen, A. Juhazs, J.P. Kirk, and S. Fedoroff, Astroglia and plasminogen activator activity: differential activity level in the immature, mature and "reactive" astrocytes. In: Current Issues In Neural Regeneration Research (Reier, P.J., Bunge, R.P. and Seil, F.J. eds.) Alan R. Liss Press, New York, pp. 271-280 (1988).

PLASMINOGEN ACTIVATORS IN DEVELOPMENT, INJURY AND PATHOLOGY OF THE NEUROMUSCULAR SYSTEM

DANIEL HANTAÏ[1], CLAUDINE SORIA[2], JEANNETTE SORIA[3],
BRIGITTE BLONDET[1,4], GEORGIA BARLOVATZ-MEIMON[4], JASTI S. RAO[5]
AND BARRY W. FESTOFF[5]

[1]INSERM U.153, [2]Hôpital Lariboisière
and [3]Hôtel Dieu, Paris, France;
[4]Université Paris-Val de Marne,
Créteil, France;
and [5]Department of Veterans Affairs
Medical Center, Kansas City, MO, USA

INTRODUCTION

The development of the neuromuscular junction entails a complex, cascade-like series of events that include the directed extension of neuronal processes, the sequential localized deposition, removal and redeposition of extracellular matrix (ECM) components, and, ultimately, stabilization of the macromolecular associations between the participating cells.[1,2] The nature of these events suggests that the formation, maintenance and elimination of neuromuscular synapses may be regulated by locally expressed proteases and protease inhibitors acting on synaptic basement membrane (BM) associated molecules. *In vitro* experiments have shown that certain serine proteases, plasminogen activators (PAs), are released from neurite growth cones[3,4] and that protease inhibitors profoundly affect neurite outgrowth.[5] However *in vivo* evidence for the proteolytic control of synaptic junctions is lacking.

For several years our interest has been in exploring the balance of PAs and inhibitors in the neuromuscular system. PAs are present *in vivo* in mouse muscle, but after experimental denervation a dramatic and early rise in muscle PA activity occurs.[6] Using the nerve crush paradigm PA activity rise returns to baseline closely paralleling the muscle reinnervation suggesting regulation of muscle PA by the nerve (Hantaï et al., submitted). Using an *in situ* immunoassay the PAs present in muscle after denervation were found to degrade some, if not all, muscle BM zone components.[7] In other studies, during muscle development urokinase-like PA (uPA) activity, initially high at birth, drops significantly coinciding temporally with the wave of neuromuscular synapse elimination in neonatal muscle.[8] Exploring the possibility that an alteration in serine proteases may have a role in the pathogenesis of some degenerative neuromuscular diseases,[9] we studied patients suffering from the fatal amyotrophic

Serine Proteases and Their Serpin Inhibitors in the Nervous System
Edited by B. W. Festoff
Plenum Press, New York, 1990

219

lateral sclerosis (ALS) as well as mutant Wobbler mice which approximate the genetic human lower motor neuron diseases. Increased activity of plasminogen activators was detected in the more affected muscles of these mutant mice. Preliminary data show that PA and PA-inhibitor levels in plasma, before and after venous occlusion, were abnormal in patients suffering from ALS supporting the hypothesis for a role of the fibrinolytic system in the pathogenesis of this disorder.[9]

RELEASE OF PAs FROM MUSCLE AFTER AXOTOMY OR NERVE CRUSH

When muscle is deprived of its innervation, experimentally or by disease, the nerve terminals degenerate, are phagocytized and are completely engulfed by Schwann cells in less than 3 days.[10,11] Muscle fibers are, thereby, left denervated. They respond to this condition with a variety of metabolic alterations, the most dramatic being atrophy.[12] However, much earlier changes after denervation are located at the muscle cell surface and include the loss of endplate-specific 16S acetylcholinesterase[13] and reduction of fibronectin.[14] After a few days lag intact axons make new connections with the denervated end-plates by a process of collateral or ultraterminal outgrowth which comes from the junction or the subterminal nodes of the surviving axons and functional neuromuscular connections begin to reform.[10,15-18] The process of reinnervation, and hence, regeneration begins with the clearing away of surface components of the myofiber after denervation or injury. Extracellular, neutral proteases were considerable suitable candidates for surface degradation.

Muscle PAs Increase after Muscle Axotomy

Previous studies indicated that PA was the predominant neutral protease produced and secreted by cultured clonal murine skeletal muscle cells.[19] The detected PA activity was developmentally regulated in these muscle cell cultures, the greatest rise in secreted PA occurring at the time of peak fusion.[20] PA activity was also

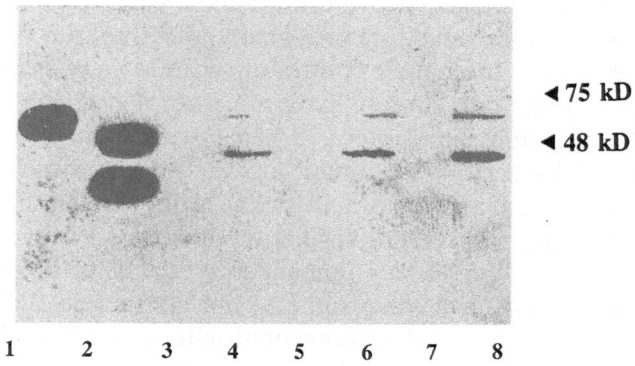

Figure 1. *Fibrin zymogram of control and denervated muscle extracts.* Electrophoresis of extracts from control (lanes 3, 5, 7) and denervated (lanes 4, 6, 8) muscles extracts at 7 (lanes 3, 4), 10 (lanes 5, 6) and 17 (lanes 7, 8) days after denervation applied to the fibrin layer, as described by Granelli-Piperno and Reich.[24] After hydrolysis at 37°C, the gel was photographed under darkfield conditions. In lane 1 is human tPA (75 kD) and in lane 2, human uPA (33 and 55 kD) for molecular weight comparison.

detected in muscle organ culture media.[21] These muscle PAs were able to degrade the extracellular matrix synthesized by the same cells in culture,[22] but little or no information was available whether PAs were present in muscle *in vivo*. We addressed this by performing axotomy of mouse leg muscles, and removed these after a time course from 2 to 17 days. Denervated and control muscles were homogenized, extracted, centrifuged and then assayed using a sensitive colorimetric method with a chromogenic substrate relatively specific for plasmin[23] and able to distinguish uPA from tissue PA (tPA). Extracts were also run on SDS-PAGE, after which fibrin zymography[24] was performed. Amidolytic activity in extracts of skeletal muscle (both control and denervated) was totally plasminogen dependent. This PA activity increased 10-fold between 0 and 7 days after denervation, after which it plateaued. Activity occurred in the absence of fibrin but some potentiation was found when fibrin monomer was added.[6] By zymography (Figure 1) the first PA detected at 2 days had a molecular mass of 48 kD, consistent with mouse uPA, while after 7 days a PA at 75 kD appeared,[6] consistent with mouse tPA (see ref. 25). These studies suggested neural regulation of muscle PA, either at transcriptional, translational or post-translational levels.

The Nerve Crush Paradigm

If the nerve was able to regulate plasminogen activator in muscle, would this coincide with evidence of stabilization of the newly formed endplates and reinnervation?[8] We turned to experiments where the sciatic nerve crush was crushed instead of the more definitive axotomy in order to allow the nerve to reinnervate the muscle. Muscle contraction on nerve stimulation and return of choline acetyltransferase activity[26,27] were used to monitor reinnervation. Coincident with muscle denervation and reinnervation uPA activity rises and declines both in soluble and membrane-bound muscle fractions as shown by amidolytic activity of chromogenic

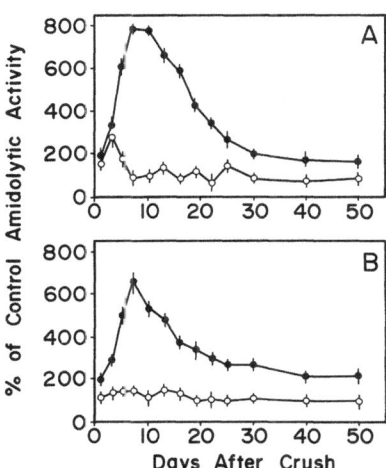

Figure 2. *PA activity in muscle cytosol and membranes after sciatic crush.* (A) uPA activity (•) and tPA activity (○) in 100,000 x g supernatants of sciatic-innervated mouse muscle extracts from day 1 to day 50 after sciatic crush. (B) uPA and tPA activity in Triton-extracted pellets from 100,000 x g centrifugation of muscle extracts over the same period. The results are expressed as the mean percentage of the absorbance at 405 nm per mg protein of contralateral (control) leg muscle extracts plus/minus standard deviation.

synthetic substrate (Figure 2) and zymography (Hantaï et al., submitted). These neurally-regulated variations in PA activity predominently involve uPA,[6,8] similar to uPA of other mouse tissues,[28] and less so tPA.[25] Although membrane-bound activity is 5-fold higher than cytosol activity, there is no shift between cellular compartments during the entire time-course. We concluded that plasminogen activation may have a role early in muscle denervation. One mechanism might be by reducing and releasing 16S acetylcholinesterase,[13] reducing sarcolemmal fibronectin[15] and acting on basement membrane remodeling.[7] These changes may be necessary for the activation of satellite cells as well as for the effective reinnervation of denervated muscle fibers.[29] These results show tight regulation of PA levels in muscle by some neural influence. The mechanism of this influence is currently under study.

DEGRADATION OF MUSCLE ADHESIVE MACROMOLECULES BY PAs

Since PA activity was high in development, low in the normal adult but rose dramatically after denervation, we asked what might be the role of the increased PA in denervated and embryonic muscle? We attempted to show the action of uPA directly in normal and denervated muscle. We developed an *in situ* immunoassay in which antibodies to BM components (i.e., laminin, type IV collagen and fibronectin) were applied either in the presence or absence of denervated or control muscle extracts, in the presence or absence of plasminogen. Qualitative immunofluorescence was performed as well as laser-directed image analysis to quantitate results. These studies indicated a clear plasminogen-dependent degradation of fibronectin > type IV collagen > laminin.[7] The effect of denervation was to increase plasminogen activation to plasmin locally, which then resulted in BM zone degradation. A representative experiment with type IV collagen is shown in Figure 3, similar patterns were found with fibronectin antisera. Thus, extracts of denervated, but not control, muscle contained PA(s) powerful enough to produce BM degradation. In addition, this activity was found locally in the muscle fibers increasing BM degradation progressively after denervation.

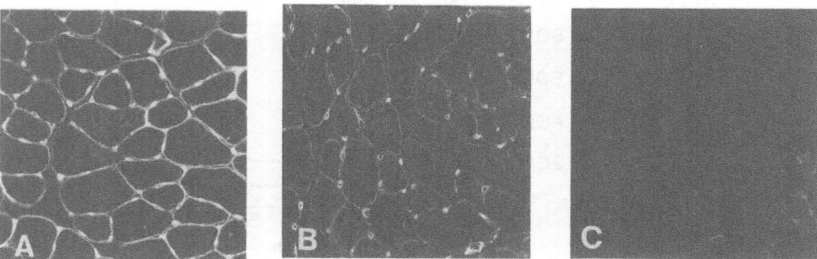

Figure 3. *Denervation-induced degradation of basement membrane antigens.*[7] Type IV collagen immunofluorescence after 2 hr incubation at 37°C with plasminogen. A reduction of fluorescence was seen during the entire time course of denervation. Normal muscle (A) and muscle denervated for 4 (B) and 7 (C) days.

Is it possible that the changes in PA activity "induced" by experimental denervation of adult muscle, might also occur, perhaps less obviously, during *in vivo* development? Programmed changes involving elimination of polyneuronal synapses occur during mammalian embryonic and neonatal development. This is particularly obvious in mammalian, especially rodent, skeletal muscle where synapses, which are still polyneuronally innervated at birth, change over a very narrow period of time to the more adult situation of mononeuronal innervation.[30,31] The mechanism involved in the elimination of polyneuronal innervation is still unknown, but O'Brien et al.[32] have raised the hypothesis that neuromuscular activity might stimulate the release of proteolytic enzymes at the endplate. These enzymes could then attack the attachment of nerve terminals causing the withdrawal of all but the most resistant terminal from adhesion to post-synaptic membranes. We asked whether changes in PA levels or activity might reflect their possible role in this programmed developmental change. We showed that in homogenates of mouse leg muscle both forms of PA, uPA as well as tPA, are present during the first few days after birth.[8] Very quickly over the next few days these enzymes markedly decrease, reaching the adult stage at 27 days. In fact, by measuring the plasminogen-dependent amidolytic activity,[23] we found a

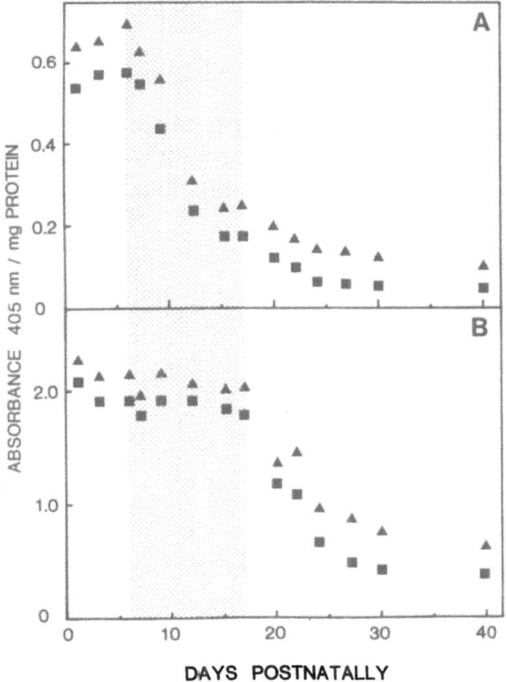

Figure 4. *PA activity in muscle cytosol and membranes during postnatal development.* (A) PA activity without fibrin monomer (■) and with fibrin monomer (▲) in 100,000 x g supernatants of mouse leg muscle extracts. (B) PA activity in Triton-extracted pellets from 100,000 x g centrifugation of muscle extracts. Activity is expressed as absorbance at 405 nm per mg of protein and represents duplicate wells of three experiments. Replicates agreed within 5-10%. The vertical stippled column represents the range of time for change from polyneuronal to mononeuronal innervation of muscle fibers in rodent leg muscles.

dramatic drop, primarily of uPA activity, from day 6 to day 17, i.e. during the period of maximum regression of polyneuronal innervation (Figure 4). Although membrane-bound activity was higher than cytosol activity, there was no shift between cellular compartments. These experiments suggest that there is a regulation of PA activity during normal neuromuscular development. This occurs temporally coincident with elimination of polyneuronal innervation and the differentiation and stabilization of the mature neuromuscular synapse, is just opposite to the situation following denervation which results in neuromuscular dedifferentiation,[33] and parallels what happens during muscle reinnervation (i.e., a kind of re-differentiation).

These studies, together with the discovery that protease nexin I is highly concentrated at the neuromuscular junction[34] (see also Festoff et al. in this volume), suggest an important role for the PA-plasmin system in neuromuscular junction formation and maintenance by acting on synaptic basement membrane, or basement membrane associated components. In this context, it was previously hypothesized[9] that in the fatal disease, ALS, de-regulation of this PA-plasmin system may explain the progressive random denervation observed in this disease.

PAs IN LOWER MOTOR NEURON DISEASES

We decided to further explore this question by using two different approaches. First, we used the neurological mutant Wobbler mouse as a model which approximates the human lower motor neuron diseases, such as the ALS. Secondly we initiated a study in patients suffering from ALS by measuring their PA and PA-inhibitor plasma levels which appeared to be abnormal.

The Mutant Wobbler Mouse

The neurological mutant Wobbler mouse carries an autosomal recessive gene (wr) and has been studied as a model of lower motor neuron disorders with associated muscle atrophy and denervation/reinnervation phenomena.[35] In addition a recent

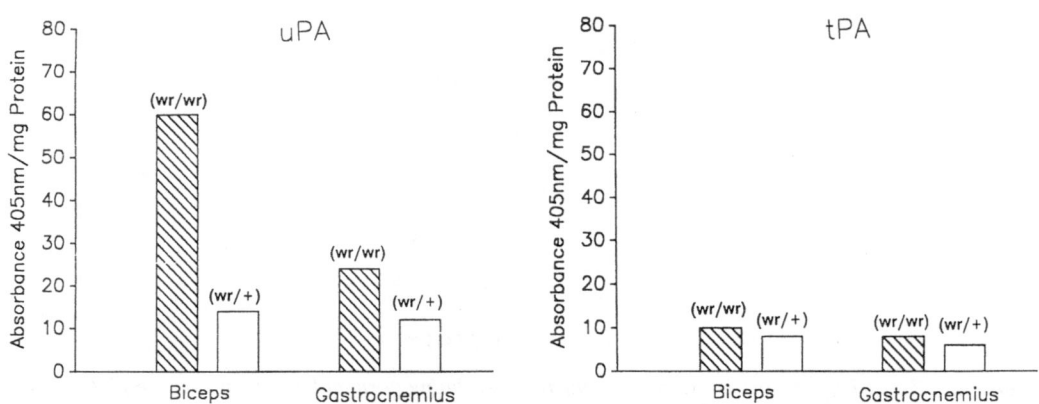

Figure 5. *PA activity in biceps and gastrocnemius total muscle extracts of 4 month-old Wobbler mice.* Specific PA activity is increased in Wobbler muscle versus unaffected littermate muscle and more in biceps than in gastrocnemius muscle, which is less impaired by the mutation. The predominant PA is uPA more than tPA.

Table 1. PA and PA-Inhibitor levels in ALS patients and controls.

Patient	Age	Sex	MND/Duration (yr)	tPA activity ng/ml		euglobulin lysis time hr		tPA (ELISA) ng/ml		PA-inh. activity IU/ml		uPA immuno-act. ng/ml	
				0min	5min	0min	5min	0min	5min	0min	5min	0min	5min
1. C.G.	75	F	PBP	3	2.7	5.0	3.2	>4	11.0	7.5	1.2	1.7	3.3	5.2
2. A.M.	48	M	ALS	2	1.8	1.7	1.7	3.5	3.8	3.0	1.0	1.6	6.0	6.0
3. A.Q.	72	M	ALS	1	0.6	4.3	2.0	3.0	3.5	3.0	1.6	2.7	2.0	7.2
4. J.L.	53	M	ALS	2	0.1	0.3	3.2	3.2	8.5	11.0	2.3	2.7	2.4	4.4
5. J.L.	54	M	ALS	2.5	ND	ND	3.2	0.8	11.0	17.0	0.8	0.5	2.7	5.0
6. M.C.	56	F	ALS	0.5	0.8	0.6	3.5	3.2	3.0	3.0	0.8	1.8	15.6	16.0
7. E.S.	67	M	PBP	2	4.2	9.0	1.1	3.5	9.5	5.0	1.3	1.2	24.0	10.0
8. M.F.	82	F	PBP	2	2.5	5.5	1.8	3.2	4.3	3.5	0.4	0.4	4.0	3.2
9. V.R.	48	M	ALS	0.5	ND	ND	1.8	0.5	4.0	10.5	0.6	1.0	4.8	4.8
10. B.L.	75	F	ALS	2	ND	ND	>4	>4	18.0	25.5	2.2	2.1	5.2	8.8
11. R.E.	55	M	PBP	0.5	ND	ND	3.5	1.7	6.0	7.3	0.9	2.0	2.8	3.6
12. B.R.	52	M	ALS	1	ND	ND	0.8	1.1	15.0	10.0	0.8	0.8	4.4	4.4
X	61.4				1.8	1.8	2.5	2.6	8.1	8.9	1.2	1.5	6.8	6.6
SEM	±3.4				±0.5	±1.2	±0.3	±0.3	±1.4	±2.0	±0.2	±0.2	±1.9	±1.1
CONTROLS					0.2	15.0	3.0	<1	6.0	8.0	0.8		3.0	3.0
					±.06	±5.0			±3.0	±4.0	±0.2		±0.4	±0.4

report has identified neuromuscular junction defects and, hence, it may also be a model for defects of synapse formation.[36]

In order to determine the possible involvement of PAs in the basement membrane degradation associated with denervation in Wobbler, specific PA activities in biceps and gastrocnemius muscle extracts were estimated using both zymography and the synthetic substrate S-2251.[23] These activities were compared to those of the corresponding muscles in unaffected mice (wr/+) from the same litter. PA activity was increased in Wobbler mouse muscles versus control muscles and more in biceps muscle than in gastrocnemius muscle, which is less impaired by the mutation (Figure 5). Here again, muscle PA was mostly uPA (Mr=48 kD) more than tPA (Mr=75 kD). In addition, kinetic characteristics of the PA assay obtained with Wobbler mouse muscles lacked the conventional lag-phase, thus evidencing the presence of already formed plasmin.

Study of PA and Inhibitors in ALS

An activation of serine proteases at the synapse has been postulated to be part of the mechanism underlying the pathogenesis of the fatal human ALS.[9] In this scheme ALS would be a "dying-back" neuropathy as Charcot first described.[37]

In order to determine if the fibrinolytic enzymes in plasma were altered in ALS, we studied twelve patients and collected their plasma using the venous stasis/exercise protocol.[38] These patients had an average age of 61. Seven had "classical" and 5 had the progressive bulbar palsy (PBP) form of ALS.[39]

Using the S-2251 (Kabi) chromogenic assay,[23] we measured tPA activity in plasma taken before venostasis and after 5 minutes ischemic exercise. We compared the results obtained to those of a group of 50 "control" subjects (Table 1). We found, before venostasis and exercise, an increase in tPA activity in all patients investigated (seven) except in one patient (No4) who was receiving both baclofen and dantrolene sodium. The mean value of tPA for the ALS patients compared to a control group was almost 10 times that of the control group using the zero time point (prior to venous stasis/ischemic exercise). In addition to this dramatic increase in baseline the ALS patients tPA levels did not rise after 5 minutes of ischemic exercise. Of interest we have found in all but 2 patients an absence of shortening of the euglobulin lysis time after venous occlusion and exercise, suggesting an anomaly in the tPA pathway. ELISA measuring both active and inactive tPA[40] revealed slightly higher levels of tPA in ALS patients than in controls but these results were not statistically significant. There was also no difference in PA-inhibitory activity between control and ALS patients. Interestingly uPA immuno-activity (i.e., uPA first immunoabsorbed and then tested for activity) in 3 ALS patients was dramatically increased to values previously not observed by us.

These results show an augmentation and/or impairment of the PA-PA inhibitor system in the plasma of ALS patients. They support the hypothesis that plasminogen activators may play a role in the pathogenesis of ALS by acting at the synapse, although the levels of activity observed in plasma may not reflect precisely what is going on in impaired muscles. A study of PAs and their inhibitors in muscle biopsies of ALS patients is in progress.

CONCLUSION

These results bring converging lines of evidence for a possible role of muscle plasminogen activators during developmental elimination of synapses, in stabilization and plasticity of adult neuromuscular junctions, in the reinnervation phenomena following denervation, and in the pathogenic events which underly lower motor neuron diseases.

Several questions can, and should, be raised regarding these results. For instance what are the cells of origin of PA (and especially uPA) in muscle? These could be denervated muscle fibers, satellite cells, but also macrophages, Schwann cells and, less probably, nerve terminals.

What regulates the activation of PA in muscle? In normal, adult muscle PA activity is present but quite low, consistent with the idea that differentiation decreases PA. This also suggests that the nerve might negatively regulate PA synthesis, at either transcriptional or translational levels, and/or may also function posttranslationally, to inactivate the PA after it is released (see Botteri et al., this volume). One or more inhibitors, such as protease nexin I [34] may also function to regulate PA, at post-translational levels, (see also Festoff et al. in this volume).

Finally, what might be the physiologic function of uPA in muscle? It might be similar to its proposed role in tissue degradation in cancer.[25] The rationale for such degradation might be in denervated muscle to remove certain BM components of the muscle fiber to allow orderly regeneration. In developing muscle, one of the possible actions in the nerve/muscle attachment/disattachment phenomena would be through the local degradation of these BM components. The low muscle PA level remaining present throughout life would allow a slow turnover of these BM macromolecules and, thus, bring a kind of inherent plasticity to the neuromuscular junction. Activated in some neuro-degenerative diseases such as ALS, this PA-plasmin system might take part in the progressive, random and uncontrollable denervation of muscle.

Acknowledgements: We are indebted to Professor Jean Émile who referred the patients and to Professor Michel Fardeau for his continuing support and encouragement. This work was supported in part by INSERM, the Association Française contre les Myopathies, the Fondation Philippe and the Medical Research Service of the Veterans Administration.

REFERENCES

1. R.W. Sperry, Chemoaffinity in the orderly growth of nerve fiber patterns and connections. **Proc. Natl. Acad. Sci. USA** 50: 703 (1963).

2. M.J. Dennis, Development of the neuromuscular junction: inductive interactions between cells. **Ann. Rev. Neurosci.** 4: 43 (1981).

3. A. Krystosek, and N.W. Seeds, Peripheral neurons and Schwann cells secrete plasminogen activator. **J. Cell. Biol.** 98: 773 (1984).

4. R.N. Pittman, Release of plasminogen activator and a calcium-dependent metalloprotease from cultured sympathetic and sensory neurons. **Dev. Biol.** 110: 91 (1985).

5. D. Monard, Neuronal cell behaviour: modulation by protease inhibitors derived from non-neuronal cells. **Cell Biol. Int. Rep.** 9: 297 (1985).

6. B.W. Festoff, D. Hantaï, J. Soria, A. Thomaïdis, and C. Soria, Plasminogen activator in mammalian skeletal muscle: characteristics of effect of denervation on urokinase-like and tissue-activator. **J. Cell Biol.** 103: 1415 (1986).

7. D. Hantaï, and B.W. Festoff, Degradation of muscle basement membrane by locally generated plasmin. **Exp. Neurol.** 95: 44 (1987).

8. D. Hantaï, J.S. Rao, C. Kahler, and B.W. Festoff, Decrease in plasminogen activator correlates with synapse elimination during neonatal development of mouse skeletal muscle. **Proc. Natl. Acad. Sci. USA** 86: 362 (1989).

9. B.W. Festoff, Role of neuromuscular junction macromolecules in the pathogenesis of amyotrophic lateral sclerosis. **Med. Hypotheses** 6: 121 (1980).

10. D. Barker, and M.C. Ip, Sprouting and degeneration of mammalian motor axons in normal and de-afferentated skeletal muscle. **Proc. Roy. Soc. London** B 163: 538 (1966).

11. R. Miledi, and C.R. Slater, On the degeneration of rat neuromuscular junctions after nerve section. **J. Physiol. (London)** 207: 507 (1970).

12. E. Gutmann, and J. Zelená, Morphological changes in the denervated muscle, in: "The denervated muscle" Gutmann E. (ed.), Czech. Acad. Sci. Publishing House, Prague, p 57 (1962).

13. H.L. Fernandez, M.J. Duell, and B.W. Festoff, Neurotrophic regulation of 16S acetylcholinesterase at the vertebrate neuromuscular junction. **J. Neurobiol.** 10: 442 (1979).

14. B.W. Festoff, K.L. Oliver, and N.B. Reddy, *In vitro* studies of muscle membranes. Effects of denervation on the macromolecular components of cation transport in red and white skeletal muscle. **J. Membr. Biol.** 32: 345 (1977).

15. S. Ramon y Cajal, Degeneration and regeneration of the nervous sytem, ed. May, R.M. Oxford University Press, London (1928).

16. M.V. Edds, Collateral nerve regeneration. **Quart. Rev. Biol.** 28: 260 (1953).

17. A. Gorio, Sprouting and regeneration of peripheral nerve, in: "The Node of Ranvier" eds. J.C. Zagoren & S. Fedoroff, Academic Press, Orlando. p 353 (1984).

18. G.T. Vrbová, T. Gordon, and R. Jones, in: "Nerve-muscle interaction" Chapman and Hall, London p 105 (1978).

19. B.W. Festoff, M.R. Patterson, and K. Romstedt, Plasminogen activator: The major secreted neutral protease of cultured skeletal muscle cells. **J. Cell Physiol.** 110: 190 (1982).

20. B.W. Festoff, M.R. Patterson, D. Eaton, and J.B. Baker, Plasminogen activator and protease nexin in myogenesis. **J. Cell Biol.** 91: 43a (1981).

21. B.W. Festoff, and D. Hantaï, Plasminogen activators and inhibitors: roles in muscle and neuromuscular regeneration. **Prog. Brain Res.** 71: 423 (1987).

22. R.L. Beach, W.V. Burton, W.J. Hendricks, and B.W. Festoff, Extracellular matrix synthesis by skeletal muscle in culture: proteins and effect of enzyme degradation. **J. Biol. Chem.** 257: 11437 (1982).

23. M. Rånby, B. Norrman, and P. Wallén, A sensitive assay for tissue plasminogen activator. **Thromb. Res.** 27: 743 (1982).

24. A. Granelli-Piperno, and E. Reich, A study of protease and protease inhibitor complexes in biological fluids. **J. Exp. Med.** 148: 223 (1983).

25. K. Danø, P.A. Andreasen, J. Grøndahl-Hansen, P. Kristensen, L.S. Nielsen, and L. Skriver, Plasminogen activator, tissue degradation and cancer. **Adv. Cancer Res.** 44: 139 (1985).

26. M.W. McCaman, Biochemical effects of denervation on normal and dystrophic muscle: acetylcholinesterase and choline acetyltransferase. **Life Sci.** 5: 1459 (1966).

27. S. Tucek, Choline acetyltransferase activity in skeletal muscles after denervation. **Exp. Neurol.** 40: 23 (1973).

28. V. Kielberg, P.A. Andreasen, J. Grøndahl-Hansen, L.S. Nielsen, L. Skriver, and K. Danø, Proenzyme to urokinase type plasminogen activator in the mouse *in vivo*. **FEBS Lett.** 182: 441 (1985).

29. A. Mauro, Satellite cells of skeletal muscle fibers. **J. Biophys. Biochem. Cytol.** 9: 493 (1961).

30. R.A.D. O'Brien, A.J.C. Östberg, and G. Vrbová, Observations on the elimination of polyneuronal innervation and the developing mammalian skeletal muscle. **J. Physiol.** 282: 571 (1978).

31. D.C. Van Essen, Neuromuscular synapse elimination, *in:* "Neuronal Development" Spitzer N.C. (ed.) p 333 Plenum, New York (1982).

32. R.A.D. O'Brien, A.J.C. Östberg, and G. Vrbová, The effect of acetylcholine on function and structure of the developing mammalian neuromuscular junction. **Neurosci.** 5: 1367 (1980).

33. M.C. Brown, R.L. Holland, and W.G. Hopkins, Motor nerve sprouting. **Ann. Rev. Neurosci.** 4: 17 (1981).

34. D. Hantaï, J.S. Rao, and B.W. Festoff, Serine proteases and serpins: their possible roles in the motor system. **Rev. Neurol. (Paris)** 144: 680 (1988).

35. J.M. Andrews, M.B. Gardner, F.J. Wolfgram, G.W. Ellison, D.D. Porter, and W.W. Brandkamp, Studies on a murine form of spontaneous lower motor neuron degeneration: the wobbler *(wr)* mouse. **Am. J. Path.** 76: 63 (1974).

36. J. H. La Vail, and K.P. Irons, Abnormal neuromuscular junctions in the lateral rectus muscle of wobbler mice. **Brain Res.** 463: 78 (1988).

37. J.-M. Charcot, and A. Joffroy, Deux cas d'atrophie musculaire progressive avec lésions de la substance grise et des faisceaux antérolatéraux de la moelle épinière. **Arch. Physiol. Norm. Path.** 2: 354 (1869).

38. B. Wiman, G. Mellbring, and M. Rånby, Plasminogen activator release during venous stasis and exercise as determined by a new specific assay. **Clin. Chem. Acta** 127: 279 (1983).

39. B.W. Festoff, and H.L. Fernandez, Plasma and red cell acetylcholinesterase in amyotrophic lateral sclerosis. **Muscle Nerve** 4: 41 (1981).

40. D.C. Rijken, and D. Collen, Purification and characterization of the plasminogen activator secreted by human melanoma cells in culture. **J. Biol. Chem.** 256: 7035 (1981).

RELATIONSHIP BETWEEN PLASMINOGEN ACTIVATORS AND REGENERATION CAPACITIES OF RAT SKELETAL MUSCLES

GEORGIA BARLOVATZ-MEIMON, ÉRIC FRISDAL, YANN BASSAGLIA, DANIEL HANTAÏ[1], EDUARDO ANGLÉS-CANO[2] AND JEAN GAUTRON

MYREM, Université Paris XII
94010 Créteil, France;
[1]INSERM U.153, 75005 Paris, France;
and [2]INSERM U.143
94275 Le Kremlin-Bicêtre, France

INTRODUCTION

Mammalian skeletal muscle regeneration may be divided in two major parts. The first is a degenerative period leading to muscle atrophy as a result of the catabolic conditions due to tissue breakdown, cell death, and subsequent presence of large amounts of lysosomal and non lysosomal proteinases.[1,2] The second step is characterized by tissue remodelling resulting in a relatively complete regeneration.[3-8] This step was described by Bischoff [9] as dependent upon the capacity of neutral proteinases to digest the basement membrane. It involves changes in the amount and distribution of various components of the basement membrane[10] such as fibronectin[2], laminin and type IV collagen.[12,13] The participation of the plasminogen activator (PA) system in these modifications has been described.[14] In addition, a large amount of information links tissue remodelling to the activity of serine proteinases such as PAs.[15]

Therefore we decided to address the following questions:
- do rat skeletal muscles contain PAs, and if so what kind?
- do slow and fast contracting muscles have different amount of these enzymes and do these correlate with their behavior in degenerative and regenerative processes?
- as these regenerative processes were correlated to the density and/or the myogenic capacities of the myogenic stem cells (i.e., satellite cells) of these muscles, could these cells, in culture, express the same differences in the amount of PAs?
Briefly, could we correlate the muscle regenerative capacities with the PAs amounts of the satellite cells?

The results reported here show that, *in vivo*, rat skeletal muscle contain both types of PAs with molecular weights of 38,000 for urokinase (uPA) and 65,000 for tissue type PA (tPA), and that the slow contracting muscle contains more activity, of both types, than does the fast contracting one.

Serine Proteases and Their Serpin Inhibitors in the Nervous System
Edited by B. W. Festoff
Plenum Press, New York, 1990

In vitro, the satellite cells isolated from slow or fast muscle are different in number and in myogenicity. They differ also in their content in PAs. Moreover, the cell fusion process can be correlated with increasing amount of uPA.

In addition, during post-traumatic regeneration, the *extensor digitorum longus* muscle (EDL), undergoes modifications of the amount of uPA showing a very slight increase during the first days after injury, with a dramatic increase of this enzyme at the time of the start of the remodeling process.

MATERIALS AND METHODS

Chemicals and Reagents

The following products were used: ethylenediaminetetraacetic acid (EDTA), bovine serum albumin (BSA), and Tween 20 (Serva, Heidelberg, FRG); Ultrogel, AcA 44, DEAE-Trisacryl and CM-Trisacryl (IBF, Villeneuve-la-Garenne, France); chromogenic synthetic substrate D-Val-Leu-Lys-p-nitroanilide, S-2251 (Kabi Vitrum, Mölndal, Sweden) and D-norleucyl-cyclohexyl alanyl-arginine-p-nitroanilide, CBS 3308 (Diagnostica Stago, Asnières, France); lysine- and gelatine-Sepharose 4B (Pharmacia, Uppsala, Sweden); polyvinylchloride U-shaped micro-titration plates and adhesive plate sealers (Dynatech, Marne-La-Coquette, France); glutaraldehyde 25% aqueous solution (TAAB Laboratories, Reading, Berks, UK). Immunoplates for plasminogen determination were provided by Behringwerke AG (Marburg, FRG). Other chemicals and reagents were of the best analytical grade commercially available.

Buffers

uPA buffer: 0.1 M Tris HCl and 2 mM EDTA pH 7.6. tPA buffers: (A) 0.05 M sodium phosphate buffer, pH 7.4, containing 0.08 M NaCl; (B) buffer A, pH 6.8. Assay-buffer: buffer A containing 2 mg/ml BSA and 0.01% Tween 20. Binding-buffer: buffer B containing 4 mg/ml BSA, 0.01% Tween 20 and 2 mM EDTA.

Proteins

[Glu1]-plasminogen was purified from diisopropylphospho-fluoridate-treated fresh frozen human plasma by affinity chromatography on lysine-Sepharose in the presence of Trasylol-, ion-exchange chromatography on DEAE-Trysacryl and gel filtration chromatography on Ultrogel AcA44. The concentration of plasminogen was determined by measuring the absorbance at 280 nm using $E_{1cm}^{1\%}$, 280 nm = 16.8.[16] A human melanoma tPA preparation (> 99% sc-tPA) purified from the conditioned medium of Bowes melanoma cell cultures was kindly provided by Dr. M. Rånby (Biopool, Umea, Sweden). Its specific activity was 680,000 IU/mg as determined by reference to the First International Standard for tPA. This standard preparation, coded 83/517, was provided by the National Institute for Biological Standards and Control (London, UK). UPA purified from human urine, also used as a standard, was a kind gift of Institut Choay (Paris, France).

Preparation of Muscle Extracts

Slow (white) *soleus*, fast (red) EDL and mixed (white and red) *sternomastoideus* (white SM and red SM) muscles were obtained from 11 to 14 weeks old (5 male and 3 female) white Wistar rats. The rats were anesthetized with ether and perfused with

40 ml of Ca^{2+}-and Mg^{2+}-free cold phosphate buffer, pH 7.4, containing 0.15 M NaCl, to avoid interference by circulating proteases and inhibitors. The four muscles were then dissected out and individually sectioned, weighed and pounded in an ice-cold Potter tube with 0.1 M Tris-HCl buffer, pH 7.6, containing 2 mM EDTA and 0.4% Triton X-100. The resulting four extracts were centrifuged for 20 min at 4°C at 12,000 g, and the supernatants were stored in aliquots at -80·C until use.

Cell Cultures

Satellite cells were prepared as following : after dissection, rat muscles were minced in small pieces and incubated for 2 hours at 37°C in Ham's F 12 medium containing 15% pronase and buffered with 10 mM HEPES. Thereafter, the remaining fragments were centrifuged for 3 mn at 90 g. The supernatant containing cells in suspension was centrifuged. The cellular pellet was washed again 3 times in Dulbecco Modified Eagle's Medium (DMEM). The last washing was performed in DMEM containing 10 % fetal calf and 10 % horse serum (Gibco). The cellular suspension was then adjusted in the same medium in order to seed 2×10^3 cells per cm^2 on gelatinized Petri dishes. Cultures were incubated in 5 % CO^2 at 37°C. Medium was renewed every 4 days.

Electron Microscopy

Cultures grown in Petri dishes were fixed with 2.5 % glutaraldehyde in 0.1 M phosphate buffer, postfixed with 2 % OsO_4 in buffer, dehydrated in graded alcohol, and embedded in Epon 812. They were then photographed. Small pieces of the Epon sheet corresponding to the photos were cut and fixed on prepolymerized blocs. Ultrathin sections were obtained, contrasted with uranyl acetate/lead citrate, and observed in a Philips 410 electron microscope at 80 KV.

Regeneration Studies

Regeneration studies were performed on adult rats of 11 to 14 weeks old. After anaesthesia, the EDL muscle was denervated and then crushed from one tendon to the other. The wounded muscle was left in its bed, with its tendons attached. The regenerative muscle was taken off after 2, 4, 7, 8 or 14 days after injury. The contralateral was used as a control after testing there was no contralateral effect. The enzymatic activities were then assessed as described above.

Assay of PAs

All spectrophotometric assays were done in triplicate in microtitration plates at 37°C and were calibrated with either purified uPA or tPA. A chromogenic substrate selective for plasmin (S-2251 or CBS 3308) was used to follow the initial rate of plasminogen activation by measuring the rate of p-nitroaniline generation. Plasminogen was activated by uPA-like activators as follows: 40 μl of muscle extracts or standard samples were mixed with 160 μl uPA buffer containing 400 nM plasminogen and 1.6 mM of S-2251 as substrate. The generation of plasmin was detected by measuring the p-nitroaniline release from the substrate as indicated below.

To determine tPA activity, a solid-phase fibrin assay was utilized.[17] In this assay, the tPA in the samples was separated from the other proteins using solid phase fibrin prepared in microtitration plates. To separate tPA, 100 μl per well of each sample was added, incubated for 1 hr at 37°C and thoroughly washed with binding buffer to eliminate unbound proteins. Fibrin-bound tPA was determined according to the above assay. Briefly, 100 μl per well of assay buffer containing 0.2 μM plasminogen and 0.6 mM CBS 3308 was added, and the amidolytic activity of the plasmin generated by the muscle extracts activators was followed by measuring the absorbance at double wave length (405/630 nm) at regular intervals using a microtiter plate counter (MR 610, Dynatech) equipped with a thermostatic device to maintain a constant temperature of 37°C. Complete automation was obtained by coupling the MR 610 apparatus to an Apple IIe microcomputer through an RS232 serial interface using a computer program (Cinenzymic-log INSERM) developed at the INSERM U.152. An excess of the plasminogen and chromogenic substrates accelerated the reaction to a degree proportional to the activation of plasminogen. Since the activation progression curve is described by a parabolic relationship, the reaction rates were determined from the time course of the overall reaction by calculating the second degree equation for linear motion with constant acceleration, as indicated elsewhere.[18] Linear standard curves were drawn for the initial rate of p-nitroaniline release vs known concentrations of uPA or tPA standards. The levels of free enzyme were then determined from the standard curve. The results were expressed as specific activities by relating the initial activity of each PA ($OD_{405nm} \times 10^{-3} \times min^{-1}$) to the protein concentration of each sample.

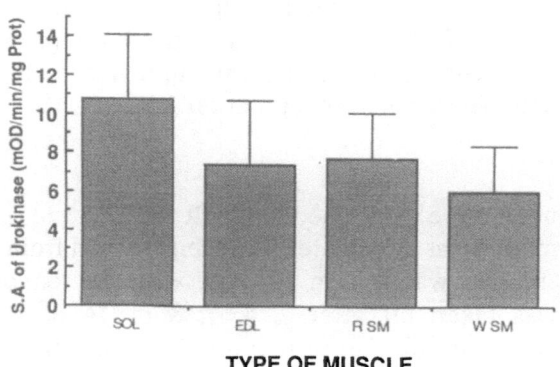

Figure 1. *uPA content of skeletal muscle extracts.* Two slow-twitch muscles the *soleus* (SOL) and the red part of the *sternomastoideus* (R SM) and of two fast-twitch muscles the *extensor digitorum longus* (EDL) and the white part of *sternomastoideus* (W SM) were extracted as described in materials and methods. The assay was performed in the presence of 400 nM plasminogen, and 1.6 mM of the synthetic substrate S-2251. The generation of plasmin was detected by measuring the p-nitroaniline release from the substrate. The values are expressed in specific activities (mOD/min/mg prot). Higher activities can be seen in the slow muscles compared to fast ones.

Miscellaneous

Protein concentration in the muscles extracts was measured according to Lowry[19] using bovine serum albumin as standard. Plasminogen was measured using a double immunodiffusion test according to the indications of the manufacturer (Behringwerke AG).

RESULTS

In Vivo Activities

In the slow and fast muscles extracts we found PAs activities of both uPA and tPA. In absolute terms, tPA activity was much lower than uPA.

The *soleus* (slow muscle) showed a higher amount of uPA than did the EDL (fast muscle) (Figure 1). The slow part of the *sternomastoideus* (red SM) showed the same kind of difference with the fast part (white SM). These activities could be entirely inhibited by aprotinin, the well known plasmin inhibitor, when added at the concentration of 10,000 U/ml confirming the PA nature of these activities.

The activities of the tPA, in the different muscle extracts, display the same differences as those of uPA (Figure 2). The assay we used is the solid phase fibrin assay (SOFIA) developed by one of us,[18] which ensures the measurement of only the PA linked to the fibrin network. Moreover, the samples' preparation procedure avoids any circulating tPA. The difference in tPA amounts between the slow part of the sternomastoideus (red SM) and the fast part (white SM) is even greater than that existing between the *soleus*, and the EDL. And indeed, the fast part of the *sternomastoideus* contains only type II fibers. It has the lowest amount of tPA. On the contrary, slow muscles that contain more type I fibers, contains more tPA. These findings reinforce the relationship between the fiber type of a muscle and the PAs content of this muscle.

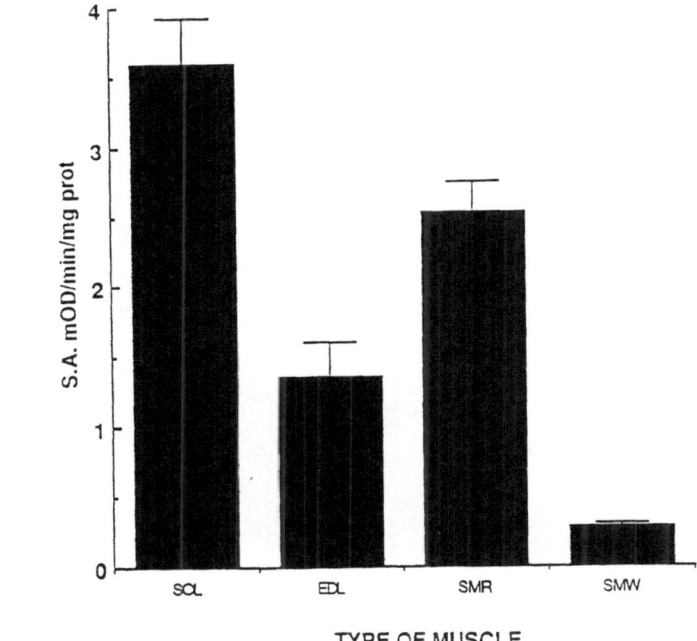

TYPE OF MUSCLE

Figure 2. *tPA content of skeletal muscles extracts.* Extracts of two slow-twitch muscles (*soleus* (SOL) and red part of the *sternomastoideus* (R SM) and of two fast-twitch muscles (*extensor digitorum longus* (EDL) and white part of the *sternomastoideus* (W SM) were performed as described in materials and methods. Fibrin-bound tPA was determined in the presence of 0.2 μM of plasminogen and 0.6 mM of the synthetic substrate CBS 3308. The amidolytic activity of the plasmin generated was then followed. The values are expressed in specific activities (mOD/min/mg prot). The obvious differences between slow and fast muscles can be observed with higher amount of tPA in slow muscles (SOL and R SM) than in fast muscles (EDL and W SM); but the level of this enzyme in muscle extracts is, in general, lower than that of uPA.

In Vitro Activities

In order to understand the nature of the physiological difference between slow fast muscles, and to attribute these activities to one or the other type of cells that can be found in muscles, we cultivated satellite cells descending from a mixture of muscles, from *soleus* (slow muscle), or· from EDL (fast muscle). These satellite cells are the myogenic stem cells that ensure the mature muscle's growth and regeneration.[4] They can be isolated and cultivated *in vitro* where they mimic some steps of the *in vivo* myogenic differentiation. In this process they undergo first a proliferation phase and then a differentiation phase. Proliferation peaks after day 4 of culture for rat satellite cells, in our culture conditions.[20] These conditions allow the cells to grow and then, to differentiate into myotubes i.e. multinucleated cells. Therefore, satellite cells have to fuse either with one another or to fuse into existing myotubes. For this process to occur, important membrane modifications are probably needed.

The micrograph of a cell that is going to fuse with an already formed myotube is shown in Figure 4. The plasma membrane of the cell seems to be undergoing profound modifications as already observed by others.[21] *In vitro*, this differentiation phase begins after day 6 and goes on for several days. At day 8, numerous fused cells can be seen.[20] This is the moment where a dramatic increase in their amount of uPA will be seen as shown in Figure 3, whereas during proliferation the level stays quite low. These results establish the relationship between cellular PA's content and differentiation i.e. fusion.

When satellite cells are isolated from slow or fast muscles, differential quantitative and qualitative data are obtained (Table 1). The two types of muscles

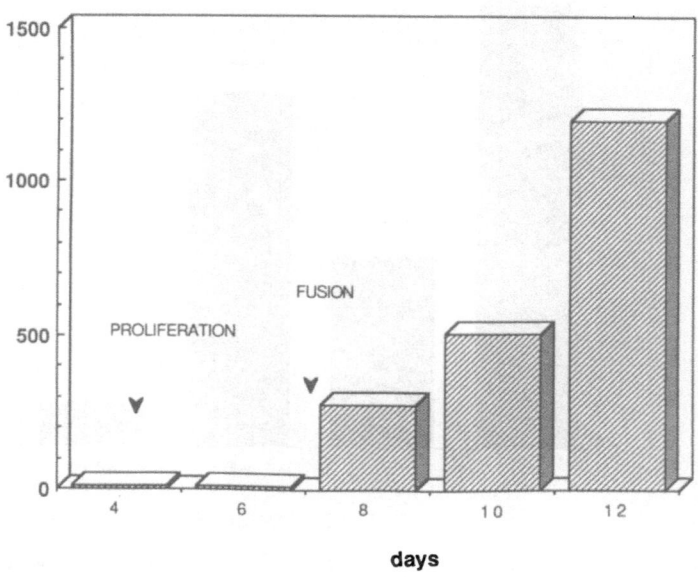

Figure 3. *Intracellular uPA content as a function of in vitro myogenesis.* Rat satellite cells in culture show a peak of proliferation at day 4. The fusion process begins between day 6 and day 8. Before extraction, the cells were rinsed and left 24 hours without serum in medium supplemented with 0.1 % BSA. The extraction was performed as described in material and methods. Even if low, uPA is present during the first days of culture, then it increases in a dramatic manner at day 8. This increase parallels the period of cell fusions.

Table 1. Characteristics of satellite cells isolated from adult rat muscles

	SOLEUS	EDL
PROLIFERATION		
number of cells $\times 10^6$	1.5	0.6
MYOGENICITY :		
Number of myotubes formed for 10^4 cells seeded	68	20
Nucleus number per myotube	10	6
ENZYMATIC CONTENT (ratio)*		
uPA	1.86	1
tPA	1.82	1

*EDL is considered equal to 1, *soleus* is then compared

differ in the number of satellite cells that can be isolated from them. These cells differ also in their myogenicity expressed as the number of myotubes formed for the same number of cells seeded in a culture dish, and in the number of nuclei per myotube.[22] We checked their content in PAs and we found the same ratio for uPA and tPA. Satellite cells isolated from the *soleus* contain nearly twice as much of both PAs then cells isolated from EDL.[23]

These results confirm that the nerve influence on muscle is kept in a sort of memory by the satellite cells this muscle contains. We propose that, *in vivo*, these cells, wedged between the plasma membrane and the basement membrane of the myofiber, have to use adapted enzymatic equipment to be mobilized and allow their proliferation and differentiation. This equipment includes PAs, which could be used for the activation/mobilization and the fusion processes *in vivo* and *in vitro*. This hypothesis is supported by the presence of uPA in proliferating cells and by its dramatic increase when these cells enter a fusing process. In addition, it is also supported by the presence of other proteinases (elastase-like proteinases) as well as inhibitors in muscle extracts (our unpublished results). The PAs, together with other proteinases, and perhaps in synergy, could play a key-role in either or both the "mobilization" of the satellite cells and the fusion process. The inhibitors could be kept out of the fusion area and uPA specific receptors could be the membrane support for these directional events. This has been supposed in other models (see Blasi et al., this volume) and is summarized in Figure 5. The existence of receptors might be important in ensuring a directional proteolytic process.

Regeneration Studies

After denervating and crushing the muscle, leaving the tendons intact, the muscle degenerates in a few days. The degeneration was controlled histologically; no intact fiber persisted. Then, the first steps of regeneration that are known to be nerve-independent[24] were observed rapidly. We checked the presence of PAs of both types in post-traumatic regenerating EDL muscle comparing it to the contralateral (same undamaged muscle of the other leg). The values of both uPA and tPA for these controls, fitted in the usual ranges of the corresponding muscles, confirming the absence of any "contralateral" effect.

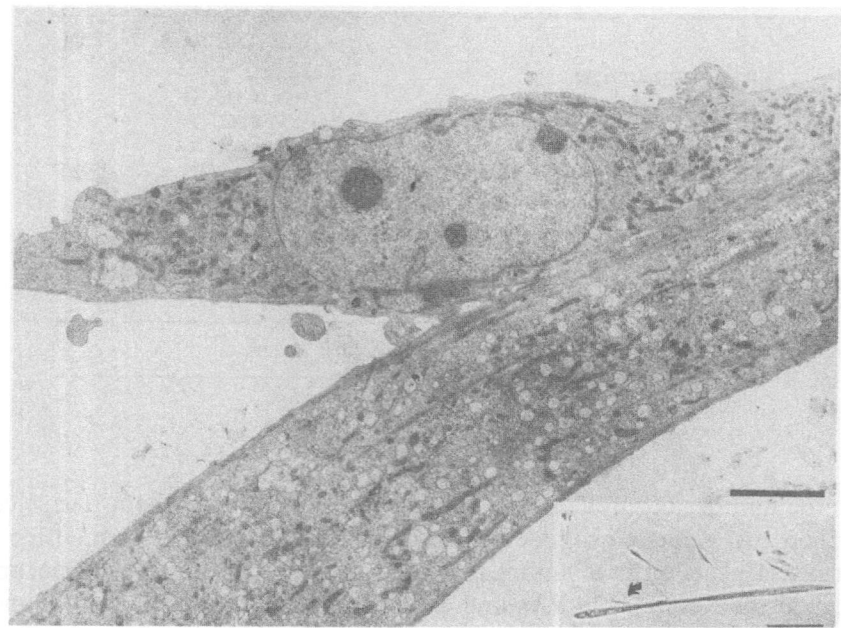

Figure 4. *Electron microscopy of a probably fusing cell.* A myoblast and a myotube just before fusion (Scale bar = 5 μ). Insert: general view. The arrow shows the concerned cell (Scale bar = 100 μ).

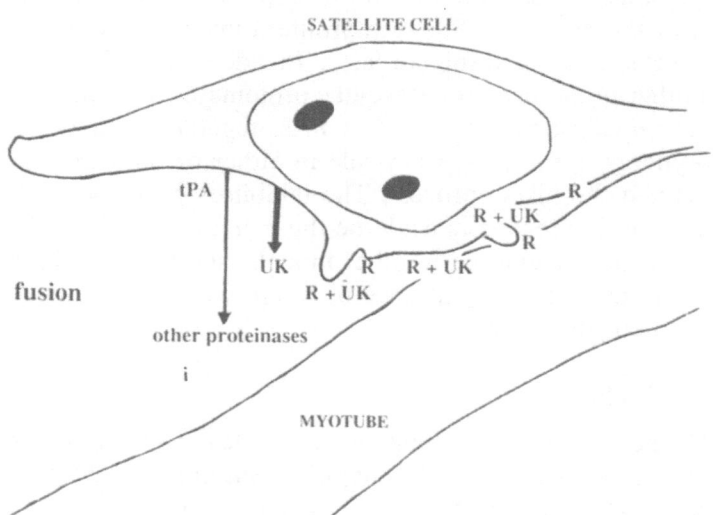

Figure 5. *A possible interpretation...* PAs and other proteinases, could play a key-role in the mobilization of the satellite cells and then in the fusion processes through directional proteolysis. The inhibitors could be kept out of the fusion area and uPA specific receptors could be the membrane support of these directional events.

tPA. Days 2, 4, 7 and 8 were compared (Figure 6) for tPA activities. No significant difference in the amount of tPA was seen between the injured muscles and the contralaterals, in either the degenerative or the regenerative phase. The values for both kinds (injured and not) of muscles were similar to the controls in previous studies.

uPA. On the contrary, the amounts of uPA found in the injured muscles was always higher then the control (Figure 7). A slight, but significant increase, was observed during the first days after muscle crush (days 2 and 4). Day 7 shows the beginning of a dramatic increase in the amount of uPA, while the control remains unchanged. This increase ranges from 3 to more then 15 fold (day 8). Day 14 is marked by the beginning of a decrease of this phenomenon. Again, contralateral values ranged inside the usual control values showing the absence of a contralateral effect. The variation of uPA amounts in the degenerative/regenerative extracts compared to the contralateral muscle as a control, strongly suggests an involvement of this enzyme in muscle regeneration.

DISCUSSION

Our previous results indicate that *in vivo*, normal rat skeletal muscles possess measurable levels of PAs of both urokinase and tissue-type with molecular weights of 38,000 and 65,000 respectively.[23] These extracts contain inhibiting capacities as well.[25] Moreover, supporting previous studies[26] we found in the same extracts, amidolytic activities (measured on SLAPNA) that can be considered as elastase-like proteinases, and their inhibitors as well (our unpublished results).

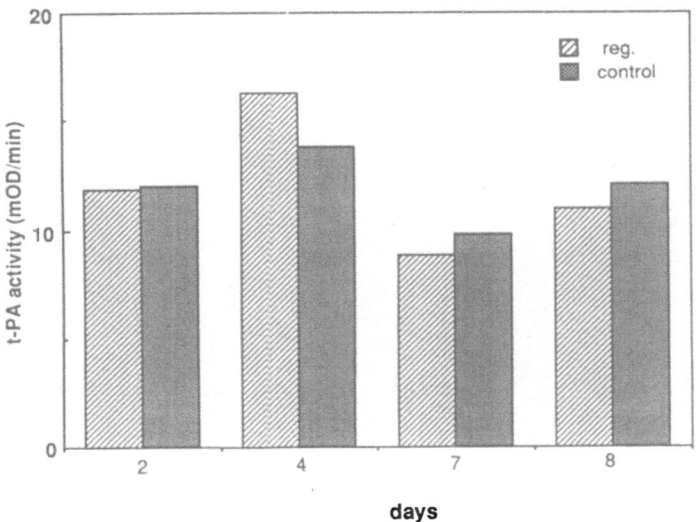

Figure 6. *tPA activities during muscle regeneration.* Extracts of post-traumatic regenerating EDL muscles were compared in their tPA activities with the contralateral as a control. Values are expressed in specific activities (mOD/min/mg prot). No significant differences were observed for 2, 4, 7 and 8 days of regeneration in these experiments.

When compared in terms of PAs activities, slow and fast muscle extracts differ with both amidolytic (S-2251), solid phase fibrin (SOFIA) assays or zymography[27] In particular, uPA is present always in higher amounts, and slow muscles contain more uPA than fast muscles. These results are expressed in terms of specific activities. They thus reflect the synthesis of these proteinases by a greater number of cells or a greater synthesis by an equal number of cells, or both. Briefly, they account for a different quantity or a different quality (or both) of cells in the different types of muscles. The main difference *in vivo* between myogenic cells of slow and fast muscle is the type of innervation of the muscle that has lead it to become slow or fast contracting with the accompanying consequences, i.e. type of fibers etc.. Also the total population of satellite cells differ: slow twitch oxidative muscle fibers have more satellite cells than fast-twitch glycolytic fibers [28,29] Could these cells differ in terms of enzymatic activities also?

We thus asked the question whether the myogenic cells i.e. satellite cells, isolated from slow or fast muscle and cultured *in vitro* would also behave differently. Would these cells keep any memory, in term of proliferative or differentiating capacities and of PAs activities, of the type of muscle they descend from?
Myogenic cells can be cultivated.[30] They then mimic important phases of myogenesis i.e. proliferation and fusion.

Our results show that, slow muscle myogenic cells, that have higher proliferative and myogenic potentialities differ from fast muscle myogenic cells,[23] in the amount of PAs of both types. These differences were seen in proliferative as well as in differentiating cells.

To better investigate this process, we studied the role of uPA in the *in vitro* myogenesis. Our results show a strong correlation between the increase of uPA and

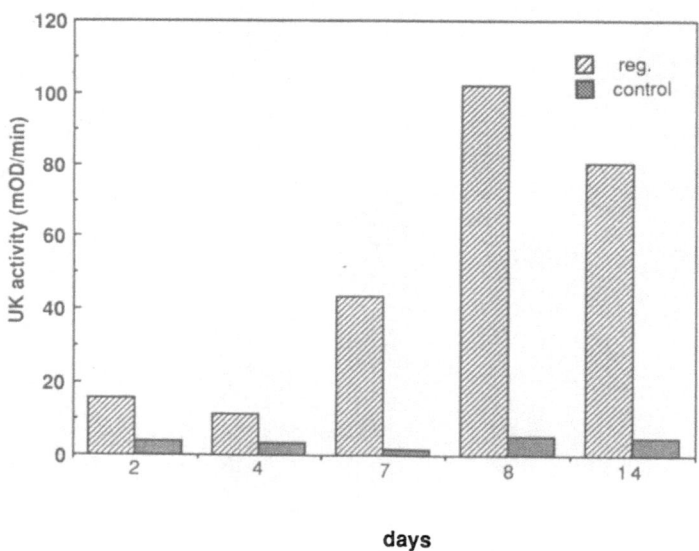

days

Figure 7. *uPA activities during muscle regeneration.* Extracts of post-traumatic regenerating EDL muscles were compared in their uPA activities with the contralateral as a control. Values are expressed in specific activities (mOD/min/mg prot). Early after injury the amounts of uPA are slightly increased in comparison with those found in control muscles. But at day 7, a dramatic increase of these activities can be observed which slows down at day 14. This result shows the correlation between uPA and the regenerative process in muscle.

the *in vitro* timing of fusion initiation. These results give an argument to consider uPA more as a "reconstructive" enzyme than a destructive one. It might be more relevant to consider the timing and amount of uPA secretion and its proportion towards others enzymes and/or inhibitors, rather than simply its synthesis or secretion. Analyzing these results, we have to keep in mind that the activities must eventually be considered in terms of how they affect the balance of plasminogen activation.

The *in vivo* situation that recapitulates foetal myogenesis and involves proliferation and fusion of myoblasts, is the post-traumatic regeneration of the skeletal muscle. The experimental method used here for these studies, leaves the tendons in place, but severe the blood circulation and the nerve. It has been demonstrated that the early phases of regeneration of an entire muscle, do not need the nerve at all. [25] This regeneration is characterized, among others, by the proliferation of the myogenic cells and their differentiation. The enormous increase of uPA in muscle regeneration, establishes clearly the involvement of this enzymatic system in these processes. While we cannot exclude the participation of other cells in this increase, the importance of the myogenic cells is suggested by the following arguments: a) the degenerative phase where an invasion of macrophages occurs, does not lead to any important modifications of the uPA activity; b) a strong correlation exists between uPA activity and the *in vitro* fusion process. These facts lead us to suggest, at least, a strong participation of these cells to the observed phenomenon. Moreover, reinforcing our hypothesis, an increase of the satellite cells number is expected as a regenerative response to degeneration (as it is the case in Duchenne muscular dystrophy).[31]

These findings together with those establishing the relationship between differentiated muscle fibers and their satellite cells (relationship that makes these cells secrete different amounts of uPA), allow us to correlate PAs, satellite cells and muscle regeneration capacities.

The necessity of PAs presence to influence the turnover of muscle basement membrane has been stressed before.[32] On this basis, we hypothesize a more complex utilization of PAs in muscle: on one hand, this enzymatic system could act as "helper" in the necessary mobilization of the myogenic stem cells i.e. satellite cells, and on the other hand, as a "reconstructive" system in the regeneration processes; in other words, as differentiation controllers. This could be done through an action on cell membranes, and would be modulated by the presence, the number and the quality of specific receptors.

CONCLUSION

These results, together with previous studies, support the hypothesis for a key-role of the proteinase/anti-proteinase system in muscle, especially in experimental situations as denervation or degeneration/regeneration. Even if this role seems to be more complex than it appeared, and is far from explained at the molecular level, it is now known that the cellular basis appears to reside within the muscle itself. Satellite cells are the reservoir myogenic cells, they seem to account for important proteinases activities in muscle. The role of these myogenic cells in regeneration was already demonstrated.[6,7] We showed here the involvement of PAs and particularly uPA, in muscle regeneration processes where satellite cells play a crucial role. Therefore these results warrant further studies to uncover the molecular basis and the physiological function of these phenomena.

Acknowledgements: We thank Professor F. Blasi for constructive discussion. This research has been partly supported by INSERM (grant n°876015) and AFM (Association Française contre les Myopathies).

REFERENCES

1. R.J. Pennington, Proteinases in Muscle, *in:* "Proteinases in Mammalian Cells an Tissues", ed. A.J. Barett, Elsevier North Holland Biochemical Press, p 515 (1977).

2. K. Takahasi, Y. Ichihara, and K. Sogawa, Membrane bound neutral proteinase in the microsomal fraction of skeletal muscle; its occurrence and properties, *in:* "Muscular Dystrophy : Biomedical aspects", eds. S. Ebashi and E. Osawa, Japan Sci. Soc. Press, Tokyo, p 305 (1983).

3. D. Allbrook, Skeletal muscle regeneration. **Muscle and Nerve** 4: 234 (1981).

4. B.M. Carlson, The regeneration of skeletal muscle: a review. **Am. J. Anat.** 137: 119 (1973).

5. B.M. Carlson, and J.A. Faulkner, The regeneration of skeletal muscle fibers following injury: a review. **Med. Sci. Sports Exercise** 15: 187 (1983).

6. A. Mauro, Satellite cell of skeletal muscle fibers, **J. Biophys. Cytol.** 9: 493 (1961).

7. A. Mauro, *in:* "Muscle Regeneration" Raven Press, New-York (1979).

8. L.C. Maxwell, Muscle regeneration in nerve-intact and free skeletal muscle autografts in cats. **Am. J. Physiol.** 246: C96 (1984).

9. R. Bischoff, Tissue culture studies on the origin of myogenic cells during muscle regeneration in the rat, *in:* "Muscle Regeneration", ed. A. Mauro, Raven Press, New-York, p 13 (1979).

10. A.K. Gulati, A.H. Reddi, and A.A. Zalewski, Changes in the basement membrane zone components during skeletal muscle fiber degeneration and regeneration. **J. Cell Biol.** 97: 957 (1983).

11. R. Hynes, and K. Yamada, Fibronectins: multifunctional modular glycoproteins. **J. Cell Biol.** 95: 369 (1982).

12. R. Timpl, H. Rohde, P. Gehron Robey, S. Rennard, J.M. Foidart, and G. Martin, Laminin: a glycoprotein from basement membranes. **J. Biol. Chem.** 254: 9933 (1979).

13. R. Timpl, R. Glanville, G. Wick, and G. Martin, Immunocytochemical study on basement membrane (type IV) collagens. **Immunology** 38: 109 (1979).

14. D. Hantaï, and B.W. Festoff, Degradation of muscle basement membrane by locally generated plasmin. **Exp. Neurol.** 95: 44 (1987).

15. K. Danø, P. Andreasen, J. Grøndahl-Hansen, P. Kristensen, L. Nielsen, and L. Skriver, Plasminogen activators, tissue degradation and cancer. **Adv. Cancer Res.** 44: 139 (1985).

16. P. Wallén, and P. Wiman, Characterization of human plasminogen. II. Separation and partial characterization of different molecular weight of human plasminogen. **Biochem. Biophys. Acta** 257, 122 (1972).

17. E. Anglés-Cano, A spectrometric solid-phase fribrin-tissue plasminogen activator activity assay (SOFIA-tPA) for high-fibrin-affinity tissue plasminogen activators. **Analyt. Biochem.** 153: 201 (1986).

18. M. Rånby, B. Norrman, and P. Wallén, A sensitive assay for tissue plasminogen activator. **Thromb. Res.** 27: 743 (1982).

19. O.H. Lowry, N.J. Rosenborough, A.L. Farr, and R.J. Randall, Protein measurement with the Folin phenol reagent. **J. Biol. Chem.** 194: 265 (1951).

20. B. Lassalle, J. Gautron, I. Martelly, and A. Le Moigne, Image analysis of rat satellite cell proliferation *in vitro*. **Cytotechnology** 2:213 (1989).

21. R. Bischoff, and M. Lowe, Cell surface components and the interaction of myogenic cells, *in:* "Exploratory concepts in muscular dystrophy", ed. Milhorat, Excerpta Medica, Amsterdam, Vol II, p 17 (1974).

22. G. Barlovatz-Meimon, Y Bassaglia, É. Frisdal, J. Gautron, A. Le Moigne, and I. Martelly, Do satellite cells isolated from fast and slow muscles and cultured *in vitro* remember their origin. I. Relationship between plasminogen activator expression, myogenic capacity and muscle regeneration. **Cell Diff. Dev.** 27: S66 (1989).

23. G. Barlovatz-Meimon, J. Lebrazi, and J. Gautron, Plasminogen activators in slow and fast Muscles. **Anticancer Research** 7: 895 (1987).

24. B.Carlson, Nerve-muscle interrelationships in mammalian skeletal muscle regeneration, *in:* "Control of cell proliferation and differentiation during regeneration", ed. H.J. Anton, Monographs in Dev. Biol., Karger, Vol 21 p 49 (1988).

25. G. Barlovatz-Meimon, Mesure de la capacité d'inhibition rapide du tPA, *in:* "Progrès en hématologie: physiopatologie de l'hématologie", eds. Y. Sultan and A.M. Fischer, Douin, Paris (1986).

26. M. Kidron, M. Dudai, I. Nachson, and M. Mayer, Protease inhibitor activity in human skeletal muscle. **Biochem. Med. Met. Biol.** 36: 136 (1986).

27   G. Barlovatz-Meimon, É. Frisdal, D. Hantaï, E. Anglés-Cano, and J. Gautron, Slow and fast rat skeletal muscle differ in their plasminogen activator activities. **Eur. J. Cell Biol.** In press.

28. H. Schmallbruch, and U. Hellhammer, The number of nuclei in adult rat muscles with special reference to satellite cells. **Anat. Rec.** 189:169 (1977).

29. M.C. Gibson, and E. Schultz, Age-related differences in absolute numbers of skeletal muscle satellite cells. **Muscle and Nerve** 6: 574 (1983).

30. D. Yaffe, Retention of differentiated potentialities during prolonged cultivation of myogenic cells. **Proc. Natl. Acad. Sci. USA** 61: 477 (1968).

31. P. Heinman, and U. Herbort, Satellite cells densities in neuromuscular mutants of the mouse: lack of correlation with fiber degeneration, *in:* "Control of cell proliferation and differentiation during regeneration", ed. H.J. Anton, Monographs in Dev. Biol., Karger, Vol 21 p 57 (1988).

32. D. Hantaï, J.S. Rao, C.B. Kahler, and B.W. Festoff, Plasminogen activator decline correlates with synapse elimination during neonatal development of skeletal muscle. **Proc. Natl. Acad. Sci.** USA 86: 362 (1989).

SECTION IV

Balance of serine proteases and serpins in the nervous system

A CASCADE APPROACH TO SYNAPSE FORMATION
BASED ON THROMBOGENIC AND FIBRINOLYTIC MODELS

BARRY W. FESTOFF, JASTI S. RAO, BOKKA R. REDDY
AND DANIEL HANTAÏ

Neurobiology Research Laboratory (151)
Department of Veterans Affairs Medical Center
4801 Linwood Boulevard
Kansas City, MO 64128 USA
and
Unité 153, I.N.S.E.R.M.
75005 Paris, FRANCE

INTRODUCTION

Cell-cell interaction in the nervous system has been most intensively studied at the peripheral cholinergic synapse, the neuromuscular junction. The interaction of the motor neuron's pre-synaptic axonal twigs with the sub-synaptic portion of the muscle fiber surface is influenced both by humoral factors as well as by specific adhesive components of the extracellular matrix (ECM). Neuromuscular synapse formation is initiated with the outgrowth of a neurite destined to be the axon of the motor neuron. This initial ramification begins a complex cascade of steps which include the directed extension of this axonal process through an inhospitable extracellular milieu, the recognition of an appropriate site(s) on the muscle fiber's surface, the localized deposit, removal and subsequent re-deposition of components of the muscle fiber's ECM. Eventually, in the neonatal period, a stabilization or linkage of macromolecules of the motor neuron and muscle fiber with the synaptic ECM occurs, which may allow for plasticity or remodelling over the life of the organism. The very nature of these cascade events suggests the well-studied biochemical events associated with regulated extracellular proteolysis underlying thrombogenesis and fibrinolysis. Our laboratory has been actively testing the hypothesis that synapse formation, more specifically neuromuscular junction formation, may be based on molecular mechanisms similar to clot formation and retraction. This MARATEA CONFERENCE has resulted from theoretical and empirical observations involving several members of the fibrinolytic cascade in muscle and neural cells in culture and *in vivo*. Some of this information has been presented elsewhere in this volume (see Hantaï et al.). In this paper we present evidence for the presence, and probable involvement, of a serine protease inhibitor with anti-thrombin and anti-plasminogen activator (PA) activity in one or more steps that might lead to stabilization of mature neuromuscular synapses.

Serine Proteases and Their Serpin Inhibitors in the Nervous System
Edited by B. W. Festoff
Plenum Press, New York, 1990

245

PROTEASES AT THE NEUROMUSCULAR JUNCTION

The basis for this hypothesis originated with studies of experimental denervation in animals to determine the neurotrophic basis for the localization, but not enzyme activity, of acetylcholinesterase (AChE) at the neuromuscular junction.[1] It was subsequently presented as an hypothesis for metabolism of the ECM under normal conditions and as might be found in the fatal, enigmatic neuromuscular disorder, amyotrophic lateral sclerosis (ALS).[2] A significant literature suggesting the presence of proteases at the neuromuscular junction already existed (Table 1). These early were concerned with proteases released under nerve stimulation and were postulated to be metalloproteases, largely calcium-activated. In addition to hypothesizing the participation of neutral extracellular proteases in the degradation of ECM components it also introduced the concept of a protease inhibitor produced by one or more of the cell types present at the neuromuscular junction.

This has subsequently been tested with the resultant demonstration of localized concentration of a specific protease inhibitor, protease nexin I (PNI) at adult murine junctions.[3,4] Recent data on the production of PNI isoforms by skeletal muscle cells in culture and its localization on myotubes, the binding of protease:PNI complexes to myotubes and expression of PNI protein in adult muscle have been submitted. These and other data allow us to propose a model for the involvement of thrombogenic and fibrinolytic cascade components and inhibitors in the formation and stabilization of the neuromuscular synapse, the testing of which is in progress.

Table 1. Chronology of protease regulation in neuromuscular system

Release of lysosomal enzymes at NMJ by nerve stimulation	(49,50)
Decrease in lysosomal enzymes with end of polyneuronal innervation	(51)
ACh causes release of lysosomal enzymes in developing muscle	(52)
Cultured chick muscle secretes PA after Rous sarcoma virus infection	(53)
PA:plasmin affects AChRs on Rous-infected chick muscle	(54)
Denervation causes early release of A12 AChE from NMJ BM	(1,55)
Hypothesis of extracellular neutral proteases in synaptic disorders such as ALS starting with loss of A_{12} AChE and BM and prediction of inhibitor	(2)
Finding of protease nexin in clonal mouse muscle cells	(40)
PA:plasmin involved in muscle BM degradation in culture	(33)
PA major secreted neutral protease of cultured muscle	(36)
PA:plasmin increases internalization of AChRs in clonal mouse muscle	(56)
uPA followed by tPA activated early in denervated muscle	(37)
uPA:plasmin degrades BM components in muscle *in situ* and increases early after denervation	(35)
PNI localizes to NMJ	(34)
uPA and tPA decline rapidly from birth as polyneural innervation is eliminated	(38)
PNI prevents uPA-driven myotube BM degradation	(57)
uPA, not tPA, in muscle rapidly regulated by nerve	(39)

INHIBITORS OF SERINE PROTEASES

Kunitz-Type Inhibitors (Kunins)

This group of polypeptides and glycoproteins represents the oldest-known and best studied of the serine protease inhibitors.[5] Members of this class include the bovine pancreatic trypsin inhibitor (BPTI; Kunitz inhibitor; aprotinin) as well as the inter-α-trypsin inhibitor ($I\alpha I$),[6] lipoprotein-associated coagulation inhibitor (LACI)[7] and the Kunitz protease inhibitor (KPI)-domain-containing forms of the β amyloid precursor protein, βAPP_{751} and βAPP_{770} (see Enghild et al., and Tanzi, this volume). Kunins may be single, double or tri-headed. Of interest to the concept of synapse formation involving a balance between one or more of the serine proteases and the peptide/polypeptide serine proteinase inhibitors is the very close association of kunins with glycosaminoglycans (GAGs), especially heparan sulfate and chondroitin sulfate, as well as the disulfide links based on their intra- and inter-chain cysteines. It is apparent from studies of $I\alpha I$ and the $\beta APPs$, that the kunins, apparently unlike the serpins, have been embedded in larger precursors.

Serpins

These include circulating and cellular inhibitors and homologous molecules for which a serine protease inhibiting function has not, as yet, been found. Members in the former sub-group are α_1-protease inhibitor ($\alpha_1 PT$) and α_1-anti-chymotrypsin ($\alpha_1 ACT$). The serpins without known inhibitory activity include angiotensinogen and ovalbumin.

Protease nexin I (PNI) is a cell-associated member of the serpin superfamily of serine protease inhibitors.[8,9] It is a heparin-activatable thrombin inhibitor with reported molecular masses (Mr) of 38-54 kDa.[8-10] The variability in Mr is presumably due to differences in glycosylation. In addition to inhibiting thrombin at rates faster than any other mammalian inhibitor, (in the presence of heparin, $1 \times 10^9 \cdot mol^{-1} \cdot s^{-1}$) PNI is also a "respectable" inhibitor of urokinase (uPA) and tissue (tPA) plasminogen activators. It is a much less potent inhibitor of plasmin and trypsin. Like other serpins, PNI is "single-headed" and forms 1:1 molar complexes with serine proteinases. Most serpins have a primary, physiologic target protease and for PNI it is likely to be thrombin but this may vary during development. Although originally found in human foreskin fibroblasts, PNI is synthesized by a number of anchorage-dependent, extravascular mammalian cells. Important to its possible role in cholinergic neuromuscular synapse formation is the fact that PNI binds to and is localized to the ECM.[11] This finding has prompted Cunningham and colleagues to postulate that cell surfaces can regulate the amount and type (i.e, target protease) of inhibitory activity of PNI (see Cunningham et al, this volume).

Plasminogen activator inhibitors 1, 2 and 3 (PAI-1,2,3) are more specific cellular serpins than PNI and have little or no inhibitory specificity towards serine proteases other than the PAs.[12-14]

Protease nexin II (PNII), another serine protease inhibitor was first detected by its ability to "link" the epidermal growth factor (EGF)-binding protein[15] and/or the gamma subunit of nerve growth factor (NGF).[16] Once isolated and purified it was found to be an effective inhibitor of chymotrypsin.[17] Recently, while THE MARATEA CONFERENCE was being held, work demonstrating that the sequence of PNII was identical to the β amyloid precursor protein (βAPP_{751}) isolated from senile amyloid plaques from brains of patients with Alzheimer's disease (AD) was submitted and eventually published.[18,19] A controversy currently exists as to whether

Table 2. Partial reactive center of serpin inhibitors

	Sequence	P2	P1	P1^{1}	P2^{1}	P3^{1}
Rodent						
	cos3E-46	Phe-	Arg-	Ser-	Arg-	Arg
	subC	Pro-	Leu-	Ser-	Ala-	Lys
	subD	Leu-	Arg-	Cys-	***-	Gly
	2A1	Gln-	Cys-	Cys-	Gln-	Gly
	2A2	Gly-	Cys-	Cys-	Ala-	Val
	subE	Phe-	Met-	Ser-	Ala-	Lys
	3E2	Phe-	Gln-	Ser-	Ser-	Lys
	Mouse α1-PI	Pro-	Tyr-	Ser-	Met-	Pro
	Contrapsin	Gly-	Arg-	Lys-	Ala-	Ile
Human						
	α1-PI	Pro-	Met-	Ser-	Ile-	Pro
	PCI	Phe-	Arg-	Ser-	Ala-	Arg
	α1-ACT	Leu-	Leu-	Ser-	Ala-	Leu
	ATIII	Gly-	Arg-	Ser-	Leu-	Asn
	HCII	Pro-	Leu-	Ser-	Thr-	Gln
	PAI-1	Ala-	Arg-	Met-	Ala-	Pro
	PAI-2	Gly-	Arg-	Thr-	Gly-	His
	PNI	Ala-	Arg-	Ser-	Ser-	Pro
	I-αI	Cys-	Arg-	Ala-	Phe-	Ile
	APP	Cys-	Arg-	Ala-	Met-	Ile

PNII is a serpin or, perhaps, a kunin (see above), based upon its ability to form 1:1 molar ratio complexes with EGF-binding protein or chymotrypsin that are resistant to boiling SDS.

The representative kunins and serpins discussed above are presented in tabular form, depending on their reactive center (P$_1$) residue, in Table 2.

SERPINS IN THE NERVOUS SYSTEM

Serpins in the central nervous system (CNS) were first evaluated by Monard and his co-workers (this volume), who have focused on a polypeptide initially isolated from rat C6 glioma conditioned medium that promoted neurite outgrowth from mouse neuroblastoma cells cultured in the absence of NGF.[20] The isolated and purified molecule was shown to be a protease and fibrinolytic inhibitor and, subsequently, to be virtually identical to PNI. When the cDNAs for the glioma factor,[21] subsequently termed the glial derived nexin (GDN) and human fibroblast PNI[22] were compared GDN differed from PNI by only three amino acids. Peripheral tissues, fibroblasts and muscle (Festoff et al, unpublished data), produce two forms, termed alpha and beta, which are in a 2:1 ratio, while CNS cells produce only, or primarily, the β form.

Neurite outgrowth promotion was the bioassay used by Monard's group to define, isolate and, ultimately, to clone the GDN. This effect has been attributed to PNI's inhibition of a thrombin-like enzyme.[23] This has been confirmed[24] for PNI's equipotent promotion of neurite outgrowth while the thrombin present in the culture medium was found to be inhibitory in the outgrowth bioassay. However, to date the presence of thrombin in the nervous system, or mRNA for prothrombin, has not been reported.

Neuronal survival has been stated to be unaffected by GDN[20,21] (see Monard et al., this volume). However, similar experiments have not been reported for PNI or the PAIs. In previous experiments with day 8 chick embryo mixed spinal cord cultures it was found that an extract of adult chicken ischiatic (sciatic) nerve, after transferrin was removed by affinity chromatography, markedly enhanced neuronal survival, as well as neurite outgrowth.[25] Preliminary characterization of this extract showed that anti-thrombin and anti-urokinase-like PA (uPA) activity was present in a heparin-agarose affinity elution peak. This peak has recently been shown to contain PNI (Festoff, unpublished). Thus, PNI and/or other serpins may also have neuronal survival properties, along with their effects on neurite outgrowth, further suggesting their roles as trophic factors in the nervous system.

Distribution of serpins in the CNS has recently been reported by Monard and his colleagues who have shown that GDN is enriched in the olfactory system: bulb, tract and lobe.[26] This finding is of considerable interest and importance since the olfactory system undergoes almost continuous degeneration and regeneration in the adult mammalian nervous system. In addition, in AD several groups of investigators have recently focused attention on the olfactory system as a region susceptible to aluminum intoxication and where amyloid plaques are found in high concentration.[27,28] PNI has recently been shown to be present in amyloid plaques in AD brain.[29]

SERPINS IN THE PERIPHERAL NERVOUS SYSTEM (PNS)

The study of serpins and their target proteases as trophic factors in the PNS has been an interest of our group over the last several years. We were led to studies on protease inhibitors in the neuromuscular system by experiments on the neurotrophic localization of end-plate acetylcholinesterase (AChE),[1,30-32] muscle ECM metabolism,[33-35] and serine proteases in muscle.[36-39] This resulted in the demonstration of PNI synthesis in clonal and primary muscle cells in culture[40,41] and the localization of PNI to adult neuromuscular junctions.[3,4] Recently, GDN has been shown to increase in the distal stump of rat sciatic nerve after axotomy.[42] In crush injury of mouse sciatic nerve PNI mRNA also increases in the distal stump (Festoff et al., in preparation) over the same time course as that of uPA increase in the denervated muscle.[39]

The junctional concentration of PNI is of significance for several reasons. First, although several antigens have been localized to the synaptic ECM in muscle, the only known proteins so far identified have been A_{12} AChE and a form of heparan sulfate proteoglycan (HSPG). Secondly, forms of HSPG have been shown to be associated with both the high concentration of membrane acetylcholine receptors[43] (AChRs; see Anderson, this volume) and A_{12} AChE[44] present at the junction and appear to undergo nerve-induced remodelling in culture. PNI binds to heparin in solution and a component(s) of the ECM, most likely HSPG (see Cunningham et al., this volume). Thirdly, several proteins are known to be selectively expressed by nuclei under the end-plate area, so-called sub-synaptic nuclei. PNI localization to this region suggests that it, too, may be selectively expressed by these nuclei, likely to be strongly influenced by specific signals emanating from the nerve.[45]

THROMBOGENIC AND FIBRINOLYTIC CASCADE MODELS OF SYNAPSE FORMATION AND PLASTICITY

Based on the foregoing information, we have proposed a model for synapse formation at the neuromuscular junction, but testable for central synapses as well.[3] In Figure 1, the principal features of the model are diagrammatically represented. Here seemingly opposing forces, or activities, favoring synapse formation (left-hand side) or synapse plasticity (as might occur in programmed elimination of polyneuronal synapses) or disintegration (as might occur in synaptic disorders such as the fatal amyotrophic lateral sclerosis) on the right-hand side, are presented.

Prominent components of the ECM (i.e, fibronectin, laminin, type IV collagen and HSPG) are displayed along with thrombospondin (TSP), a platelet α-granule glycoprotein found to be a constituent of a number of basement membranes, including muscle. In muscle, TSP has a unique developmental appearance, compacting in the ECM only after the period of polyneuronal synapse elimination is over.[45] TSP has

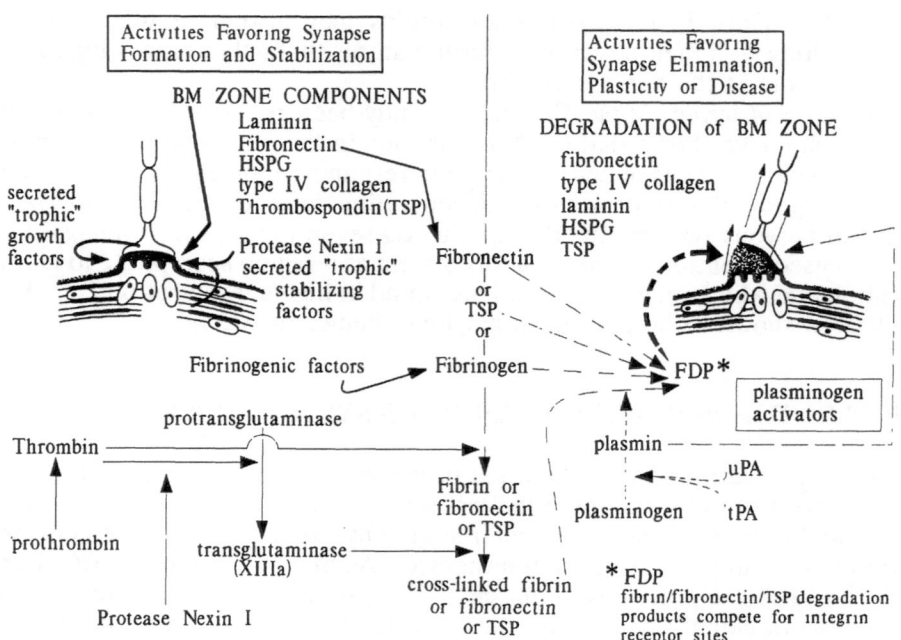

Figure 1. *Model of neuromuscular synapse formation and plasticity based on concept of thrombogenic and fibrinolytic cascades operating with ECM components and RGDS receptors (integrins).* Balance of serine proteases (secreted "trophic" factors of nerve) and inhibitors (serpins, such as protease nexin I, secreted "trophic" stabilizing factors of target muscle fiber) essential for this model.

been found to be increased in muscle of patients with ALS, in contrast to other components of the ECM.[47] In addition, PNI is depicted as one of several secreted trophic stabilizing factors from muscle that interact with one or more secreted trophic growth factors from the nerve. One or more PAs, secreted by axonal growth cones, may fall into this latter category. In forming a synapse the focus is on adhesion and stabilization. At the level of the membrane are the ECM components receptors, the integrins[48] which allow for the multi-functional glycoproteins of the ECM to interact with both nerve terminal and post-synaptic muscle fiber membranes. These elements are shown overlying several zymogen components of the thrombogenic cascade. Activation of prothrombin to active serine protease, in turn activating protrans-glutaminase to Factor XIIIa, capable of cross-linking fibronectin, TSP or other multi-functional ECM glycoproteins to effectively stabilize the nascent neuromuscular junction, is shown as well.

On the opposite side, activities that favor plasticity as might occur in programmed synapse elimination, in experimental denervation or in natural disease processes are shown. Here the focus is on dis-adhesion and important to this are the RGDS peptides generated by the action of fibrinolytic enzyme activities, themselves generated from precursor zymogens by specific activating enzymes regulated both by their own zymogen precursor state, by hormonal activity and by the action of specific cellular serpins such as the PAIs and the protease nexins. Degradation products, FDP (fibrin or fibronectin), TDP (thrombospondin) or LDP (laminin) could compete for binding of the intact ECM component for its specific integrin receptor, thereby enhancing the possibility of dis-adhesion.

Several of the factors and components have already been studied in neonatal muscle during polyneuronal synapse elimination and in experimental denervation, both axotomy and crush injury (see Hantaï et al., this volume). Work is in progress to determine the roles of thrombin activation and Factor XIIIa activation in developing muscle. In addition, work demonstrating the production of several isoforms of PNI by skeletal muscle in culture suggest that glycosylation of this serpin may influence some or all of its functions as a secreted stabilizing trophic factor in neuromuscular junction formation.

SUMMARY

A review of the various members of the kunin and serpin superfamilies of serine protease inhibitors has been presented from the perspective of their potential participation in the formation and stabilization of neuromuscular synapses. In addition, evidence for involvement of neutral, extracellular proteases at the neuromuscular junction has been focused primarily on components of the fibrinolytic cascade, in particular, urokinase-like plasminogen activator. A critical balance of serine proteases and inhibitors is suggested by the finding that the heparin-activatable cellular serpin, PNI, is produced by muscle cells in culture, binds both thrombin and uPA in the fluid phase, as well as a component of the myotube surface. In addition, complexes of PNI with either of these proteases are recognized by receptors on myotubes which are distinct from those that bind PNI alone. PNI has been localized, by monospecific antibody, to the adult neuromuscular junction co-localizing with AChRs. Since it is a bi-functional molecule, with binding sites not only for its target proteases but also for heparin-like molecules on the cell surface, a model accounting for these and other data has been proposed that relates serine proteases of the thrombogenic and fibrinolytic cascades with components of the ECM (substrates), the cellular serpins (such as PNI) and integrin receptors for both "native" ECM multi-domain glycoproteins as well as for their degradation products, the RGDS peptides. This model has a number of attractive features which may be readily tested. Prominent among these involves the several levels of control and regulation by the interacting cells, the growing nerve and elongating muscle fiber, their respective surfaces and secreted trophic factors.

REFERENCES

1. H.L. Fernandez, M.J. Duell, and B.W. Festoff, Neurotrophic regulation of 16 S acetylcholinesterase at the neuromuscular junction. J. Neurobiol. 10, 442-454 (1979).
2. B.W. Festoff, Role of neuromuscular junction macromolecules in the pathogenesis of amyotrophic lateral sclerosis. Med. Hypothesis 6, 121-131 (1980).
3. D. Hantaï, J.S. Rao, and B.W. Festoff, Serine proteases and serpins: their possible roles in the motor system. Rev. Neurol. 144, 680-687 (1988).
4. B.W. Festoff, D. Hantaï, and J.S. Rao, Plasminogen activators and inhibitors in the neuromuscular system III. Protease nexin I, a neurite-outgrowth promoting serine protease inhibitor (serpin) synthesized by muscle, co-localizes with acetylcholine receptors at neuromuscular synapses. J. Cell. Biol. submitted (1990).
5. R.W. Carrell, and D.R. Boswell, Serpins: the superfamily of plasma serine proteinase inhibitors. In: Proteinase Inhibitors, A. Barret, G. Salvesen, eds., Elsevier, Amsterdam. 403-420.
6. M. Laskowski, Jr., and I. Kato, Protein inhibition of proteinases. Ann. Rev. Biochem. 49, 593-626 (1980).
7. T.-C. Wun, K.K. Kretzmer, T.J. Girard, J.P. Miletech, and G.T. Broze, Cloning and characterization of a cDNA coding for the lipoprotein-associated coagulation inhibitor shows that it consists of three tandem Kunitz-type inhibitory domains. J. Biol. Chem. 263, 6001-6004 (1988).

8. J.B. Baker, D.A. Low, R.L. Simmer, and D.D. Cunningham, Protease nexin: A cellular component that links thrombin and plasminogen activator and mediates their binding to cells. Cell 21, 37-45 (1980).

9. D. Knauer, J.T. Thompson, and D.D. Cunningham, Protease nexins: cell secreted components that mediate the binding, internalization and degradation of regulatory serine proteases. J. Cell. Physiol. 117, 385-396 (1983).

10. J.B. Baker, D.J. Knauer, and D.D. Cunningham, Protease nexins: secreted protease inhibitors that regulate protease actions at and near the cell surface. 153-172. In: The Receptors, vol. 3, P.M. Conn (ed.), Academic Press, NY (1986).

11. D.H. Farrell, and D.D. Cunningham, Glycosaminoglycans on fibroblasts accelerate thrombin inhibition by protease nexin I. Biochem. J. 245, 543-550 (1988).

12. D.J. Loskutoff, and T.S. Edgington, Synthesis of a fibrinolytic activator and inhibitor by endothelial cells. J. Biol. Chem. 256, 4142-4145 (1982).

13. L. Holmberg, I. Lecander, B. Persson, and B. Astedt, An inhibitor from placenta specifically binds urokinase and inhibitor plasminogen activator released from ovarian carcinoma in tissue culture. Biochim. Biophys. Acta 544, 128-137 (1978).

14. D.C. Stump, M. Theinpont, and D. Collen, Purification and characterization of a novel inhibitor of urokinase from human urine quantitation and preliminary characterization. J. Biol. Chem. 261, 12759-12766 (1986).

15. D.J. Knauer, and D.D. Cunningham, Epidermal growth factor carrier protein binds to cells via a complex with released carrier protein nexin. Proc. Natl. Acad. Sci. USA 79, 2310-2314 (1982).

16. D.J. Knauer, K.M. Scaparro, and D.D. Cunningham, The subunit of 7S nerve growth factor binds to cells via complexes formed with two cell-secreted nexins. J. Biol. Chem. 257, 15098-15104 (1982).

17. W.E. Van Nostrand, and D.D. Cunningham, Purification of protease nexin II from human fibroblasts. J. Biol. Chem. 262, 8508-8514 (1987).

18. T. Oltersdorf, L.C. Fritz, D.B. Schenk, J. Lieberburg, K.L. Johnson-Wood, E.C. Beattie, P.J. Ward, R.W. Blacher, H.F. Dovey, and S. Sinha, The secreted form of the Alzheimer's amyloid precursor protein with the Kunitz domain is protease nexin II. Nature 341, 144-147 (1989).

19. W.E. Van Nostrand, S.L. Wagner, M. Suzuki, B.M. Choi, J.S. Farrow, J.W. Geddes, C.W. Cotman, and D.D. Cunningham, Protease nexin II, a potent antichymotrypsin, shows identity to amyloid β-protein precursor. Nature 341, 546-569 (1989).

20. D. Monard, K. Stockel, and R. Goodman et al, Distinction between nerve growth factor and glial factor. Nature 258, 444 (1975).

21. S. Gloor, K. Odink, J. Guenther, H. Nick, and D. Monard, A glial-derived neurite promoting factor with protease inhibitory activity belongs to the protease nexins. Cell 47, 687-693 (1986).

22. M.P. McGrogan, J. Kennedy, M.P. Li, C. Hsu, R.W. Scott, C. Simonsen, and J.B. Baker, Molecular cloning and expression of two forms of human protease nexin I. Biotechnology 6, 172-177 (1988).

23. D. Monard, E. Niday, A. Limat et al, Inhibition of protease activity can lead to neurite extension in neuroblastoma cells. Prog. Brain Res. 58, 359 (1983).

24. D. Gurwitz, and D.D. Cunningham, Thrombin modulates and reverses neuroblastoma neurite outgrowth. Proc. Natl. Acad. Sci. USA 85, 3440 (1988).

25. H. Popiela, T. Porter, R.L. Beach, and B.W. Festoff, Peripheral nerve extract promotes long-term survival and neurite outgrowth in cultured spinal cord neurons. Cell. Molec. Neurobiology 4, 67-77 (1984).

26. E. Reinhard, R. Meier, W. Halfter et al, Detection of glia-derived nexin in the olfactory system of the rat. Neuron 1, 387 (1988).

27. D.P. Perl, and P.F. Good, Uptake of aluminum into central nervous system along nasal-olfactory pathways. Lancet 2, 1028 (1987).

28. D.P. Perl, and P.F. Good, The association of aluminum, Alzheimer's disease and neurofibrillary tangles. J. Neurol. Transmission (Suppl.) 24, 205-211 (1987).

29. D.E. Rosenblatt, C. Geula, and H.H. Mesulam, Protease nexin immunostaining in Alzheimer's disease. Annals of Neurol. 26, 628-634 (1989).

30. H.L. Fernandez, M.J. Duell, and B.W. Festoff, Cellular distribution of 16S acetylcholinesterase. J. Neurochem. 32, 581-585 (1979).

31. H.L. Fernandez, M.J. Duell, and B.W. Festoff, Bidirectional axonal transport of 16S acetylcholinesterase in rat sciatic nerve. J. Neurobiol. 10, 31-39 (1980).

32. B.W. Festoff, H. Popiela, and R.L. Beach, Neurocrine regulation of acetylcholinesterase in health and disease, 317-330. In: Cholinesterases: Fundamental and Applied Aspects, M. Brzin, E.A. Barnard and D. Sket (eds.), de Gruyter, Berlin (1984).

33. R.L. Beach, W.V. Burton, W.J. Hendricks, and B.W. Festoff, Extracellular matrix synthesis by skeletal muscle in culture: proteins and effect of enzyme degradation. J. Biol. Chem. 257, 11437-11442 (1982).

34. J.S. Rao, R.L. Beach, and B.W. Festoff, Extracellular matrix synthesis in muscle cell cultures: quantitative and qualitative studies during myogenesis. Biochem. Biophys. Res. Comm. 130, 440-446 (1985).

35. D. Hantaï, and B.W. Festoff, Degradation of muscle basement membrane by locally generated plasmin. Exp. Neurol. 95, 44-55 (1987).

36. B.W. Festoff, M.R. Patterson, and K. Romstedt, Plasminogen activator: The major secreted neutral protease of cultured skeletal muscle cells. J. Cell. Physiol. 110, 190-195 (1982).

37. B.W. Festoff, D. Hantaï, J. Soria, A. Thomaïdis, and C. Soria, Plasminogen activator in mammalian skeletal muscle: characteristics of effect of denervation on urokinase-like and tissue activator. J. Cell Biol. 103, 1415-1421.

38. D. Hantaï, J.S. Rao, C.B. Kahler, and B.W. Festoff, Decrease in plasminogen activator correlates with synapse elimination during neonatal development of mouse skeletal muscle. Proc. Natl. Acad. Sci. USA 86, 362-366 (1989).

39. D. Hantaï, J.S. Rao, and B.W. Festoff, Rapid neural regulation of muscle urokinase-like plasminogen activator as defined by nerve crush. Proc. Natl. Acad. Sci. 87, 2926-2930 (1990).

40. B.W. Festoff, M.R. Patterson, D. Eaton, and J.B. Baker, Plasminogen activator and protease nexin in myogenesis. J. Cell. Biol. 91, 43a (1981).

41. B.W. Festoff, D. Hantaï, A. Rayford, and J.S. Rao, Plasminogen activators and their inhibitors in the neuromuscular system. II. Serpins and protease: serpin complex receptors increase during in vitro myogenesis. J. Cell. Physiol. 140, 272-279 (1990).

42. R. Meier, P. Spreyer, R. Ortmann, A. Harel, and D. Monard, Induction of glial-derived nexin after lesion of a peripheral nerve. Nature 342, 548-550 (1989).

43. M.J. Anderson, Nerve-induced remodelling of muscle basal lamina during synaptogenesis. J. Cell Biol. 102, 863-877 (1986).

44. M. Vigny, G.R. Martin, and G.R. Grotendorst, Interactions of asymmetric forms of acetyl-cholinesterase with basement membrane components. J. Biol. Chem. 258, 8794-8798 (1983).

45. D. Hantaï, J.S. Rao, and B.W. Festoff, Fibrinolytic enzymes and neuromuscular synapses. International ALS MND Update 20, 9-11 (1988).

46. D. Hantaï, J.S. Rao, R.R. Miller, and B.W. Festoff, Developmental appearance of thrombospondin in neonatal mouse skeletal muscle parallels elimination of polyneuronal innervation. Submitted J. Cell Biol. 1990.

47. J.S. Rao, D. Hantaï, and B.W. Festoff, Thrombospondin, but not collagen types I, III, IV, laminin or fibronectin increases in muscle of patients with amyotrophic lateral sclerosis. Submitted J. Clin. Invest. (1990).

48. E. Ruoshlahti, and M.D. Pierschbacher, New perspectives in cell adhesion: RGD and integrins. Science 238, 491-496 (1987).

49. M. Poberai, G. Sávay, and B. Csillik, Function-dependent proteinase activity in the neuromuscular synapse. Neurobiology 2, 1-7 (1972).

50. M. Poberai, and G. Sávay, Time course of proteolytic enzyme alterations in the motor end-plates after stimulation. Acta Histochem. (Jena) 57, 44-48 (1976).

51. E. Gutmann, J.A. Melichna, A. Herbrychová et al, Different changes in contractile and histochemical properties of reinnervated and regenerated slow soleus muscles of guinea pigs. Pflüegers Arch. 364, 191-194 (1976).

52. R.A.D. O'Brien, A.J.C. Östberg, and G. Vrbová, Observations on the elimination of polyneuronal innervation and the developing mammalian skeletal muscle. J. Cell Physiol. 282, 571-582 (1978).

53. R. Miskin, T.G. Eaton, and E. Reich, Plasminogen activator in chick embryo muscle cells: Induction of enzyme by RSV, PMA and retinoic acid. Cell 15, 1301-1312 (1978).

54. J. Hatzfeld, R. Miskin, and E. Reich, Acetylcholine receptor: effects of proteolysis on receptor metabolism. J. Cell Biol. 92, 176-182 (1982).

55. S.G. Younkin, C. Rosenstein, P.L. Collins, and T.L. Rosenberry, Cellular localization of the molecular forms of acetylcholinesterase in rat diaphragm. J. Biol. Chem. 257, 13630-13637 (1982).

56. K. Romstedt, R.L. Beach, and B.W. Festoff, Studies of acetylcholine receptor in clonal muscle cells: role of plasmin and effects of protease inhibitors. Muscle & Nerve 6, 283-290 (1983).

57. J.S. Rao, C.B. Kahler, J.B. Baker, and B.W. Festoff, Protease nexin I, a serpin, inhibits plasminogen-dependent degradation of muscle extracellular matrix. Muscle & Nerve 12, 640-646 (1989).

LOCALIZED EXTRACELLULAR PROTEOLYSIS MAY CONVEY INDUCTIVE SIGNALS BETWEEN NERVE AND MUSCLE CELLS DURING SYNAPTOGENESIS

M. JOHN ANDERSON, LAUREN E. SWENARCHUK, AND
SHASIKANT CHAMPANERIA

Department of Pharmacology and Therapeutics
The University of Calgary
Calgary, Alberta, Canada T2N 4N1

SHORT-RANGE CELLULAR INTERACTIONS

A variety of fundamental regulatory processes seems to be dependent upon short-range cellular interactions with immediate neighbors, and with the organized structural scaffolding of the extracellular matrix (ECM). These interactions contribute to the maintenance and repair of tissue organization in adult animals, and appear to be particularly important for conveying the positional information that is essential for the control of cellular migration, proliferation and differentiation during embryogenesis. While all of these processes can be influenced by exogenous humoral factors, and by the adhesive substances of the ECM, it is not yet clear how the relevant regulatory signals are generated and distinguished at appropriate sites of close cell contact *in vivo* (for reviews, see references 1-3). Our ignorance stems primarily from the fact that the most striking 'inductive' cellular interactions occur only transiently, at discrete sites within intact living organisms, a situation which restricts the use of conventional biochemical methods. If these limitations could be overcome by cytochemical techniques, therefore, it should become possible to determine the sequence of biochemical events that are initiated during inductive interactions between individual cell-pairs, and perhaps to identify the key molecular events that are most directly affected by close cell contact.

One relatively simple inductive interaction has, in fact, been found to be particularly suited to cytochemical analysis. In this process, contact between embryonic neurons and their target cells results in the focal assembly of organized synaptic structures, thereby directing the final stages of cellular differentiation in the nervous system. Due to the unusual size and abundance of skeletal muscle fibers, this class of cellular interaction has been most intensively studied during the establishment of the vertebrate neuromuscular junction. In fact, motor neurites have been shown to interact with skeletal myocytes in culture, eventually forming morphologically and physiologically intact neuromuscular junctions at their sites of contact.[4-7] In this system, therefore, specific regulatory signals which are generated only by an appropriately matched cell-cell contact control the several dynamic processes that assemble synaptic structures in both nerve and muscle cells.

It has been found, for example, that the motor neurite directs an orderly redistribution of mobile acetylcholine receptors (AChR) throughout the sarcolemma, leading both their aggregation within the developing postsynaptic membrane, and to the disappearance of any pre-existing extrajunctional AChR clusters.[8,9] Even though this action is independent of the relatively well understood electrophysiological phenomena that are unique to excitable cells,[10] it seems to be restricted exclusively to the cholinergic neurites which are destined to innervate skeletal muscle fibers.[5,7] Due to its cell-specificity, and to the striking effects it has on muscle surface organization, nerve-muscle interaction offers a particularly accessible model system in which to analyze the molecular mechanisms of short-range inductive signaling between developing cells (for reviews, see references 11,12).

There have, in fact, been several attempts to identify humoral agents, which function as 'paracrine' neural signals that evoke postsynaptic differentiation. Several substances have thus been proposed to act as 'inductive factors', based upon their common ability to cause increases in the number of AChR aggregates present on the surface of embryonic muscle cells, when they are used to supplement culture medium.[13-24] Unfortunately, we still have no evidence that synapse induction is mediated by a conventional 'paracrine' mechanism, or that an inductive neural factor would evoke widespread AChR clustering, following its non-physiological bath-application. Furthermore, these agents differ chemically (ranging almost 5000-fold in molecular weight), and include a number of ubiquitous vitamins, metabolic cofactors and structural proteins, that seem surprising candidates for a quasi-hormonal role in the control of synaptic differentiation.[20,21,24] It thus remains unclear whether these substances will eventually be seen as functional elements in an orderly mechanism that directs synaptic differentiation throughout the nervous system, or simply discounted as artifacts of an ambiguous bioassay.

There is another striking experimental phenomenon, however, which seems more revealing than the formation of discrete surface specializations on denervated muscle cells, or its dependence upon the vitamins and adhesion-promoting substances present in culture medium. The topical application on the muscle surface of (positively-charged) polystyrene microspheres has been found to duplicate precisely the complex changes in muscle surface organization that are evoked only by the motor neurite under physiological conditions.[25-29] Similar polymer microbeads also induce the focal development of presynaptic organelles in developing neurites.[27,30] Since these simple 'inductive' agents function in the absence of the complementary cell-type (and its biosynthetic products), their potent biological activity is unlikely to result from (a) the binding or release of 'paracrine' regulatory factors, (b) the transmission of regulatory substances through gap-junctions, or (c) direct molecular interactions with stereo-specific surface receptors. The inductive properties of polymer microbeads are particularly significant, therefore, since they demonstrate that the formation of conventional ligand-receptor complexes (between substances from nerve and muscle) is not obligatory for the induction of 'synaptic' differentiation in either cell. They are experimental anomalies, however, because no established mechanism of intercellular communication can yet explain why a regulatory cellular interaction should be mimicked so effectively by contact with a positively charged polymer microbead.

Our recent efforts have been directed at resolving this puzzling situation, through a systematic analysis of the processes through which nerve-muscle interaction regulates the *de novo* formation of a specialized basal lamina at the developing synapse. This strategy was initially based upon earlier findings that the accumulation of synaptic components occurs preferentially near surviving rudiments of the 'synaptic' basal lamina, during nerve and muscle regeneration in adult frogs.[31,32] It thus seemed reasonable to suppose that the mechanisms of synapse induction might be revealed in the process through which the synaptic basal lamina first becomes assembled during the development of embryonic neuromuscular junctions in cell culture.

One possible explanation for the apparent 'inductive' properties of the adult junctional basal lamina, for example, is that this structure contains stable 'paracrine' factor(s) that modulate corresponding surface receptors on nerve and muscle cells. The secretion (and anchorage) of these 'inductive' ECM substances could then direct both the formation and maintenance of chemically specialized pre- and postsynaptic structures at the site of cell contact.[22] Alternatively, the junctional basal lamina could simply be an extracellular extension of the postsynaptic apparatus, and be induced by another distinct form of 'inductive' signal. In this latter case, the junctional basal lamina could still contribute to synapse formation and regeneration, but would do so as a 'second messenger' for another biochemical process or substance.

Since the basal lamina that covers the adult muscle fiber shows no obvious morphological specialization at the neuromuscular junction, we used hybridoma methods to generate monoclonal antibodies to one of the several molecular markers which distinguish the 'synaptic' basal lamina. This is a (Mr $\sim$700,000) heparan sulfate proteoglycan (HSPG), that is concentrated at least 5-fold at the frog neuromuscular junction.[33] The antibodies could then be used to study (a) how nerve-muscle interaction causes the formation of a specialized (high HSPG) basal lamina during synaptogenesis, and (b) how basal lamina deposition correlates with the accumulation of other established synaptic components.

In our initial experiments, we found several significant differences in phenomenology between synapse formation and regeneration, which together seem inconsistent with the idea that these processes are mediated simply by the deposition of 'inductive factors' into the synaptic basal lamina. The organized (and chemically specialized) junctional basal lamina was induced by nerve-muscle contact, for example, and appeared coincidently with other postsynaptic structures. These included accumulations of AChR, and an organized 'postsynaptic density' (a thickening of the muscle cell surface that reflects the assembly of an organized complex of membrane and cytoskeletal proteins).[11,34] In fact, the developing synaptic basal lamina was cospatial with the 'postsynaptic density', and extended well beyond the immediate site of nerve-contact. It thus appeared merely to be an extracellular extension of the muscle fiber's synaptic apparatus (see Figure 1).

When developing synapses were denervated to eliminate any ongoing 'inductive' signals provided by the nerve cell, the postsynaptic accumulation of AChR dispersed within a day. This rapid disappearance of junctional AChR occurred despite the survival of the junctional accumulation of basal lamina HSPG.[34] This contrasts with the striking stability of junctional AChR in denervated adult frog muscle, a key phenomenon which ultimately led to the hypothesis that the (stable) basal lamina acts as an 'inductive' agent during synapse formation.[31,32] If the embryonic induction of postsynaptic differentiation is mediated simply by the deposition of stable basal lamina components, which retain their ability to direct synaptic specialization even in the

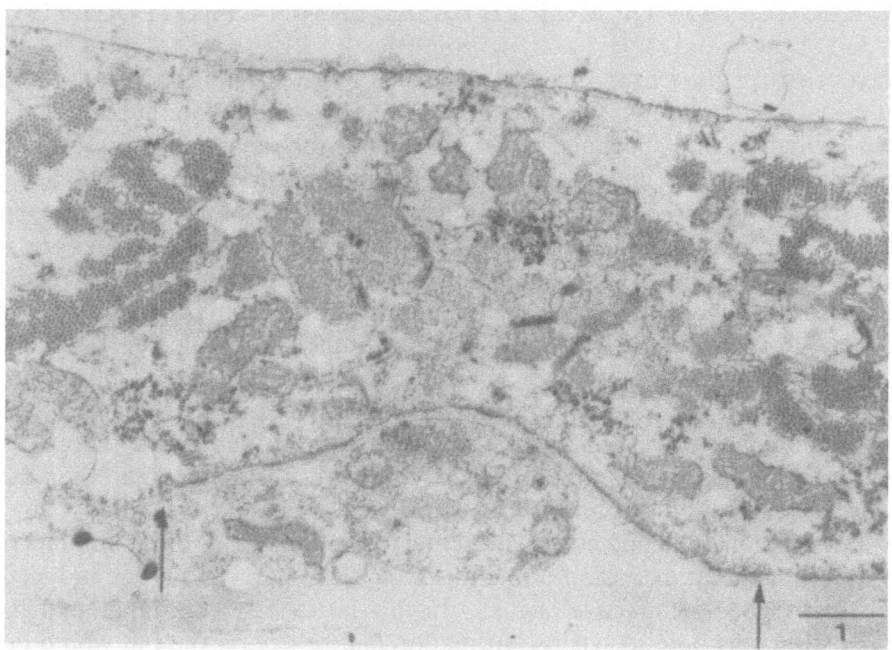

Figure 1. *Neuromuscular junction developed in cell culture. Xenopus laevis* neurons and muscle cells were co-cultured for two days. Presumptive neuromuscular junctions were selected on the basis of organized 'synaptic' staining with α-bungarotoxin, and submitted to ultrastructural reconstruction after semi-serial sectioning (see reference 5). Note the full complement of synaptic organelles in both the nerve and muscle cell, and the broad extent of the "post-synaptic density" and synaptic basal lamina (arrows) over the muscle surface. Similar views in many such sections suggest that the synaptic basal lamina is simply an extracellular extension of the muscle fiber.

adult,[22] it is not clear why these constituents fail to maintain similar specializations for more than a few hours on embryonic muscle fibers. This result shows, instead, that the formation and maintenance of postsynaptic differentiation is initially dependent upon a labile (ongoing) 'inductive' action of the nerve terminal, and does not correlate with the presence of stable basal lamina components like HSPG.

Further analysis of the nerve-induced deposition of junctional HSPG has continued to strengthen this impression, that the junctional basal lamina is simply part of the muscle fiber's postsynaptic apparatus. These recent experiments have employed species-specific monoclonal antibodies, to identify the cellular origin of the HSPG accumulations at chimeric frog-salamander junctions developing in cell culture (Swenarchuk, Champaneria and Anderson, submitted for publication). Our results demonstrate that junctional HSPG is almost exclusively of muscle origin (Figure 2), like the corresponding accumulation of junctional acetylcholinesterase.[35] It is already clear from other work that denervated muscle cells can synthesize all of the known components of the synaptic basal lamina, and assemble them in organized quasi-synaptic surface specializations.[33,36-39] This recent evidence is significant, therefore, since it shows that the chemical specialization of the junctional basal lamina can be accounted for most simply, by the demonstrated ability of the nerve-muscle contact to control the site of deposition of muscle basal lamina precursors.

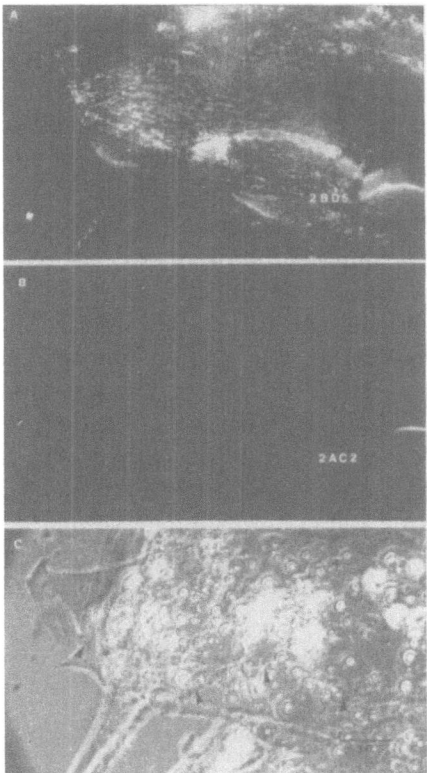

Figure 2. *Post-synaptic origin of basal lamina proteoglycan at chimeric frog-salamander neuromuscular junction in culture.* Synapses were formed between neurons of *Rana Pipiens* and muscle cells of *Ambystoma mexicanum* (and vice versa - not shown). These were stained with monoclonal antibody 2BD5, which react with both species-forms of HSPG (A), and counterstained with an anuran-specific HSPG antibody 2AC2 that was labeled with a contrasting fluorochrome (B). Note that the junctional accumulation of HSPG (compare A with phase-contrast image of the cell in (C) was almost exclusively derived from the muscle cell, like that over the remainder of the muscle surface. Bar in C is 30 μm

Taking all current evidence into consideration, therefore, we find no experimental support for the idea that synaptic differentiation is dependent upon the neural secretion of paracrine (basal lamina or other) substances. Instead, we have compelling experimental evidence that paracrine (nerve and muscle) factors are irrelevant to the experimental mimicry of synapse induction by polymer microbeads (see above). It thus seems appropriate to consider the possibility that synapse induction is mediated by another molecular mechanism. In fact, our recent results support an unexpected alternative that is consistent with the data on synapse development and regeneration, and can rationalize the (otherwise) anomalous 'inductive' properties of polymer microbeads.

NEURAL REMODELING OF THE MUSCLE BASAL LAMINA

We have found that innervation by motor neurites has another striking effect on the biochemical organization of the muscle basal lamina. Before the elaboration of synaptic structures, the developing motor neurite first eliminates any endogenous accumulations of muscle HSPG from the presumptive synaptic surface of the muscle fiber. Figure 3 shows an example of this phenomenon on an innervated muscle cell in culture, whose basal lamina was labeled with anti-HSPG antibody before the establishment of nerve-muscle contact. The developing nerve fiber has caused a conspicuous local reduction in HSPG staining intensity along the path of cell contact. This focal elimination of muscle HSPG deserves particular consideration, because it is both specific to developing junctions, and the earliest detectable neural perturbation of muscle surface organization during synaptogenesis.[40]

To determine whether the neural removal of HSPG could result from a local increase in the level of pericellular proteolysis, we also examined the 'subtractive' actions of developing synapses on adsorbed ECM-derivatives, attached to the adjacent culture dish. These experiments compared the elimination of ECM-derivatives at developing neuromuscular junctions with that at morphologically equivalent muscle contacts with other neurites, which do not undergo synapse differentiation. Our results show that there is an enhanced elimination of gelatin-derivatives along the path of the nerve fiber at the developing synapse (Figure 4). This can, moreover, be distinguished from a corresponding elimination of laminin and fibronectin, which (a) is equally prevalent at synaptic and non-synaptic nerve-muscle contacts (Table 1), and (b) also occurs at a variety of other cell substratum contacts in culture (not shown, see references 41-45). These results show (a) that nerve-muscle interaction results in

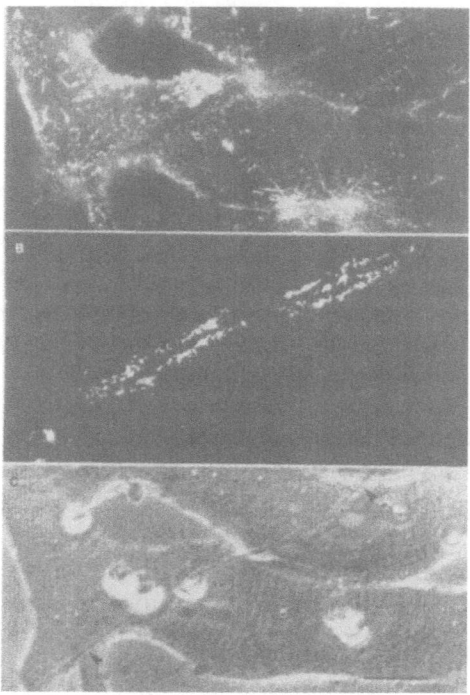

Figure 3. *Neural elimination of muscle surface proteoglycan during synapse development.* Cultures of *Xenopus* muscle cells were decorated with antiproteoglycan antibody, immediately before adding dissociated nerve cells. After synapses (see above) had developed, 2 d later, the cultures were stained with fluorescent goat anti-mouse IgG, and counterstained with α-bungarotoxin to allow identification of synaptic sites (see above). Synaptic accumulations of AChR (B) were then seen to be associated preferentially with local regions of HSPG-elimination (compare A with the path of the nerve shown by arrows in the phase-contrast image in (C). Similar removal was not seen at other nerve-muscle or nerve-ECM contact sites (see 1). The bar in C is 20 μm.

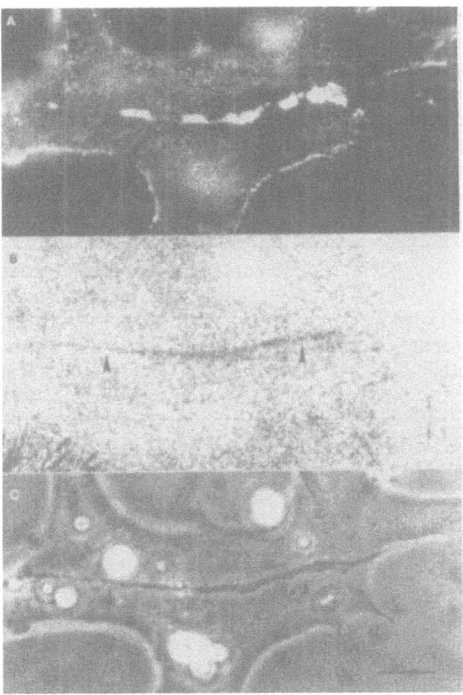

Figure 4. *Localized removal of fluorescent gelatin at a developing neuromuscular junction.* Nerve-muscle cultures were grown on coverslips coated with Texas Red-labeled gelatin. Synaptic sites were identified 2 d later, after counterstaining with Lucifer Yellow-labeled antibody to basal lamina HSPG. Junctional accumulations of HSPG (A) were preferentially associated with sites of enhanced substrate-degradation (B) along the path of nerve-substratum contact (see phase-contrast image in C). These observations were quantitated under different surface-coating conditions, and are summarized in Table 1. Bar in C: 20 μm. ECM substrates during synaptogenesis, and (b) that analogous changes in proteinase activity occur on both the nerve and muscle surfaces.

a selective increase in the degradation of some ECM substrates during synaptogenesis, and (b) that analogous changes in proteinase activity occur on both the nerve and muscle surfaces.

This conclusion raises obvious questions about the function of a focal increase in proteolytic activity that is restricted to developing synapses, independent of synaptic transmission, and begins before any synaptic differentiation becomes apparent.[40] A local increase in degradative activity could, for example, be one of the several secondary processes (like AChR redistribution and basal lamina deposition, see above) which together result in the assembly of the synaptic apparatus. Alternatively, proteolytic activity could transmit the 'inductive' signals that direct synaptic differentiation. Localized extracellular proteolysis could, for example, modulate receptor function on both nerve and muscle cells, either by the liberation of humoral factors sequestered on the adjacent cell surfaces, or by the disruption of stable ligand-complexes (see below). This latter possibility seemed particularly promising to us, because it offered a means of rationalizing the anomalous 'inductive' properties of polystyrene microbeads.

Table 1. Preferential elimination of gelatin-derivatives adjacent to synaptic nerve-muscle contacts. Etching of the fluorescent substratum was compared at synaptic and non-synaptic nerve-muscle contacts, identified by counterstaining with fluorescent derivatives of α-bungarotoxin or anti-proteoglycan IgG. Similar comparisons were made under a variety of surface-coating conditions with each matrix substrate. Approximately 50 sites of nerve-muscle contact were scored in each culture, 56 ± 13 % of which were synaptic. Results are presented as (a) percentages of the identified cell contacts in each culture that showed detectable substrate-elimination (see Figure 4), and (b) the corresponding ratios between synaptic and non-synaptic sites, within individual cultures (± standard deviation; n = number of cultures scored).

Note the distinct preference of developing cholinergic junctions for gelatin-derivatives, and the absence of substrate-selectivity on both laminin and fibronectin, even when different surface-coating conditions substantially changed the fraction of sites displaying substrate-etching.

| Fluorescent substrate | Percentage nerve-muscle contacts that etch substrate | | |
	synaptic	non synaptic	Ratio synaptic to non-synaptic
Laminin 10-40μg/ml (n=6)	89 ± 14	86 ± 8	1.04 ± 0.14
Laminin 80-200μg/ml (n=5)	47 ± 13	54 ± 15	0.87 ± 0.18
Fibronectin 100-200μg/ml (n=4)	93 ± 5	94 ± 6	0.99 ± 0.07
Fibronectin 400-500μg/ml (n=5)	99 ± 3	98 ± 2	1.00 ± 0.02
Gelatin 100-400μg/ml (n=7)	93 ± 6	13 ± 6	8.64 ± 3.75

DESORPTIVE ACTIONS OF INDUCTIVE POLYMER MICROBEADS

We explored the interactive properties of several microbead preparations, to determine which might correlate with the striking ability of some polymer beads to mimic the cell-specific 'inductive' signals that regulate synaptogenesis. We thereby determined, for example, that native polystyrene microbeads are intrinsically bioactive under low-serum culture conditions, and can evoke both the accumulation of AChR and the formation of a specialized quasi-synaptic muscle basal lamina (see Figure 5). This 'inductive' activity is easily blocked by serum proteins, however, in a quenching-reaction that is partially inhibited by pretreatment of the beads with polycations such as polylysine (data not shown). Such experiments showed, therefore, that the 'inductive' ability of polymer microbeads is unrelated to their surface charge, and instead correlates with the well established (saturable) protein-adsorbing properties of polystyrene.

Since polystyrene beads mimic inductive signals on muscle in the virtual absence of soluble (or nerve-derived) proteins in the culture medium (see above), their biological activity must somehow be dependent upon an ability to bind other proteins, provided by the adjacent muscle cell. Control experiments showed, however, that 'inductive' beads did not evoke a passive aggregation of adjacent ECM proteins like

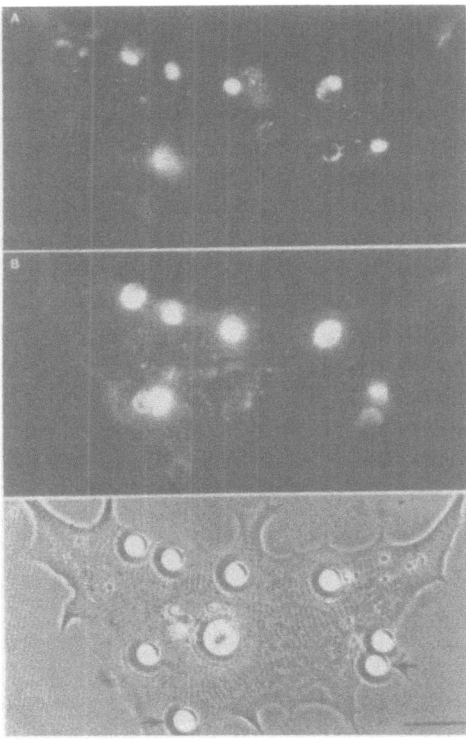

Figure 5. *Inductive properties of native polystyrene microbeads.* Polystyrene microbeads (diameter ~ 10 μm) were allowed to interact with *Xenopus* muscle cells for 2 d under serum-free conditions. The cells were then stained with monoclonal anti-HSPG antibodies (see above) and counterstained with α-bungarotoxin for AChR. Note that polystyrene beads induce focal accumulations of both HSPG (A) and AChR (B) at sites of bead-muscle contact (see phase contrast image in C). They also eliminate ectopic accumulations of AChR, but not those of HSPG (compare A,B). Bar in C represents 20 μm.

HSPG (not shown), just as they do not seem to evoke a passive aggregation of mobile AChR.[25] Instead, both of these muscle surface components seem to accumulate actively, via the endogenous physiological processes that normally generate both synaptic and ectopic surface specializations.[4,8,40] This leaves us with two other simple possibilities. Each adsorptive microbeads trap secreted muscle products at the sites of bead-contact, or they act as competitive diffusion-traps, that cause a local removal of weakly bound proteins from the adjacent cell surface.

This latter desorptive property of polystyrene beads is illustrated in Figure 6, which shows the effect of bead-substratum contact on adsorbed films of fluorescent ECM proteins bound to glass coverslips (similar to those used above to reveal local degradative activity). Polystyrene beads cause a gradual removal of the fluorescent protein at the site of closest contact between the flat coverslip and the spherical microbead. This becomes visible (after ~ 20 min in our experiments) as a small dark spot of reduced fluorescence near the bead center, when viewed from below in the light microscope (Figure 6A,B). Similar desorptive actions were not observed, for example, with polyacrylamide beads (Figure 6C,D), nor after quenching surface binding-sites with soluble serum proteins (not shown). Significantly, this "passive" desorptive property (observed under cell-free conditions) correlates closely with the ability of the same bead-preparation to mimic inductive signalling between nerve and muscle (see above).

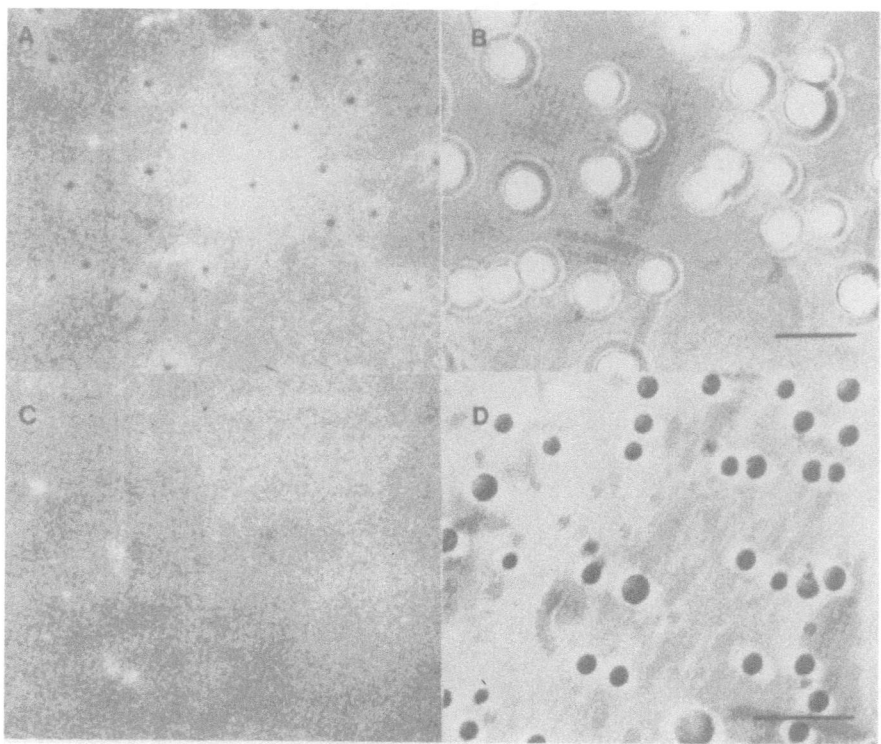

Figure 6. *Desorptive actions of polymer microbeads: correlation with their inductive effect on AChR distribution.* Coverslips were coated with fluorescent gelatin, laminin or fibronectin (as above). Polystyrene beads (see phase-contrast image in B) removed fluorescent proteins at sites of bead-substratum contact, producing foci of reduced fluorescence intensity (compare arrows in A and B). This action was duplicated neither by protein-quenched polystyrene beads (not shown), nor by polyacrylamide beads (see fluorescence and phase-contrast images in C,D). Only desorptive polystyrene beads mimicked the abilities of motor neurites to remove matrix proteins from adjacent surfaces, and to induce the redistribution of fluorescently stained AChR on muscle (see above). Bars in B,D: 20 μm.

Adsorptive microbeads could thus assume 'inductive' properties either (a) because they remove weakly-bound protein ligands from their existing receptor sites on an adjacent cell surface, or (b) because they adsorb and accumulate secreted cell products, which then act like autocrine 'inductive factors' on the cell which made them. Neither of these alternatives resembles a conventional 'paracrine' signalling mechanism, where only the appropriate target cell should respond to the secreted regulatory gland. Therefore, unless we are prepared to accept the *ad hoc* alternative that nerve and muscle cells both synthesize and secrete 'autocrine' synapse-inducing factor(s), we must consider the possibility that the local removal of cell surface proteins is somehow perceived by both the neuron and muscle fiber as a close analog of the physiological inductive signal.

This would have considerable conceptual utility, of course, since we have already shown (a) that the earliest detectable change in surface organization during synaptogenesis is the (presumably degradative) elimination of an endogenous muscle

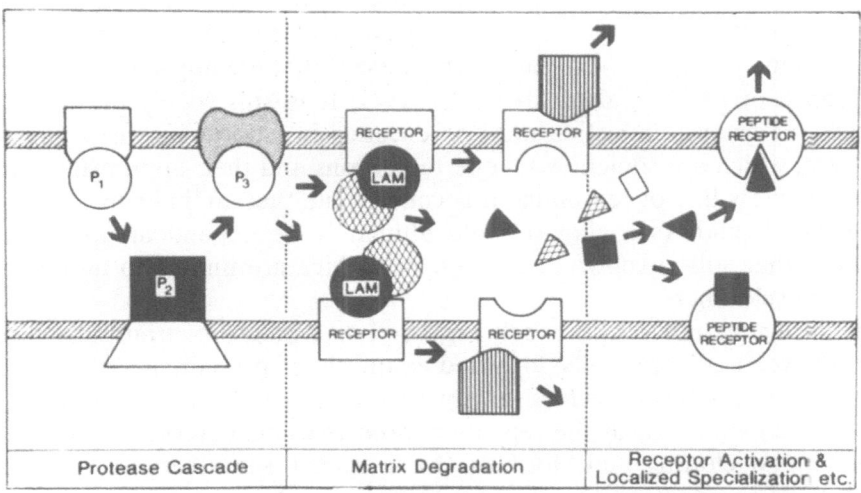

Figure 7. *Diagrammatic representation of molecular mechanisms through which extracellular proteinase-activation could lead indirectly to the modulation of signal-transduction systems on both the nerve and muscle surfaces.* In this hypothetical process, interactions between surface-bound proteinases, zymogens (and inhibitors) could activate latent proteinases at sites of close cell contact, leading to the focal degradation of ECM and other surface components. This could then result in (a) the dissociation of degraded ECM molecules from their receptor sites, and (b) the release of proteolytic fragments and sequestered regulatory peptides. These latter events should indirectly modulate the function of both constitutive ligand-receptor complexes (such as those of 'integrins') and conventional peptide (such as 'growth factors') receptors on adjacent cell surfaces.

ECM protein, and (b) that similar site-specific degradative mechanisms also become active on the adjacent nerve-surface (see above). Taken together, therefore, our results demonstrate that cholinergic neurites, skeletal muscle cells and 'inductive' polymer microbeads all share one distinctive property, which is not available to their biologically inert homologs. Their inductive properties thus seem likely to result from this common ability, to cause the removal of constitutive surface components at sites of close cell contact. We therefore propose that contact with polystyrene microbeads conveys quasi-inductive signals to cultured *Xenopus* nerve and muscle cells, because it mimics the 'subtractive' consequence of the relevant inductive process. Synapse induction could thus be mediated by localized increases in the rate of pericellular proteolysis, which evoke the deposition of the synaptic basal lamina, and the elaboration of other pre- and postsynaptic organelles.

ALTERNATIVE MODES OF INDUCTIVE CELL COMMUNICATION

In the absence of direct evidence about its molecular mechanism(s), it has long been presumed that short-range cellular communication is routinely mediated by the action of more or less cell-specific paracrine (secreted or surface-bound) agonists, which regulate corresponding receptor systems on the adjacent 'target cells' (see above). A central problem in evaluating the contribution of any physiological process to 'inductive' cellular communication, is the need to account for the remarkable cellular specificity of 'inductive' events such as synapse formation. Given the extraordinary variety of synaptic contacts present in many vertebrates, for example, this is a difficult issue for all hypothetical inductive mechanisms, particularly for those that must employ unique 'recognition substances' that identify each cell-pair. Due to the absence of realistic alternatives, however, it is still widely presumed that the specificity of synapse induction must be encoded in (still unknown) agonist-receptor complexes, which are somehow unique to neurons and their appropriate target cells. If we follow this line of reasoning, it is conceivable that an 'inductive' elimination of cell surface ligands (see above) could ultimately be dependent upon the primary action of other still-unknown agonist factors, which are unique to motoneurons and skeletal muscle fibers.

While it may not yet be widely regarded as a means of transmitting regulatory signals between adjacent cells, localized extracellular proteolysis, is, nevertheless, an established biochemical mechanism (for reviews, see references 46,48). It has been studied most extensively as the regulatory process which determines the site and rate of both blood clotting and clot-lysis (for a review see reference 49), but is also implicated in other physiological and pathological processes, such as cell growth, cell migration and metastatic invasion (see also references 47,50,51). In fact, a variety of cell-types, including the developing neurite,[52,53] releases plasminogen-activators and other proteinase zymogens in culture. These zymogens appear to become concentrated and activated in regions of enhanced pericellular degradation, such as fibroblast adhesion-sites.[42-44,54,55] The molecular mechanisms which determine the local rates of extracellular proteolysis under physiological conditions still remain only partly understood, however, and may turn out to be as elaborate as those which are already known to control fibrinogenesis and fibrinolysis.

It is already clear from this body of work, nevertheless, that local increases in the rate of extracellular proteolysis commonly occur under a unique form of biochemical regulation, which is not dependent upon conventional agonist-receptor interactions (for reviews, see references 46-48). As illustrated in the formation and breakdown of fibrin, extracellular proteolysis is instead controlled by a temporal sequence of zymogen-activation reactions, which is modulated by a variety of complementary proteinases, inhibitors and substrates.[49] Based upon these precedents, therefore, proteolytic activity at the developing synapse is also likely to be regulated by interactions between complementary proteinases, inhibitors and substrates. In fact, the most parsimonious way to explain our present results would be to propose that neural proteinases interact reciprocally with corresponding muscle surface zymogens (and/or inhibitors), thereby initiating a localized cascade-reaction which results in (a) the activation of zymogens on both cell surfaces, and (b) a corresponding breakdown and elimination of adjacent ECM constituents.

This possibility has significant implications, since zymogen activation-cascades are self-regulating processes, and could thus be initiated by the close proximity of complementary proteinases, brought about simply by the establishment of cell-cell contact. Our results do not, therefore, require the proposition of a unique (but hypothetical) agonist-receptor complex, merely to account for the site-specificity of the degradative actions we have observed at the developing neuromuscular junction. In fact, based upon current evidence (see below), at least two families of surface receptors could respond indirectly to the degradative consequences of localized extracellular proteolysis. Proteinase-activation may thus function as an 'inductive' signal in its own right. This would obviate any need to postulate the existence of unique 'inductive substances,' simply to account for the cellular specificity of synapse formation.

POSSIBLE TRANSDUCTION OF PERICELLULAR PROTEOLYSIS BY ECM RECEPTORS.

It is already known, for example, that the proteolytic removal of ECM substances from the cell surface disrupts cell adhesion, and induces a striking rearrangement of cytoskeletal structures. These properties are (a) routinely exploited during the preparation of cell cultures from intact tissues, and (b) can both be mimicked by the action of a single monoclonal antibody that displaces fibronectin from its receptor-sites on the cell surface.[56,57] Replacement of the degraded fibronectin reverses these changes in cytoskeletal organization, and promotes the formation of new cell-substratum adhesion sites. These opposing actions of proteinases and fibronectin thus seem to be mediated by the same transmembrane 'receptors,' whose ligands include the adhesive proteins of the ECM.

The best known member of this receptor family, now called 'integrins,' has several unusual biochemical properties which distinguish it from receptors that interact with conventional hormone-like agonists. Integrin is known, for example, to (a) be a relatively abundant membrane protein, (b) establish stable but moderate-avidity complexes with a number of related ECM components (such as fibronectin, vitronectin and laminin), and (c) interact also with cytoplasmic tyrosine-kinases and cytoskeletal proteins such as talin (for reviews, see references 56,58). Most significantly, however, even though integrin can be studied as a 'fibronectin receptor', it does not simply make transient complexes that inform the cell of local changes in the concentration of its soluble ligand, like conventional hormone or 'growth factor' receptors. Instead, it seems to be present at cell-adhesion sites as one component of a constitutive transmembrane complex, which reversibly joins the cytoskeletons and ECM of a remarkable variety of cell types.[59,60]

Under experimental conditions, therefore, these constitutive complexes seem to function as both ECM and proteolysis receptors, which respond reversibly to the association and dissociation of their ligands in the ECM, thereby conveying directive signals to and from the cytoplasmic enzyme systems that determine rearrangements of membrane and cytoskeletal proteins. Localized proteolysis of endogenous pericellular matrices (such as basal laminae) may thus be able to transmit analogous regulatory signals under physiological conditions. While the results of this proteolytic process might appear similar to those afforded by the secretion of more conventional paracrine agents, the underlying chemical signal would be (a) more comparable to the local elimination of an endogenous inhibitor, and (b) mimicked by desorptive agents like polystyrene microbeads.

INDIRECT PROTEOLYTIC REGULATION OF GROWTH FACTOR RECEPTORS

While ECM receptors currently seem the most likely recipients of 'substractive' inductive signals at sites of nerve-muscle interaction, they are not the only membrane transduction-system that could convey regulatory signals in response to localized pericellular proteolysis. There is, for example, increasing evidence that other well established agonist-receptors could also be modulated indirectly by proteolytic interactions on adjacent cell surfaces. These receptors react with peptide 'growth factors' such as the epidermal growth factor (EGF), transforming growth factor (TGFβ), and fibroblast growth factor (FGF - for reviews, see references 46,61,62).

The most rigorous means of identifying individual molecular constituents, which contribute directly to short-range inductive cellular interactions, currently involves the cloning and sequencing of mutant gene-products that disrupt the differentiation of restricted cell populations. Several such mutations have already been characterized, in what appear to be short-range 'inductive' interactions between adjacent *Drosophila* cells. Molecular analysis of the relevant cDNA clones has, however, implicated a number of surprisingly ubiquitous regulatory substances. In fact, the developmental lesions evoked by each of these mutations commonly appear in only a limited subset of the cells that actually express the mutant gene-product.[63,64]

Several of these 'homeotic' loci seem particularly revealing. It is known, for example, that a (presumptive) plasma membrane protein-kinase, related in amino-acid sequence to the vertebrate EGF receptor, is directly implicated in the *Drosophila* "Sevenless" mutation, suggesting a role for this well-established signal-transduction system in cell contact-mediated induction of the insect photoreceptor (for a review, see reference 64). Furthermore, analogous point-mutations in separate (but similar) EGF-like domains, arranged linearly within a individual 'Notch' gene-product, can each prevent the differentiation of distinct sub-populations of (Notch-expressing) embryonic nerve cells.[63,65] Recent evidence also suggests that other common 'growth factors,' such as FGF and TGFβ, may act like paracrine inductive substances ("morphogens") during early amphibian morphogenesis.[61,62] It is clear, therefore, that substances which are closely related to peptide 'growth factors' (or their precursors and receptors) seem to contribute to the transmission of a striking variety of specific inductive signals during embryogenesis (for reviews, see references 63,65).

While it is well established that these classical 'growth factors' modulate the activity of cytoplasmic 'second messengers' under experimental conditions, ultimately regulating biosynthetic processes and structural reorganizations which are of particular importance during development (see reference 66 for a review), it is also clear that they do not function as conventional hormones *in vivo*. The classical form of EGF, for example, appears to be secreted by exocrine cells (notably in the submandibular glands of the male mouse), but has no obvious access to most ECF receptors, which are present on a wide variety of cells.[67,68] It has recently become evident, moreover, that common 'growth factors' (notably TGFβ and FGF) may also be ubiquitous constituents of the cell surface and ECM.[61,69] Taken together, these findings lead us into another puzzling situation, where specific 'inductive' signals seem to be transmitted by ubiquitous regulatory ligands that are present constitutively on the cell surface and ECM.

It is thus interesting to note that a proteolytic 'inductive' mechanism (similar to that suggested by our recent results - see above), could also resolve this puzzle, since it should lead to a site-specific liberation of any humoral regulatory ligands sequestered (by proteinase sensitive agents) on the cell surface, or in the ECM. Analogous proteolytic signals at other sites of close cell-contact could thus mediate the processing of active peptide fragments from membrane-associated EGF-precursors (such as Notch), and the liberation of FGF and TGFβ from the degraded ECM. Furthermore, since these 'growth factors' would be liberated by site-specific enzymatic processes, they would be acting merely as "second messengers" for the primary proteolytic inductive signal. In this role, peptide 'inductive factors' have no greater need for precise cell-specificity than conventional peptide neurotransmitters, which also convey the effects of another more specific biological signal (localized synaptic activity).

REGULATION OF CELLULAR ADHESION, PROLIFERATION AND DIFFERENTIATION

In view of this potential to transmit (what have usually been presumed to be) 'autocrine' and 'paracrine' regulatory signals, it is also instructive to reconsider the observation that regions of enhanced pericellular proteolysis seem to occur routinely at sites of adhesive interaction between neighboring cells, or with the ECM (see above). This correspondence between local adhesion-sites and the enhanced degradation of the ECM substances which mediate cell-adhesion is superficially paradoxical. It thus seems plausible that this enhanced local degradation of the ECM actually reflects only a single component process in a higher-order mechanism which controls adhesive cell-cell and cell-matrix interactions.

It is conceivable, for example, that the strength of local cell-adhesion may be modulated by the relative rates of several dynamic but antagonistic physiological processes. Proteolysis of the ECM obviously offers an appropriate biochemical means for weakening the strength of localized adhesive interactions on the cell surface, as does the modulation of integrin ligand-affinity by cytoplasmic enzymes. Likewise, the local deposition of recently synthesized (adhesive) matrix substances should serve to enhance the strength of cell-adhesion, and to foster the assembly of local attachment-sites with the cytoskeleton. Spatial and temporal differences in the local rates of these dynamic processes could thus alternately foster and oppose adhesive molecular interactions between the cell surface and its immediate environment. While it has yet to be demonstrated experimentally, some mechanism along these lines is clearly required to explain the transient formation and detachment of highly localized adhesion-sites during cell migration.

In this context, it is useful to note that nerve-muscle interaction leads first to the local removal of muscle HSPG, and then to its redeposition as part of the basal lamina that joins these cells at their specialized adhesion site, the neuromuscular junction.[35,40] While this process is restricted to contacts between the motor neuron and its appropriate target-cell, it has obvious parallels at other sites of adhesive interaction between cells and their environment, such as the fibroblast adhesion-plaque (for reviews see references 48,59). Furthermore, the ECM constituents (and their receptors) which mediate cell adhesion, and the components of the membrane signal-transduction systems that could respond to the biochemical consequences of localized extracellular proteolysis (see above), are all ubiquitous cellular components. The proteolytic mechanism which seems to mediate synapse development (see above) may thus represent one example of a generic means of conferring site-specificity into the modulation of 'inductive' and 'adhesive' interactions between cells and their immediate surroundings.

The particular value of pericellular proteolysis in such regulatory processes would be expected to derive from the separation it could allow between the generation and transduction of regulatory signals at contacts with adjacent cells, and with ECM. As illustrated in the proteolytic cascades of the clotting system (for reviews see references 47,49), the regulation of proteinase-activity is encoded in a temporal sequence of zymogen-activation reactions. By incorporating different lineage-specific enzymes and/or inhibitors, such activation-sequences may be subject to a host of subtle permutations that could allow proteolytic degradation to have unique biochemical properties at different sites of cell interaction (as it does at different sites of nerve-muscle contact - see above).

This could have important implications for many regulatory processes that are dependent upon specific site-recognition, since permutations in the reaction-sequence of zymogen-activation cascades could permit a small family of regulatory gene-products to encode the generation of a vast number of unique proteolytic signals, which could then be used to specify different sites of cell contact. Once initiated, however, proteolytic degradation could (a) modulate cell adhesion (and the signal-transduction systems coupled to 'adhesion-receptors') on each of the interacting cells, and (b) affect any neighbors that possess receptors for the humoral substances that are liberated by proteolysis. These properties would be of particular significance, since they could allow zymogen activation-cascades to confer site-specificity into the control of cell migration, proliferation and differentiation.

SUMMARY

To determine whether a localized proteolytic remodeling of the muscle surface could mediate the neural induction of synaptic differentiation, we examined the degradative actions of developing *Xenopus* nerve and muscle cells on thin films of extracellular matrix substances, bound to the glass surface of a culture chamber. Connective tissue cells, skeletal myocytes and paths of nerve-muscle contact were all found to remove fluorescent derivatives of fibronectin and laminin from the substratum, immediately adjacent to regions of close cell-surface contact. In addition to this, however, developing synapses also displayed a uniquely enhanced rate of gelatin-elimination, when compared with non-innervated myocytes or non-synaptic nerve-muscle contacts. This localized degradative process seems to transmit important directive signals, since an ability to eliminate bound matrix proteins from adjacent substrata is the only distinctive property we find to be shared by both 'inductive' polymer microbeads and motor neurites, but missing from their biologically inactive homologs.

These observations are most consistent with the ideas that (a) extracellular proteinases become activated on developing motor neurites and muscle cells, adjacent to their sites of contract, and (b) the resulting removal of constitutive cell surface components transmits inductive positional signals during synapse formation. Our findings thus raise the possibility that site-specific proteinase activation-cascades may have ubiquitous directive roles in the coordination of cellular migration and differentiation, modulating both the formation of adhesive cell-contacts and the release of regulatory peptide factors at appropriate sites of cell-cell and cell-matrix interaction.

REFERENCES

1. P. Ekblom, D. Vestweber, and R. Kemler, Cell-matrix interactions and cell adhesion during development. **Ann. Rev. Cell Biol.** 2:27-47 (1986).

2. J. Gurdon, Embryonic induction - molecular prospects. **Development** 99:285-306 (1987).

3. A.G. Jacobson, and A.K. Slater, Features of embryonic induction. **Development** 104:341-359 (1988).

4. M.J. Anderson, M.W. Cohen, and E. Zorychta, Effects of innervation on the distribution of acetycholine receptors on cultured amphibian muscle cells. **J. Physiol. (Lond.)** 268:731-756 (1977).

5. M.W. Cohen, and P.R. Weldon, Localization of acetylcholine receptors and synaptic ultrasstruc-ture at nerve-muscle contacts in culture: dependence on nerve type. **J. Cell. Biol.** 86:388-401 (1980).

6. E.L. Frank, and G.D. Fischbach, Early events in neuromuscular junction formation *in vitro*. Induction of acetylcholine receptor clusters in the postsynaptic membrane and morphology of newly formed synapses. **J. Cell Biol.** 83:143-158 (1979).

7. Y. Kidokoro, M.J. Anderson, and R. Gruener, Changes in synaptic potential properties during acetylcholine receptor accumulation and neurospecific interactions in *Xenopus* nerve-muscle cell culture. **Devel. Biol.** 78:464-483 (1980).

8. M.J. Anderson, and M.W. Cohen, Nerve-induced and spontaneous redistribution of acetylcholine receptors on cultured muscle cells. **J. Physiol. (Lond.)** 268:757-773 (1977).

9. L. Ziskind-Conhaim, I. Geffen, and Z.W. Hall, Redistribution of acetylcholine receptors on developing rat myotubes. **J. Neurosci.** 4:2346-2349 (1984).

10. D.F. Davey, and M.W. Cohen, Localization of acetylcholine receptors and cholinesterase on nerve-contacted and non-contacted muscle cells grown in the presence of agents that block action potentials. **J. Neurosci.** 6:673-680 (1986).

11. S.J. Burden, The extracellular matrix and subsynaptic sarcoplasm at nerve-muscle synapses. In: **The Vertebrate Neuromuscular Junction,** ed. M. M. Salpeter, pp 163-186, A.R. Liss, Inc., New York (1987).

12. M.M. Salpeter, Developmental and neural control of the neuromuscular junction and of the junctional acetylcholine receptor. In: **The Vertebrate Neuromuscular Junction,** ed M.M. Salpeter, pp. 55-115. A.R. Liss, New York (1987).

13. K.F. Barald, G.D. Phillips, J.C. Jay, and I.F. Mizukami, A component in mammalian muscle synaptic basal lamina induces clustering of acetylcholine receptors. **Prog. Brain Res.** 71:397-408 (1987).

14. H.C. Bauer, M.P. Daniels, P.A. Pudimat, L. Jacques, H. Sugiyama, and C.N. Christian, Characterization and partial purification of a neuronal factor which increases acetylcholine receptor aggregation on cultured muscle cells. **Brain Res.** 209:395-405 (1981).

15. C.N. Christian, M.P. Daniels, H. Sugiyama, Z. Vogel, L. Jacques, and P.G. Nelson, A factor from neurons increases the number of acetylcholine receptor aggregates on cultured muscle cells. **Proc. Natl. Acad. Sci. USA** 75:4011-4015 (1978).

16. J.A. Connolly, P.A. St. John, and G.D. Fischbach, Extracts of electric lobe and electric organ from *Torpedo californica* increase the total number as well as the number of aggregates of chick myotube acetylcholine receptors. **J. Neuroscience** 2:1207-1213 (1982).

17. B. Fontaine, A. Klarsfeld, T. Hökfelt, and J. P. Changeux, Calcitonin gene-related peptide, a peptide present in spinal cord motoneurons, increases the number of acetylcholine receptors in primary cultures of chick embryo myotubes. **Neurosci. Lett** 71:59-65 (1986).

18. T.M. Jessel, R.E. Siegel, and G.D. Fischbach, Induction of acetylcholine receptors on cultured skeletal muscle by a factor extracted from brain and spinal cord. **Proc. Natl. Acad. Sci. USA** 76:5397-5401 (1979)

19. C. Kalcheim, Z. Vogel, and D. Duksin, Embryonic brain extract induces collagen biosynthesis in cultured muscle cells: involvement in acetylcholine receptor aggregation. **Proc. Natl. Acad. Sci. USA** 79:3077-3081 (1982).

20. D. Knaack, I. Shen, M.M. Salpeter, and T.R. Podleski, Selective effects of ascorbic acid on acetylcholine receptor number and distribution. **J. Cell Biol.** 102:795-802 (1986).

21. G.J. Markelonis, R.A. Bradshaw, T.H. Oh, J.L. Johnson, and O.J. Bates, Sciatin is a transferrin-like polypeptide. **J. Neurochem.** 39:315-320 (1982).

22. R.M. Nitkin, M.A. Smith, C. Magill, J.R. Fallon, Y.-M. M. Yao, B.G. Wallace, and U.J. McMahan, Identification of Agrin, a synaptic organizing protein from *Torpedo* electric organ. **J. Cell. Biol.** 105:2471-2478 (1987).

23. T.B. Usdin, and G.D. Fischbach, Purification and characterization of a polypeptide from chick brain that promotes the accumulation of acetylcholine receptors in chick myotubes. **J. Cell Biol.** 103:493-507 (1986).

24. Z. Vogel, C.N. Christian, M. Vigny, H.C. Bauer, P. Sonderegger, and M.P. Daniels, Laminin induces acetylcholine receptor activity of a neuronal factor. **J. Neurosci.** 3:1058-1068 (1983).

25. H.B. Peng, Participation of calcium and camodulin in the formation of acetylcholine receptor clusters. **J. Cell Biol.** 92:550-557 (1984).

26. H.B. Peng, Elimination of preexistent acetylcholine receptor clusters induced by the formation of new clusters in the absence of nerve. **J. Neurosci.** 6:581-589 (1986).

27. H.B. Peng, Q. Chen, M.W. Rochlin, D. Zhu, and B. Kay, Mechanisms of neuromuscular junction development studied in tissue culture. In: **Developmental Neurobiology of the Frog,** ed. E.D. Pollack and H.D. Bibb, pp. 103-119, A.R. Liss, New York (1988).

28. H.B. Peng, and P. Cheng, Formation of postsynaptic specializations induced by latex beads in cultured muscle cells. **J. Neurosci.** 2:1760-1774 (1982).

29. H.B. Peng, P.C. Cheng, and P.W. Luther, Formation of Ach receptor clusters induced by positively charged beads. **Nature (Lond.)** 292:831-834 (1981).

30. R.W. Burry, R. Ho, and W.D. Matthew, Presynaptic elements formed on polylysine-coated beads contain synaptic vesicle antigens. **J. Neurocytol.** 15:409-419 (1986).

31. S.J. Burden, P.B. Sargent, and U.J. McMahan, Acetylcholine receptors in regenerating muscle accumulate at original synaptic sites in the absence of nerve. **J. Cell Biol.** 82:412-425 (1979).

32. J.R. Sanes, L.M. Marshall, and U.J. McMahan, Reinnervation of muscle fiber basal lamina after removal of myofibers. Differentiation of regenerating axons at original synaptic sites. **J. Cell Biol.** 78:176-198 (1978).

33. M.J. Anderson, and D.M. Fambrough, Aggregates of acetylcholine receptors are associated with plaques of a basal lamina heparan sulfate proteoglycan on the surface of skeletal muscle fibres. **J. Cell Biol.** 97:1396-1411 (1983).

34. M.J. Anderson, F.G. Klier, and K.E. Tanguay, Acetylcholine receptor aggregation parallels the deposition of a basal lamina proteoglycan during development of the neuromuscular junction. **J. Cell Biol.** 99:1769-1784 (1984).

35. C.B. Weinberg, and Z.W. Hall, Junctional forms of acetyl-cholinesterase restored at nerve-free endplates. **Devel. Biol.** 68:631-635 (1979).

36. E.K. Bayne, M.J. Anderson, and D.M. Fambrough, Extracellular matrix organization in developing muscle: correlation with acetylcholine receptor aggregates. **J. Cell Biol.** 99:1486-1501 (1984).

37. M.P. Daniels, M. Vigny, P. Sonderegger, H.C. Bauer, and Z. Vogel, Association of laminin and other basement membrane components with regions of high acetylcholine receptor density on cultured myotubes. **Int. J. Develop. Neurosci.** 2:87-99 (1984).

38. J.R. Fallon, and C.E. Gelfman, Agrin-related molecules are concentrated at acetylcholine receptor clusters in normal and aneural developing muscle. **J. Cell Biol.** 108:1527-1535 (1989).

39. F. Moody-Corbett, and M.W. Cohen, Localization of cholinesterase at sites of high acetylcholine receptor density on embryonic amphibian muscle cells cultured without nerve. **J. Neurosci.** 1:596-605 (1981).

40. M.J. Anderson, Nerve-induced remodeling of muscle basal lamina during synaptogenesis. **J. Cell Biol.** 102:863-877 (1986).

41. Z. Avnur, and B. Geiger, The removal of extracellular fibronectin from areas of cell-substrate contact. **Cell** 25:121-132 (1981).

42. J.-M. Chen, and W.-T. Chen, Fibronectin-degrading proteases from the membranes of transformed cells. **Cell** 48:193-203 (1987).

43. W.-T. Chen, J.-M.K. Chen, S.J. Parsons, and J.T. Parsons, Local degradation of fibronectin at sites of expression of the transforming gene product pp60src. **Nature (Lond.)** 316:156-158 (1985).

44. W.-T. Chen, K. Olden, B.A. Bernard, and F.-F. Chu, Expression of transformation-associated protease(s) that degrade fibronectin at cell contact sites. **J. Cell Biol.** 98:1546-1555 (1984).

45. F. Grinnell, Focal adhesion sites and the removal of substratum-bound fibronectin. **J. Cell Biol.** 103:2697-2706 (1986).

46. F. Blasi, J.-D. Vassalli, and K. Danø, Urokinase-type plasminogen activator: Proenzyme, receptor, and inhibitors. **J. Cell Biol.** 104:801-804 (1987).

47. O. Saksela, Plasminogen activation and regulation of pericellular proteolysis. **Biochem. Biophys. Acta** 823:35-65 (1985).

48. O. Saksela, and D.B. Rifkin, Cell-associated plasminogen activation: regulation and physiological functions. **Ann. Rev. Cell. Biol.** 4:93-126 (1988).

49. B. Furie, and B.C. Furie, The molecular basis of blood coagulation. **Cell** 53:505-518 (1988).

50. L.A. Liotta, C.N. Rao, and U.M. Wewer, Biochemical interactions of tumour cells with the basement membrane. **Ann. Rev. Biochem.** 55:1037-1057 (1986).

51. G.K. Scott, Proteinases and eukaryotic cell growth. **Comp. Biochem. Physiol.** 87B:1-10 (1987).

52. A. Krystosek, and N.W. Seeds, Peripheral neurons and Schwann cells secrete plasminogen activator. **J. Cell. Biol.** 98:773-776 (1984).

53. R.N. Pittman, Release of plasminogen activator and a calcium-dependent metalloprotease from cultured sympathetic and sensory neurons. **Devel. Biol.** 110:91-101 (1985).

54. J. Pöllänen, K. Hedman, L.S. Nielsen, K. Danø, and A. Vaheri, Ultrastructural localization of plasma membrane-associated urokinase-type plasminogen activator at focal contacts. **J. Cell Biol.** 106:87-95 (1988).

55. J. Pöllänen, O. Saksela, E.M. Salonen, P. Andreasen, L. Nielsen, K. Danø, and A. Vaheri, Distinct localization of urokinase-type plasminogen activator and its type 1 inhibitor under cultured human fibroblasts and sarcoma cells. **J. Cell Biol.** 104:1085-1096 (1987).

56. C.A. Buck, and A.F. Horwitz, Cell surface receptors for extracellular matrix molecules. **Ann. Rev. Cell Biol.** 3:179-205 (1987).

57. A. Horwitz, K. Duggan, R. Grecos, C. Decker, and C. Buck, The cell substrate attachment (CSAT) antigen has properties of a receptor for laminin and fibronectin. **J. Cell Biol.** 101:2134-2144 (1985).

58. R.O. Hynes, Integrins: A family of cell surface receptors. **Cell** 48:549-554 (1987).

59. K. Burridge, K. Faith, T. Kelly, G. Nuckolls, and C. Turner, Focal adhesions: transmembrane junctions between the extracellular matrix and the cytoskeleton. **Ann. Rev. Cell Biol.** 4:487-525 (1988).

60. C. Damsky, K.A. Knudsen, D. Bradley, C.A. Buck, and A.F. Horwitz, Distribution of the cell substratum attachment (CSAT) antigen on myogenic and fibroblastic cells in culture. **J. Cell Biol.** 100: 1528-1539 (1985).

61. D. Rifkin, and D. Moscatelli, Recent developments in the cell biology of basic fibroblast growth factor. **J. Cell Biol.** 109:1-6 (1989).

62. A. Rizzino, Transforming growth factor-β - multiple effects on cell differentiation and extracellular matrices. **Devel. Biol.** 130:411-422 (1988).

63. S. Artavanis-Tsakonas, The molecular biology of the Notch locus and the fine tuning of differentiation in *Drosophila*. **Trends Genetics** 4:95-100 (1988).

64. A. Tomlinson, Cellular interactions in the developing *Drosophila* eye. **Development** 104:183-193 (1988).

65. W. Bender, Homeotic gene products as growth factors. **Cell** 43:559-560 (1985).

66. D.R. Sibley, J.L. Benovic, M.G. Caron, and R.J. Lefkowitz, Regulation of transmembrane signaling by receptor phosphorylation. **Cell** 48:913-922 (1987).

67. G. Carpenter, Receptors for epidermal growth factor and other polypeptide mitogens. **Ann. Rev. Biochem.** 56:881-914 (1987).

68. G. Carpenter, and S. Cohen, Epidermal growth factor. **Ann. Rev. Biochem.** 48:193-216 (1979).

69. K.C. Flanders, N.L. Thompson, D.C. Cissel, E. van Obberghen-Schilling, C.C. Baker, M.E. Kass, L.R. Ellingsworth, A.B. Roberts, and M.B. Sporn, Transforming growth factor-β1: histochemical localization with antibodies to different epitopes. **J. Cell Biol.** 108:653-660 (1989).

STEPS IN ESTABLISHING A BIOLOGICAL RELEVANCE FOR GLIA-DERIVED NEXIN

DENIS MONARD, EVA REINHARD, ROLAND MEIER, JUERG SOMMER, LYNNE FARMER, GIORGIO ROVELLI AND RAINER ORTMANN

Friedrich Miescher-Institut
P.O. Box 2543
4002 Basel, Switzerland

INTRODUCTION

The development and maintenance of the nervous system requires a multiplicity of interactions between the neuroblast and the neuron and their respective environments. After a phase of cellular migration, the first step in the differentiation program of the neuroblast is to extend neurites which will become either dendrites capable of receiving information from the afferent cells or axons which will establish contact with specific targets in order to propagate this information. Today, neurite outgrowth is considered to be a complex phenomenon which can be influenced by many different molecules. Neurotrophic factors, neurotransmitters, and components of the extracellular matrix trigger or modulate this event *in vitro* and, in some cases, *in vivo*.[1-6] Some of the same molecules are probably also involved in the regenerative events whereby neurite outgrowth is reinitiated following a lesion.

During development the burst of glial cell proliferation precedes the explosion of neuritic growth.[7] In many instances, lesion of the established nervous system triggers mitosis of the surrounding glial cells.[8] Glial cells have therefore been thought to release molecules able to initiate or modulate neurite outgrowth.[9]

NEURITE OUTGROWTH MEDIATED BY SERINE PROTEASE INHIBITION

A neurite-promoting activity has been detected in the medium conditioned by an established cell line of glial origin, the rat C6 glioma cells.[10] This activity was also released by rat brain primary cultures established after a critical developmental stage.[11] This glia-derived neurite-promoting factor was shown to be distinct from the well-known Nerve Growth Factor.[12] During establishment of our purification procedure, it was seen that the neurite-promoting activity was always associated with a potent serine protease inhibitor activity. Nanograms of thrombin antagonized the neurite-promoting activity, and hirudin, a potent thrombin inhibitor, promoted neurite

Serine Proteases and Their Serpin Inhibitors in the Nervous System
Edited by B. W. Festoff
Plenum Press, New York, 1990

outgrowth at concentrations similar to those needed to obtain the same effect as that of the glia-derived protein.[13] The glia-derived neurite-promoting factor was finally purified to homogeneity; it turned out to be a 43 kDa protein which was also able to inhibit serine proteases such as thrombin, urokinase, plasminogen activator and trypsin. The 43 kDa glia-derived protein formed SDS-resistant complexes with these proteases.[14]

The dissociation constants of the equilibrium complexes of the factor with trypsin, urokinase, and thrombin are 17, 280 and 18 pM, respectively.[15] The glia-derived protein inactivates thrombin about 200-fold faster than does antithrombin III. Kinetic experiments have also indicated that the rate at which the glia-derived inhibitor reacts with thrombin increases by more than 40-fold in the presence of heparin.[16] Heparin also decreases the dissociation of the complex with thrombin by more than 80-fold, to 0.3 pM. Different heparin types, fractionated on the basis of their affinity for antithrombin III, give an optimal rate of α-thrombin inhibition by the glia-derived protein. At optimal heparin concentrations, the rate of inactivation is 0.5-1.2 $nM^{-1}s^{-1}$, which suggests that under these conditions the interaction is diffusion-controlled.

Demonstration that the glia-derived neurite-promoting factor was in fact a potent serine protease inhibitor provided the first experimental evidence that glial cells could influence the behaviour of neuronal cells by modifying the amount of the proteolytic activity associated with their membrane or localized in their vicinity.

CHARACTERIZATION BY DETERMINATION OF THE PRIMARY SEQUENCE

A cDNA library constructed from the mRNA of rat C6 glioma cells was screened by hybridization-selected translation. The identity of the cDNA clone which was isolated was established by the fact that tryptic peptides of the purified protein matched with the amino acid sequence deduced from the sequencing of the cDNA.[17] This rat cDNA clone was used to screen a cDNA library made from a human glioma cell line for the purpose of obtaining the sequence of the corresponding human protein.[18]

There is an 83% homology between the rat and human protein sequences.[17] The sequences obtained show regions of homology with known protease inhibitors: endothelial plasminogen activator inhibitor, antithrombin III and α_1-protease inhibitor. Best alignment studies revealed that the arginine[345] and the serine[346] are the reactive center of this glia-derived protease inhibitor. With this a given, the P_{17} residue is a glutamic acid and the P_{69} residue a lysine. This is of interest because the formation of a salt bridge between the Glu^{342} and the Lys^{290} (and thus also at a 52-residue equidistance) has been reported to stabilize the three-dimensional structure of α_1-proteinase inhibitor.[19] These sequence properties in fact fulfil the criteria proposed by Carell and Travis for definition of the serpin (*serine protease inhibitors*) superfamily.[20] In addition, the first 28 amino acids of the human protein were found to be identical to the partial amino terminal sequence known at the time for protease nexin I.[21] This sequence data disclosed two important facts: 1) the glia-derived proteins belong to the serpins, and 2) the nexins belong to the serpins. The glia-derived neurite-promoting factor can thus be termed glia-derived nexin (GDN).

In contrast to rat GDN, two distinct cDNA subclones have been identified for human GDN.[17] These clones are identical except that, in the ß form, the two residues Thr^{310} and Gly^{311} replace the residue Arg^{310} of the α form. This reveals the presence of two corresponding mRNAs in the human glioma cells. It is worth noting that two of the three cysteine residues are conserved at exactly the same positions (117 and 131, respectively) in rat and human GDN. It remains to be seen if a disulfide bridge between these residues is of importance for the inhibitory and the neurite-promoting activity of the protein. The putative glycosylation sites (Asn-X-Thr/Ser) conserved in the rat and human sequences are located at the same positions (364-366; 365-367 for the ß form). Again, further studies should establish if glycosylation at these residues can influence the biological properties of GDN.

These data have recently been corroborated by results demonstrating that human protease nexin I is identical to human GDN at the level of the primary structure.[22] The two forms of mRNA are also found in the human fibroblasts where Northern hybridization indicates that these cells produce about twice as much α form as ß form. Transfection with the cDNA coding for either of the two forms revealed that they have the same protease inhibitor characteristics.

GDN BIOLOGICAL EFFECTS *IN VITRO*

Since GDN has been purified as a protein which promotes neurite outgrowth in a mouse neuroblastoma cell line of tumor origin, it was necessary first to establish whether cultured primary neurons would also respond to it. GDN was shown to promote neurite outgrowth in cultured dissociated neurons of the chick sympathetic ganglion.[23] GDN does not support the survival of those neurons, which require Nerve Growth Factor (NGF). The dose-dependent GDN effect took place when the neurons were seeded at low density on petri dishes coated with polyornithine and incubated in serum-free medium, that is, under conditions which do not allow optimal neurite outgrowth. When the neurons of the chick sympathetic ganglia were seeded on laminin, one of the best substrates for support of neurite outgrowth, the potentiation of GDN remained marginal. This data suggests a permissive role for GDN and indicates that GDN could become important when environmental conditions do not permit an optimal rate of neurite outgrowth.

Recombinant GDN was recently added to astrocyte-free cultures of rat hippocampal neurons.[24] After 48 h of incubation GDN promoted a significant increase in axon length, although dendrite length and total number of neurites were not affected. In chick sympathetic ganglion cell cultures, hirudin and synthetic tripeptides which specifically inhibit thrombin did not affect neurite outgrowth.[23] In contrast, in hippocampal cultures, hirudin was found to enhance neurite outgrowth at concentrations similar to those required for GDN. Nerve Growth Factor (NGF) did not influence the morphology of these cultured pyramidal cells and there was no additional effect when NGF was added together with GDN. This result, together with the one obtained with cultured chick sympathetic neurons whose neurite outgrowth was potentiated by the combination of NGF and GDN, indicates different degrees of sensitivity in different types of neuronal cells. It also further supports the hypothesis that GDN is rather to be considered as a permissive factor. Along with the fact that nanograms of thrombin antagonize[13] or even reverse[25] neurite outgrowth in mouse neuroblastoma cells, the effect of GDN in cell culture experiments indicated that a delicate balance between cell-derived or cell-associated proteases and protease inhibitors can modulate the rate of neurite outgrowth under certain conditions.[6]

The results of these *in vitro* experiments are insufficient to demonstrate a biological relevance for GDN *in vivo*.

An initial step in this demonstration was to learn where GDN is localized *in vivo* and whether GDN synthesis can be triggered following an *in vivo* lesion at a site where regeneration can take place.

LOCALIZATION OF GDN

Polyclonal antibody to purified and paraformaldehyde-fixed GDN were raised in two rabbits. Only immunization with prefixed antigen led to immunoglobulins that could be used for immunochemistry. The specificity of these anti-GDN antibodies was demonstrated on immunoblots performed with either concentrated C6 rat glioma cell-conditioned medium or crude homogenate of the adult rat olfactory bulb. Of the myriad protein bands revealed on the strips of nitrocellulose by Ponceau red staining, only a single 43 kDa band co-migrating with purified rat GDN was detected by the immunoglobulin preparation.[26]

Immunocytochemical analysis of these GDN-specific antibodies demonstrated the prominence of GDN in the olfactory system of the rat. GDN was found in the olfactory epithelium, in the submucosa and in the olfactory nerve layer of the olfactory bulb, exclusively restricted to the primary olfactory projection.

In contrast to earlier Northern blot analysis indicating that in rat brain the amount of GDN mRNA increases up to postnatal day 12 and then decreases with age,[18] the level of GDN mRNA in the olfactory system did not fluctuate significantly during development and remained high throughout adulthood. The amount of GDN mRNA was high both in the peripheral portion of the olfactory system (epithelium and submucosa) and in the olfactory bulb, indicating that the GDN immunostaining in the submucosa and the olfactory nerve layer could not be solely attributed to a mechanism associated with the axonal transport of the protein to the central nervous system. Recent analysis by in situ hybridization has confirmed synthesis of GDN in the peripheral and central parts of the olfactory system. However, at this stage, neither immunocytochemistry nor in situ hybridization have provided the resolution required for precise identification of the types of cells which synthesize GDN in these structures.

The localization and constitutive expression of GDN in the olfactory system, the only structure in the mammalian central nervous system in which degeneration, neurogenesis and axogenesis take place throughout life, represent an initial support for but not yet a definitive demonstration of the relevance of GDN *in vivo*.

UP-REGULATION OF GDN FOLLOWING LESION OF THE RAT SCIATIC NERVE

Another step in establishment of the *in vivo* relevance of GDN was to consider its regulation following a lesion in a structure where regeneration is possible. Heumann et al. have shown that transection of the sciatic nerve leads to a rapid accumulation of NGF in the distal segment immediately adjacent to the transection site.[27] The increase in NGF level, detected as early as 6 h after the transection, reached a maximum at 24 h. The NGF content decreased after three days but remained at a steady-state level which was seven-fold higher than the control unlesioned nerve for up to two weeks. These changes in NGF were confirmed by a biphasic increase in the level of NGF mRNA. Crushing of the rat sciatic nerve leads to degeneration of the axon stump distal to the injury. It was thus of interest to investigate whether the synthesis of GDN would also be affected following this type of lesion.

In a first set of preliminary experiments segments of rat sciatic nerve were incubated in serum-free Dulbecco modified Eagle's medium and the release of proteolytic activity monitored at 24-hour intervals using a microtiter assay based on the plasminogen-dependent degradation of milk casein. The amount of proteolytic activity released was high during the first 24-hour incubation period and gradually decreased, becoming undetectable during the fifth subsequent 24-hour incubation period. At a later stage the released proteolytic activity again increased. This result suggested that the amount of GDN released by the explants could increase during the first five days of incubation but that at a later stage it would be insufficient to inhibit the increasing amount of proteolytic activity released by the degenerating explant.

Experiments were then performed *in vivo*.[28] The total RNA from the sciatic nerve was analyzed at one-day intervals following crushing of the nerve. Northern blot quantitation revealed an approximately seven-fold induction of the GDN transcript, with the maximum reached six days after injury. No induction of the transcript could be detected in sham-operated animals. To look at how synthesis of GDN protein followed the increase in GDN mRNA, a 5-mm nerve segment just distal to the lesion site was removed from the animal 1.5, 7, 10 and 14 days after the crushing. The nerve pieces were incubated for three hours in serum-free modified Eagle's medium containing 1 mg BSA/ml. The amount of GDN released by the tissue was monitored by a sandwich ELISA which specifically detects as little as 0.1 ng of GDN per sample. In the fragments removed 1.5 days after injury, the amount of GDN released was the same as that released by a similar fragment from intact sciatic nerve. A three- to four-fold increase in the amount of GDN released by the fragment removed on days 7 and 10 following the lesion was measured. The fragments removed from the animal 14 days after the lesion did not release more GDN than those originating from control intact nerve, indicating as shown by the estimation of the GDN mRNA, that the *in vivo* lesion induces a transient increase in GDN synthesis. Similar data illustrating the *in vivo* increase of GDN was obtained by immunoblot analysis of nerve fragments homogenized just after removal from the animal seven days after lesion. Fifty μg of protein were loaded on a polyacrylamide gel, electrophoretically separated and transferred onto nitrocellulose paper. A low amount of GDN was detected in the sample derived from the contralateral, unlesioned control nerves with or without sham-operation. A marked increase in the amount of GDN was shown in the samples derived from crushed or transected nerves. This latter result indicates that the increased level of GDN is not exclusively linked to regeneration, since transection does not permit this process to take place.

Immunohistochemistry with anti-GDN antibodies revealed that only a few cells were weakly positive in sections of non-injured nerve. After injury both the intensity and the number of GDN-positive cells increased markedly over those of the contralateral control nerve. Most interestingly, these GDN-positive cells were scattered throughout the nerve fragment distal to the lesion and could not be detected in the proximal stump where the nerve fibers do not degenerate.

Double immunofluorescent experiments with macrophage, fibroblast and Schwann cell markers did not permit definition of the type of cells synthesizing GDN in the sciatic nerve in this post-lesion situation.

It should be pointed out that the kinetics of GDN induction are much slower than the increase in NGF mentioned above. Since GDN potentiates the effect of

NGF in cultured sympathetic neurons, the somewhat delayed synthesis of GDN lends further support to a permissive effect for this protein on neurite outgrowth.

Nonetheless, these results clearly show that GDN synthesis is induced following lesion, and this represents an important step in establishment of an *in vivo* function for GDN.

CONCLUSION

Protease nexin I (or human GDN) is localized in human brain; the amount of free nexin able to form high-molecular-weight complexes with thrombin is reduced in homogenates derived from Alzheimer's brain.[29] These results represent another argument for the relevance of GDN *in vivo*.

The prominence of GDN in the olfactory system of the rat, its induction following lesion of the sciatic nerve and its occurrence in human brain nevertheless do not demonstrate that GDN has an *in vivo* function.

GDN is one of the best, if not the best, natural thrombin inhibitors known today. Thrombin antagonizes,[13] even reverses,[25] neurite elongation *in vitro*. It is thus possible that GDN is involved in the mechanisms which should protect nerve cells from degenerative events triggered by an abnormal localized invasion of proteases such as thrombin, urokinase or plasminogen activator. On the other hand, GDN was first isolated as a neurite promoting factor and free GDN or GDN complexed with a protease can interact with components of the extracellular matrix such as vitronectin or laminin.[30] It is thus possible to envisage a dual *in vivo* function whereby GDN would take care of an excess of abnoxious proteolytic activity and at the same time establish an appropriate balance between proteases and their inhibitors, in this way creating a permissive environment for neurite outgrowth.[6]

It will be difficult to demonstrate the validity of such an hypothesis *in vivo*. A study of the consequences of the *in vivo* application of GDN will certainly be one of the next steps required. GDN has recently been expressed in yeast,[31] and the amounts of recombinant protein now available will unquestionably represent an important tool for attaining this goal.

REFERENCES

1. H. Thoenen, and D. Edgar, Neurotrophic factors. **Science** 229: 238 (1985).
2. D. Bray, Growth cones: do they pull or are they pushed? **TINS** 10: 431 (1987).
3. Y. Dodd, and T.M. Jessell, Axon guidance and the patterning of neuronal projections in vertebrates. **Science** 242: 692 (1988).
4. Y.A. Barde, Trophic factors and neuronal survival. **Neuron** 2: 1525 (1989).
5. S. Carbonetto, The extracellular matrix of the nervous system. **TINS** 7: 382 (1984).
6. D. Monard, Cell-derived proteases and protease inhibitors as regulators of neurite outgrowth. **TINS** 11: 541 (1988).
7. R.P. Skoff, D.L. Price, and A. Stocks, Electron microscopic autoradiographic studies of gliogenesis in rat optic nerve. I. Cell proliferation. **J. Comp. Neurol.** 169: 291 (1976).
8. J.L. Salzar, and R.P. Bunge, Studies of Schwann cell proliferation. **J. Cell Biol.** 84: 739 (1980).
9. M.E. Hatten, and C.A. Mason, Neuron-astroglia interactions *in vitro* and *in vivo*. **TINS** 9: 168 (1986).
10. D. Monard, F. Solomon, M. Rentsch et al., Glia-induced morphological differentiation in neuroblastoma cells. **Proc. Natl. Acad. Sci. USA** 70: 1894 (1973).
11. Y. Schurch-Rathgeb, and D. Monard, Brain development influences the appearance of glial factor-like activity in rat brain primary cultures. **Nature** 273: 308 (1978).

12. D. Monard, K. Stockel, R. Goodman et al., Distinction between nerve growth factor and glial factor. **Nature** 258: 444 (1975).

13. D. Monard, E. Niday, A. Limat et al., Inhibition of protease activity can lead to neurite extension in neuroblastoma cells. **Prog. Brain Res.** 58: 359 (1983).

14. J. Guenther, H. Nick, and D. Monard, A glia-derived neurite-promoting factor with protease inhibitory activity. **EMBO J.** 4: 1963 (1985).

15. S.R. Stone, H. Nick, J. Hofsteenge et al., Glia-derived neurite-promoting factor is a slow-binding inhibitor of trypsin, thrombin, and urokinase. **Arch. Biochem. Biophys.** 252: 237 (1987).

16. A. Wallace, G. Rovelli, J. Hofsteenge et al., Effect of heparin on the glia-derived-nexin-thrombin interaction. **Biochem. J.** 257: 191 (1989).

17. J. Sommer, S.M. Gloor, G.F. Rovelli et al., cDNA sequence coding for a rat glia-derived nexin and its homology to members of the serpin super family. **Biochemistry** 26: 6407 (1987).

18. S. Gloor, K. Odink, J. Guenther et al., A glia-derived neurite promoting factor with protease inhibitory activity belongs to protease nexins. **Cell** 47: 687 (1986).

19. H. Loebermann, R. Tokuoka, J. Deisenhofer et al., Human α_1-proteinase inhibitor. **J. Mol. Biol.** 177: 531 (1984).

20. R. Carrell, and J. Travis, α_1-Antitrypsin and the serpins: variation and countervariation. **Trends Biochem. Sci.** 10: 20 (1985).

21. R.W. Scott, B.L. Bergman, A. Bajpai et al., Protease nexin. **J. Biol. Chem.** 260: 7029 (1985).

22. M. McGrogan, J. Kennedy, M. Ping Li et al., Molecular cloning and expression of two forms of human protease nexin I. **Biotechnology** 6: 172 (1988).

23. A.D. Zurn, H. Nick, and D. Monard, A glia-derived nexin promotes neurite outgrowth in cultured chick sympathetic neurons. **Dev. Neurosci.** 10: 17 (1988).

24. L. Farmer, J. Sommer, and D. Monard, Glia-derived nexin potentiates neurite extension in hippocampal pyramidal cells *in vitro*. **Dev. Neuroscience**. *In press*.

25. D. Gurwitz, and D.D. Cunningham. Thrombin modulates and reverses neuroblastoma neurite outgrowth. **Proc. Natl. Acad. Sci. USA** 85: 3440 (1988).

26. E. Reinhard, R. Meier, W. Halfter et al., Detection of glia-derived nexin in the olfactory system of the rat. **Neuron** 1: 387 (1988).

27. R. Heumann, S. Korsching, C. Bandtlow et al., Changes of nerve growth factor synthesis in non-neuronal cells in response to sciatic nerve transection. **J. Cell Biol.** 104: 1623 (1987).

28. R. Meier, P. Spreyer, R. Ortmann et al., Glia-derived nexin, a serine protease inhibitor with neurite promoting activity, is induced after lesion of a peripheral nerve. **Nature**. *In press*.

29. S.L. Wagner, J.W. Geddes, C.W. Cotman et al., Protease nexin-I, an antithrombin with neurite outgrowth activity, is reduced in Alzheimer's disease. **Proc. Natl. Acad. Sci. USA** 86: 8234 (1989).

30. G. Rovelli, K.T. Preissner, and D. Monard, Specific interaction of vitronectin with the cell-secreted protease inhibitor glia-derived nexin (GDN) and the GDN-thrombin complex. Submitted for publication.

31. J. Sommer, B. Meyhack, G. Rovelli et al., Expression of glia-derived nexin in yeast. **Gene**. *In press*.

SECTION V

Serpins in degenerative and malignant neurologic diseases

PROTEASE INHIBITORS IN NEUROLOGIC DISEASES

NORIAKI ADACHI

Department of Medicine (Neurology)
Shinshu University School of Medicine
3-1-1, Asahi Matsumoto
390 Japan

INTRODUCTION

Proteases and their inhibitors are known or suspected to be implicated in many aspects of normal physiology and disease processes. Although the etiologies of all neurological diseases are not yet known, many are fatal and can be classified as hereditary. Among them are amyotrophic lateral scleroses (ALS) and familial amyloidotic polyneuropathy (FAP). FAP is encountered often in our clinical work at Shinshu University School of Medicine because our hospital is located in one of the areas reporting the highest incidence of FAP in the world, second only to Portugal. Therefore, myself and my colleagues have taken a special interest in investigating the states of proteases and their inhibitors in neurologic diseases. This chapter briefly summarizes our current state of knowledge concerning the implications of protease and protease inhibitors in ALS and FAP. These developments should provide insights for subsequent steps toward clarifying the etiologies of these diseases.

Protease Inhibitors in Amyotrophic Lateral Sclerosis

ALS is characterized by a catabolic process for which abnormal activity of proteases and/or protease inhibitors may be responsible. Antel and colleagues showed increased acid protease activity in muscle of patients with ALS, suggesting that muscle tissue is actively digested.[1] In other studies, increased collagenase activity has been shown in skin fibroblasts of Guamanian ALS patients.[2] In studies related to protease activity in plasma of patients with ALS, Festoff measured the levels of α_2-macroglobulin in ALS patients and found that they were significantly decreased $(p < 0.001)$.[3,4] The interpretation of these studies was that lowered levels of this protease inhibitor may play a role in the destructive process characteristic of ALS.

Recently, in a study conducted at our hospital, we measured the serum levels of four protease inhibitors: α_1-antitrypsin (α_1AT), C1-inactivator (C1I), α_2-macroglobulin (α_2M) and antithrombin-III (ATIII). These inhibitors were measured in 11 patients with ALS (6 males and 5 females), and in a control group of patients with non-neurologic diagnoses.[5] Our results indicated no significant differences in the levels of serum α_2M between the two groups. However, we did find slight, but significantly, lower levels of serum ATIII in patients with ALS. Although the reduction of ATIII

Serine Proteases and Their Serpin Inhibitors in the Nervous System
Edited by B. W. Festoff
Plenum Press, New York, 1990

285

was slight (approximately 25% less than controls), this may be a potential clue to the understanding of the disease mechanism involved. Further study of this and other protease inhibitors in connection with the systemic biochemical changes in ALS is needed.

Protease Inhibitors in FAP

Direct measurement of inhibitors in the sera of FAP patients. FAP is an autosomal dominant systemic amyloidosis that has been described in several kinships throughout the world.[6-8] Recent work on the analysis of amyloid protein in this disease revealed that a single base transmutation in the genomic DNA underlies its pathogenesis. Specifically, the mutation of guanine to adenine in the codon corresponding to valine causes a single amino acid residue to mutate from valine to methionine. This results in a change in the protein structure of transthyretin (TTR, formerly called prealbumin) in plasma and leads to the deposition of amyloid, characteristic of this disease.[9-14] Both normal and abnormal transthyretin are present in the serum of FAP patients, and the latter is now believed to be the main cause of the disease. It has been postulated that a protective mechanism in some individuals harboring the mutant transthyretin gene delays or prevents the development of clinical FAP.[15] Very recently, Holmgren et al. reported the case where a patient possessed homozygosity for the TTR-met30-gene, while the sister of the patient presented the same homozygous RFLP pattern without any of the typical symptoms of FAP, nor any amyloid deposits in a skin biopsy specimen.[16] These cases indicate that the mutation of the protein involved in amyloid formation may be necessary but is clearly not sufficient for the clinical symptoms. On the other hand, in patients with secondary amyloidosis a component of normal serum is able to abolish the Congo red-staining properties of amyloid fibrils.[17] Inhibition of this activity by α_1-antitrypsin and α_2-macroglobulin suggests that the component is an enzyme whose activity is lower in patients with amyloidosis secondary to rheumatoid arthritis or familial Mediterranean fever than in patients who have had these diseases but have not developed amyloidosis.[18] These findings indicate that the balance between enzymes with amyloid degrading activity and their inhibitors may be important to the development of amyloidosis, as proposed by Skogen et al.[19] In a study to investigate the state of protease and protease inhibitors in FAP, serum levels of six protease inhibitors were evaluated: α_1-antitrypsin, C1-inactivator, α_2-macroglobulin, antithrombin-III, α_1-antichymotrypsin, and inter-α-trypsin inhibitor in patients with FAP and a control group with non-neurologic patients.[20] No significant differences were observed between the study group and controls. It may be that other undefined protease inhibitors may be responsible so that further investigation is required.

Indirect assumption of the state of protease and protease inhibitors from the clinical symptoms of FAP. The state of protease and protease inhibitors in serum, especially coagulation factors, are easily investigated through clinical studies because patients with all forms of amyloidosis are prone to focal or generalized hemorrhage with acquired factor X deficiency,[21-24] and combined factor IX and X deficiency[25] have also been postulated. We reported a retrospective study of the frequency of the bleeding problem and its relationship to clotting abnormalities in patients with FAP.[26] In 24 FAP patients, five had experienced one or more bleeding episodes. In four of these five patients, bleeding occurred in the terminal stage. The incidence of hemorrhage in familial amyloidotic polyneuropathy previously reported for this disease is lower than in other types of amyloidosis. In our experience clotting abnormalities are rare; a deficiency of clotting factor does not appear to exist in FAP.

The Possibility of Application of Proteases to the Management of FAP

Intrinsic proteases and protease inhibitors have been the subject matter of many studies for the purpose of investigating their utilization in disease management. Recently we have taken on this question, and would like to briefly comment on relevant background information. Despite the delineation of most of the biochemical pathogenesis of FAP, and the fact that amyloid fibril proteins have been identified and characterized in affected patients, effective treatment of this disease has not yet been established. Various agents have been used in attempts to relieve the diverse symptoms of FAP which include progressive polyneuropathy and such autonomic nervous system disturbances as orthostatic hypotension, digestive irregularities and sweating disturbances, as well as ulcerative or bullous skin lesions and sexual impotence. Unfortunately, therapies aimed at relieving the symptoms of FAP have not altered its progression. As genetic modulation of the disease does not appear to be realistic at this time, the attention of many investigators has turned to degradation of the amyloid deposits. Dimethyl sulphoxide (DMSO) has been tested for over 10 years. DMSO, an organic solvent, is an inexpensive by-product of wood pulp manufacture in the paper and allied industries. Being highly polar, it is a powerful solvent of most aromatic and unsaturated hydrocarbons, organic compounds, and other substances. Its chief commercial applications are as a solvent of acetylene, sulfur dioxide, and other gases, as an antifreeze or hydraulic fluid when mixed with water, and as a paint and varnish remover. Its important biologic activities include the ability to penetrate plant and animal tissues and to preserve living cells during freezing. DMSO has been used experimentally as a topical anti-inflammatory agent and as a carrier to enhance drug penetration. Also, it is available as a SOB solution for direct instillation into the bladder in the treatment of interstitial cystitis. Isobe and colleagues were the first to propose the use of DMSO for the treatment of amyloidosis.[27] Although most clinical trials have shown DMSO to be effective when administered orally, there have been no reports of complete recovery or control of FAP with this substance. In secondary amyloidosis, the effect of DMSO is believed to be mediated through a reduction of amyloid precursors rather than through the degradation of existing amyloid substance.[18] The anti-inflammatory action of DMSO reduces the acute-phase serum proteins presumed to be involved in secondary amyloidosis.[28] The degradation of amyloid material by normal serum is under study, and hypotheses concerning the mechanism of natural amyloid degrading factor (ADF) have been advanced.[29] Although it is not known whether ADF may be a protease, its activity is strongest at 37°C, diminished at 4°C, and is abolished after boiling for several minutes. In 1982 Skogen et al. found that brinase, a proteolytic enzyme produced by *Aspergillus oryzae*, degraded amyloid fibrils in sections of liver tissue obtained from patients with secondary amyloidosis.[30] We obtained a similar result when we applied brinase solution *in vitro* to kidney sections from a patient who died of FAP.[20] Since brinase is no longer commercially available, we subsequently investigated the *in vitro* effects of four other enzymes: α-chymotrypsin (EC 3. 4. 24. 3), bromelain (EC 3. 4. ,22. 4.), collagenase (EC 3. 4. 24. 3), and lysozyme (EC 3. 2. 1. 17).[31] Using amyloid tissue sections from a patient with FAP in study, we found that the degradation of amyloid fibrils was extensive with the solution of α-chymotrypsin, moderate with bromelain and collagenase, and only slight with lysozyme. α-chymotrypsin, bromelain, and lysozyme are widely used clinically as oral mucolytics and anti-inflammatory agents, especially in Japan, and their efficacy is well established. Chymotrypsin has also been used topically for debriding necrotic lesions and given orally, buccally, or intramuscularly for reducing inflammation and edema.

Concerning its side effects, in 1964 Kirsch reported transient ocular hypertension following cataract surgery associated with the use of chymotrypsin. However, this effect is not known to follow oral administration of this enzyme. It is interesting that most recent reports which show the efficacy of the oral use of chymotrypsin are from outside USA[32-35] and reports from USA are few although some interest was previously shown in the early 1960's.[36] Bromelain is a protein-digesting and milk-clotting enzyme found in the stem and juice of the pineapple (*Ananas comosus var. Cayenne*). This enzyme is identified as being either fruit or stem bromelain. The available tablets contain stem bromelain having a molecular weight of about 33,000 and probably the first proteolytic enzyme of plant origin to be identified as a glycoprotein. Bromelain is considered to be nontoxic and to have a wide range of potential clinical applications: as an antitumor agent, a blood coagulant and also as an anticoagulant, an anti-inflammatory agent, in the debridement of third-degree burns and in the enhancement of drug absorption. The mechanism of action of bromelain is related in part to its modulation of the arachidonic cascade. Few side effects have been reported with bromelain, which has been used as a folk remedy in the tropics, and no reports of toxicity related to this are known. Although the chemical structures of the active components of bromelain have not been fully determined, it shows distinct pharmacologic promise. It is possible, therefore, that they can be administered orally for treating FAP. However, there remain two major obstacles to advancing from laboratory investigations to clinical trials with these enzymes First, the observed decrease in fluorescence of the amyloid substance in tissue sections *in vitro* does not mean that the amyloid material has been completely degraded, or that it will be flushed into the blood stream. To prove these events beyond doubt would require considerable study. Even then, *in vitro* findings with these enzymes may not predict the *in vivo* results. The task of amassing sufficient evidence to warrant clinical study appears formidable. Second, the molecular weights of α-chymotrypsin and bromelain are high, 25 kDa and 33 kDa, respectively. There exists the possibility that they might be digested in the stomach, thereby losing their proteolytic activity, and that their high molecular weights may prevent absorption. However, the contention that only low molecular weight substances can be absorbed from the intestine may not be correct. There is evidence that these enzymes may be absorbed in active form through the intestinal wall. The phenomenon of passive immunity in mammalian species involves the absorption of specific immunolactoglobulins from maternal milk by the proximal intestine.[37] Ferritin (molecular weight 500,000),[38] insulin, and resin particles larger than many bacteria (1 to 5 microns)[39] are reported to be readily absorbed from the gastrointestinal tract. Some researchers have documented the presence of orally administered, radiolabeled protease in the blood or tissues. However, it is possible that only the degraded enzyme is absorbed through the intestine; such degraded substances may lack the desired activity on amyloid. Table 1 summarizes three reports concerning the absorption and tissue distribution of α-chymotrypsin.[40-42] All three studies showed a direct activity of protease in blood or tissue via measurement of the proteolytic effect on specific substrates. Bruce et al. using a special viscometer, objectively demonstrated a decrease in the viscosity of tracheal mucus after the oral administration of α-chymotrypsin.[40] Bromelain and other enzyme preparations have also been reported to be absorbed in the active form via the intestine.

Table 1. Absorption of α-chymotrypsin through the intestinal wall summarizes three reports concerning the absorption and tissue distribution of α-chymotrypsin. All three studies showed a direct activity of protease in blood or tissue via measurement of the proteolytic effect on specific substrates. Bruce et al. using a special viscometer, objectively demonstrated a decrease in the viscosity of tracheal mucus after the oral administration of α-chymotrypsin. Bromelain and other enzyme preparations have also been reported to be absorbed in the active form via the intestine.

Species (Route)	Observation	Time to Therapeutic Effect	Author
human (oral)	decreased sputum viscosity	given for 4 weeks; effect increased every week	Bruce et al. (1962)
rabbit (rectal)	degradation by α-chymotrypsin of specific substrate, ATFE,	maximum at 20 min after dose (50 min after administration)	Kabacoff et al. (1963)
human (oral)	same as above	maximum at 4 hours after administration	Avakian et al. (1964)

Protease Inhibitors and Other Neurologic Diseases

Serine protease inhibitors may also have an important role in the pathogenesis of other neurological diseases. Among them the following two studies seem to be especially promising in this field. Abraham et al.[43] postulated the relationship between serine protease inhibitors and A-4 amyloid protein in Alzheimer's disease (see Abraham, this volume). In experimental allergic encephalomyelitis (EAE), a model of multiple sclerosis, Koh and colleagues showed that clinical signs of EAE were significantly suppressed in AMCA (trans-4-aminomethyl-cyclohexanecarboxylic acid) treated rats.[44] This synthetic inhibitor of plasminogen activator may have potential efficacy in the therapy of multiple sclerosis.

CONCLUDING COMMENTS

The fact that protease and protease inhibitors are known or suspected to be implicated in many aspects of normal physiology and disease processes has generated much interest in their potential for clarifying the etiology of amyotrophic lateral sclerosis and as pharmacological agents for the treatment or prevention of familial amyloidotic polyneuropathy. As discussed in this paper proteases facilitate the degradation of amyloid material in the tissue of FAP in the laboratory. I am confident that exciting new developments will continue to emerge in the field of protease and protease inhibitors in neurologic diseases.

ACKNOWLEDGMENTS

I would like to express my strong appreciation to the many talented and also encouraging colleagues who have participated in the research on protease and protease inhibitors in neurological diseases which were carried out in our department during the past five years. It is a privilege for me to act as their spokesman through the invitation to this honorable workshop. The research in our laboratory in this field has been supported mainly by grants from the Intractable Disease Division of the Public Health Bureau of Japan and by a Grant-in-Aid for scientific research from the

Ministry of Education, Science and Culture of Japan (grant Nos. 58770577 and 62770202) and also by the Uehara Memorial Foundation.

REFERENCES

1. J.P. Antel, E. Chelmicka Schorr, M. Sportiello, K. Stefansson, R.L. Wollman, and B. G. W. Arnason, Muscle acid protease activity in amyotrophic lateral sclerosis: correlation with clinical and pathologic features. **Neurology**, 32:901-903 (1982).
2. R.L. Beach, L.S.Rao, B.W.Festoff, E.T. Reyes, R.Yanagihara, and D. C.Gajdusek, Collagenase activity in skin fibroblasts of patients with amyotrophic lateral sclerosis. **J. Neurol. Sci.**,72:49-60 (1986).
3. B.W. Festoff, Circulating protease inhibitors in amyotrophic lateral sclerosis: reduced α_2-macroglobulin. **Ann. Neurol.**, 8:121 (1980).
4. B.W. Festoff, Occurrence of reduced α_2-macroglobulin and lowered protease inhibiting capacity in plasma of amyotrophic lateral sclerosis patients. **Ann. N.Y. Acad Sci.**, 421:369-376 (1983).
5. N. Adachi, and S. Shoji, Studies of protease inhibitors in the sera of patients with amyotrophic lateral sclerosis. **J. Neurol. Sci.**, 89:165-168 (1989).
6. C. Andrade, A peculiar form of peripheral neuropathy:familial atypical generalized amyloidosis with special involvement of the peripheral nerves. **Brain**, 75:408-427 (1952).
7. S. Araki, S. Mawatari, M. Ohta, A. Nakajima, and Y. Kuroiwa, Polyneuritic amyloidosis in a Japanese family. **Arch. Neurol.**, 18: 593-602 (1968).
8. M.D. Benson, and A.S. Cohen, Generalized amyloid in a family of Swedish origin. **Ann. Intern. Med.**, 86:419-424 (1977).
9. P.P. Costa, A.S. Figueira, and F. R. Bravo, Amyloid fibril protein related to prealbumin in familial amyloid polyneuropathy. **Proc. Natl. Acad. Sci. USA**, 75:4499-4503 (1978).
10. M. Pras, F. Prelli, E.C. Franklin, and B. Fragione, Primary structure of an amyloid prealbumin variant in familial polyneuropathy of Jewish origin. **Proc. Natl. Acad. Sci. USA**, 80:539-542 (1983).
11. S. Tawara, M. Nakazato, K. Kangawa, H. Matsuo, and S. Araki, Identification of amyloid prealbumin variant in familial amyloidotic polyneuropathy (Japanese type). **Biochem. Biophys. Res. Commun.**, 116:880-888 (1983).
12. M.J.M. Saraiva, S. Birken, P.P. Costa, and D.S. Goodman, Amyloid fibril protein in familial amyloidotic polyneuropathy, Portuguese type. **J Clin Invest** 74:104-19 (1984).
13. F.E. Dwulet, and M.D. Benson, Primary structure of an amyloid prealbumin and its plasma precursor in a heredo-familial polyneuropathy of Swedish origin. **Proc Nat Acad Sci USA** 81:694-98 (1984).
14. S. Mita, S. Maeda, K. Shimada, and S. Araki, Cloning and sequence analysis of cDNA for human prealbumin. **Biochem Biophys Res Commun** 124:558-64 (1984).
15. M.J.M. Saraiva, P.P. Costa and D.S. Goodman, Genetic expression of a transthyretin mutation in typical and late-onset Portuguese families with familial amyloidotic polyneuropathy. **Neurology**, 36: 1413-1417 (1986).
16. G. Holmgren, E. Haettner, I. Nordenso, O.L. Sandgren, L. Steen, and E. Lundgren, (1988) Homozygosity for the transthyretin-met30-gene in two Swedish sibs with familial amyloidotic polyneuropathy. **Clin Genet** (Denmark) 34 (5) p333-8.
17. I. Kedar, M. Ravid, and E. Sohar, Demonstration of amyloid degrading activity in normal human serum. **Proc. Soc. Exp. Biol. Med.**, 145:343-345 (1974).
18. O. Wegelius, A.M. Teppo, and C.P.J. Maury, Reduced amyloid A-degrading activity in serum in amyloidosis associated with rheumatoid arthritis. **Br. Med. J.**, 284:617-619 (1982).
19. B. Skogen, and L. Natvig, Degradation of amyloid proteins by different serine proteases. **Scand. J. Immunol.**, 14:389-396 (1981).
20. N. Adachi, S.Shoji, S.Nakagawa, C-S.Koh, N.Tsukada and N.Yanagisawa, Studies of protease and protease inhibitors in familial amyloidotic polyneuropathy. **J. Neurol. Sci.**, 81:79-84 (1987).
21. M. Howell, Acquired factor X deficiency associated with systemic amyloidosis: report of a case. **Blood** 21:739-744 (1963).
22. J.R. Krause, Acquired factor X deficiency and amyloidosis. **Am. J. Clin. Path.** 67:170-173 (1977).
23. M. Quitt, E. Aghai, D. Miriam, R. Kohan, Y.B. Ari, and P. Froom, Acquired factor X and antithrombin III deficiency in a patient with primary amyloidosis and nephrotic syndrome. **Scand. J. Haematol.** 35:155-157 (1985).
24. B. Furie, E. Greene, and B.C. Furie, Syndrome of acquired factor X deficiency and systemic amyloidosis: *in vivo* studies of the metabolic fate of factor X. **New Engl. J. Med.** 297:81-85 (1977).
25. R.A. McPherson, J.W. Onstad, R.J. Ugoretz, and P.L. Wolf, Coa-gulopathy in amyloidosis: combined deficiency of factors IX and X. **Am. J. Hematol.** 3:225-235 (1977).
26. N. Adachi, C.S. Koh, N. Tsukada, S. Shoji, and N. Yanagisawa, In vitro degradation of amyloid material by four proteases in tissue of a patient with familial amyloidotic polyneuropathy. **J.Neurol. Sci.**, 84:295-299 (1988).

27. T. Isobe, and E.F. Osserman, Effects of dimethyl sulphoxide (DMS0) on Bence-Jones proteins, amyloid fibrils and casein-induced amyloidosis. In: Wegelius O, Pasternack A, eds. **Amyloidosis**. London, New York and San Francisco: Academic Press, 1976:247-57 (1976).

28. H. Falck, and O. Wegelius, Treatment of secondary renal amyloidosis with dimethyl sulphoxide. In: **Proceedings of European Amyloidosis Research Symposium**. Bristol, Sept 10-12, 1981.

29. I. Kedar, E. Sohar, and M. Ravid, Degradation of amyloid by a serum component and inhibition of degradation. **J. Lab. Clin. Med.**, 99:693-700 (1982).

30. B. Skogen,and E. Amundsen, Degradation of amyloid proteins with protease I from Aspergillus oryzae. *In vivo* increase in SAA clearance rate after enzyme infusion. **Scand. J. Immunol** 16, 509-514 (1982).

31. N. Adachi, S. Shoji, and N. Yanagisawa, Bleeding Manifestations in 24 Patients with Familial Amyloidotic Polyneuropathy. **Europ. Neurol.** 28: 115-116 (1988).

32. T. Kobayashi, H. Ozone, H. Kamei, and K. Ishimaru, Chymoral for inflammatory diseases in the orodental area. **Shikai Tenbo.** 64(4):813-8 (1984).

33. P.H. Brakenbury, and J. Kotowski, A comparative study of the management of ankle sprains. **Br J Clin Pract.** 37(5);181-5 (1983).

34. A.D. Roberts, and D.M. Hart, Polyglycolic acid and catgut sutures, with and without oral proteolytic enzymes, in the healing of episiotomies. **Br J Obstet Gynaecol.** 90(7);650-3 (1983).

35. V.G. Glozman, and I.S.Anchupane, Use of chymotrypsin in inflammatory diseases of the scrotal organs. **Urol Nefrol (Mosk).** (5):44-6 (1982).

36. S. Avakian, Current concepts in therapy. Chymotrypsin and trypsin. **N Engl J Med.** (1961).

37. R.D. Rodewald, Selective antibody transport in the proximal small intestine of the neonatal rat. **J Cell Biol** 45:635-40 (1970).

38. A. Stochino, G. Tecce, and G.G. Tedeschi, Sull'assorbimento intestinale delle proteine omologhe ed eterologhe. **Boll Soc Ital Biol Sper** 27:1672-4 (1951).

39. J.M. Payne, B.F. Sansom, R.J. Garner, A.R. Thomson, and B.J. Miles, Uptake of small resin particles (1-5 microns diameter) by alimentary canal of calf. **Nature (London)** 188:586 (1960).

40. R.A. Bruce, and K.C. Quinton, Effect of oral α-chymotrypsin on sputum viscosity. **Br Med J**, 1:282-284 (1962).

41. B.L. Kabacoff, A. Wohlman, M. Umkey, and S. Avakian, Absorption of chymotrypsin from the intestinal tract. **Nature**, 199:815 (1963).

42. S. Avakian, Further studies on the absorption of chymotrypsin. **Clin Pharmacol Ther**, 5:712-5 (1964).

43. C.R. Abraham, D. J. Selkoe and H. Potter, Immunochemical identification of the serine protease inhibitor, α_1-antichymotrypsin, in the brain amyloid deposits of Alzheimer's disease. **Cell**, 52:487-501 (1988).

44. C.S. Koh, and P.Y. Paterson, Suppression of Clinical Signs of Cell Transferred Experimental Allergic Encephalomyelitis and Altered Cerebrovascular Permeability in Lewis Rats Treated with a Plasminogen Activator Inhibitor. **Cellular Immunology** 107, 52-63 (1987).

45. L.G. Millard, and N.R. Rowell, Primary amyloidosis and myelomatosis associated with excessive fibrinolytic activity. **Br. J. Derm.** 94: 569-571 (1976).

46. L.S. Perlin, P. Brakman, H.S. Berg, P.J. Kirchner, R.B. Moguin, T. Astrup, Enhanced blood coagulation and fibrinolysis in a patient with primary fibrinolysis. **Thromb Haemostasis** 26:9-14 (1971).

PRESENCE AND SIGNIFICANCE OF α_1-ANTITRYPSIN IN HUMAN BRAIN TUMORS

RAYMOND SAWAYA

Department of Neurosurgery
University of Cincinnati College of Medicine
and Veterans Administration
Cincinnati, Ohio 45267

INTRODUCTION

It is increasingly recognized that proteinases are involved in the basic biological events associated with the neoplastic growth. Tumor invasiveness and the ability of tumor cells in culture to degrade extracellular proteins are indirect evidence of the involvement of the proteolytic enzymes. More direct evidence has been provided by the detection of increased levels of proteinases capable of digesting connective tissue matrix component within the tumor environment.[1-7] Studies involving brain tumor tissues have demonstrated the production of plasminogen activators (PA) in high concentrations by cultured brain tumor cells[8] and by a variety of fresh surgical specimen of human brain tumors.[9]

As a consequence of these studies, interest has arisen in regard to the presence and distribution of proteinase inhibitors in tumors. A plasmin inhibitory activity associated with brain tumors was previously reported[10] and based on a literature review it was postulated that α_1-antitrypsin (A_1AT) was the most likely protease inhibitor to be associated with brain tumors.

α_1-antitrypsin has been found in normal liver, pancreas, lung and gastrointestinal tract.[11-13] In addition, several neoplasms of those organs have been shown to contain A_1AT.[14-16] With the exception of hemangioblastoma,[15] A_1AT has not been described in association with brain tumors.

These studies were undertaken to confirm the presence of A_1AT in homogenates of fresh human tumor samples and to localize the expression pattern of this serpin at the cellular level using immunohistochemical methods.[17] Additionally, an attempt was made to evaluate the significance of its presence in the brain tumor environment.[18]

CLINICAL MATERIAL AND METHODS

Seventy-seven consecutive adult patients with various brain tumors were entered in this study. Clinical and biological parameters were obtained on all study subjects

Serine Proteases and Their Serpin Inhibitors in the Nervous System
Edited by B. W. Festoff
Plenum Press, New York, 1990

293

including age, sex, duration of symptoms, Karnofsky score, plasma fibrinogen level, prothrombin time, partial thromboplastin time, platelet count, sedimentation rate, and histological diagnosis.

Computerized tomography (CT) scans were used to identify tumor volume according to the following formula: Volume = π (a.b.c)/6 where a , b, and c represent the three axes of the tumor mass.[19] Peritumoral brain edema was defined as a low- attenuation zone surrounding the mass lesion[20] and was graded from 0 (no edema) to 5+ (marked edema) where the low-density area exceeded the size of the mass lesion.

Biopsies obtained from fresh surgical specimens during craniotomy were immediately labeled, placed on dry ice, and transferred to a -80°C freezer prior to homogenization. Similar samples were fixed in 10% buffered formalin and routinely embedded in paraffin. Four-micron thick sections were cut and used for hematoxylin and eosin staining for histological diagnosis and for immunohistochemical studies.

Biochemical Studies

Tumor tissue was thawed and rapidly weighed, minced, and homogenized in 37.5 mM Tris-HCl, 0.75 mM ethylenediaminetetraacetic acid, 75 mM NaCl, and 15 mM lysine buffer at a final pH of 9.5 in a Teflon pestle tissue homogenizer. For each specimen, 50 mg of tissue was homogenized in 1 ml of Tris buffer with 20 strokes. The A_1AT and $\alpha_2$macroglobulin (A_2MG) contents of the tumor extracts were qualitatively assessed by Ouchterlony immunodiffusion techniques.[21] Sheep anti-human A_1AT and sheep anti-human A_2MG antibodies and appropriate controls were obtained (Behring Diagnostic, Inc. Somerville, New Jersey). The PA activity was assayed electrophoretically on sodium dodecyl sulfate gels as described in detail elsewhere.[9] Statistical analysis of this data was done by means of chi-square test, student's t-test and analysis of variance for a significance level of $p < 0.05$.

Immunohistochemical Methods

α_1-antitrypsin was demonstrated by the regular peroxidase antiperoxidase (PAP) method of Sternberger et al.[22] In brief, paraffin sections of 21 tissue samples were trypsinized and incubated with specific A_1AT antibody (rabbit anti-human A_1AT, Dako, Santa Barbara, CA). After washing a second incubation was carried out with goat anti-rabbit immunoglobulin (DAKO). The sections were then treated with a preformed PAP complex (DAKO) and the final reaction was developed with diaminobenzidine (DAB, Sigma Chemical Co., St. Louis, MO) hydrogen peroxide reaction. Various control tests were applied as previously described.[12,16] The specificity of PAP reaction was verified by absence of positive reaction after absorption of anti-A_1AT with nonimmune rabbit serum. Positive immunoreactivity for A_1AT was graded semiquantitatively as follows: negative (-), slightly positive (+), moderately positive (+ +), and strongly positive (+ + +).

RESULTS

Biochemical Results

Marked differences in A_1AT positivity were noted among the various histological groups of brain tumors (Table 1). The difference in A_1AT reactivity reached statistical significance for the following groups: acoustic neuroma versus meningioma,

Table 1. α_1-Antitrypsin (A_1AT) in Various Tumors.

Diagnosis	Number of Samples	A_1AT positivity %
Acoustic Neuroma	10	100
Metastasis	12	91.6
Glioblastoma	20	78.6
Low-grade glioma	7	71.4
Meningioma	22	50.0
Miscellaneous*	6	33.3

*includes pituitary adenoma, ependymoma, sarcoid granuloma, lymphoma, neuroblastoma, and demyelinating process.

metastasis versus meningioma, and acoustic neuroma versus miscellaneous brain tumors. No statistical differences in A_1AT positivity was found between malignant glioma and low-grade glioma. The difference in A_1AT positivity between acoustic neuroma and meningioma is surprising and should be emphasized because of the many common features to both tumors, such as their benign, noninvasive nature and their location in the extraaxial space.

The patients were then separated into two groups according to the reactivity of their tumors to sheep anti-human A_1AT. Group I was composed of 53 patients with positive anti-A_1AT (68.6%), and Group II of 24 patients with negative anti-A_1AT reactivity (31.4%). All specimen were negative for A_2MG. No statistically significant differences were observed between the two groups with regards to age, sex, Karnofshy score, duration of symptoms, size of tumor, PT, PTT, platelet count and sedimentation rate (Table 2). Statistically significant differences were noted in the extent of the peritumoral edema on CT scans and the preoperative serum fibrinogen level, both parameters being higher in patients with A_1AT reactive tumors. However, most importantly, the PA activity of the A_1AT positive tumors was significantly higher (p=0.001) than that of the tumors in Group II.

Immunohistochemical Results

Four normal brain samples showed A_1AT positivity that varied from + to + +. Positive staining was restricted to neurons and was primarily intracytoplasmic in the form of fine granules (Table 3). Cell nuclei appeared to be negative for A_1AT. Weak staining was seen in a few vascular structures.

Table 2. Summary of Clinical and Biochemical Results*

Parameter	A_1AT (+)	A_1AT (-)	P
Age (years)	52.01 ± 2.4	54.59 ± 2.68	NS
Sex (% males)	47	59	NS
Symptoms (days)	327.89 ± 69	222.13 ± 64.99	NS
Karnofsky Score	81.22 ± 2.36	83.63 ± 3.05	NS
Tumor Size (mm)	26.09 ± 4.2	39.2 ± 12.0	NS
PT (sec)	11.85 ± 0.20	11.39 ± 0.22	NS
PTT (sec)	25.83 ± 0.56	26.23 ± 1.07	NS
Platelets (1 cu mm)	24.857 ± 15.85	290.556 ± 23.04	NS
ESR (mm/hr)	9.8 ± 1.44	9.28 ± 1.80	NS
PAA (units/mg)	125 ± 8.21	73 ± 8.48	0.001
Fibrinogen (mg/100ml)	374.43 ± 30.38	263.46 ± 19	0.025
Edema Score (0 to 5)	2.05 ± 0.25	1.076 ± 0.32	0.049

*Values are means ± standard error.
PAA = plasminogen activator activity.

Seventeen brain tumor samples were studied, and of these, 15 showed A_1AT positivity. All 15 samples exhibited granular intracytoplasmic staining sparing the nuclei. In 12 cases, a concomitant extracellular positive reaction was seen. No samples showed A_1AT positivity in the extracellular space and A_1AT negativity in the cytoplasm.

Malignant tissues demonstrated the strongest intracellular and extracellular staining. Two low-grade glioma were completely negative for A_1AT whereas two others exhibited mild intracytoplasmic and extracytoplasmic staining and an additional sample showed only slight intracytoplasmic staining.

The two acoustic schwannoma samples showed moderate intracytoplasmic A_1AT immunoreactivity; only one of the two samples showed extracellular positivity.

Finally, three of four cases of meningioma exhibited slight extracytoplasmic and intracytoplasmic staining, whereas only one showed a moderate intracytoplasmic A_1AT positivity without extracellular staining.

DISCUSSION

α_1-antitrypsin is a major protease inhibitor present in mammalian blood. This glycoprotein is known as an acute phase reactant and is increased in serum of cancer patients as well as in those with acute and chronic inflammation, trauma, and autoimmune diseases.[23-25]

Patients with tumors of the central nervous system were also found to have significant increases in their serum A_1AT compared to normal controls.[26,27] High levels of A_1AT were reported in cerebrospinal fluid of patients with brain tumors,[28] and hyaline globules similar to those present in meningiomas and glioblastomas[29,30] have recently been shown to contain A_1AT in cerebellar hemangioblastomas.[15]

Table 3. Histochemical localization of A_1AT

| Patient no. | Sex | Histologic type | Immunoreactivity | | |
			Nuclear	Cytoplasmic	Extracellular
1	M	Normal brain	−	++	−
2	M	Normal brain	−	++	−
3	F	Normal brain	−	++	−
4	M	Normal brain	−	+	−
5	F	Metastasis	−	++	+++
6	M	Metastasis	−	+	++
7	M	Glioblastoma	−	++	+++
8	F	Glioblastoma	−	+	+++
9	M	Glioblastoma	−	+++	+
10	M	Glioblastoma	−	+++	+
11	F	Low-grade glioma	−	−	−
12	M	Low-grade glioma	−	−	−
13	M	Low-grade glioma	−	+	−
14	F	Low-grade glioma	−	++	+
15	M	Low-grade glioma	−	+	+
16	F	Ac. neuroma	−	+	++
17	F	Ac. neuroma	−	+	−
18	F	Meningioma	−	++	−
19	M	Meningioma	−	+	+
20	F	Meningioma	−	+	+
21	F	Meningioma	−	+	+

-: negative; +: slightly positive; + +: moderately positive; + + +: strongly positive
$A_1AT = \alpha_1$-antitrypsin

The studies described herein have demonstrated that immunologically detectable A_1AT is present in both malignant and non-malignant brain tumors. Its role and significance in brain tumors are not known with certainty. A_1AT has a broad range of enzyme inhibitory activity and is the major serum protease inhibitor of human plasma. As several lines of evidence suggest that proteases may play an important role in the tumor-host relationship, maintenance of a proper balance between proteases and protease inhibitors might be essential in the regulation of tumor growth. In fact, in the group of tumors with detectable A_1AT in our study, the mean value for the PA activity was significantly higher than that for the group with nondetectable A_1AT. Local concentrations of protease inhibitors might therefore be important in modulating tumor proteases, and in this manner affecting the neoplastic spread.[31,32] Indeed, many synthetic and natural antiproteinases have the ability to retard tumor growth.[19] Conversely, evidence suggests that A_1AT is also an important factor in the inflammatory and immune responses. It inhibits the cytotoxic reactions of lymphocytes including antibody-dependent cell-mediated cytotoxicity[33,34] T cell-mediated cytotoxicity[22,35] and natural killer activity.[33,36] It also modulates the activity of a number of inflammatory pathways controlling an excessive inflammatory response. Additionally, because A_1AT inhibits serine esterases released from host leukocytes, some of which participate in immune cytolysis of tumor cells, the production of A_1AT may compromise a vital host defense mechanism.[37]

The source of A_1AT in brain tumors is unknown. Akatsuka et al.,[25,38] using immunoelectron microscopy, have localized A_1AT to the endoplasmic reticulum and golgi apparatus of human tumors transplanted in nude mice. It is our hypothesis that A_1AT is actively produced in the region of the tumor either by the neoplastic cells or by the supportive and/or reactive cells. In favor of this hypothesis is the fact that A_2MG was not found in any of the samples tested, thus excluding the likelihood of a passive transfer of A_1AT from the serum. Similarly, the lack of differences in several blood parameters between the two groups of patients is evidence against the possibility of a non-specific acute-phase reaction. In addition, the observations derived from the immunohistochemical studies have supported a local origin including the finding of intracytoplasmic A_1AT in the absence of any extracellular immunoreactivity in several samples, the morphologic appearance of diffusely dispersed intracytoplasmic fine granules suggestive of an increased synthesis,[39] and the intensity of the A_1AT immunoreactivity seen in some of the malignant samples.

Finally, it remains to be determined whether A_1AT is the product of the tumor itself or of the reactive cells in response to the abnormally high proteolytic activity generated by the tumor. It appears likely that A_1AT is primarily produced by the tumor cells in a protective role and as an immunosuppressant to the host.

SUMMARY

Proteases and protease inhibitors play an important role in tumor biology. Their role in brain tumors is unknown because of lack of data. In this study, we have measured α_1-antitrypsin (A_1AT) and α_2-macrogrobulin (A_2MG) in 77 consecutive brain tumor extracts obtained freshly in the operating room. The results of these measurements were correlated with several clinical and laboratory parameters (age, Karnofsky, tumor size, PT, PTT, Fibrinogen, etc.) as well as with the content of the tissue in plasminogen activator (PA), a major proteolytic enzyme. A_1AT and A_2MG were qualitatively assessed by ouchterlony immunodiffusion and PA was assayed electrophoretically on SDS gels. Appropriate controls were also included.

Sixty-eight percent of the samples were positive for A_1AT, while all specimens were negative for A_2MG. The frequency of A_1AT positivity varied with the histological type and ranged from 100% for acoustic schwannoma to 50% for meningioma. Metastatic tumors and glioblastoma were 91.6 and 78.9% positive respectively. Clinical and biological parameters failed to show statistically significant differences between the group of patients with positive A_1AT and the group with negative A_1AT with the exception of the following three parameters:

PA activity ($P=0.001$), peritumoral brain edema as quantitated on CT scan ($P=0.05$) and the preoperative serum fibrinogen level ($P=0.025$), all three values being higher in the group with positive A_1AT.

This study supports the hypothesis that A_1AT is produced locally by tumor cells in proportion to the regional proteolytic activity. To further support this hypothesis, we have immunohistochemically demonstrated A_1AT intracellularly in various brain tumors.

ACKNOWLEDGEMENT

This work was made possible by a grant from the Veterans Administration. Data was managed and analyzed using the GCRC Data Management and Analysis System, the University of Cincinnati General Clinical Research Center Grant No. M01 RR00068, supported by the Division of Research Resources of NIH.

Dr. Mario Zuccarello has assisted in the data collection and Dr. Mukunda Ray in the immunohistochemical studies.

The secretarial assistance was provided by Saundra K. Eversole.

REFERENCES

1. W.E. Laug, Y.A. DeClerck, and P.A. Jones: Degradation of the subendothelial matrix by tumor cells. **Cancer Res.** 43:1827-1834, (1983).
2. L.A. Liotta, S. Abe, P.G. Robey, et al.: Preferential digestion of basement membrane collagen by an enzyme derived from a metastatic murine tumor. **Proc. Natl. Acad. Sci. USA** 76:2268-2271, (1979).
3. L. Ossowski, J.C. Unkeless, A. Tobia, et al.: An enzymatic function associated with transformation of fibroblasts by oncogenic viruses. II. Mammalian fibroblast cultures transformed by DNA and RNA tumor viruses. **J. Exp. Med.** 137:112-126, (1973).
4. A.R. Poole, K.J. Tiltman, A.D. Recklies, et al.: Differences in secretion of the proteinase cathepsin B at the edges of human breast carcinomas and fibroadenomas. **Nature** 273:544-547, (1978).
5. B.F. Sloane, J.R. Dunn, and K.V.Honn: Lysosomal cathepsin B: correlation with metastatic potential. **Science** 212:1151-1153, (1981).
6. J. Unkeless, K. Dano, G.M. Kellerman, et al.: Fibrinolysis associated with oncogenic transformation. Partial purification and characterization of the cell factor, a plasminogen activator. **J. Biol. Chem.** 249:4295-4305, (1974).
7. J. Unkeless, A. Tobia, L. Ossowski, et al.: An enzymatic function associated with transformation of fibroblasts by oncogenic viruses. I. Chick embryo fibroblast cultures transformed by avian RNA tumor viruses. **J. Exp. Med.** 137:85-112, (1973).
8. W.S. Tucker, W.M. Kirsch, A. Martinez-Herandez, et al.: *In vitro* plasminogen activator activity in human brain tumors. **Cancer Res.** 38:297-302, (1978).
9. R. Sawaya, and R. Highsmith: Plasminogen activator activity and molecular weight patterns in human brain tumors. **J. Neurosurg.** 68:73-79, (1988).
10. R. Sawaya, C.J. Cummins, and P.L. Kornblith: Brain tumors and plasmin inhibitors. **Neurosurgery** 15:795-800, (1984).
11. K. Geboes, M.B. Ray, P. Rutgeerts, et al.: Morphological identification of α_1-antitrypsin in the human small intestine. **Histopathology** 6:55-60, (1982).
12. M.B. Ray, and V.J. Desmet: Immunohistochemical demonstration of α_1-antitrypsin in the islet cells of human pancreas. **Cell Tissue Res.** 187:69-77, (1978).

13. W.C. Tuttle, and R.K. Jones: Fluorescent antibody studies of α_1-antitrypsin in adult human lung. **Am J. Clin. Pathol.** 64:477-482, (1975).

14. P.J. Klemi, L. Meurman, M. Gronroos, et al.: Clear cell (mesonephroid) tumors of the ovary with characteristics resembling endodermal sinus tumor. **Int. J. Gynecol. Pathol.** 1:95-100, (1982).

15. B.S. Mann, and J.F. Geddes: The nature of cytoplasmic inclusions in cerebellar haemangioblastomas. **Acta Neuropathol.** (Berl) 67:174-176, (1985).

16. M.B. Ray, K. Geboes, F. Callea, et al.: α_1-antitrypsin immunoreactivity in gastric carcinoid. **Histopathology** 6:289-297, (1982).

17. M. Zuccarello, R. Sawaya, M. Ray: Immunohistochemical demonstration of α1-proteinase inhibitor in brain tumors. **Cancer** 60:804-809, (1987).

18. R. Sawaya, M. Zuccarello, and R. Highsmith: α1-antitrypsin in human brain tumors. **J. Neuro Surg.** 67:258-262, (1987).

19. R. Sawaya, T. Mandybur, I. Ormsby, et al.: Antifibrinolytic therapy of experimentally grown malignant brain tumors. **J. Neurosurg.** 64:263-268, (1986).

20. J.J. Gilbert, J.E. Paulseth, R.K. Coates, et al.: Cerebral edema associated with meningicmas. **Neurosurgery** 12:599-605, (1983).

21. O. Ouchterlony: Diffusion-in-gel methods for immunological analysis. **Prog. Alleregy** 5:1-78, (1959).

22. L.A. Sternberger, P.H. Hardy, and J.J Culis: The unlabeled antibody enzyme method of immunohistochemistry. **J. Histochem. Cytochem.** 18:315-333, (1970).

23. E.H. Cooper, R. Turner, A. Geekie, et al: α-globulins in the surveillance of colorectal carcinoma. **Biomedicine** 24:171-178, (1976).

24. C.C. Harris, A. Primack, and M.H. Cohen: Elevated α_1-antitrypsin levels in lung cancer patients. **Cancer** 34:280-281, (1974).

25. W.K. Mueller, R. Handschumacher, and M.E. Wade: Serum haptoglobin in patients with ovarian malignancies. **J. Obstet. Gynecol.** 38:427-435, (1971).

26. H. Matsuura, and S. Nakazawa: Prognostic significance of serum α1-acid glycoprotein in patients with glioblastoma multiforme: A preliminary communication. **J. Neurol. Neurosurg. Psychiat.** 48:835-837, (1985).

27. J.F. Weiss, R.A. Morantz, and W.P. Bradley: Serum acute-phase proteins and immunoglobulins in patients with gliomas. **Cancer Res.** 39:542-544, (1979).

28. S. Galvez, A. Farcas, and M. Monari: The concentration of α_1-antitrypsin in cerebospinal fluid and serum in a series of 40 intracranial tumors. **Clin. Chim. Acta** 91:191-196, (1979).

29. M.N. Hart: Hyaline globules (Letter). **Arch. Pathol.** 96:144, (1973).

30. T. Kubota, A. Hirano, and S. Yamamoto: Electron microscopic study of hyaline inclusions in meningioma. **No. Shinkei. Geka.** 10:521-528, (1982).

31. D.E. Mullins, and S.T. Rohrlich: Role of proteinases in cellular invasiveness. **Biochim. Biophys. Acta** 695:177-214, (1983).

32. D.B. Rifkin, and R.M. Crowe: Isolation of a protease inhibitor from tissues resistant to tumor invasion. **Hoppe Seylers Z. Physiol. Chem.** 358:1525-1531, (1977).

33. E.W. Ades, A. Hinson, C. Chapuis-Cellier, et al: Modulation of the immune response by plasma protease inhibitors. I. α_2-macroglobulin and α_1-antitrypsin inhibit natural killing and antibody-dependent cell-mediated cytotoxicity. **Scand. J. Immunol.** 15:109-113, (1982).

34. D. Redelman, and D. Hudig: The mechanism of cell-mediated cytotoxicity. I. Killing by murine cytotoxic T lymphocytes requires cell surface thiols and activated proteases. **J. Immunol.** 124:870-878, (1980).

35. A. Akatsuka, S. Yoshimura, J. Hatu, et al: Intracellular localization of human plasma proteins revealed by peroxidase-antibody method in human tumors transplanted in nude mice. **J. Electron Microsc.** 28:93-99, (1979).

36. D. Hudig, T. Haverty, C. Fulcher, et al.: Inhibition of human natural cytotoxicity by macromolecular antiproteases. **J. Immunol.** 126:1569-1574, (1981).

37. S.N. Breit, D. Wakefield, J.P. Robinson, et al: The role of α_1-antitrypsin deficiency in the pathogenesis of immune disorders. **Clin. Immunol. Immunopathol.** 35:363-380, (1985).

38. S. Yoshimura, N. Tamaoki, Y. Ueyama, et al.: Plasma protein production by human tumor xenotransplanted in nude mice. **Cancer Res.** 38:3474-3478, (1978).

39. I. Reintoft, and I. Hagerstrand: Demonstration of α_1-antitrypsin in hepatomas. **Arch. Pathol. Lab Med.** 103:495-498, (1979).

SERPINS AND BRAIN TUMORS: ROLES IN PATHOGENESIS

JASTI S. RAO, RIICHIRO SUZUKI, AND BARRY W. FESTOFF

University of Kansas and Department of Veterans Affairs Medical Center
Neurobiology Research Laboratory (151)
Kansas City, MO 64128

INTRODUCTION

Brain tumors constitute a group of heterogeneous neoplasms. Pathophysiological mechanisms associated with the neoplastic process involve a complex balance of fibrinolytic and coagulation enzymes. The fibrinolytic enzyme system covers a wide spectrum ranging from tumor growth to tumor invasiveness, tumor hemorrhage and tumor host interactions leading to serious coagulopathies and thromboembolic complications.[1-4] The primary malignant brain tumors are also rich in fibrinolytic activity and are recognized for their low potential to metastasize, while they invade locally. The low potentiality of brain tumors to metastasize may be due to behavior of enzymes and the overproduction of inhibitors. Tumor spread correlates directly with tumor activation of fibrinolysis and inversely with inhibitors of fibrinolysis. Both proteolysis and anti-proteolytic activity have been associated with primary brain malignancies.

Tissue culture pioneers noted that cancer explants were capable of dissolving plasma clots,[5,6] and suggested that this phenomena was related to tissue degradation accompanying invasive growth of tumors. The process of tumor metastasis involves a sequence of events which include the making and breaking of cell-cell interactions, the movement of cells through tissue and movement into and out of blood or lymph vessels. These events may require the production of various proteolytic enzymes by the tumor cells.[7-9] Serine proteases are important to a discussion of limited proteolysis since they are involved in multi-enzyme cascade reactions such as thrombogenesis and fibrinolysis and often mediate these processes by converting inactive zymogens to active enzymes. Plasminogen activators, namely uPA and tPA, belong to a group of serine proteases that have been associated with neoplastic transformation, including tumor invasiveness and metastasis.[1-4]

Post-translational regulation of PA activity is accomplished by complex formation with certain specific polypeptide inhibitors which are members of a large family of protease inhibitors now known as serpins[10] and subsequent internalization of these complexes by specific cell surface receptors.[11] Four distinct PA inhibitors are known, protease nexin I (PNI),[1,12] plasminogen activator inhibitors, PAI-1,[13] PAI-2,[14] and PAI-3.[15]

Serine Proteases and Their Serpin Inhibitors in the Nervous System
Edited by B. W. Festoff
Plenum Press, New York, 1990

301

α_1-anti-trypsin or α_1 protease inhibitor is a model serpin,[10] a circulating inhibitor, upon which much of the chemistry of serine protease inhibition has been based. Interestingly, trypsin inhibitors were reported in human gliomas 20 years ago. Differences were found in subcellular localization of the trypsin inhibitor from normal brain when compared to the glioma trypsin inhibitor.[16,17] Recently, a qualitative study demonstrated the presence of α_1-anti-trypsin in a variety of human brain tumors.[18] The possibility of serpin involvement in degenerative nervous system diseases was postulated several years ago.[19] With the advent of molecular genetics, it has now been shown that serpin sequences are present in β-amyloid isolated from senile neuritic plaques of Alzheimer disease victim's brains.[20] Independently, three separate laboratories have found Kunitz-type serpin sequences in the A_4 (amyloid precursor protein or APP) segment of β-amyloid.[21-23] Another group has found substantial evidence for the presence of α_1-antichymotrypsin, another serpin, in neuritic plaques of Alzheimer disease brains.[24] Still another serpin, the glial derived nexin (GDN), has also been detected in rat and human gliomas.[25] This serpin is virtually identical to the β form of PNI.[26] Antibodies to GDN and PNI cross-react[25] and PNI is almost as active in neurite outgrowth promotion as is GDN,[27] most likely due to its inhibition of thrombin.[27]

Because of the potentially important biological roles of fibrin and fibrinolysis in brain tumors, and the possible specific functions of serpins in human gliomas[18] we determined the levels of these proteins in model rat 9L tumor cells[28] in vitro and in vivo and estimated the PNI mRNA levels from various human brain tumors obtained at time of specific neurosurgical procedure.

EXPERIMENTAL PROCEDURES

Materials

Human α thrombin was a gift of Dr. John Fenton II, Public Health Labs., Albany, New York. Plasminogen was prepared from fresh human plasma obtained from the Community Blood Center of Greater Kansas City (courtesy Dr. David Jenkins) and purified on a lysine-agarose column as described.[29] Fetal bovine serum (FBS) was from Hazleton - KC Biologicals (Lenexa, Kansas). Eagle's basal medium and antibiotics were from GIBCO (Grand Island, New York). Sea-Kem agarose and GelBond were from FMC Marine Colloids (Rockland, ME). Tissue culture plates were from Falcon (Oxnard, CA) and roller bottles from Corning (Fisher Scientific, St. Louis, MO). Trypsin (Type III, bovine pancreas Triton X-100 and Coomassie Brilliant Blue-G were from Sigma (St. Louis, MO). Plasminogen-free human fibrinogen was from KABI VITRUM (Helena Diagnostics, Beaumont, TX). Iodogen (diphenylglycouril) was from Pierce Biochemicals (Rockford, IL). ^{125}I-Na 1 (Amersham, Arlington Heights, IL) and ^{35}S-α-methionine (Dupont New England Nuclear) were all used according to manufacturer's instructions. Human urokinase was purchased from American Diagnostica (New York, NY).

Methods

Cell culture. 9L cells were grown in Eagle's basal medium (MEM) supplemented with vitamins, non-essential amino acids, antibiotics and 10% fetal or newborn calf serum at pH 7.2-7.4. Stock cultures were maintained by removing cells from a flask with 0.25% trypsin in Mg^{++} and Ca^{++} free Hank's basal salt solution (HBSS) containing 20.0 mg sterile Na_2 EDTA. The trypsin activity was inactivated by addition of complete MEM and the cells were counted and re-seeded into either 75 cm² flasks (1-2x10⁻⁶ cells) or 25 cm² (1-2x10⁻⁵ cells).

Preparation of conditioned media from 9L cells. 9L cells were grown in dishes as described above. At the 90% confluence level the cell layers were rinsed 3 times with serum-free media and then incubated for 24 hrs in the same media. Medium was collected and centrifuged at 5,000 x g to remove cellular debris.

Tissue processing. Normal rat brain tissue and 9L tumor was collected, placed immediately on dry ice, and transferred to a -80°C freezer until used. The tissue was thawed, weighed and homogenized in Tris-buffer (50 mM Tris-Cl, pH 9.5, containing 0.75 Na₂ EDTA and 75 mM NaCl) and centrifuged at 5,000 x g, for 30 min and discarded the pellet and aliquot supernatant.

Concanavalin-A chromatography. 9L CM and C2 CM were purified by chromatography on a Concanavalin-A-Sepharose column. The bound material was eluted with 0.01 M sodium phosphate pH 7.4, containing 0.5 M α-methyl D-mannoside[13] and the inhibitor-containing fractions were pooled, concentrated and stored at -80°C.

Inhibition or release of labeled extracellular matrix (ECM). Metabolically radiolabeled ECM was prepared using clonal muscle cells as described.[30,31] Myotube labeled matrices were shown previously to be excellent models to study ECM degradation.[30,31] This preparation is a modification of those used to study ECM degradation by metastatic cells.[32,33] Labeled ECM was incubated with the indicated enzymes (trypsin, uPA and plasminogen) in the presence or absence of diluted 9L CM, for 5 hrs at 37°C. The release, or inhibition of release, of ECM was assessed by removing 50 μl aliquots for measurement of radioactivity in a liquid scintillation counter (Packard model 2200CA, efficiency 95% for ³⁵S).

Fibrin zymography. SDS-PAGE was performed according to Laemmli,[34] in 10% gels using pre-stained standards and was followed by fibrin-plate zymography according to the method of Granelli-Piperno and Reich.[35] A current modification was binding the fibrinogen to Gelbound (FMC Bioproducts, Rockland, ME) for casting the fibrin gels.[36] After SDS-PAGE and removal of SDS by washing the acrylamide gels in 2.5% Triton X-100 the gels were transferred onto the pre-cast fibrin plates, incubated in a moist atmosphere at 37°C for 24 hours and then stained with amidoblack. Photography of the fibrin plates was performed with transillumination.

Reverse zymography. This technique, although similar to zymography[35,42] utilizes agar gels that also contain uPA, in addition to fibrinogen and plasminogen. The gels are rapidly digested leaving the inhibitor band clearly visualized, which is then marked with Indian ink and photographed.

Radioiodination of proteins. Purified human α thrombin or uPA was iodinated using the Iodogen method as described previously.[11,37] The fractions containing radioiodinated proteins were pooled and dialyzed against 2 liter of dd H₂0 for 6 hrs at 5°C. After dialysis, 50 μl were precipitated by TCA to determine the extent of incorporation of ¹²⁵I. The remainder of the radioiodinated protein preparation was stored at -80°C.

Complex formation studies. Serum-free 9L conditioned media or tissue homogenates from normal rat brain or 9L tumor were incubated with 1.7 mM SDS (or buffer only) at 37°C for 30 min and dialyzed against PBS containing 0.01% Tween

80 and 0.02% sodium azide at 4°C for 18 hrs. The pre-incubated samples, with and without SDS, were then incubated with ^{125}I-α thrombin or ^{125}I-uPA in the presence or absence of non-labeled thrombin at 37°C for 30 min. The reaction was stopped by the addition of SDS-PAGE sample buffer. ^{125}I-protease:inhibitor complexes were detected by SDS-PAGE and analyzed by autoradiography. The quantity of Th-PNI or uPA-PNI was determined by measuring radioactivity in gel slices after electrophoresis.

Extraction and purification of RNA. Total RNA were extracted from various grades of human brain tumors with SET buffer in the presence of proteinase K[38] or by guanidine thiocyanate as described previously.[39] The purity of the RNA is established by monitoring absorbance at 260-280 nM and ethidium bromide staining of RNA electrophoresed on agarose gels. Poly(A)-containing RNA is selected by oligo (dT) cellulose chromatography.[40]

Northern hybridization. A nick-translated PNI probe was used to estimate the levels of PNI by Northern hybridization.[41]

RESULTS

Inhibition of ECM Degradation by 9L Conditioned Medium

We investigated the effect of 9L cell serum-free conditioned medium (CM) on the destruction of extracellular matrix by urokinase and plasminogen to confirm whether 9L tumor cells secrete inhibitors of fibrinolysis (the plasmin/PA system). Metabolically-labeled mouse muscle cell ECM was used as previously described.[31] The amount of radioactivity released by trypsin was then taken as the 100% for calculation of total releasable activity while uPA + plasminogen released 90% (Table 1). The buffer control, as well as 9L CM plus plasminogen, released only 12-13% of label. It indicates that 9L CM might contain one or more protease inhibitors, or that these cells did not produce and/or secrete PA. To determine which was the case, we added 9L CM to trypsin, as well as to uPA plus plasminogen, and studied ECM degradation. Table 1 shows that 9L CM inhibited matrix release 60% and 85% in the case of trypsin and uPA plus plasminogen, respectively.

Fibrin Zymography

To further characterize the plasminogen activators produced by 9L cells *in vitro* and *in vivo* we utilized zymography, to provide directly the relative molecular weight of the plasminogen-dependent fibrin-degrading enzymes. Fibrin zymography of cell extracts and conditioned media showed a predominant zone of lysis around 48 Kd (uPA) in both medium and cell extract while 75 Kd (tPA) faintly in cell extract

Table 1. Release of label from [^{35}S]methionine-labeled extracellular matrix. Incubation of enzymes or media with metabolically-labeled muscle cell matrix as described in Materials and Methods.

Enzyme	%Release
Trypsin	100
Buffer control	12
uPA + plasminogen	90
9L CM + plasminogen	13
Trypsin + 9L CM	40
9L CM + uPA + plasminogen	15

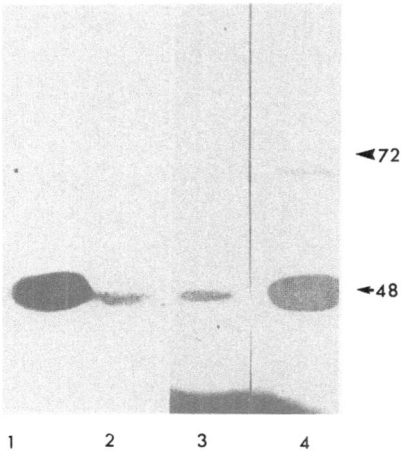

Figure 1. *Fibrin zymogram from 9L cells in vitro and in vivo.* Serum free medium (lane 1), and cell homogenate (lane 2) of 9L cells grown *in vitro.* Normal rat brain tissue homogenate (lane 3), and *in vivo* tumor homogenate (lane 4) run on SDS-PAGE for fibrin zymography as described in methods.

(Figure 1). Both normal and tumor brain extracts showed the presence of 48 Kd uPA and 75 Kd tPA; the level of uPA was increased in brain tumors with no change in tPA activity (Figure 1) when compared to normal brain tissue.

Reverse Fibrin Zymography

Reverse fibrin zymography was used to determine the molecular weights of fibrinolytic inhibitors present in the 9L CM. For comparison we also used CM from a clonal, murine skeletal muscle cell line, C-2. We concentrated and partially purified any PAI-1 that was present by absorption to Con A Sepharose. Figure 2 shows the reverse zymogram in which the presence of a single 51 Kd inhibitor in the concanavalin A-purified material from C2 cell CM (lane 1). On the other hand, concanavalin A-purified material from 9L CM shows inhibitory zones at both 51 Kd (a narrow zone) and a broader zone migrating at 55 Kd.

Complex Formation Studies with Labeled Urokinase or Thrombin

Complex formation studies with labeled urokinase or thrombin were used to estimate the level of active serpin. Incubation of ^{125}I-thrombin or ^{125}I-urokinase with tissue homogenates of normal rat brain or 9L tumor growing in rat brain, with and without SDS pre-treatment, resulted in the formation of unique high molecular weight complexes. Figure 3A shows a complex at 78 Kd with labeled thrombin in lanes 1 and 3. No complexes were detected in lanes 2 or 4 when incubation was in the presence of excess unlabeled thrombin, suggesting PNI. More importantly, the data in Figure 3A suggests that the active PNI-like factor was present in much larger quantity in the tumor (lane 3) than in normal brain tissue (lane 1). The complex bands with labeled thrombin or urokinase were excised and analyzed in a gamma counter. These results indicate that the complex formation increased four times in 9L tumor in comparison to normal brain tissue on a per μg protein loaded basis (Table 2).

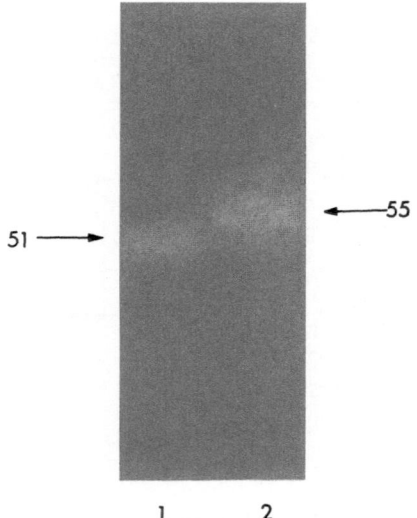

Figure 2. 9L CM and C2 CM were purified by chromatography on a concanavalin-A column, fractionated by SDS-PAGE and then analyzed by reverse zymography as described in Methods [reverse zymogram from C2 (lane 1) and 9L conditioned media (lane 2)].

Strikingly, Figure 3B reveals that the tumor (lane 3), but not the normal brain tissue (lane 1), contained a <u>latent</u> SDS-activated factor that formed 88 Kd complexes with thrombin. The figure suggests that this thrombin inhibiting factor is relatively abundant in the 9L tumor. As shown in Figure 3B, complex formation was inhibited (lane 4) by excess non-labeled thrombin demonstrating saturability and specificity.

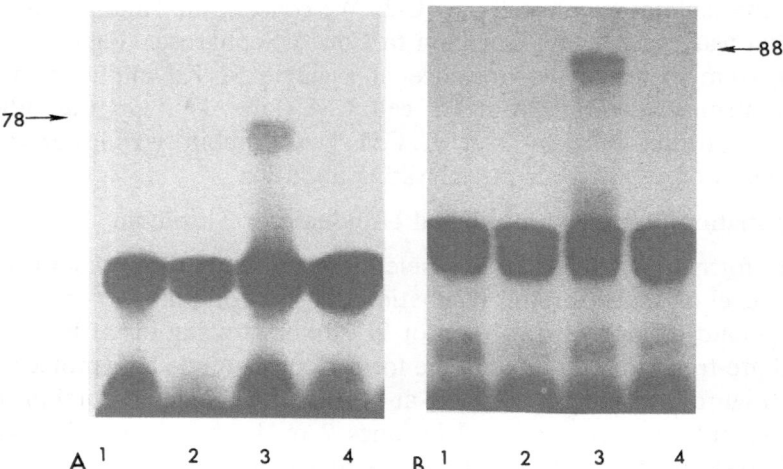

Figure 3. *Complexes formed with iodinated thrombin by extracts of normal brain and 9L tumor with and without SDS pre-treatment.* Samples from normal brain and tumor tissue homogenates were incubated with labeled thrombin for 30 minutes at 37°C in the presence (even numbered lanes) or absence (odd numbered lanes) of excess non-labeled thrombin without (Figure 3A) and with (Figure 3B) SDS pre-treatment. The samples were analyzed by SDS-PAGE and autoradiography. Lanes 1 and 2, normal brain homogenate; 3 and 4, tumor homogenate.

Table 2. Complex Formation Studies with Rat Normal Brain and Tumor Extracts. Tissue extracts were incubated with labeled Th or uPA at 37°C for 30 min in the presence or absence of cold thrombin. The samples were analyzed by 10% SDS-PAGE and autoradiography. The gels were sliced and counted.

Condition	cpm/μg of protein	
	^{125}I-Th-PNI	^{125}I uPA-PNI
Normal brain	1460 ± 115	2765 ± 190
Tumor	5894 ± 215	12432 ± 368

Northern Hybridization

Total RNA was extracted from various human brain tumors and hybridized with a full-length ^{32}P-cDNA to human PNI. Figure 4 shows the presence of 2.7 Kb PNI mRNA in human fibroblasts, tissues of normal and human brain tumors. Compared to normal brain tissue PNI mRNA levels were increased in astrocytoma and glioblastoma.

DISCUSSION

Our present results indicate (Table 1) that serum-free 9L CM inhibits the release of labeled ECM in the presence of trypsin or uPA plus plasminogen. This demonstrates that the 9L CM contained powerful and sufficient serine protease inhibitor(s) capable of inhibiting both exogenous trypsin and PAs. Zymography confirmed that these cells did, indeed, secrete activatable uPA (Mr of 48 Kd, Figure 1) with the molecular weight of rodent urokinase.[9,42] The protease inhibitor activity detected by reverse zymography (Figure 2) is consistent with the factor(s) being responsible for the relative absence of matrix degradation when plasminogen, with or without uPA, was added to 9L CM (Table 1).

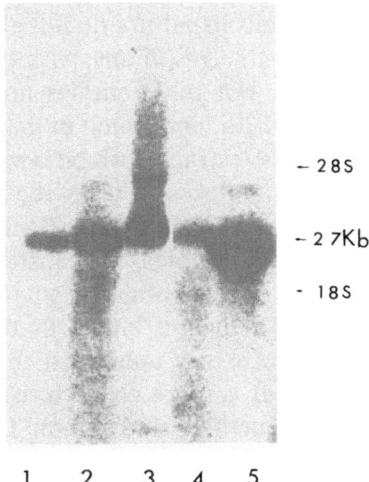

Figure 4. Northern hybridization with a full length ^{32}P labeled human PNI cDNA hybridized to total RNA from normal human brain, and brain tumor tissues. Total RNA from normal human brain (lane 1). Total RNA from human brain tumors (lanes 2-4); glioblastoma, anaplastic astrocytoma, and meningioma, respectively. Total RNA from human fibroblast used as a control (lane 5).

High levels of plasminogen-dependent fibrinolytic activity occur in cells derived from an ethylnitrosourea-induced glioma of rat brain when compared to normal rat brain.[43] An earlier report demonstrated that fresh human brain tumors showed lysis zones on fibrin plates.[44] Fibrinolytic and thromboplastic activities were variable in different grades of tumors. Recent reports indicated that PA activity was highest in metastatic tissue and acoustic neuroma, followed by glioblastoma, meningioma and low grade glioma and the Mr pattern for each type of brain tumor was entirely different with few exceptions.[45] Our results also showed the same pattern of PA activity (data not shown) in human brain tumors.

To date, four PA inhibitors, PNI, PAI-1, PAI-2 or PAI-3 have been identified and well-characterized.[11-15] PNI from human fibroblasts effectively inhibited fibrosarcoma (HT-1080) cell-mediated degradation of ECM.[33] Our own earlier results indicated that a PNI-like protein purified from a clonal muscle cell line also inhibits the hydrolysis of labeled ECM by the uPA/plasminogen system.[31] When concanavalin A-purified material from 9L CM was analyzed by reverse zymography two inhibitory zones at 51 Kd (a narrow zone) and a broader zone migrating at 55 Kd were found (Figure 2). It is known that human PAI-1 yields an inhibitory zone in reverse zymography at 51 Kd.[13] The 55 Kd broader inhibitory band might represent a novel inhibitor, also showing affinity for concanavalin A, or a somewhat larger variant of PAI-1.

We further characterized these protease inhibitors by complex formation studies with ^{125}I-labeled urokinase or thrombin. Pre-treatment with SDS was used to determine SDS sensitivity or reactivation of latent inhibitors.[11,13] Incubation was also performed to assess specificity and saturability in the presence or absence of excess non-labeled thrombin or urokinase. A factor in 9L CM had the characteristics of PNI,[11] an SDS-sensitive inhibitor of thrombin and urokinase (data not shown). However, SDS-pre-treated 9L CM samples (data not shown) also revealed a separate and distinct 78 Kd complex with labeled urokinase which was not inhibited by excess non-labeled thrombin. These are the characteristics of PAI-1, and SDS-activated inhibitor of PAs, but not of thrombin.[13] It have previously been reported that SDS and excess of non-labeled thrombin or urokinase inhibit the complex formation by PNI with labeled thrombin or urokinase.[11,12]

Since the characteristics of cells in culture do not always mimic in vivo conditions, we were interested to learn what changes, if any, with these inhibitors occurred when 9L cells grew as tumors in vivo. For these studies normal rat brain and 9L tumor extracts were pre-treated with SDS and incubated in the presence or absence of excess non-labeled thrombin. Brain tissue extracts formed typical PNI-type 78 Kd complexes with labeled thrombin. These complexes were increased in the tumor compared to normal brain (Table 2). A unique 88 Kd (Figure 3B, lane 3) complex was detected, but only in SDS-pre-treated tumor samples. Exhaustive studies failed to detect the presence of 88 Kd complexes in normal brain or when 9L was grown in vitro. It is of interest that this antithrombin activity is latent, that is, it is only detected in SDS-pre-treatment, and specific, since excess unlabeled thrombin prevents detection of labeled 88 Kd complexes. This Mr is more characteristic of resistant complexes with ^{125}I-thrombin to antithrombin III or heparin cofactor II. However, this brain tumor thrombin inhibitory factor exhibits novel behavior, since it is inactive until treated with SDS, in contrast to other thrombin inhibitors, ATIII or heparin cofactor II, which are inactivated by SDS. Typical PNI-type complexes (72 Kd) were found with labeled urokinase, inhibited by excess non-labeled thrombin or pre-treatment with SDS, and these were also increased in the 9L tumor (Table 2).

Figure 4 showed that increased levels of PNI mRNA were found in brain tumors when compared to normal tissue, more prominent in astrocytoma and glioblastoma. This serpin has been found to be more abundant in neonatal compared to adult rat brain suggesting a possible link between expression of this protein and brain, predominantly glial, growth.[25] Many studies have reported increased deposition of amyloid precursor protein (APP) in Alzheimer's disease victims' brains but decreased levels of PNI.[46] Recent reports indicate that the APP protein sequence is identical to protease nexin II (PN-II),[47,48] a 105 Kd chymotrypsin inhibitor.

SUMMARY

The plasminogen activators, urokinase type (uPA) and tissue type (tPA), are members of a group of proteases and key fibrinolytic enzymes which have been repeatedly implicated as mediators of tumor cell invasion and cellular destruction. Regulation of these proteases may occur by surface associated inhibitors. Four potential physiological inhibitors, namely protease nexin I (PN-I), and the plasminogen activator inhibitors, PAI-1, PAI-2, PAI-3 have been purified and cloned. As with other malignancies, plasminogen activators (PAs) and inhibitors (serpins) appear to play important roles in brain tumor biology, since primary brain tumors locally invade but do not metastasize. We have studied the 9L gliosarcoma, as a model for brain tumor therapy, as well as various human tumors obtained at biopsies, to evaluate the balance of PAs and serpins in brain tumor pathogenesis. Our results indicate that cells cultured from the tumor *in vitro* express PA inhibitory activities of both PN-I and PAI-1. However, the serpin PA inhibitory activity in extracts from both the normal brain and brain tumors are of the PN-I type, and this activity is found to be higher in the tumor than in the normal tissue. In addition, we present evidence and exciting new data for a novel thrombin inhibitor which a) is only present in the tumor growing in the rat brain, since it is undetectable in either the normal brain tissue or the cells *in vitro*; b) this inhibitor is in a latent state, and may be activated by SDS to form 88 kD complexes with thrombin; c) this inhibitor appears to be distinct from the other thrombin inhibitors such as antithrombin III (AT III), or heparin co-factor II (HC-II); d) the presence of excess quantity of unlabeled thrombin inhibits the formation of the inhibitor:[125]I-thrombin complexes; e) this inhibitor appears not to bind uPA. Our results with human brain tumors also indicate increased levels of PN-I, by Northern blot hybridization with a full-length, nick-translated PN-I cDNA probe when compared to normal brain tissue. These results suggest that serpins may play an important role in brain tumor pathogenesis.

ACKNOWLEDGMENTS

Support for these studies was provided, in part, by the Speas Foundation (JSR) (University of Kansas Medical Center), the National Science Foundation (BWF), and the Medical Research Service of the Department of Veterans Affairs (BWF). We thank Joyce Capps and Lori Peterson for preparation of the manuscript.

REFERENCES

1. G. Girmann, H. Pees, and G. Schwarze, Immunosuppression by micromolecular fibrinogen degradation products in cancer, Nature 259:339-401 (1976).
2. E.F. Plow, D. Freaney, and T.S. Edgington, Inhibition of lymphocyte protein synthesis by fibrinogen-derived peptides, J. Immunol. 128:1595-1599 (1982).
3. R. Ng, and J.A. Kellen, The role of plasminogen activators in metastasis, Med. Hypothesis 10:291-293 (1983).
4. N.A. Booth, R. Bennett, and G. Wijngaards, A new lifelong hemorrhage disorder due to excess plasminogen activator, Blood 61:269-275 (1983).
5. A. Carrel, and M.T. Burrows, Cultivation *in vitro* of malignant tumors, J. Exp. Med. 13:571-575 (1911).
6. P. Goldhaber, I. Cornman, and R.A. Ornsbee, Experimental alteration of ability of tumor cells to lyse plasma clots in vitro, Proc. Soc. Exp. Biol. Med. 66:590-595 (1947).
7. D.E. Mulling, and S.T. Rohrlich, The role of proteinases in cellular invasiveness. Biochem. Biophys. Acta 695:177-214 (1983).
8. T. Kalebic, S. Garbisa, B. Glaser, and L.A. Liotta, Basement membrane collagen: degradation by migrating endothelial cells. Science 221:281-283 (1983).
9. K. Danø, P.A. Andreasen, J. Grøndahl-Hansen, J.P. Kristensen, L.S. Nielsen, and L. Skriver, Plasminogen activators, tissue degradation and cancer, Adv. Cancer Res. 44:139-266 (1985).
10. R.W. Carrel, α_1-antitrypsin molecular pathology, leukocytes and tissue damage, J. Clin. Invest. 78:1427-1431 (1986).
11. J.B. Baker, D.A. Low, R.L. Simmer, and D.D. Cunningham, Protease nexin: A cellular component that links thrombin and plasminogen activator and mediates their binding to cells, Cell 21:37-47 (1980).
12. R.W. Scott, and J.B. Baker, Purification of human protease nexin. J. Biol. Chem. 258:10439-10444 (1983).
13. D.J. Loskutoff, and T.S. Edgington, Synthesis of a fibrinolytic activator and inhibitor by endothelial cells. J. Biol. Chem. 256:4142-4145 (1982).
14. L. Holmberg, I. LeCander, B. Persson, and B. Astedt, An inhibitor from placenta specifically binds urokinase and inhibitor plasminogen activator released from ovarian carcinoma in tissue culture, Biochem Biophys. Acta 544:128-137 (1978).
15. D.C. Stump, M. Thienpont, and D. Collen, Purification and characterization of a novel inhibitor of urokinase from human urine. Quantitation and preliminary characterization. J. Biol. Chem. 261:12759-12766 (1986).
16. A. A. Brecher, and N.M. Quinn, The distribution of a particulate trypsin inhibitor in brain. J. Neurochem. 14:203-206 (1967).
17. B.B. Oliphant, J.B. Suszkiw, and A.S. Brecher, The presence of distinct soluble and particulate trypsin inhibitors in human astrocytoma tissue. Proc. Soc. Exp. Biol. Med. 135:754-756 (1970).
18. R. Sawaya, M. Zuccarello, and R. Highsmith, α_1-antitrypsin in human brain tumors. J. Neurosurg. 67:258-262 (1987).
19. B.W. Festoff, Neuromuscular junction macromolecules in the pathogenesis of Amyotrophic Lateral Sclerosis, Med. Hypothesis 6:121-132 (1980).
20. R.W. Carrell, Alzheimer' disease: Enter a protease inhibitor. Nature 331:478-479 (1988).
21. N. Kitaguchi, Y. Takahaski, Y. Tokushima, S. Shiotiri, and H. Ito, Novel precursor of Alzheimer's disease amyloid protein shows protease inhibitory activity, Nature 331:530-532 (1988).
22. P. Ponte, D.P. Gonzales, J. Schilling, T. Miller, D. Hsu, B. Greenberg, K. Davis, G.W. Wallace, T. Lieberburg, F. Fuller, and Crodell, A new A_4 amyloid mRNA contains a domain homologous to serine proteinase inhibitors, Nature 331:525-527 (1988).
23. R.E. Tanzi, A.I. McClatchey, E.D. Lamperti, L. Villa-Komaroff, J.F. Gusella, and R.L. Neve, Protease inhibitor domain encoded by an amyloid protein precursor mRNA associated with Alzheimer's disease, Nature 33:528-530 (1988).
24. C.R. Abraham, D.J. Selkoe, and H. Potter, Immunochemical identification of the serine protease inhibitor α_1-antichymotrypsin in the brain amyloid deposits of Alzheimer's disease, Cell 52:487-501 (1988).
25. S. Gloor, K. Odink, J. Guenther, H. Nick, and D. Monard, A glia derived neurite promoting factor with protease inhibitory activity belongs to the protease nexins, Cell 47:687-693 (1986).
26. M. McGrogan, J. Kennedy, M.P. Li, C. Hsu, R.W. Scott, C. Simonsen, and J.B. Baker, Molecular cloning and expression of two forms of human protease nexin I, Biotechnology 6:172-177 (1988).

27. D. Gurwitz, and D.D. Cunningham, Thrombin modulates and reverses neuroblastoma neurite outgrowth, Proc. Natl. Acad. Sci. 85:3440-3444 (1988).

28. M.L. Rosenblum, K.D. Knebel, K.T. Wheeler, and C.B. Wilson, Development of an *in vitro* colony formation assay for the evaluation of *in vivo* chemotherapy of a rat brain tumor. In Vitro 11:264-273 (1975).

29. D.G. Deutsch, and E.T. Mertz, Plasminogen: purification from human plasma by affinity chromatography, Science 170:1095-1096 (1970).

30. R.L. Beach, W.V. Burton, W.J. Hendricks, and B.W. Festoff, Production of extracellular matrix by cultured muscle cells: Basal lamina proteins, J. Biol. Chem. 257:11437-11442 (1982).

31. J.S. Rao, C.B. Kahler, J.B. Baker, and B.W. Festoff, Protease nexin I, a serpin, inhibits plasminogen-dependent degradation of muscle extracellular matrix, Muscle & Nerve 12:640-646 (1989).

32. L. Werb, M.J. Banda, and P.A. Jones, Degradation of connective tissue matrices by macrophages. III. Morphological and biochemical studies on extracellular, pericellular, and intracellular events in matrix proteolysis by macrophages in culture, J. Exp. Med. 152:1537-1553 (1980).

33. B.L. Bergman, R.W. Scott, A. Bajpai, S. Watts, and J.B. Baker, Inhibition of tumor-cell mediated extracellular matrix destruction by a fibroblast proteinase inhibitor protease nexin I, Proc. Natl. Acad. Sci. USA 83:996-1000 (1986).

34. U.K. Laemmli, Cleavage of structural proteins during the assembly of the bed of bacteriophage T-4, Nature (London) 227:680-685 (1970).

35. A. Granelli-Piperno, and E. Reich, A study of proteases and protease-inhibitor complexes in biological fluids, J. Exp. Med. 148:223-224 (1978).

36. D. Hantaï, J.S. Rao, C.B. Kahler, and B.W. Festoff, Developmental regulation of plasminogen activators in post-natal skeletal muscle, Proc. Natl. Acad. USA 86:362-366 (1989).

37. B.W. Festoff, M.R. Patterson, and K. Romstedt, Plasminogen activator: the major secreted neutral protease of cultured skeletal muscle cells, J. Cell Physiol. 110:190-195 (1982).

38. D.W. Rowe, R.C. Moen, J.M. Davidson, P.H. Byers, and P.D. Palmiter, Correlation of procollagen mRNA levels in normal and transformed chick embryo fibroblasts with different rates of procollagen synthesis, Biochemistry 17:1581-1589 (1978).

39. J.H. Han, C. Stratowa, and W.J. Rutter, Isolation of full-length putative rat lysophospholipase cDNA using improved methods for mRNA isolation and cDNA cloning, Biochem. 26:1617-1625 (1987).

40. H. Aviv, and P. Leder, Purification of biologically active globin messenger RNA by chromatography on oligothymidylic acid-cellulose, Proc. Natl. Acad. Sci. USA 69:1408-1413 (1972).

41. P.S. Thomas, Hybridization of denatured RNA and small DNA fragments transferred to nitrocellulose, Proc. Natl. Acad. Sci. USA 77:5201-5205 (1980).

42. B.W. Festoff, D. Hantaï, J. Soria, A. Thomaidis, and C. Soria, Plasminogen activator in mammalian skeletal muscle: Characterization of effects of denervation on urokinase-like and tissue activators. J. Cell Biol. 103:1415-1421 (1986).

43. T.A. Hince, and J.P. Roscoe, Differences in pattern and level of plasminogen activator production between a cloned cell line from an ethylnitrosourea-induced glioma and one from normal adult rat brain. J. Cell Physiol. 106:199-207 (1980).

44. H. Kraus, Fibrinolytic activity of cerebral tumors and the treatment of a tendency to hemorrhage. Arch. Neurol. Neurochir. Psychiatr. 108:27-31 (1971).

45. R. Sawaya, and R. Highsmith, Plasminogen activator activity and molecular weight patterns in human brain tumors. J. Neurosurg. 68:73-79 (1988).

46. S.L. Wagner, J.W. Geddes, C.W. Cotman, A.L. Lau, D. Gurwitz, P.J. Isackson, and D.D. Cunningham, Protease nexin I, an antithrombin with neurite outgrowth activity, is reduced in Alzheimer's disease. Proc. Natl. Acad. Sci. USA 86:8284-8288 (1989).

47. T. Oltersdorf, L.C. Fritz, D.B. Sehenk, I. Lieberburg, K.L. Johnson-Wood, E.C. Beattie, P.J. Ward, R.W. Blacher, H.F. Dovey, and S. Sinha, The secreted form of Alzheimer's amyloid precursor protein with the Kunitz domain is protease nexin-II. Nature 341:144-147 (1989).

48. W.E. Van Nostrand, S.L. Wagner, M. Suzuki, B.H. Choi, J.S. Farrow, J.W. Geddes, C.W. Cotman, and D.D. Cunningham, Protease nexin-II, a potent antichymotrypsin, shows identity to β-amyloid precursor protein. Nature 341:546-549 (1989).

A SERINE PROTEASE INHIBITOR DOMAIN ENCODED WITHIN THE ALZHEIMER DISEASE-ASSOCIATED AMYLOID β-PROTEIN PRECURSOR GENE

RUDOLPH E. TANZI

The Molecular Neurogenetics Laboratory
Massachusetts General Hospital
Boston MA 02114

INTRODUCTION

A crucial event in the pathogenesis of AD involves the aggregation of a 4.2 kiloDalton (kd) hydrophobic peptide referred to as the amyloid β- protein (ABP) into insoluble extracellular proteinaceous fibers. These aggregates take the form of both senile (neuritic) plaques (SP) and cerebrovascular amyloid (CVA) deposits.[1] Besides AD, amyloid deposits also occur in Dutch cerebral amyloidosis,[2] older patients with Down syndrome (DS; trisomy 21), and to a limited extent in normal aged individuals. The formation of amyloid plaques and blood vessel CVA deposits appears to correlate well, although not perfectly, with the degree of dementia in AD patients.[3] Whether amyloidogenesis is a primary or secondary event in AD-related neuropathology is not known. However, given the association between neuronal cell death and the presence of amyloid in AD and DS, multiple laboratories have focused their efforts on delineating the mechanism by which the 40 amino acid peptide, ABP, is generated and subsequently aggregated into insoluble amyloid fibrils. ABP is derived from a much larger precursor protein (APP) encoded by a gene located on chromosome 21.[4-7] This precursor protein may consist of 695, 751 , or 770 amino acids depending on the alternate splicing of two exons of 168 and 57 basepairs.[5,8-10] The larger exon encodes a functional serine protease inhibitor domain in the Kunitz family. The effect of this inhibitor on the proteolytic processing of APP and the generation of amyloid is not yet known.

The Relationship Between the APP Gene and Familial Alzheimer's Disease

Physical mapping techniques employing somatic cell hybrid panels and *in situ* hybridization place the APP gene immediately proximal to the obligate Down syndrome region at the border of bands 21q21.3 and 21q22.[11] In other studies, APP has been mapped more proximally around 21q21.1.[12,13] The localization of the APP gene on chromosome 21 implies that the presence of β-amyloid deposits in the brains of patients with DS could be most easily attributed to gene dosage as a result of the third copy of the APP gene. Increased levels of APP mRNA have been recently dem-

Serine Proteases and Their Serpin Inhibitors in the Nervous System
Edited by B. W. Festoff
Plenum Press, New York, 1990

313

onstrated in DS fetal brain when compared to a control sample (Figure 1. lanes 1 and 2). Interestingly, the amount of APP751/770 mRNA in the DS brain is approximately eight-fold higher than in the DS brain. This value is considerably higher than the expected 1.5 fold increase predicted for a linear gene dosage effect. Therefore, either three copies of the APP gene may somehow deregulate APP gene expression, or, alternatively, other genes on chromosome 21 may somehow upregulate the APP gene when in a trisomic state. APP gene dosage at the germline level does not appear to account for the generation of amyloid in AD since gene duplication of APP has not been consistently observed in brain and lymphocyte DNA of AD patients.[14-16]

Recently, a locus for familial Alzheimer's disease (FAD) has been localized on chromosome 21 by genetic linkage analysis of four large FAD kindreds and DNA markers specific for chromosome 21.[17] One of the linked markers, D21S1/S11, is also genetically linked to the APP gene, thus, implying that the APP and FAD loci reside in the same general vicinity of chromosome 21.[7] Subsequent studies have demonstrated that the APP and FAD genes are not identical but separated by a considerable genetic distance.[18] Whether it is merely coincidental that the FAD and APP genes reside on the same chromosome in the same general vicinity, or the two loci interact in some way at the transcriptional or translational levels is unknown. One possibility would involve a *cis* or *trans* effect of the FAD gene or it's product on the expression of the APP gene.

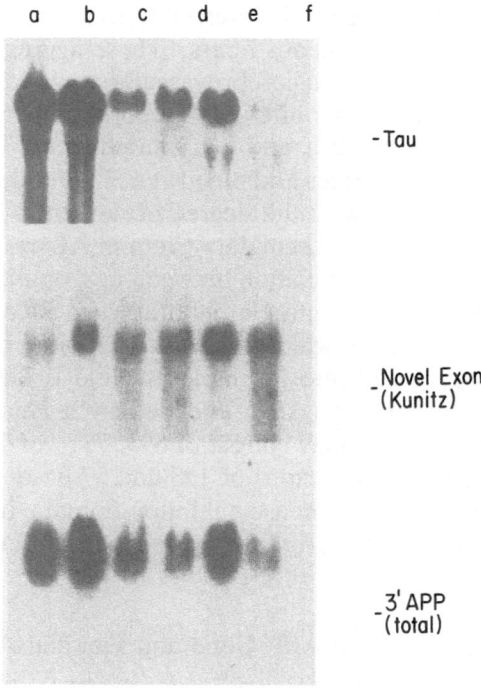

Figure 1. *Northern blots of hybridizations to total RNA using a protease inhibitor-encoding exon and a 1.1 kilobase probe from the 3' end of APP.* 25 μg of RNA were loaded from 19-week normal (lane a) and trisomy 21 (lane b) brains, adult normal (lane c) and AD (lane d) cerebellum and adult normal (lane e) and AD (lane f) frontal cortex. Control hybridization with a cDNA for the microtubule-associated protein tau is also shown to compare RNA load. The three autoradiograms are the results of independent hybridizations with the same filter.

A Kunitz-type Serine Protease Inhibitor Domain within APP

The APP gene produces at least three different messages as a result of alternative splicing of two adjacent exons. The two largest transcripts, APP751 and APP770 contain a 56 amino acid domain that interrupts residue 289 of APP695.[8-10] This domain represents a Kunitz-type serine protease inhibitor and has been shown to inactivate trypsin *in vitro*.[8] In this experiment, Cos cells were transfected with APP751, and conditioned media was used to successfully inhibit trypsin activity. The full-length amino acid sequence of APP resembles an integral membrane-bound protein[5] with the Kunitz protease inhibitor (APP-KPI) domain located in the large extracellular portion of the molecule. The extracellular portion of APP undergoes both O-linked, and N-linked glycosylation and is cleaved and released into the extracellular space.[19] Sinha et al. (these proceedings) have reported that APP751-derived secreted polypeptides possess strong trypsin inhibitory activity capable of preventing their own degradation, as well as the proteolysis of synthetic substrates when challenged with exogenous trypsin. Significantly, thrombin activity was not inhibited by the APP-KPI domain. It is, therefore, unlikely that APP-KPI plays a role in the nervous system analogous to that of the serine protease inhibitor class of neurite promoting factors referred to as nexins.

The demonstration of trypsin inhibitory activity by APP-KPI agrees with the observation that the reactive site in this domain consists of an arginine flanked by a cysteine and an alanine indicating a specificity for trypsin and trypsin-like proteases. This same reactive site motif is employed in another member of the Kunitz family present in human, the so-called HI-30 portion of the serum protease inhibitor, inter-α-trypsin inhibitor (ITI).[20] HI-30 is actually composed of two interconnected KPI domains (I and II), each with a relative molecular weight of 7 kd, but with two carbohydrate groups on the N-terminus of domain I. The APP-KPI domain exhibits a greater degree of homology to Kunitz domain II of HI-30 which is highly selective for trypsin-like proteases.[21] These data suggest that in considering proteases that may be involved in processing APP, special emphasis should be placed on trypsin-like serine proteases that are inactivated by ITI *in vitro* (e.g. cathepsin G, chymotrypsin, and plasmin). Additional support for the involvement of serine proteases and their inhibitors in plaque formation derives form the finding that α_1-antichymotrypsin is present in SP.[22]

Given the similarities between APP-KPI and ITI, it is also worthwhile to consider the available data on the function of ITI HI-30 domain. HI-30 domain II is identical to the endothelial cell growth factor ECGF 2b.[23] This proliferative effect may be somewhat non-specific since other serine protease inhibitors have been shown to stimulate endothelial cell growth *in vitro* as well. It has also been observed that HI-30-protease complexes are cleared from the bloodstream at a much slower rate than other protease-protease inhibitor complexes presumably due to deposition in extravascular spaces.[24] It would be interesting to explore whether these observations bear any similarity to the process by which the amyloid β peptide is concentrated in the cerebrovascular spaces of AD patients. Interestingly, HI-30 levels in plasma are increased up to 500-fold in response to certain states of neoplasia and inflammation suggesting that it may function as an acute-phase reactant protein.[25] Given that APP751 and 770 (KPI⁺) are the major forms of APP in the peripheral organs including liver, kidney, and spleen,[10] it might be useful to test for upregulation of APP in the peripheral organs and plasma of AD patients. If AD, and especially, FAD, begin as systemic disorders centralized around the cerebral blood vessels, a set of abnormal conditions involving the upregulation of APP751 and 770 as acute phase reactant-like proteins might be important to consider in the formation of cerebrovascular amyloid.

Expression of Alternate Transcripts for the APP Gene

Northern blot analysis reveals that the KPI-containing APP transcripts, APP751 and 770, are present in both neuronal and non-neuronal tissues, while the APP695 transcript, lacking the inhibitor, appears to be more limited to the brain.[9,16] In addition, while transcripts for APP695 appear to be highly abundant in fetal tissues, APP751 RNA is present at higher levels in adult tissues. This suggests that the precursor lacking the KPI domain may be required in larger amounts during fetal brain development. The ability of the brain to alternatively splice out the exon encoding the KPI domain from APP RNA suggests that the protease inhibitor domain may regulate some activity of APP that is more crucially needed during fetal brain development (e.g. synaptogenesis). The simplest explanation would be that the protease inhibitor domain prohibits specific cleavages of APP (or other proteins associated with plasma or intracellular membranes) that are vitally needed during periods of neuronal plasticity.

The presence of a functional protease inhibitor in APP might have profound effects on the process of amyloidogenesis in AD (Figure 2). On one hand, amyloid formation might be prevented by the KPI domain by its ability to inhibit certain proteases capable of releasing ABP or its penultimate proteolytic intermediate. On the other hand, the KPI domain could potentially accelerate the generation of ABP by impeding specific proteases from degrading this peptide or its penultimate fragment. Given the ability of the KPI domain to protect APP from trypsin degradation *in vitro* (Sinha, these proceedings), it is conceivable that the proteolytic intermediates of APP might differ appreciably depending upon the presence or absence of the KPI domain and that only one of these sets proceeds toward subsequent modification into amyloid plaques and deposits. It is therefore an important task to determine whether APP695, 751, or 770 gives rise to amyloid fibrils.

If APP molecules destined for amyloidogenesis are neuronally-derived, it is important to determine which forms of APP transcripts (with or without the KPI encoding exon) are prevalent in affected regions of AD brain. Preliminary results

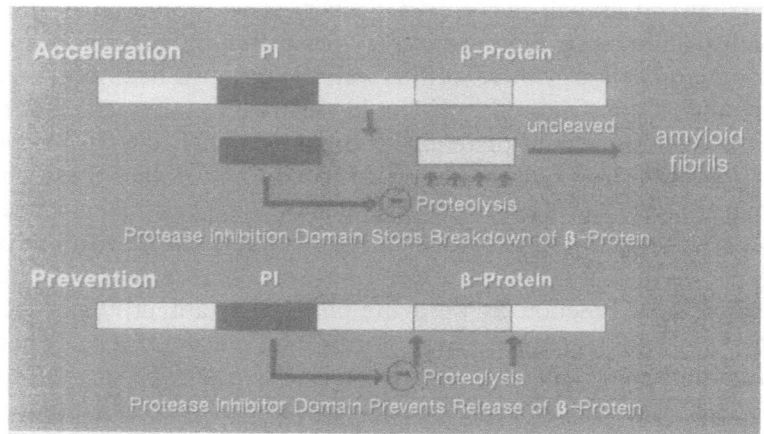

Figure 2. *Possible Effects of APP Protease Inhibitor Domain on β-Amyloid Formation.*

have been somewhat contradictory. Northern blot analysis indicates a selective reduction of APP695 (lacking the inhibitor) in the frontal cortex of patients with AD, with relatively little change in the amount of KPI-containing transcripts (Figure 1., lanes 5 and 6).[10,26] Message levels for virtually all neuronal genes tested on Northern blots of mRNA derived from affected regions of AD brains are drastically reduced and this is commonly attributed to neuronal cell loss. Unlike other neuronal mRNA's, the messages for KPI-containing APP are relatively normal in affected regions of AD brain (Figure 1., lane 6). Therefore, one interpretation of this result is that the neurons which primarily express APP695 are reduced in number resulting in lower levels of APP695 message. If APP751 and 770 in the brain are produced by non-neuronal cell populations such as glial cells and microglia, their levels would not be as drastically affected. An alternate explanation for the selective reduction in APP695 is that since overall protein synthesis is diminished in neurons from affected brain regions, protein factors necessary for splicing out the KPI-encoding exon (to produce APP695) would also be limited, resulting in greater levels of APP751 and/or APP770 transcripts. In support of this latter possibility, Tanaka et al. have reported an increase in affected regions of AD brain in the ratio of APP770 to all other forms of APP.[27] APP770, the largest transcript would require presumably no alternative splicing. A final conceivable explanation for the sustained levels of APP751/770 in affected regions of AD brain is that this species of mRNA may be particularly resistant to degradation by ribonucleases.

In situ hybridization of APP cRNA probes to sections of affected brains of eleven Alzheimer's patients reveals a different story. Palmert et al report 2-3 fold increases in overall APP mRNA in the nucleus basalis of Meynert and in the locus coeruleus but not in the subiculum, basis pontis, or occipital cortex of brains from AD patients.[28] Using a cRNA probe specific for KPI-encoding APP transcripts, no differences are observed in APP mRNA levels in these two same sub-cortical regions. The authors conclude that the increase in APP must be due to APP transcripts lacking the KPI domain (e.g. APP695). Alternatively, given the extent of nucleic acid homology between APP-KPI and other Kunitz type serpins (e.g. >50% with HI-30), it is possible that the KPI probe is able to detect other homologous KPI-encoding mRNA's present in the nBM and LC neurons, thereby masking the effect of any increase in APP751 and 770. Confirmation of these results will, therefore, require *in situ* hybridization using junctional oligonucleotide probes specific for the APP695 transcript.

FINAL CONSIDERATIONS

The expression studies described above make the assumption that amyloid deposits are derived from neuronally-produced APP. It is equally possible that systemic or blood-derived APP may be the actual source of cerebrovascular and senile plaque amyloid. APP polypeptides have been observed in both human plasma and CSF (D. Selkoe, personal communication; and unpublished results). To address this possibility it will be necessary to study APP mRNA levels in peripheral tissues of AD and control individuals, as well as compare APP polypeptides in these tissues and in plasma. Finally, although it is important to determine the actual source of amyloid (e.g. neuronal or systemic, APP695 or APP751), it may be also worthwhile to identify the common denominator(s) in specific regions of the brain where amyloidogenesis occurs (e.g. cerebral cortex, hippocampus, nucleus basalis). These factors might reside in either the blood-brain barrier or the neuronal microenvironment and include proteases, inhibitors, or even certain cell populations (e.g. microglial cells) that specifically play a role in amyloid formation.

REFERENCES

1. G.G. Glenner, and C.W. Wong, Alzheimer's disease: initial report of the purification and characterization of a novel cerebrovascular amyloid protein. **Biochem. Biophys. Res. Commun.**, 120: 885-890 (1984).
2. S.G. van Duinen, E.M. Castano, F. Prelli, G.T. Bots, W. Luyendijk, and B. Frangione, Hereditary cerebral hemmorrhage with amyloidosis in patients of Dutch origin is related to Alzheimer disease. **Proc, Nat. Acad Sci (U.S.A.)**, 84: 5991- 5994 (1987).
3. G. Blessed, B.E. Tomlinson, and M. Roth, The association between quantitative measures of dementia and of senile change in the cerebral grey matter of elderly subjects. **Brit. J. Psychiatry**, 114: 797-811 (1968).
4. D. Goldgaber, J.I. Lerman, O.W. McBride, U. Saffiotti, D.C. and Gajdusek, Characterization and chromosomal localization of a cDNA encoding brain amyloid of fibril protein. **Science**, 235: 877-880 (1987).
5. J. Kang, H.G. Lemaire, A. Unterbeck, J. Salbaum, L. Masters, K.H. Grzeschik, G. Multhaup, K. Beyreuther, and B. Muller-Hill, The precursor of Alzheimer's disease amyloid A4 protein resembles a cell-surface receptor. **Nature**, 325: 733-736 (1987).
6. N.K. Robakis, N. Ramakrishna, G. Wolfe, and H.M. Wisniewski, Molecular cloning and characterization of a cDNA encoding the cerebrovascular and the neuritic plaque amyloid peptides. **Proc. Nat. Acad. Sci. (U.S.A)**, 84: 4190-4194 (1987a).
7. R.E. Tanzi, J.F. Gusella, P.C. Watkins, G.A. Bruns, P. St. George-Hyslop, M.L. VanKeuren, S.P. Patterson, D.M. Kurnit, and R.L. Neve, Amyloid β protein gene: cDNA, mRNA distribution, and genetic linkage near the Alzheimer locus. **Science**, 235: 880-884 (1987a).
8. N. Kitaguchi, Y. Takahashi, Y. Tokushima, S. Shiojiri, H. Ito, Novel precursor of Alzheimer's disease shows protease inhibitory activity. **Nature**, 331: 530-532 (1988).
9. P. Ponte, P. Gonzalez-DeWhitt, J. Schilling, J. Miller, D. Hsu, B. Greenberg, K. Davis, W. Wallace, I. Lieberburg, F. Fuller, and B. Cordell, A new A4 amyloid mRNA contains a domain homologous to serine proteinase inhibitors. **Nature**, 331: 525-527 (1988).
10. R.E. Tanzi, A.I. McClatchey, E.D. Lamperti, L. Villa-Komaroff, J.F. Gusella, and R.L. Neve, Protease inhibitor domain encoded by an amyloid protein precursor mRNA associated with Alzheimer's disease. **Nature**, 331: 528-530 (1988).
11. D. Patterson, K. Gardiner, F-T. Kao, R. Tanzi, P. Watkins, and J. Gusella, Mapping of the gene encoding the β-amyloid precursor protein and its relationship to the Down syndrome region of chromosome 21. **Proc. Nat. Acad. Sci. (U.S.A.)**, 85: 8266-8270 (1988).
12. J.R. Korenberg, R. West, and Pulst S.M., The Alzheimer protein precursor maps to chromosome 21 sub-bands q21.15-q21.2. **Neurology** (suppl.), 38: 265 (1988).
13. N.D. Robakis, H. M. Wisniewski, E.C. Jenkins, E.A. Devine-Gage, G.E. Houck, X.L. Yao, N. Ramakrishna, G. Wolfe, W.P. Silverman, and W.T. Brown, Chromosome 21q21 sublocalization of gene encoding β-amyloid peptide in cerebral vessels and neuritic (senile) plaques of people with Alzheimer disease and Down syndrome. **Lancet** ii: 384 (1987b).
14. M. Podlisney, G. Lee, and D. Selkoe, Gene dosage of the amyloid β precursor protein in Alzheimer's disease. **Science**, 238: 669-671 (1987).
15. P.H. St. George-Hyslop, R.E. Tanzi, R.J. Polinsky, R.L. Neve, D. Pollen, D. Drachman, J. Growdon, L.A. Cupples, L. Nee, R.H. Myers, D. O'Sullivan, P.C. Watkins, J.A. Amos, C.K. Deutsch, J.W. Bodfish, M. Kinsbourne, R.G. Feldman, A. Bruni, L. Amaducci, J-F. Foncin, and J.F. Gusella, Absence of duplication of chromosome 21 genes in familial and sporadic Alzheimer's disease. **Science**, 238: 664-666 (1987a).
16. R.E. Tanzi, E.D. Bird, S.A. Latt, and R.L. Neve, The Amyloid β protein gene is not duplicated in brains from patients with Alzheimer's disease. **Science**, 238:666-669 (1987b).
17. P.H. St. George-Hyslop, R.E. Tanzi, R.J. Polinsky, J.L. Haines, L. Nee, P.C. Watkins, R.H. Myers, R.G. Feldman, D. Pollen, D. Drachman, J. Growdon, A. Bruni, J-F. Foncin, D. Salmon, P. Frommelt, L. Amaducci, S. Sorbi, S. Piacentini, G.D. Stewart, W.J. Hobbs, P.M. Conneally, and J.F. Gusella, The genetic defect causing familial Alzheimer's disease maps on chromosome 21. **Science**, 235: 885-889 (1987b).
18. R.E. Tanzi, P.H. St.George-Hyslop, J.L. Haines, R.J. Polinsky, L. Nee, J-F. Foncin, R.L. Neve, A.I. McClatchey, P.M. Conneally, and J.F. Gusella, The genetic defect in familial Alzheimer's disease is not tightly linked to the amyloid β-protein gene. **Nature**, 329: 156-157 (1987c).
19. A. Weidemann, G. Konig, D. Bunke, P. Fischer, M.J. Salbaum, C. Masters, and K. Beyreuther, Identification, biogenesis, and localization of precursors of Alzheimer's disease A4 Amyloid protein. **Cell**, 57: 115-126 (1989).
20. J.F. Kaumeyer, J.O. Polazzi, and M.P. Kotick, The mRNA for a proteinase inhibitor related to the HI-30 domain of inter-α-trypsin inhibitor also encodes α_1-microglobulin (protein HC) **Nuc. Acid Res.**, 14: 7839- 7850 (1986).
21. E. Wachter, and K. Hochstrasser, Kunitz type proteinase inhibitors derived proteolysis of the Inter-α-trypsin, IV. **Hoppe-Seyler's Z. Physiol. Chem.**, 360: 1351-1355 (1981).

22. C.R. Abraham, D.J. Selkoe, H. Potter, Immunochemical identification of the serine protease inhibitor α_1-antichymotrypsin in the brain amyloid deposits of Alzheimer's disease. **Cell**, 52: 487-501 (1988).

23. W.L. McKeehan, Y. Sakgami, H. Hoshi, and K.A. McKeehan, Two apparent human endothelial cell growth factors from hepatoma cells are tumor associated protease inhibitors. **J. Biol. Chem.**, 261: 5378-5383 (1986).

24. C.W. Pratt, and S.V. Pizzo, *In vivo* metabolism of inter-α-trypsin inhibitor and its proteinase complexes: Evidence for proteinase transfer to α_2- macroglobulin and α_1-proteinase inhibitor. **Arch. Biochem. Biophys.**, 248: 587-596 (1986).

25. R.K. Chawla, D.J. Roushe, F.W. Miller, W.R. Vogler, and D.H. Lawson, Abnormal profile of serum proteinase inhibitors in cancer patients. **Cancer Res.**, 44: 2718-2723 (1984).

26. S.A. Johnson, G.M. Pasinetti, P.C. May, P.A. Ponte, B. Cordell, C.E. Finch, Selective reduction of mRNA for the β-amyloid precursor protein that lacks a Kunitz-type protease inhibitor motif in cortex from Alzheimer brains. **Exp. Neurol.**, 102: 264-268 (1988).

27. S. Tanaka, S. Nakamura, K. Ueda, M. Kameyama, S. Shiojiri, Y. Takahashi, N. Kitaguchi, and H. Ito, Three types of amyloid protein precursor mRNA in human brain: their differential expression in Alzheimer's disease. **Biochem. Biophys. Res. Commun.**, 157: 472-479 (1988).

28. M.R. Palmert, T.E. Golde, M.L. Cohen, D.M. Kovacs, R.E. Tanzi, J.F. Gusella, M.F. Usiak, L.H. Younkin, and S.G. Younkin, Amyloid protein precursor messenger RNAs: differential expression in Alzheimer's disease. **Science**, 241: 1080-1084 (1988).

THE SERPIN, α_1-ANTICHYMOTRYPSIN, IN BRAIN AGING AND DISEASES OF THE NERVOUS SYSTEM

CARMELA R. ABRAHAM* AND HUNTINGTON POTTER

Department of Neurobiology
Harvard Medical School
Boston, MA 02115, U.S.A.

INTRODUCTION

One of the main neuropathological lesions that characterize the brains of Alzheimer disease patients is the neuritic or senile plaque, which consists of a spherical core of extracellular protein filaments surrounded by a halo of degenerating nerve cell processes. Extracellular protein filaments similar to those in the cores of neuritic plaques also occur in the walls of meningeal and intracortical blood vessels. The proteinaceous deposits in the cores of neuritic plaques and in blood vessels are referred to by the generic term "amyloid" generally defined as an aggregate of extacellular 6-10 nm protein filaments that has certain tinctorial properties.

The first identified constituent of Alzheimer amyloid deposits was purified and sequenced in 1984 by George Glenner and Caine Wong from the amyloid in meningeal blood vessels.[1] This protein, termed the β-protein, also proved to be the major constituent of the cores of the neuritic plaques of Alzheimer's disease and Down's syndrome[2-4] and served as the beachhead for cloning the gene encoding this protein. Subsequently, four groups, independently and at nearly the same time, succeeded in using oligonucleotide probes designed on the basis of the β-protein sequence to screen human brain cDNA libraries and obtain the corresponding gene for the Alzheimer's β-amyloid protein precursor (βAPP).[5-8] Analysis of an apparently full-length clone showed the precursor of the β-protein to be a 695 acid protein which resembles a cell surface receptor.[6] The β-protein is a small portion of this polypeptide near its carboxy terminus, and includes part of the putative membrane spanning region and part of the adjacent extracellular domain.

α_1-Antichymotrypsin (ACT) in Alzheimer Amyloid Deposits

A second, parallel line of experimentation indicated that another component of the amyloid deposits found in the cores of neuritic plaques and in the walls of selected meningeal and cortical blood vessels in Alzheimer's disease brains was the serine

---

*Present address: Boston University Medical School
            Boston, MA 02118, USA

Serine Proteases and Their Serpin Inhibitors in the Nervous System
Edited by B. W. Festoff
Plenum Press, New York, 1990

321

protease inhibitor, α_1-antichymotrypsin.[9] This conclusion was derived from two approaches: molecular cloning and immunochemical analysis. An antiserum raised gainst insoluble Alzheimer amyloid filaments identified cDNA clones which, upon sequencing, were found to code for α_1-antichymotrypsin (ACT). Conversely, several antisera raised against pure ACT specifically recognized virtually all light-microscopically visible amyloid deposits in six Alzheimer brains. The fact that the amyloid antiserum recognized at least two distinct epitopes on purified ACT strongly suggests the presence in the amyloid of the inhibitor itself, rather than an immunologically-related protein. Immunogold visualization of SDS/β-mercaptoethanol-purified Alzheimer's amyloid deposits in the electron microscope indicates that the ACT protein is integral to, or is very tightly associated with, the amyloid filaments. RNA encoding ACT was found in liver, as expected, and also in Alzheimer brain. The data indicated that expression appears to be higher in those areas of Alzheimer brain known to develop many neuritic plaques. The possibility that ACT plays a role in the formation of Alzheimer amyloid deposits was suggested by the specificity of its localization and the tightness of its association with the amyloid fibers: (1) only ACT, but not other related protease inhibitors, was detected in Alzheimer amyloid, (2) ACT was not present in the cerebral amyloid of Creutzfeldt-Jakob disease, and (3) immunogold electron microscopy localized ACT on extensively purified and extracted amyloid fibers.

Amyloid Deposition During Normal Aging

Like aged humans, some animals show certain Alzheimer-like neuropathological changes as they reach advanced age.[10-12] In particular, aged (> 20 year-old) rhesus monkeys develop amyloid deposits in blood vessels and in neuritic plaques that are highly similar to those seen in aged humans and patients with Alzheimer's disease. Thus, non-human primates provide a very useful animal model for the study in which the appearance and relative protein composition of amyloid deposits (β-protein, α1-antichymotrypsin and other potential components) can be followed over time.

We have examined aged monkey brains for the presence of ACT immunoreactivity and have found that this protein is also present in the brain amyloid deposits of aged monkeys, as it is in humans. Specifically, amyloid deposition associated with neuritic plaques and with the brain vasculature was analyzed in aged monkeys using three probes: thioflavin S, a cytochemical stain classically used to identify amyloid and believed to recognize protein arrays with extensive β-pleated sheet conformation; and specific antibodies recognizing ACT or the β-protein, respectively. Based on these experiments, we conclude that (1) both ACT and β-protein are components of the amyloid deposits arising in aged rhesus monkey brain, as they are in the similar deposits found in Alzheimer's disease and normal aged human brain; (2) amyloid deposition in the meningeal vessels and the cores of neuritic plaques arises approximately contemporaneously (about age 25 years), but within an individual animal, plaques may be present in the absence of detectable microvascular amyloid; (3) amyloid deposition in the cortical vasculature is detected only in the oldest monkeys (30, 31 and 37 years) and by all three probes; (4) white matter vasculature appears to be free of amyloid deposition, and (5) ACT immunoreactivity is found in perivascular cells around both grey and white matter blood vessels (Abraham, Selkoe, Potter, Price and Cork, **Neuroscience**, in press).

These findings demonstrate that two of the brain amyloid components of human senescence and Alzheimer's disease the β-protein and the protease inhibitor α_1-antichymotrypsin are also present in the amyloid deposits of normal aged monkey brain. The extended molecular parallels between normal brain aging and Alzheimer's disease suggest that similar biochemical mechanisms may underlie progressive amyloid deposition in both situations.

Another issue emphasized by this study is that the aging monkeys develop vascular amyloid deposits only in the grey matter and in meningeal vessels. White matter is entirely spared. Brains from patients with AD also show little or no amyloid in white matter vasculature.[13-15] These results suggest that vessels of the grey matter and meninges, in contrast to the white matter, share some characteristic which makes them preferred sites for amyloid deposition, despite the fact that blood (containing ACT, for instance) circulates through all vessels.

The Origin of Amyloid Proteins in Alzheimer's Disease

Whether proteins contributing to amyloid deposits in Alzheimer's disease are synthesized locally, or are derived from a blood-borne precursor, has been a matter of some debate because it may bear on the pathogenesis of the disease (for discussion, see Abraham et al., 1988).[9] With respect to the β-protein precursor, its expression can be found not only in neurons in the brain but in cells of various tissues, including spleen, kidney and heart.[7] The structure of the protein and some elegant pulse-chase experiments by Weidemann, Beyreuther and their colleagues indicate that it is a cell surface glycoprotein with the ability to be proteolytically cleaved to generate a smaller secreted form.[16] This free protein can be detected in normal CSF but not serum, suggesting that it is generated in brain (Palmert and Younkin, personal communication, 1988, Society for Neuroscience meeting). The C-terminal portion left in the cell membrane that includes the β-protein sequence may provide the immediate precursor of the amyloid β-protein. Thus, the brain appears to be the most likely source of the β-protein found in Alzheimer amyloid deposits, but it is not yet possible to exclude the possibility that the βAPP also circulates in the blood and reaches the brain through a defective blood-brain barrier.

The presence of ACT in Alzheimer amyloid deposits also opened the question of the origin of this protease inhibitor. Because ACT is present in high concentration in the blood, a likely source of the inhibitor in Alzheimer's amyloid deposits is the circulation. However, the finding that both ACT messenger RNA and soluble ACT protein were greatly elevated in Alzheimer's disease grey matter, especially those areas most affected by neuropathology, indicates that the brain itself is also a potential source. This contrasts with our finding of almost undetectable levels of ACT-RNA and protein in normal grey matter and is consistent with the results of Justice and her co-workers, who also concluded from absorption experiments and immuno-histochemistry that ACT is only found in small amounts in normal brain.[17] We next wished to identify those cells producing ACT in Alzheimer's disease. Besides the extracellular ACT staining found in amyloid deposits of normal aging, Alzheimer's disease and Down's syndrome, ACT was detected in three types of cells: astrocytes, neurons and pericytes. In general, as compared to normal brain, the areas of grey matter around damaged tissue always exhibited more numerous ACT-positive astrocytes. Astrocytes in the subcortical and subependymal white matter were equally immunolabeled in control and diseased brain. This staining was generally cytoplasmic, but in some of the cells was also seen in the nucleus. In all cases, the positive neurons were few and tended to be large. The other cell type that was strongly immuno-positive for ACT in all control and disease brains was the epithelium of the choroid plexus. *In situ* hybridization studies confirm the immunocytochemical findings (Pasternak, Abraham, Potter and Younkin, Am. J. Pathol., in press).

Role of Proteases and Inhibitors in Amyloid Formation

Proteases and their inhibitors, including α_1-antichymotrypsin, exist in a dynamic equilibrium in many tissues of the body (for review, see reference 9). There is a delicate balance in the organism between proteases and their inhibitors associated with various physiological processes. Any alteration in activity, amount or location of a protease inhibitor such as ACT could disrupt the dynamic maintenance of the local architecture or function of brain tissue. For example, a local excess of ACT might prevent a normal brain protease from clearing abnormal cleavage products of the β-protein precursor. The fact that ACT forms a complex with its target protease that cannot be dissociated by boiling in SDS and β-mercaptoethanol[18] could thus explain the inclusion of the inhibitor in the Alzheimer amyloid filaments. Alternatively, defective inhibitor might allow proteolytic cleavage of the β-protein precursor and the release of the self-aggregating β-protein fragment. In this regard, it is interesting to note that the N-terminal of the two cleavage sites within the precursor of the β-protein contain a methionine and could therefore result from processing by a chymo-trypsin-like protease.

These same roles for a protease inhibitor in Alzheimer's disease should also be considered in light of the recent finding that the β-protein precursor can, by alternative splicing, contain a new domain with structural, and probably functional, similarity to the Kunitz family of protease inhibitors.[19-21]

Thus two completely unrelated protease inhibitors are associated with the Alzheimer amyloid and may contribute to the resistance of the deposits to clearing by further proteolytic cleavage.

Special Association Between α_1-Antichymotrypsin and β-Protein in Amyloid

Several different types of human amyloidoses have been described at the biochemical level in the past 20 years. Amyloid is a generic name given to deposits of extracellular material that consists of 7-10 nm wide fibrils and possesses a β-pleated sheet conformation which contributes to its distinct tinctorial properties. When amyloid fibrils from various clinical forms of amyloidosis were purified and their major protein component analyzed, it was found to differ in composition among the diseases. For example, a light chain of IgG or a fragment thereof makes up the amyloid fibrils in primary amyloidosis, a portion of serum amyloid A in secondary amyloidosis, a mutated form of prealbumin in familial amyloidotic polyneuropathy, and a "prion" protein associated with an unconventional infectious agent in Creutzfeldt-Jakob disease (for review see reference 22). Except for the hereditary diseases, where a mutated protein may be more prone to form amyloid, it is still unknown what causes the aberrant deposition of an otherwise normal precursor protein or the formation of protein fragments that self-aggregate to form amyloid. The fact that ACT was tightly associated with the β-protein in Alzheimer amyloid, but not in the amyloid of Creutzfeldt-Jakob disease,[9] suggested that there is at least a special association of ACT and the β-protein and perhaps an essential interaction between the two proteins in the formation of Alzheimer-type amyloid. We therefore asked whether ACT was also associated with the amyloid deposits in other amyloid-oses.

Three procedures were used to visualize amyloid deposits: Thioflavin S or Congo red staining, immunolabeling with ACT antibodies and with two types of β-protein antibodies (Table 1). ACT immunoreactivity was only detected in those amyloid deposits which have as their major component, the β-protein. As expected, these include amyloid cores of neuritic plaques and blood vessel amyloid in AD, older

Table 1. Differential Association of α_1-Antichymotrypsin with Amyloid Fibrils

Type of Amyloidosis	Major Protein in Fibrils	TS or Congo Red Staining	Staining with ACT Antibodies	Stirring with β-Protein Antibodies
Alzheimer's disease	β-protein	+	+	+
Down's syndrome	β-protein	+	+	+
Normal aging (humans and monkeys)	β-protein	+	+	+
Hereditary Congophilic Angiopathy (Dutch)	β-protein	+	+	+
Primary	Light chain of IgG	+	−	−
Secondary	Serum amyloid A	+	−	−
Familial Amyloidic Polyneuropathy	Prealbumin	+	−	−
Creutzfeldt-Jakob disease	PrP	+	−	−

Down's syndrome, and aged controls. Amyloid deposits in primary and secondary amyloidosis, familial amyloidotic polyneuropathy and Creutzfeldt-Jakob disease were birefringent under polarized light after Congo red staining but showed no immunoreactivity with either ACT antibodies or β-protein antibodies. In a recently discovered Dutch variant of hereditary cerebral hemorrhage with amyloidosis, which shows similar angiopathy, the major component of the amyloid is the β-protein,[23] not Cystatin C, as in the Icelandic version.[24] When we immunolabeled sections from this disease with antibodies to ACT, the amyloid was positive. The conclusion that can be drawn from our analysis of the amyloidoses tested is that ACT is associated with, and only with, amyloidoses of the β-protein type, indicating a special functional or physical association between these two proteins (Abraham, Shirahama and Potter, **Neurobiol. Aging,** in press).

Is α_1-Antichymotrypsin Part of a Brain 'Acute Phase Response'?

Since the concentration of ACT in the blood can increase 2-4 fold in a few hours as part of the body's acute phase response to various types of inflammation, it was possible that the increase in ACT mRNA and protein we detected in AD represented a similar general response associated with pathological states in the central nervous system. Thus, having identified astrocytes as the primary cells responsible for ACT production in Alzheimer's disease brain, we wished to determine whether such expression is AD-specific or can also be seen in other human central nervous diseases in areas of tissue disruption and neuronal loss. We therefore asked whether ACT-positive cells also arose in areas of neuropathology associated with diseases unrelated to Alzheimer's disease or Down's syndrome. In Huntington's and Parkinson's disease, stroke and brain malignancy, astrocytes immunolabeled for $\alpha_1$antichymotrypsin were found surrounding the areas of pathology (Abraham, Shirahama and Potter, **Neurobiol. Aging,** in press). From these results we conclude that the overexpression of ACT in the brain is not *per se* a characteristic of Alzheimer's disease. Nor is it sufficient to cause ACT-associated amyloid deposition.

Instead, either ACT and its expression must be aberrant in Alzheimer's disease, or its increased expression must be coupled with aberrant processing of the β-protein-precursor protein in order for characteristic Alzheimer-like amyloid to form. Indeed, the finding that the gene encoding the β-protein precursor is not mutated or amplified in the Familial Alzheimer's Disease patients further indicates that it is the proteolytic processing of this normal protein that represents the pathological defect.

We consider it a particularly intriguing possibility that the overexpression of the protease inhibitor α_1-antichymotrypsin in the various types of brain pathology may reflect a brain 'acute phase' response not unlike the acute phase response elicited by inflammation in the periphery. One of the signals which, from *in vitro* studies, can induce liver cells to start expressing acute phase proteins such as α_1-antichymotrypsin is IL-1,[25,26] which has also been shown to induce the expression of the β-protein-precursor in transfected cells (D. Goldgaber and R.J. Donnelly, personal communication, 1988, Society for Neuroscience meeting). The possibility that astrocytes express α_1-antichymotrypsin in response to IL-1 is being investigated, and would suggest that the two protein components of Alzheimer's amyloid deposits may be up-regulated by the same signals in the brain. Indeed, the possibility must be considered that in the brain, as is often the case in the periphery, the response to inflammatory assault may in itself be more harmful to the structure and function of the tissue than the original insult itself.

SUMMARY

The purpose of this study was to characterize the nature and the origin of the Alzheimer's disease amyloid deposits. We used an amyloid antiserum to screen a human liver expression library. A positive clone was sequenced and found to code for the serine protease inhibitor α_1-antichymotrypsin, an acute phase serum protein. Thus, this protein is a second component of the brain amyloid in addition to the β-protein. In order to determine whether the inhibitor originated from the serum or was made in the brain, we performed Northern blots on tissue from control and Alzheimer brain and found that α_1-antichymotrypsin RNA is present in the brain and that the diseased brain contained larger amounts than the controls. Immunocytochemistry and *in situ* hybridization show the astrocytes to produce the inhibitor, mainly around senile plaques. α_1-antichymotrypsin is only associated with the amyloid deposits of the β-protein kind in normal aging of man and monkeys, Alzheimer's, Down's syndrome and hereditary cerebral hemorrhage with amyloidosis of Dutch origin, but not in primary and secondary amyloidosis or familial amyloidotic polyneuropathy.

The specific association between α_1-antichymotrypsin and the β-protein prompted us to suggest a role for this serine protease inhibitor in the proteolytic processing of the β-protein precursor.

REFERENCES

1. G.G. Glenner, and C.W. Wong, Alzheimer's disease: initial report of the purification and characterization of a novel cerebrovascular amyloid protein. Biochem. Biophys. Res. Commun. 122:885-890 (1984).
2. C.L. Masters, G. Multhaup, G. Simms, J. Pottgieser, R.N. Martins, and K. Beyreuther, Neuronal origin of a cerebral amyloid: neurofibrillary tangles of Alzheimer's disease contain the same protein as the amyloid of plaque cores and blood vessels. EMBO J. 4:2757 (1985).
3. C.W. Wong, V. Quaranta, and G.G. Glenner, Neuritic plaques and cerebrovascular amyloid in Alzheimer disease are antigenically related. Proc. Natl. Acad. Sci. USA 82:8729-8732 (1985).

4. D.J. Selkoe, C.R. Abraham, M.B. Podlisny, and L.K. Duffy, Isolation of low-molecular weight proteins from amyloid plaque fibers in Alzheimer's disease. J. Neurochem. 46:1820-1834 (1986).

5. D. Goldgaber, M.J. Lerman, O.W. McBride, V. Saffiotti, and D.C. Gadjusek, Characterization and chromosomal localization of a cDNA encoding brain amyloid of Alzheimer's disease. Science 235:877 (1987).

6. J. Kang, H.G. Lemaire, A. Unterback, J.M. Salbaum, C.L. Masters, K.H. Grezeschik, G. Multhaup, K. Beyreuther, and B. Muller-Hill, The precursor of Alzheimer disease amyloid A4 protein resembles a cell-surface receptor. Nature 325: 733 (1987).

7. R.E. Tanzi, J.F. Gusella, P.C. Watkins, G.A.P. Bruns, P. St. George-Hyslop, M.L. Van Keuren, D. Patterson, S. Pajan, D.M. Kurnit, and R.L. Neve, Amyloid β-protein gene; cDNA, mRNA distributions, and genetic linkage near the Alzheimer locus. Science 235:880 (1987).

8. N.K. Robakis, N. Ramakrishna, G. Wolfe, and H.M. Wisniewski, Molecular cloning and characterization of a cDNA encoding the cerebrovascular and the neuritic plaque amyloid peptides. Proc. Natl. Acad. Sci. USA 84:4190 (1987).

9. C.R. Abraham, D.J. Selkoe, and H. Potter, Immunochemical identification of the serine protease inhibitor α_1-antichymotrypsin in the brain amyloid deposits of Alzheimer's disease. Cell 52:487-501 (1988).

10. H.M. Wisniewski, and R.D. Terry, Morphology of the aging brain, human and animal. In: Progress in Brain Research, 1973; Vol. 40, Neurobiological Aspects of Maturation and Aging. ed. D.H. Ford, pp. 1108-1109. Amsterdam: Elsevier.

11. R.G.Struble, D.L. Price, Jr., L.C. Cork, and D.L. Price, Senile plaques in cortex of aged normal monkeys. Brain Res. 361:267-275 (1985).

12. D.J. Selkoe, D.S. Bell, M.B. Podlisny, D.L. Price, and L.C. Cork, Conservation of brain amyloid proteins in aged mammals and humans with Alzheimer's disease. Science 235:873-877 (1987).

13. T.I. Mandybur, The incidence of cerebral amyloid angiopathy in Alzheimer's disease. Neurology 25:120-126 (1975).

14. M. Tomonoga, Cerebral amyloid angiopathy in the elderly. J. Amer. Geriatr. Soc. 29:151-157 (1981).

15. C.L. Joachim, L.K. Duffy, J. Morris, and D.J. Selkoe, Protein chemical and immunocytochemical studies of meningovascular β-amyloid protein in Alzheimer's disease and normal aging. Brain Res., 474:100-111 (1988).

16. A. Weidemann, G. König, D. Bunke, P. Fischer, C.L. Masters, and K. Beyreuther, Identification, biogenesis and localization of precursors of Alzheimer's disease A4 amyloid protein. Cell 57:115-126 (1989).

17. D.L. Justice, R.H. Rhodes, and Z.A. Tokes, Immunohistochemical demonstration of proteinase inhibitor α_1-antichymotrypsin in normal human central nervous system. J. Cell. Biochem. 34:227-238 (1987).

18. J. Travis, J. Bowen, and R. Baugh, Human α_1-antichymotrypsin: interaction with chymotrypsin-like proteinases. Biochemistry 17:5651-5656 (1978).

19. P. Ponte, P. Gonzalez-DeWhitt, J. Schilling, J. Miller, D. Hsu, B. Greenberg, K. Davis, W. Wallace, I. Lieberburg, F. Fuller, and B. Cordell, A new A4 amyloid mRNA contains a domain homologous to serine protease inhibitors. Nature 331:525-527 (1988).

20. R.E. Tanzi, A.I. McClatchey, E.D. Lamberti, L. Villa-Komaroff, J.F. Gusella, and R.L. Neve, Protease inhibitor domain encoded by an amyloid protein precursor mRNA associated with Alzheimer's disease. Nature 331:528-530 (1988).

21. N. Kitaguchi, Y. Takahashi, Y. Tokushima, S. Shiojiri, and H. Ito, Novel precursor of Alzheimer's disease amyloid protein shows protease inhibitory activity. Nature 331:530-532 (1988).

22. E.M. Castano, and B. Frangione, Biology of disease: Human amyloidosis, Alzheimer disease and related disorders. Lab. Invest. 58:122-132 (1988).

23. S.G. Van Duinen, E.M. Castano, F. Prelli, G.T.A. Bots, W. Luyendijk, and B. Frangione, Hereditary cerebral hemorrhage with amyloidosis in patients of Dutch origin is related to Alzheimer's disease. Proc. Natl. Acad. Sci. 84:5991 (1987).

24. J. Ghiso, O. Jensson, and B. Frangione, Amyloid fibrils in hereditary cerebral hemorrhage with amyloidosis of Icelandic type is a variant of γ-trace basic protein (cystatin C). Proc. Natl. Acad. Sci. 83:2974-2978 (1986).

25. H. Baumann, G.P. Jahreis, D.N. Sauder, and A. Koj, Human keratinocytes and monocytes release factors which regulate the synthesis of major acute-phase plasma proteins in hepatic cells from man, rat and mouse. J. Biol. Chem. 259:7331 (1984).

26. H. Baumann, C. Richards, and J. Gauldie, Interaction among hepatocyte-stimulating factors, interleukin 1, and glucocorticoids for regulation of acute phase plasma proteins in human hepatoma (HepG2) cells. J. Immunol. 139:4122-4128 (1987).

PROTEASE NEXIN I IMMUNOSTAINING IN ALZHEIMER'S DISEASE

DORRIE E. ROSENBLATT,[1] CHANGIZ GEULA,[2]
AND MARSEL-MAREK MESULAM[2]

[1]*Institute of Gerontology & Div. of Geriatrics, University of Michigan*
[2]*Division of Neuroscience & Behavioral Neurology,*
Harvard Medical School

INTRODUCTION

The NATO Conference which provided the impetus for this book was focused on protease and protease inhibitor activity in the nervous system. The conference covered both the Peripheral Nervous System (PNS) and the Central Nervous System (CNS) and addressed both developmental and pathological conditions. A recurrent theme throughout the conference was that of a crucial balance between proteolytic enzyme and inhibitor activity. Another theme of the conference was that of neuron-glia interaction. This chapter will report on immunohistochemical studies on Protease Nexin I (PN I), a glial-derived serine protease inhibitor with neurite promoting activity, in Alzheimer brain and will discuss the results in light of material presented at the conference.

Alzheimer's Disease is an example of a pathological process that has been hypothesized to result from an imbalance between protease and protease inhibitor activity.[1] It is a progressive dementia that is thought to effect 15% of the US population over the age of sixty five.[2] Although the clinical diagnosis is based on the observation of increasing cognitive impairment, the definitive pathological diagnosis of Alzheimer's Disease depends on the demonstration, at post mortem, of neuritic plaques and neurofibrillary tangles in the brain.[3] Both of these pathological lesions contain abnormal proteinaceous deposits.[4] The tangles are neurite remnants that contain paired helical filaments which are related to cytoskeletal proteins. The plaques consist of an amyloid core surrounded by abnormal neurites. The amyloid core is highly insoluble and to date only two proteins have been identified in the amyloid. The first is β- amyloid, a 4Kda peptide that has been solubilized, sequenced and cloned[5,6] (for review and Tanzi in this volume). The second is is α-1-antichymotrypsin which has been identified immunohistochemically in plaques[7] (and Abraham in this volume). The hypothesis that the deposition of abnormal proteins in Alzheimer's Disease arises from an imbalance between protease and inhibitor activity was based on the demonstration of α-1-antichymotrypsin in plaques and strengthened by the finding that α-1-antichymotrypsin mRNA is increased in Alzheimer brain.[7] The hypothesis was further reinforced by the finding that the amyloid precursor protein (APP 695) can contain a 56 amino acid insert (APP 751) with functional Kunitz-type

Serine Proteases and Their Serpin Inhibitors in the Nervous System
Edited by B. W. Festoff
Plenum Press, New York, 1990

protease inhibitor activity.[8] In brain, the APP 695 form predominates and its levels appear to be developmentally regulated.[9] There is also a regional increase in APP 695 in Alzheimer brain, the implications of which are discussed by Tanzi in this volume.

These data led us to investigate the role of PN I in AD. PN I is a protease inhibitor belonging to the serpin family of inhibitors. The nexins were originally described in conditioned medium from cultured human foreskin fibroblasts and categorized by molecular weight and the pattern of proteases inhibited.[10] PN I is the best characterized of the nexins and is discussed in detail in the chapter by Baker. Briefly, PN I has a MW of 43Kda and binds thrombin, urokinase and γ NGF. Inhibition of the proteases takes place via formation of a covalent bond at the serine in the enzyme active site. Formation of this covalent complex leads to cellular binding, internalization and degradation of the complex.[11] The cellular binding is mediated by a binding site that recognizes the nexin part of the complex and the binding is calcium dependent and blocked by heparin.[12] PN I is also made by astrocytes and has been shown to be identical to Glial Derived Neurite Promoting Factor[13,14,15] (and Monard in this volume) and to have neurite promoting effects in neuroblastoma N2A[16] and sympathetic superior cervical ganglia[17] and to inhibit granule cell migration.[18] PN I concentrations have been shown to vary during development[13,19] and in response to injury (see Monard this volume). Schwann cells also make PN I.[20] As a glial-derived protease inhibitor with effects on neurites PN I appears to be a good candidate for a role in the putative imbalance of protease and protease inhibitor activity in Alzheimer's Disease.

MATERIALS AND METHODS

The study used three brains from patients with a clinical diagnosis of Alzheimer's Disease and one from a 74 year old man with no history of dementia. Details of the tissue processing and staining have been reported previously.[21] Briefly, the tissue was fixed in 4% paraformnaldehyde, processed through sucrose of increasing concentrations (10-40%) and then cut in 40 μm sections on a freezing microtome. Sections stained with thioflavin-S, hematoxylin and eosin, and cresyl violet were used to confirm the diagnosis of Alzheimer's Disease.[22]

The immunostaining utilized a rabbit polyclonal antibody to PN I purified from the rat C6 glioma line that was produced and characterized by Rosenblatt.[21] An IgG fraction purified from immune sera by Protein A chromatography was used in all experiments.[23] This antibody is specific for PN I.[21] It is chemically active in that it blocks PN I inhibition of thrombin in a Chromozyme assay.[24] It is biologically active in that it blocks cellular binding of PN I : [125]I protease complexes. The immunohistochemical studies used the peroxidase anti-peroxidase methodology of Sternberger.[25] The antibody was diluted in phosphate buffered saline (20 μg/ml) and incubated with the tissue overnight at 4° C. The sections were then treated with goat anti-rabbit IgG and PAP and the peroxidase labelling visualized with diaminobenzidine. Three basic staining conditions were used: 1) anti-PN I antibody (Ab) alone: 2) Ab preincubated with PN I (Ab + PN I): 3) tissue pre-incubated with PN I and then stained with Ab. Specificity of the staining was tested in three ways: 1) using pre-immune serum 2) using immune serum that had been passed over a PN I:heparin sepharose affinity column 3) using the related protease inhibitor anti-thrombin III (AT III) and anti-AT III antibody.

To compare the binding of PN I in Alzheimer tissue to that described in tissue culture the following conditions were tested. 1) Ab was pre-incubated with 10^{-4} M Diisopropylfluorophosphate (DFP) to rule out complex formation with a protease in the IgG preparation. 2) Tissue was pre-incubated with DFP to rule out complex formation with a tissue protease. 3) To test the ability of a PN I:protease complex to bind, PN I was pre-incubated with thrombin to form covalent complexes and these complexes were substituted for PN I in the staining process. 4) The effect of heparin on PN I binding was tested by adding heparin (5 mg/ml) simultaneously with the addition of the Ab + PN I to the tissue. 5) The effect of calcium was tested by using EDTA (1 mM) in the binding buffer. 6) The role of heparan sulfate in the binding was tested by pre-treating the tissue with the heparan sulfate specific enzyme heparinitase (100 u/5 ml).

RESULTS

Thioflavin-S staining showed dense plaque and tangle formation in the Alzheimer brains and extremely rare plaques and tangles in the hippocampus and entorhinal cortex of the control brain (2-3 per coronal section of the entire temporal lobe).

Anti-PN I antibody staining yielded positive staining of plaques in the mediotemporal limbic areas (Figure 1). Analysis of adjacent sections stained with thioflavin-S

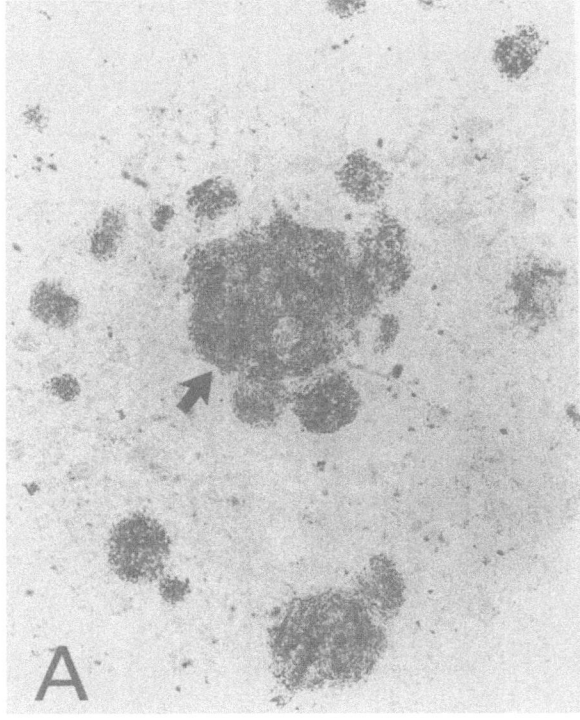

Figure 1. Immunohistochemical staining with the protease nexin I antibody in the amygdala of an AD patient. Note the absence of dense staining in plaque cores even though corresponding thioflavin-S stains showed that these plaques had amyloid cores. Some plaques (arrow) even appear to have a central area of decreased immunostaining. Magnification 236X.

showed that virtually all plaques visualized by thioflavin-S were also visualized by the immunostaining. This impression was confirmed by double staining experiments. However, while thioflavin-S gave intense staining of the amyloid core, the anti-PN I antibody staining was predominantly over the neuritic periphery of the plaques. The anti-PN I antibody staining did not visualize plaques elsewhere in the brain nor did it visualize any tangles.

When the anti-PN I antibody was pre-incubated with PN I prior to staining, the expectation was that the immunostaining would be eliminated thus confirming the specificity of antibody. However, use of the antibody + PN I aggregate gave universal staining of all the plaques and tangles in a distribution pattern that matched that seen with thioflavin-S (Figure 2). The same pattern was seen when the tissue was pre-incubated with PN I. In both cases, the PN I staining was predominantly over the neuritic periphery of the plaque, in contrast to staining with thioflavin-S or anti-β amyloid antibody both of which stained the amyloid core of the plaques. Specificity of the antibody staining was ultimately demonstrated in three ways. No plaque or tangle staining was seen with pre-immune rabbit serum, with immune serum depleted of anti-PN I antibody by passage over a PN I affinity column, or with anti-thrombin III and anti- AT III antibody.

The immunostaining with anti-PN I antibody alone was interpreted as reflecting endogenous PN I. The staining with the Ab + PN I aggregate was interpreted as reflecting PN I binding sites. The following experiments (Table 1) were done to compare the PN I binding in Alzheimer tissue with that described in fibroblast tissue culture.[11] In tissue culture, cellular binding of PN I is mediated by complex formation

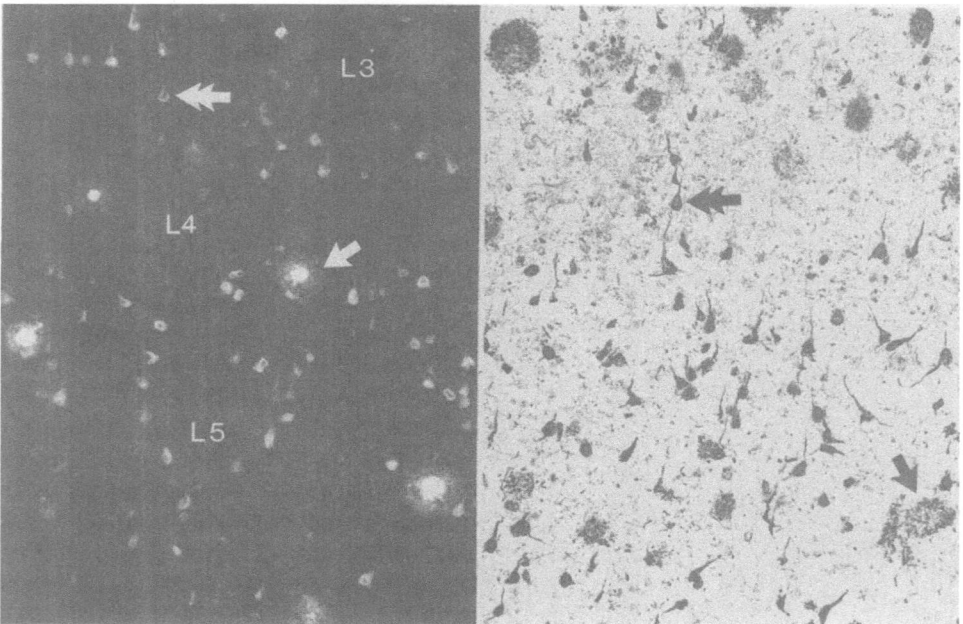

Figure 2. Thioflavin-S and PN I immunostaining of AD cortex. Prorhinal cortex from a patient with Alzheimer's disease. Magnification:140x. Left: Thioflavin-S staining and epifluorescent illumination. Plaques (single arrow) and tangles (double arrow) are fluorescent. Most plaques have a brightly fluorescent amyloid-rich core. The tangles are more frequent in layers 3 and 5 (L3 and L5) than in layer 4 (L4). Right: Immunostaining with Ab+PN I in an adjacent section. Plaques and tangles show intense immunostaining. Their density and distribution matches that shown with Thioflavin-S. Enhancement of staining over plaque cores occurs much less frequently than with thioflavin-S.

Table 1. Immunostaining conditions for characterization of the PN I binding site. [ ] represents 1 hr pre-incubation at 24°C. Reagent concentrations: Anti-PN I Antibody (Ab), 20 μg/ml. PNI, 5 μg/ml. DFP, 10^{-4} M. Heparin, 5 mg/ml. EDTA, 1 mM. Heparinitase, 100 u/5 ml

Staining Conditions	Staining Patterns
Ab	limited plaque staining
[Ab+PN I]	all plaques + tangles
Ab +[tissue +PN I]	all plaques + tangles
[Ab+ DFP] +PN I	all plaques + tangles
[Ab+ PN I] +[tissue +DFP]	all plaques + tangles
Ab +[PN I + thrombin]	all plaques + tangles
[Ab+ PN I] +heparin	no staining
[Ab+ PN I] +EDTA	all plaques + tangles
[Ab+ PN I] +[tissue + heparinitase]	all plaques + tangles

with a protease and uncomplexed PN I does not bind. To determine whether PN I was forming a complex with a protease in the antibody preparation, the antibody was pretreated with the irreversible serine protease inhibitor diisopropylfluorophosphate (DFP). This had no effect on staining suggesting that complex formation was not neccessary for binding in Alzheimer tissue. Pre-treatment of the tissue with DFP eliminated the possibility of complex formation with a tissue protease prior to binding or of a tissue protease being the binding site. Although complex formation with a protease was not a prerequisite for binding in the Alzheimer tissue, complexed protease could also bind as demonstrated by the staining obtained when PN I:thrombin complexes were substituted for PN I. In tissue culture, PN I binding is blocked by heparin. Addition of heparin to the Alzheimer tissue simultaneously with the Ab + PN I aggregate eliminated all staining. This effect was present at heparin concentrations that had no effect on other immunostaining reactions. Calcium has been shown to be important in PN I binding in tissue culture. Use of EDTA had no effect on binding in Alzheimer tissue. Binding to cell membrane and matrix proteoglycans, in particular heparan sulfate, has been shown to be important in tissue culture.[26,27] Treatment of Alzheimer tissue with the heparan sulfate specific enzyme heparinitase had no effect on staining suggesting that heparan sulfate was not involved in the binding in this tissue.

DISCUSSION

The data reported here suggest that in Alzheimer's Disease there are high concentrations of endogenous PN I localized to plaques in the amygdala and entorhinal cortex and that there are PN I binding sites associated with plaques and tangles throughout the brain. Furthermore, some of the characteristics of these binding sites differ from those described in fibroblast tissue culture. We have interpreted the immunostaining in the limbic plaques as representing abnormally high concentrations of PN I because no PN I staining is seen over cell types, such as astrocytes, which are known to make and bind PN I. This lack of staining may be due either to the fact that normal low levels of PN I are below the level of sensitivity of the immunohistochemical assay or to the fact that PN I was washed out of the tissue during tissue processing. It is also possible that the PN I associated with the limbic plaques is bound in a way that makes it resistant to washout. Studies are currently underway to determine whether the apparent increase in PN I in the plaques is due to increased synthesis or increased binding or a combination of the two.

The concept of a localized increase in PN I in Alzheimer lesions fits nicely with the concept of glial production of neurotrophic/neurite promoting factors in response to neuronal distress or death. There are many examples of localized changes in neurons, neurotransmitters and protein synthesis in aging and in Alzheimer brain.[3] The localized increases in expression of APP 695 in Alzheimer brain are an example of such changes (see Tanzi this volume). If PN I is indeed an example of a glial neurite promoting factor then it should be produced in situations associated with neuronal pathology such as the neuritic plaque. Monard et al have demonstrated an increase in PN I synthesis in injured sciatic nerve and this is apparently due to production by Schwann cells (see Monard this volume). We suggest that an increase in astrocytic production of PN I in the vicinity of limbic plaques is an equivalent finding. It is of note that Abraham (See this volume) has shown an increase in α-1-antichymotrypsin mRNA in astrocytes in similar areas. The finding of an increase in PN I in the limbic plaques is not in conflict with the report by Cunningham (see this volume) that there is a decrease in PN I activity in Alzheimer brain. Our immunohistochemical assay recognized both free (active) and complexed PN I (inactive). Cunningham reported a decrease in active PN I in brain homogenates but a relative increase in bound PN I and no changes in PN I mRNA in Alzheimer brain compared to control values. Small regional differences, such as the PN I increase in limbic plaques, may be missed or cancelled out when assaying brain homogenates.

The concept of abnormally high avidity binding permitting PN I visualization by immunohistochemical techniques is supported by the studies on PN I binding characteristics in Alzheimer brain. In fibroblast cultures two types of binding take place. One type of binding is to matrix proteoglycan binding sites.[26] The second type of binding is to a cellular receptor that has the characteristics of a class II receptor and mediates internalization and degradation of the bound PN I:protease complex.[11] It is the receptor binding that has been well characterized and which we have used as a basis of comparison for study of PN I binding in Alzheimer brain. In the fibroblast cultures, receptor binding requires complex formation with a protease, is blocked by heparin and requires calcium. The PN I binding in Alzheimer brain differs in several ways from that described in fibroblast tissue culture. Some of the differences may be attributed to the experimental conditions. For example, the lack of a calcium effect on the binding may be due to fixation effects on the receptor and the finding that PN I alone binds may be due to either the high concentrations of PN I used or to the fact that the experimental conditions could not separate receptor from matrix binding. The most interesting difference was that treatment with heparinitase did not decrease binding in Alzheimer brain. This suggests that heparan sulfate is not the crucial proteoglycan involved in the PN I binding in Alzheimer tissue. It is of note that while Snow et al[28] have immunocytochemically demonstrated heparan sulfate in plaques it was not demonstrable in tangles and they have suggested that different proteoglycans or glycosaminoglycans are present in tangles. Because the immunohistochemical techniques visualized PN I binding over plaques and tangles, and not over cells such as astrocyte and fibrocytes that are known to bind PN I,[11] we suggest that there is an abnormal binding molecule with particularly high avidity for PN I associated with the Alzheimer lesions. Changes in surface proteoglycans have been associated with development and injury in the peripheral nervous system.[27] The appearance of an abnormal proteoglycan on distressed or dying neurons in plaques and tangles may represent a similar phenomenon.

It has also been shown that the binding of PN I to matrix proteoglycan affects its complex formation with protease: speeding its interaction with thrombin and blocking its interaction with urokinase[30] (and Cunningham this volume). Binding of PN I to an abnormal binding molecule might change either the inhibitory activity or the spectrum of proteases inhibited. The resultant disturbance in PN I activity would result in an imbalance of protease and protease inhibitory activity thus fitting the hypothesis proposed for the pathogenesis of Alzheimer's Disease.

As a glial-derived protease inhibitor with neurite promoting effects PN I appeared to be a good candidate for a role in the hypothesized model for the pathogenesis of the neuritic plaques and neurofibrillary tangles in Alzheimer's Disease. The studies reported here showing high levels of PN I in limbic plaques and of PN I binding in all plaques and tangles confirm that PN I is a good candidate for a role in the pathogenesis of the Alzheimer lesions. Further studies are needed to more specifically delineate the role of PN I in the pathophysiology of Alzheimer's Disease.

REFERENCES

1. R.W. Carrell, Enter a protease inhibitor. Nature 331:478-479 (1988).
2. R. Katzman, Alzheimer's disease. N Engl J Med 314:964-973 (1986).
3. V.W. Henderson, and C.E. Finch, The neurobiology of Alzheimer's disease. J Neurosurg 70:335-353 (1989).
4. D.J. Selkoe, Altered structural proteins in plaques and tangles: what do they tell us about the biology of Alzheimer's disease. Neurobiol Aging 7:425-432 (1986).
5. G.G. Glenner, and C.W. Wong, Alzheimer's disease: initial report of the purification and characterization of a novel cerebrovascular amyloid fibril protein. Biochem Biophys Res Commun 122:1131-1135 (1984).
6. D.J. Selkoe, Biochemistry of altered brain proteins in Alzheimer's disease. Ann Rev Neurosci 12:463-490 (1989).
7. C.R. Abraham, D.J. Selkoe, and H. Potter, Immunochemical identification of the serine protease inhibitor alpha-1-anti-chymotrypsin in the brain amyloid deposits of Alzheimer's disease. Cell 52:487-501 (1988).
8. N. Kitaguchi, Y. Takahashi, Y. Tokushima, et al. Novel precursor of Alzheimer's disease amyloid protein shows protease inhibitory activity. Nature 331:530-532 (1988).
9. R.E. Tanzi, A.I. McClatchey, E.D. Lamperti, et al. Protease inhibitor domain encoded by an amyloid protein precursor mRNA associated with Alzheimer's disease. Nature 331:528-530 (1988).
10. J.B. Baker, and R.S. Gronke, Protease nexins and cellular regulation. Semin Thromb Hemost 12:216-220 (1986).
11. D.J. Knauer, J.A. Thompson, and D.D. Cunningham, Protease nexins: cell-secreted proteins that mediate the binding, internalization and degradation of regulatory serine proteases. J Cell Physiol 117:385-396 (1983).
12. E.W. Howard, and D.J. Knauer, Characterization of the receptor for protease nexin-1: protease complexes on human fibroblasts. J Cell Physiol 131:276-283 (1987).
13. S. Gloor, K. Odink, and J. Guenther, et al. A glia-derived neurite promoting factor with protease inhibitory activity belongs to the protease nexins. Cell 47:687-693 (1986).
14. D.E. Rosenblatt, C.W. Cotman, M. Nieto-Sampedro, et al. Identification of a protease inhibitor produced by astrocytes that is structurally and functionally homologous to human protease nexin-1. Brain Res 415:40-48 (1987).
15. D.J. Knauer, R.A. Orlando, and D.E. Rosenblatt, The glioma cell-derived neurite promoting activity protein is functionally and immunologically related to human protease nexin-I. J Cell Physiol 132:318-324 (1987).
16. D. Monard, F. Solomon, M. Rentsch, and R. Gysin, Glia-induced morphological differentiation in neuroblastoma cells. Proc Nat Acad Sci USA 70:1894-1897 (1973).
17. A.D. Zurn, H. Nick, and D. Monard, A glia-derived nexin promotes outgrowth in cultured chick sympathetic neurons. Dev Neurosci 10:17-24 (1988).
18. J. Lindner, J. Guenther, and H. Nick, et al. Modulation of granule cell migration by a glia-derived protein. Proc Nat Acad Sci USA 83:4568-4571 (1986).
19. Y. Schurch-Rathgeb, and D. Monard, Brain development influences the appearance of glial factor-like activity in rat brain primary cultures. Nature 273:308-309 (1978).
20. L.P. Mulligan, D.E. Rosenblatt, and D. Johnson, Protease nexin I-like activity in cultured Schwann cells. Submitted

21. D.E. Rosenblatt, C. Geula, M.-M. Mesulam, Protease Nexin I immunostaining in Alzheimer's disease. **Annals of Neurol.** 26:628-634 (1989).

22. Z.S. Khachaturian, Diagnosis of Alzheimer's disease. **Arch Neurol** 4:1097-1105 (1985).

23. H. Hjelm, K. Hjelm, and J. Sjoquist, Protein A from Staphylococcus aureus. Its isolation by affinity chromatography and its use as an immunoabsorbent for isolation of immunoglobulins. **FEBS Lett** 28:73-76 (1972).

24. U. Abildgaard, M. Lie, and O.R. Odegard, Antithrombin (heparin cofactor) assay with "new" chromogenic substrates (S-2238 and chromozym TH). **Thromb Res** 11:549-553 (1977).

25. L.A. Sternberger, P.H. Hardy Jr, J.J. Cuculis, and H.G. Meyer, The unlabeled antibody enzyme method of immunohistochemistry: preparation and properties of soluble antigen-antibody complex (horseradish peroxidase-antihorseradish peroxidase) and its use in identification of spirochetes. **J. Histochem Cytochem** 18:315-346 (1970).

26. D.H. Farrell, and D.D. Cunningham, Human fibroblasts accelerate the inhibition of thrombin by protease nexin. **Proc Natl Acad Sci USA.** 83:6858-6862 (1986).

27. D.H. Farrell, S.L. Wagner, R.H. Yuan, and D.D. Cunningham, Localization of protease nexin on the fibroblast extracellular matrix. **J Cell Physiol** 134:179-188 (1988).

28. A.D. Snow, M.A.R. Henderson, and D Nochlin et al. The presence of heparan sulfate proteoglycan in the neuritic plaques and congophilic angiopathy in Alzheimer's disease. **Am. J. Path.** 133:456-462 (1988).

29. B.W. Festoff, Proteases, their inhibitors and the extracellular matrix: factors in nerve-muscle development and maintenance. **Adv Exp Med Biol** 209:25-39 (1987).

30. S.L. Wagner, A.L. Lau, and D.D. Cunningham, Binding of protease nexin-1 to the fibroblast surface alters its target proteinase specificity. **J Biol Chem:** 2 64:611-615 (1989).

THE MARATEA CONFERENCE

Serpins at breakfast!
B. BLONDET, Y. NAGAMINE,
D. HANTAÏ, H. ALAMEDDINE,
W.-D. SCHLEUNING

Has the meeting started?
A.-J. VAN ZONNEVELD

A Serpin Dynasty?
T. CARRELL, R. CARRELL

S. WAGNER, R. PITTMAN,
D. CUNNINGHAM,
N. AHMED

Meat or fish?
J. ANDRADE, R. HILL,
D. EVANS, E. ANGLÉS-CANO

Is this a Tango?
F. SCHUMANN,
N. KALDERON

Just like Half-Moon Bay!
J. BAKER, J. BALLANCE,
E. LEVIN, D. ROSENBLATT

Fourth of July à deux!
R. SCOTT, J. ANDERSON,
C. SORIA, J. SORIA

For those folks back home!
J. FENTON, II,
N. ADACHI

Organizationally Speaking:
D. HANTAÏ, G. BARLOVATZ-
MEIMON, R. CARRELL, G.
MOONEN, B. FESTOFF, S.
FESTOFF

Protease nexin $\underline{II}$ is the........
P. GRABHAM, S. WAGNER,
D. CUNNINGHAM, J.
WISEMAN

English is not our first language!
G. DICKNETTE, N.
KALDERON, B. STELTE-
LUDWIG, A.-J. VAN
ZONNEVELD

The French Connection:
M./M. MIRSHAHI, C. SORIA,
B. FESTOFF, J. SORIA,
D. HANTAÏ

Mmmm!
D. ROSENBLATT, S.
KAYNAK, N. AHMED

Fish or meat?
R. REDDY, J. ANDERSON,
R. PITTMAN

California Dreamin'
J. BAKER, E. LEVIN,
M. McGROGAN, R. SCOTT

Nice, Francesco? Nice, Yoshi!
Y. NAGAMINE, F. BLASI

My boss, my friend!
H. POTTER, C. ABRAHAM

Pencils ready?
B. BLONDET, P. LEPRINCE,
A. BLEUEL, D. HANTAÏ,
W.-D. SCHLEUNING,
Y. NAGAMINE

R. SAWAYA, G. BARLOVATZ-
MEIMON, K. DANØ, B.
FESTOFF, D. CUNNINGHAM,
J. WISEMAN

B. FESTOFF

P. LEPRINCE, C.
DESCHEPPER, G. MOONEN,
P. DISTEFANO

G. MOONEN, P. DISTEFANO,
G. SALVESEN, R. HILL,
N. SEEDS

Nice Coffee Breaks!
J. TRAVIS, P. LEPRINCE,
R. ORTMANN, C. DeSCHEPPER,
N. SEEDS, G. MOONEN,
R. PITTMAN

PARTICIPANTS

Carmella R. Abraham, Ph.D.   Boston University School of Medicine
Boston, MA 02118 USA

Noriaki Adachi, M.D., Ph.D.   Shinshu University School of Medicine
3-1-1 Aasahi,
Matsumoto 390 JAPAN

Nahed Ahmed, Ph.D.   Marion Laboratories, Inc.
Kansas City, MO 64134 USA

Hala Alameddine, Ph.D.   I.N.S.E.R.M., U. 153,
75005 Paris, FRANCE

M. John Anderson, Ph.D.   University of Calgary,
Calgary, Alberta, CANADA T2N 4N1

José Paulo Andrade, M.D.   Oporto School of Medicine,
Porto, PORTUGAL 4200

Eduardo Anglés-Cano, M.D., Ph.D. I.N.S.E.R.M. U.143,
94275 Le Kremlin Bicêtre Cedex, FRANCE

Joffre Baker, Ph.D.   Genentech, Inc.,
S. San Francisco, CA 94080 USA

James Ballance, Ph.D.   Delta Biotechnology, Ltd.,
Nottingham NG71FD UK

Francesco Blasi, M.D.   University Institute of Microbiology
1353 Copenhagen K, DENMARK

Alicia Bleuel, Ph.D.   Friedrich Miescher-Institut,
CH-4002 Basel SWITZERLAND

Brigitte Blondet, Ph.D.   I.N.S.E.R.M. U. 153,
75005 Paris, FRANCE

Tom W.G. Carrell   University of Cambridge Clinical School
Cambridge CB2 2QL, UNITED KINGDOM

Dennis Cunningham, Ph.D.   California College of Medicine,
University of California, Irvine
Irvine, CA 92717 USA

Keld Danø, M.D.	Finsen Laboratory, Rigshospitalet DK-2100 Copenhagen, DENMARK
Christian F. Deschepper, Ph.D.	University of California, San Francisco San Francisco, CA 94143 USA
Gerhard Dickneite, Ph.D.	Behringwerke Aktiengesellschaft, Postfach 1140, D-3550 Marburg 1 WEST GERMANY
Melitta Dihanich, Ph.D.	Friedrich Miescher-Institut CH-4002 Basel SWITZERLAND
Peter Distefano, Ph.D.	Abbott Laboratories, Abbott Park, IL 60064 USA
Dyfed L. Evans, Ph.D.	University of Cambridge Clinical School Cambridge CB2 2QL, UNITED KINGDOM
John Fenton, II, Ph.D.	New York State Department of Health Albany, NY 12201 USA
H. Hugh Fudenberg, M.D., Ph.D.	Medical University of South Carolina Charleston, S.C. 29425-2230 USA
Peter W. Grabham, Ph.D.	University of Birmingham, Birmingham B15 2TJ UNITED KINGDOM
David Gurwitz, Ph.D.	University of California, Irvine Irvine, CA 92717 USA
Robert Hill, Ph.D.	Western General Hospital, Edinburgh EH4 2XUMRC, SCOTLAND
Jan Hofsteenge, Ph.D.	Friederich Miescher Institut, Basel, SWITZERLAND
Nurit Kalderon, Ph.D.	Rockefeller University, New York, NY 10021 USA
Senay Kayanak, M.D.	Iccebci, Koyluler Sokak, Ankara, TURKEY
Pierre Leprince, Ph.D.	Université de Liège, Liège, BELGIUM
Eugene Levin, Ph.D.	Scripps Clinic and Research Foundation La Jolla, CA 92037 USA
H.R. Lijnen, M.D.	Center for Thrombosis and Vascular Research B-3000 Leuven, BELGIUM
He Lu, Ph.D.	Laboratoire d'Hématologie Hôpital Lariboisière 75010 Paris, FRANCE

Beatrix Stelte-Ludwig, Ph.D. — Bayer AG, Pharmaceutical Research Center
5600 Wuppertal 1, WEST GERMANY

Michael McGrogan, Ph.D. — Invitron Corp.,
Redwood City, CA 94063 USA

Massoud Mirshahi, M.D. — Laboratoire Ste Marie, Hôtel Dieu
Parvis Notre Dame
75004 Paris, FRANCE

Denis Monard, Ph.D. — Friedrich Miescher-Institut
CH-4002 Basel SWITZERLAND

Yoshikuni Nagamine, Ph.D. — Friederich Miescher Institut
CH-4002 Basel SWITZERLAND

Rainer Ortmann, Ph.D. — CIBA-GEIGY AG,
CH-4002 Basel SWITZERLAND

Randall Pittman, Ph.D. — University of Pennsylvania School of
Medicine,
Philadelphia, PA 19104-6084 USA

Jasti Rao, Ph.D. — University of Kansas Medical Center at the
Veterans Administration Medical Center
Kansas City, MO 64128 USA

Bokka Reddy, Ph.D. — University of Kansas Medical Center at the
Veterans Administration Medical Center
Kansas City, MO 64128 USA

Eva Reinhard, Ph.D. — Friedrich Miescher-Institut
CH-4002 Basel SWITZERLAND

Dorrie Rosenblatt, Ph.D. — University of Michigan,
Ann Arbor, MI 48109-2007 USA

Guy Salveson, Ph.D. — Duke University Medical Center,
Durham, NC 27710 USA

Raymond Sawaya, M.D. — University of Cincinnati Medical Center
Cincinnati, OH 45267-0515 USA

Wolf-Dieter Schleuning, M.D. — Institute for Biochemistry, Schering AG
D-1000 Berlin 65, WEST GERMANY

Friedrich Schumann, Ph.D. — Bayer AG, Pharmaceutical Research Center
5600 Wuppertal 1, WEST GERMANY

Randy Scott, Ph.D. — Invitron Corp.,
Redwood City, CA 94063 USA

Nicholas Seeds, Ph.D. — University of Colorado Health Sciences
Center
Denver, Colorado 80262 USA

Juerg Sommer, Ph.D.

Friedrich Miescher-Institut
CH-4002 Basel SWITZERLAND

Claudine Soria, Ph.D.

Laboratoire d'Hématologie,
Hôpital Lariboisière
75010 Paris, FRANCE

Jeannette Soria, Ph.D.

Laboratoire S$^{te.}$ Marie, Hôtel Dieu
75004 Paris, FRANCE

Patrizia Stopelli, Ph.D.

Institute of Genetics and Biophysics,
80123 Naples, ITALY

Rudolph Tanzi, Ph.D.

Harvard Medical School,
Laboratory of Neurogenetics
Boston, MA 02115 USA

James Travis, Ph.D.

University of Georgia,
Athens, GA 30602 USA

Anton-Jan van Zonneveld, Ph.D.

Central Laboratory of the Red Cross
Blood Transfusion Service
1006 AK, Amsterdam,
THE NETHERLANDS

Stephen Wagner, Ph.D.

California College of Medicine
University of California, Irvine
Irvine, CA 92717 USA

Jeffrey Wiseman, Ph.D.

GLAXO Research Labs,
Morrisville, NC 27560 USA